ADVANCES IN LASER AND OPTICS RESEARCH

VOLUME 10

ADVANCES IN LASER AND OPTICS RESEARCH

ADVANCES IN LASER AND OPTICS RESEARCH

VOLUME 10

WILLIAM T. ARKIN
EDITOR

New York

For permission to use material from this book please contact us:
Telephone 631-231-7269; Fax 631-231-8175
Web Site: http://www.novapublishers.com

Library of Congress Cataloging-in-Publication Data

ISBN 978-1-62257-795-8

ISSN 1542-5126

Published by Nova Science Publishers, Inc. † New York

CONTENTS

PREFACE

This book gathers the latest research from around the globe on the subject of lasers and optics research. Topics discussed include polarization dependent effects in optical waveguides containing bragg grating structures; bacterial cell interactions with optical fiber surfaces; passively mode-locked fiber lasers with nonlinear optical loop mirrors; optical breakdown in gases induced by high-power ir co2 laser pulses and towards shaping of pulsed plane waves in the time domain via chiral sculptured thin films.

Chapter 1 - Since the discovery of photosensitivity in optical fibers by Hill et al. in 1978, Bragg gratings fabricated in optical fibers and planar waveguides have been extensively investigated over the last three decades and have been widely used in optical communication and sensor applications. In this chapter, the properties of birefringence in optical fibers and planar waveguides with Bragg grating structures are investigated experimentally, and the birefringence-induced impairments in communication systems with Bragg-grating–based components are evaluated by simulation.

Chapter 2 - Since 1840, investigations into the principles of fiber optics have continued to deliver new associated technologies and applications. The expansion of knowledge in the area of nano- and applied science, in particular, has predictably led to the introduction of novel techniques and instruments that exploit fiber optics. As such, the use of optical fibers as chemical or environmental/bio-sensors or as imaging devices has resulted in the need for a detailed exploration of possible interactions between the surface of optical fibers and environmental perturbations, including bacterial cells.

Although the impact of bacterial adhesion to a number of surfaces, such as glass, metal, or polymeric surfaces, has already been studied in considerable detail, there appears to be a lack of systematic investigation into the interaction of bacterial cells with optical fibers, and in many cases, their subsequent attachment. This study focuses on the influence of optical/imaging fiber surface characteristics and chemistry on bacterial attachment. Attachment profiles of medically significant and environmentally important bacteria on optical fibers, both as-received and eroded as a result of chemical etching, were evaluated. Bacterial cell characteristics such as surface charge and wettability were analyzed, together with optical fiber characteristics, such as surface chemistry, wettability, charge, surface tension and surface nano-roughness. A series of scanning electron microscopy (SEM), atomic force microscopy (AFM) and confocal laser scanning microscopy (CLSM) experiments were carried out to provide an insight into bacterial attachment behaviour on the optical fiber surfaces.

The results show that the as-received fiber surface was consistently more suitable for bacterial attachment than the modified surface. Therefore this study suggests that the surface treatment of optical fibers can be an important factor in the development of optical fiber sensors.

Chapter 3 - We present the operational principles of two recently introduced nonlinear optical loop mirrors (NOLMs): dispersion imbalanced NOLM (DI-NOLM) and an absorption imbalanced NOLM (AI-NOLM). These NOLMs with 3-dB fiber couplers can break the symmetry of nonlinear phase shift in a fiber loop either with two different dispersion fibers or an asymmetrically located attenuator in a fiber loop mirror. Mode-locked laser pulses were obtained utilizing these two types of NOLMs and the pulse properties of them were investigated. It is demonstrated that these lasers show different pulse operation modes where the pulsed output can be characterized as stable mode-locked pulses, broad square pulses, harmonically mode-locked pulses, bound solitons, etc. depending on the parameters of a laser system. Stable mode-locking condition for a figure-eight fiber laser with a DI-NOLM is relatively hard to obtain due to the sensitive changes of the optical dispersion parameter of a fiber with respect to parameters and external conditions such as wavelength, temperature, bending, etc. Meanwhile, it is shown that the station mode-locked output pulses can be obtained easily with an AI-NOLM because of its simple and robust operating condition.

Chapter 4 - This chapter reviews some fundamentals of laser-induced breakdown spectroscopy (LIBS) and describes some experimental studies developed in our laboratory on gases such as nitrogen, oxygen and air. LIBS of these gases at different pressures, in the spectral range ultraviolet-visible-near infrared (UV-Vis-NIR), was excited by using a high-power transverse excitation atmospheric (TEA) CO_2 laser (λ=9.621 and 10.591 μm; τ(FWHM)=64 ns; and different laser power densities). The spectra of the generated plasmas are dominated by emission of strong atomic, ionic species and molecular bands. Excitation temperatures were estimated by means of atomic and ionic lines. Electron number densities were deduced from the Stark broadening of several ionic lines. The characteristics of the spectral emission intensities from different species have been investigated as functions of the gas pressure and laser irradiance. Optical breakdown threshold intensities in different gases have been measured experimentally. The physical processes leading to laser-induced breakdown of the gases have been analyzed. Plasma characteristics in LIBS of air were examined in detail on the emission lines of N^+, O^+ and C by means of time-resolved optical-emission spectroscopy (OES) technique. The results show a faster decay of continuum and ionic spectral species than of neutral atomic and molecular ones. The velocity and kinetic energy distributions for different species were obtained from time-of-flight (TOF) OES measurements. Excitation temperature and electron density in the laser-induced plasma were estimated from the analysis of spectral data at various times from the laser pulse incidence. Temporal evolution of electron density has been used for the estimation of the three-body recombination rate constant.

Chapter 5 – The authors review their investigations concerning the propagation of ultrashort optical pulsed plane waves across chiral sculptured thin films (STFs) in the time domain. The phenomenon of pulse bleeding—i.e., the time domain manifestation of the circular Bragg phenomenon (CBP)—is explained, and its importance for the shaping of ultrashort pulses is investigated.

Chapter 6 - This chapter describes various kinds of processes to integrate optical waveguides on silicon substrates, suitable for the visible to the near infrared spectral range.

Different types of materials for the light guiding film and set ups for the waveguide structures are compared according to the propagation loss and the integration density. The link of the strip loaded rib waveguides to photo detectors, integrated in the same substrate, is performed by butt- or leaky wave coupling. Two kinds of processing for the monolithic integration of waveguides, photo detectors and CMOS circuits are presented, and an application as a pressure sensor is given. The last part of the chapter discusses optical waveguides in silicon suitable for data transmission in the 1.55 μm spectral range. Doped silicon waveguides and silicon-on insulator waveguides are presented as light guiding films. The capability of the silicon technology is demonstrated by a capacity controlled all-silicon modulator for 10 GHz.

Chapter 7 - Buildings, internal spaces and urban environments, which are not only very energy efficient, but also pleasant to occupy and look at, are needed to accelerate the rate of reduction worldwide in CO2 emissions, and to raise living standards. New materials, whose optical properties are tailored to this task, are making this goal possible. Lighting, visual, and thermal control properties of windows, skylights, paints and lighting systems are treated. These three functions must be considered together, not independently. The way a range of new polymeric and nano-structured materials enable the optimised spectral management and distribution of incoming solar or lamp radiation, and outgoing thermal radiation, is presented.

Most of the materials discussed are composites whose properties can be varied in a controlled way, if the physics is understood. The basic models linking structure, composition and optical response are presented, for example, from among four classes of optical materials: opaque, transparent, translucent, and the group – light piping, mixing and homogenisation. Issues for plasmon resonant nanoparticles in visibly transparent glazing include resonance tuning, and differences in short wavelength (Rayleigh) and plasmon resonance scattering. For energy efficient glare control and uniform daylighting, for integration of light piping and continuous illumination, and for short length light mixing and homogenisation, the use of dopants, which are close in refractive index (RI) to the host, are analysed. These are found to be ideal to use with many emerging LED applications and daylight. For LED's they enable very energy efficient white light lamps based on RGB arrays at low cost, and also, enable uniform illumination despite the small area of the source. Finally, some special spectrally selective paints are discussed; those that can be deep colours or even black and still either reflect or absorb much solar energy (whichever is desired), and those that enable efficient radiative cooling or water collection from the atmosphere.

Chapter 8 - This chapter shows mathematical expressions for various types of electro-optic modulators, such as, intensity modulators, single-sideband modulators, frequency-shift-keying modulators, phase-shift-keying modulators, etc, based on optical phase modulation. We also discuss electrode structures for high-speed operation. Resonant electrodes can enhance modulation efficiency in a particular band. Design schemes for two types of resonant electrodes: asymmetric resonant structure and double-stub structure, are given. Most important application of optical modulators is conversion of electric signals into lightwave signals. This is indispensable to high-speed optical links. However, recently, we need new functions such as optical label processing at nodes, signal timing control to prevent packet collision, etc. In addition to basic function of optical modulators, this chapter describes the photonic sideband management techniques which can manipulate optical or electrical signals in time or frequency domain, by using sideband generation at the modulators.

Chapter 9 - In the developed nations of the world, optical sensor industry is a multimillion-dollar business. This chapter is devoted to the review of the fundamentals of fiber optic sensing techniques, and describes the different types of optical sensors and their applications in different industries. Descriptions are made of intensity based sensors, distributed sensors, and interferometric sensors, their different types and specific applications. Examples of some more forms of FOSs, and their use in the context of environment are also reviewed. In addition, discussions are made of the role of integrated optics technology in sensing applications. It is noteworthy that photonic crystals and STFs have been the emerging areas of research in the present day; the chapter describes the use of such materials and/or mediums in the area of optical sensing.

Chapter 10 - This chapter describes the present state-of-the-art of fiber optic current sensing technology based on Faraday effect. The importance of the fulfillment of HICOC (homogeneous isotropic closed optical circuit) condition is addressed. The optical current sensors are categorized into three types, viz. all-fiber, bulk-optic and hybrid, and their available HICOC techniques, advantageous features, performances and applications are mentioned.

Chapter 11 - A review of the experimental realization of key high efficiency two-dimensional optical elements, built up from metal nanostructures, such as nanoparticles and nanowires to manipulate plasmon polaritons propagating on metal surfaces is reported. Beamsplitters, Bragg mirrors and interferometers designed and produced by elelectron-beam lithography are investigated. The plasmon field profiles are imaged in the optical far-field by leakage radiation microscopy or by detecting the fluorescence of an organic film deposited on the metal structures. It is demonstrated that these optical far-field methods are effectively suited for direct observation and quantitative analysis of plasmon polariton wave propagation and interaction with nanostructures on thin metal films. Several examples of two-dimensional nanooptical devices fabricated and studied in recent years are presented.

Chapter 12 - Composite materials containing metal nanoparticles (MNPs) are now considered as a basis for designing new photonic media for optoelectronics and nonlinear optics. Simultaneously with the search for and development of modern technologies intended for nanoparticle synthesis, substantial practical attention has been devoted to designing techniques for controlling the MNP size. One of the promising methods for fabrication of MNPs is ion implantation. Review of recent results on ion-synthesis and nonlinear optical properties of cupper, silver and gold nanoparticles in surface area of various dielectrics as glasses and crystals are presented. Composites prepared by the low energy ion implantation are characterized with the growth of MNPs in thin layer of irradiated substrate surface. Fabricated structures lead to specific optical nonlinear properties for picosecond laser pulses in wide spectral area from UV to IR such as nonlinear refraction, saturable and two-photon absorption, optical limiting. The practical recommendations for fabrication of composites with implanted MNPs for optical components are presented.

Versions of these chapters were also published in *Journal of Optics Research,* Volume 13, published by Nova Science Publishers, Inc. They were submitted for appropriate modifications in an effort to encourage wider dissemination of research.

In: Advances in Laser and Optics Research. Volume 10 ISBN: 978-1-62257-795-8
Editor: William T. Arkin

Chapter 1

POLARIZATION DEPENDENT EFFECTS IN OPTICAL WAVEGUIDES CONTAINING BRAGG GRATING STRUCTURES

Ping Lu,[1] Liang Chen,[2] Xiaoli Dai,[1] Stephen J. Mihailov[1] and Xiaoyi Bao[2*]

[1]Communications Research Centre Canada, Ottawa, Canada
[2]Department of Physics, University of Ottawa, Ottawa, Canada

ABSTRACT

Since the discovery of photosensitivity in optical fibers by Hill et al. in 1978 [1], Bragg gratings fabricated in optical fibers and planar waveguides have been extensively investigated over the last three decades and have been widely used in optical communication and sensor applications. In this chapter, the properties of birefringence in optical fibers and planar waveguides with Bragg grating structures are investigated experimentally, and the birefringence-induced impairments in communication systems with Bragg-grating–based components are evaluated by simulation.

1. INTRODUCTION

When the attenuation and chromatic dispersion in optical fibers are compensated, polarization mode dispersion (PMD) is the major factor limiting the capacity of fiber optic communication systems at high bit rates (>10Gb/second) [2-6]. The presence of PMD in optical networks introduces differential group delay (DGD), resulting in signal distortion and high bit error ratio (BER). The system impact of PMD has been investigated quantitatively by

* Ping Lu, Xiaoli Dai, Stephen J. Mihailov: Communications Research Centre Canada, Ottawa, ON K2H 8S2, Canada. Liang Chen, Xiaoyi Bao: Department of Physics, University of Ottawa, Ottawa, ON K1N 6N5, Canada.

examining the PMD induced system outage and eye penalty [2, 3]. The fiber optic networks are becoming more complex because of the introduction of new types of photonic components that have more functionality and can operate at higher data rates. Optical components, such as reconfigurable optical add-drop multiplexers (ROADMs), fiber Bragg gratings (FBGs), planar waveguide Bragg gratings (WBGs), etc. have both PMD and polarization-dependent loss (PDL) [4, 5]. It has been shown both theoretically and experimentally that the system impact due to the combined effect of PMD and PDL is more severe than that due to PMD alone [4, 6]. As one of the key optical components in modern networks, Bragg gratings in fibers and planar waveguides are widely used in today's communication and sensing systems. They are usually fabricated by exposure of the fiber or waveguide to the ultraviolet (UV) or ultrafast infrared (IR) fringes resulting from two-beam interference or generated by a zero-order nulled phase mask [7, 8]. The fibers and waveguides are typically hydrogen (H2)-loaded to increase photosensitivity to UV light [9]. Due to the sensitive nature of this fabrication technique, laser exposure is only on one side of the fiber or waveguide, resulting in grating spectra that exhibit polarization dependence (birefringence) [10, 11].

Compared with optical fibers, planar waveguides produced by plasma-enhanced chemical vapor deposition (PECVD) or flame hydrolysis deposition (FHD) have a very strong intrinsic birefringence that makes the Bragg wavelength and peak reflectivity sensitive to the polarization state of the propagation signal.

In this chapter, the properties and system impairments of birefringence in FBGs and WBGs are presented and discussed in detail. In section 2, the sources of birefringence in optical fibers and planar waveguides that incorporate a grating structure will be briefly introduced, and the relationship between birefringence and spectral PMD and PDL of FBGs and WBGs will be presented. This relationship will be used in the following sections to characterize the birefringence in gratings and to evaluate the birefringence-induced system impairments. In section 3, the induced birefringence in fibers by UV and femtosecond pulse duration IR lasers during the grating inscription process is examined, and its dependence on H2-loadening in fibers, the temperature while the grating is inscribed, and the polarization of the laser beam will be presented. The annealing curves of induced birefringence in FBGs are presented as well. Section 4 shows the technique for compensating the intrinsic birefringence of planar waveguide by using UV laser exposure. Finally, in section 5, the birefringence-induced penalties in optical communication systems consisting of Bragg grating based OADMs are simulated by examining the distortion of eye diagram of 10 Gb/s signal.

2. Birefringence in FBGs and WBGs

2.1. Sources of Birefringence

The birefringence of FBGs and WBGs consists of two parts: 1) the intrinsic birefringence of the fiber and waveguide, and 2) the induced birefringence by UV or ultrafast IR lasers during grating inscription. The intrinsic birefringence may come from the asymmetric geometry of the waveguide, the stress induced during waveguide fabrication, and environmental perturbations (such as bending, twisting, etc.).

The amount of birefringence in single mode fiber, such as Corning SMF-28 used in communications, is usually low (of the order of 10-7–10-6), and both the local birefringence value and the local principal axes (the fast and slow axes) vary along the fiber length. The birefringence in a planar waveguide is usually higher than that in cylindrical single mode fiber and can be of the order of up to 10-4. This high birefringence is due to the asymmetric geometry and material structures that result in larger geometrical birefringence and stress birefringence than would occur in regular single mode fibers.

Bragg gratings in fibers and waveguide are usually fabricated by exposure of the fibers and waveguides to the UV or ultrafast IR fringes resulting from two-beam interference or generated by a zero-order nulled phase mask. The fibers and waveguides are typically Hydrogen (H2) loaded to increase photosensitivity to UV light, but this is not necessary when high peak power IR irradiation is used. Due to the sensitive nature of this fabrication technique, only one side of the fiber and waveguide is exposed to the laser radiation, resulting in the induction of birefringence in addition to the intrinsic birefringence. This laser-induced birefringence has been studied both experimentally and theoretically [12-14]. The birefringent effects in FBGs, and long period gratings has been measured [15-18], and the annealing properties of the UV and ultrafast IR laser-induced birefringence have also been examined [19-21].

2.2. Relationship between Birefringence and Spectral PDL of Bragg Gratings in Fibers and Waveguides

In FBGs and WBGs inscribed by laser exposure, both the effective mode index and the index modulation can show a polarization dependence [18, 21]. The birefringence in effective mode index and the birefringence in index modulation of a grating may be characterized by measuring the spectral PDL (or PMD) at wavelengths around the grating's resonance and using the coupled-mode theory. Here we briefly introduce the relationship between a grating's spectral PDL and birefringence. The transmission and reflection of a uniform grating with grating length L and period Λ can be expressed as:

$$\begin{bmatrix} A_{out} \\ B_{out} \end{bmatrix} = \begin{bmatrix} p & q \\ q^* & p^* \end{bmatrix} \begin{bmatrix} A_{in} \\ B_{in} \end{bmatrix}, \qquad (1)$$

where A and B are the amplitudes of the forward and backward propagating waves, the subscripts of in and out refer to the waves at the grating's input and output ends respectively. In Equation (1), $p = \cosh(\sigma L) - i\delta \sinh(\sigma L)/\sigma$ and $q = i\kappa \sinh(\sigma L)/\sigma$, where $\delta = 2\pi(n_{eff}/\lambda - 1/\Lambda)$ is the detuning from Bragg resonance, n_{eff} is the effective mode index and λ is the wavelength in free space. Parameter σ is related to the detuning δ, and mode coupling constant κ, as $\sigma = \sqrt{\kappa^2 - \delta^2}$. The coupling constant, κ, is proportional to the index modulation, Δn, in the fiber core as $|\kappa| = \pi\Delta n / n_{eff}\lambda$. A grating with an apodized index profile can be treated as a concatenation of many uniform sub-gratings. The 2×2 transform matrix in Equation (1) can then be replaced by the product of the corresponding sub-matrices.

The UV exposure on one side of fiber makes both n_{eff} and Δn polarization dependent. The induced PMD and PDL in a grating can then be obtained by calculating the polarization dependent phase delays and magnitudes of A_{out} (in transmission) and B_{in} (in reflection) as follow:

$$PDL_{tran} = \left| 20\log_{10}\left(\left| \frac{A_{out}^{fast}}{A_{out}^{slow}} \right| \right) \right| \quad \text{(PDL in transmission)}$$

$$DGD_{tran} = \left| \tau_{tran}^{fast} - \tau_{tran}^{slow} \right| \quad \text{(PMD in transmission)}$$

$$PDL_{ref} = \left| 20\log_{10}\left(\left| \frac{B_{in}^{fast}}{B_{in}^{slow}} \right| \right) \right| \quad \text{(PDL in reflection)}$$

$$DGD_{ref} = \left| \tau_{ref}^{fast} - \tau_{ref}^{slow} \right| \quad \text{(PMD in reflection)} \qquad (2)$$

where A_{out}^{fast} and A_{out}^{slow} are the amplitudes of the forward waves at the grating's output end when the probe light is launched on to the fast and slow axes respectively. τ_{tran}^{fast} and τ_{tran}^{slow} are the time delays of the forward propagating wave along the fast and slow axes respectively. B_{in}^{fast} and A_{in}^{slow} are the amplitudes of backward waves at the grating's input end when the probe light is launched on to the fast and slow axes respectively. τ_{ref}^{fast} and τ_{ref}^{slow} are the time delays of the backward propagating wave along the fast and slow axes respectively.

The intrinsic birefringence in *SMF-28* fiber is on the order of 10^{-7}–10^{-6} and varies along the fiber length in both amplitude and orientation (the fast or slow axis). When the intrinsic fiber birefringence in a FBG is the major part of the total birefringence, which is likely for weakly inscribed gratings in single mode fiber, the spectral PDL shows noisy peaks due to the random variation of the intrinsic birefringence along the grating length.

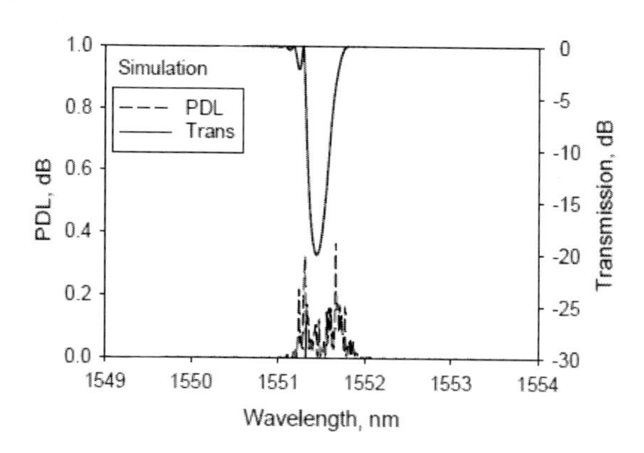

Figure 1. The simulated spectral PDL of a FBG induced by the intrinsic birefringence of the fiber.

Using Equations (1) and (2), a simulated spectral PDL induced by the intrinsic birefringence is shown in Figure 1 where the intrinsic birefringence is randomly distributed from 0 to 10^{-6} along the grating length. The grating length used for the simulation results shown in Figure 1 is 6 mm with an apodization of the Gaussian index profile with maximum index modulation of 4×10^{-4} at the grating center. The birefringence in planar waveguide is usually higher than that in single mode fiber and the fast and slow axes are usually fixed, the spectral PDL of gratings in planar waveguide is similar to that when the major part of birefringence is induced by laser irradiation that will be discussed below.

When a grating is inscribed in fiber or waveguide by exposing to UV or IR laser fringes, the "side-writing" technique produces birefringence creating fast axis and slow axes that are wavelength independent. The orientation of the fast/slow axes is dependent upon the direction of the incident beam used for the grating inscription. In the case when the laser-induced birefringence is much higher than the fiber's intrinsic birefringence, the spectral PDL of a grating can be modeled using Equations (1) and (2).

a)

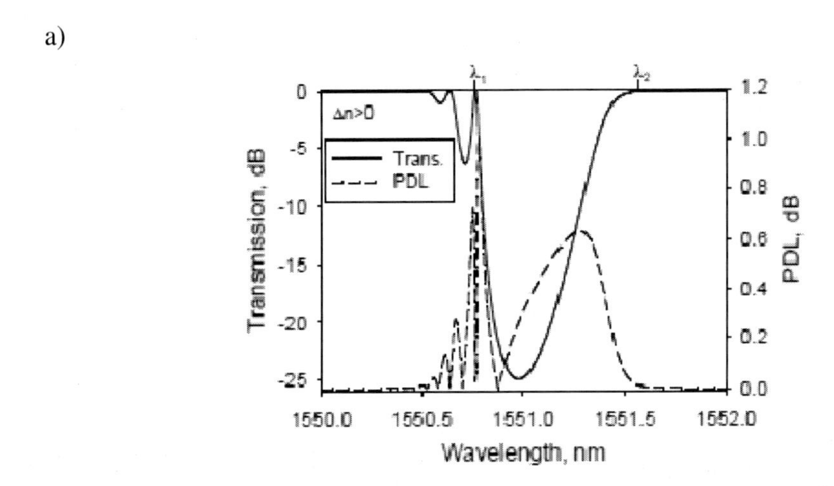

b)

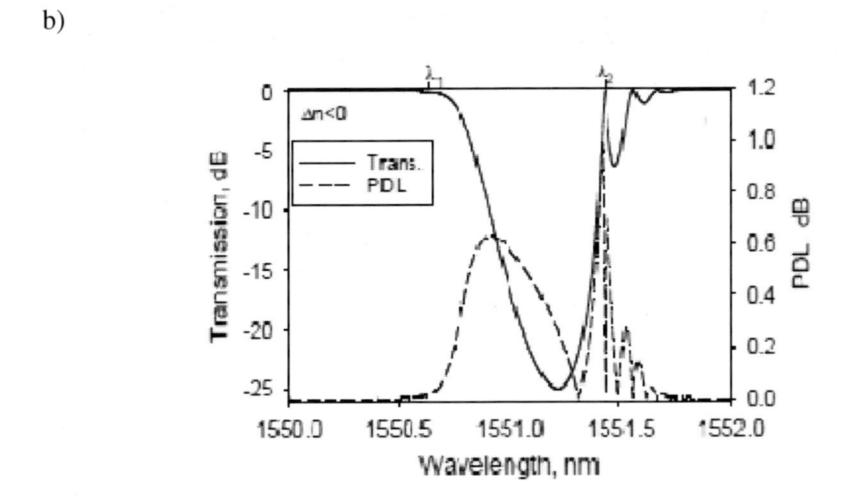

Figure 2. The simulated spectral PDLs of a FBG induced by the laser during FBG inscription. (a) the induced index is positive, (b) the induced index is negative.

Figures 2(a) and 2(b) show the simulated spectral PDLs of two gratings with the induced birefringence of 5×10^{-6}. Usually within the bandwidth of one transmission (or reflection) peak (from λ_1 to λ_2 in Figures 2(a) and 2(b)) in a grating spectrum, there are two PDL peaks. When the induced index profile is apodized with the highest index change in the middle of the grating, such as Gaussian or $\cos^2$ profile, the PDL peak on the shorter wavelength side will be narrower than that of the PDL peak in the longer wavelength side if the laser induced index is positive (Figure 2(a)). If the bandwidth of the PDL peak on the short wavelength side is broader than that of the PDL peak in the longer wavelength side, then the laser-induced index change is negative (Figure 2(b)). In the simulations of Figures 2(a) and 2(b), the FBG total length is 6 mm with an apodization of the Gaussian index profile with maximum index modulation of 5×10^{-4} at the center of the grating. Figures 1 and 2 clearly show that by modeling the grating spectral PDL, the birefringence of a grating can be obtained.

3. CHARACTERIZATION OF GRATING BIREFRINGENCE

3.1. Experiment Setup

The inscription of a grating in optical fibers and waveguides usually takes a few minutes to a few tens of minutes depending on the requirements of grating bandwidth, strength and apodization profile etc. For real time monitoring of birefringence variation during grating inscription, the measurement set-up has to be fast with reasonable wavelength resolution. The set-up to inscribe FBG and to measure birefringence used in this work is shown in Figure 3. The laser beam (either UV or ultrafast IR) is focused on the fiber by a cylindrical lens (not shown in Figure 3) through a phase mask of pitch Λ_{Bragg} that is optimized to the laser wavelength producing a Bragg resonance of $\lambda_{Bragg} = n_{eff}\Lambda_{Bragg}$. A tunable laser beam (*8164A, Agilent*) is launched into the polarization test set (*A2000 Components Analyzer, Adaptif Photonics GmbH*) into which the two ends of *SMF-28* fiber are connected in order to measure the grating's spectral PMD and PDL in transmission during grating inscription.

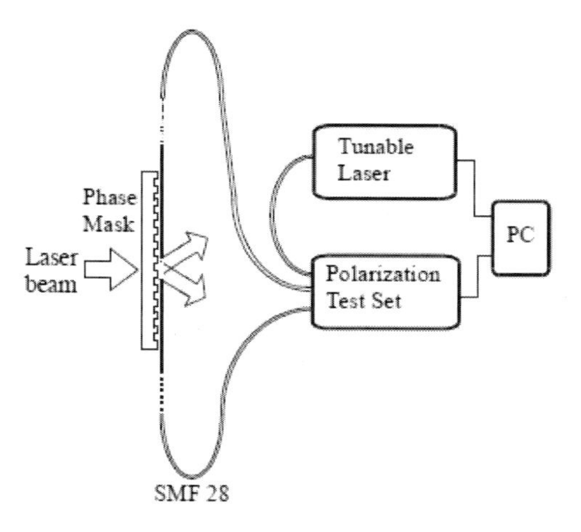

Figure 3. Set-up of writing fiber Bragg gratings and measuring the laser-induced spectral PMD and PDL.

The whole measurement set-up is controlled by a personal computer through a GPIB interface. The set-up has the ability of measuring a set of data, including grating spectrum, PMD, PDL, principal state of polarization (PSP), etc., every 2 to 3 seconds for a wavelength window of a few nanometers with a resolution of 10 picometers. The time needed for the measurement at single wavelength is less than 1 ms, which makes it possible for real-time measurements of the variations of birefringence during grating inscription. The grating's birefringence can also change very quickly when the grating is annealed at certain temperatures [19-21] as will be shown later in this section, therefore a fast measurement set-up is preferred.

3.2. UV Induced Birefringence in Fiber Bragg Gratings

3.2.1. Growth of Birefringence during Grating Inscription

The experimental set-up for writing FBGs and measuring the UV-induced birefringence is shown in Figure 3. Gratings are inscribed on *SMF-28* fibers by spot writing a focused 244 nm beam from a frequency doubled Ar+ laser through a phase mask. The focused beam spot size is 9 mm. The focal length of the cylindrical lens is 150 mm and the pitch of the phase mask is $\Lambda_{mask} = 1.07\ \mu m$, which produces a grating period of $\Lambda = 0.535\ \mu m$ in the fiber core. The polarization state of the UV beam is parallel to the fiber axis, which is preferred in order to produce minimum birefringence [12]. The fibers used in the experiment were photosensitized by hydrogen loading at room temperature and 2500 psi pressure for one week. Figure 4 and Figure 5 show the experimental results of spectral DGD and PDL of the grating with fitting curves by using the coupled mode theory. In the simulations of DGD and PDL, both the birefringence in n_{eff} and the birefringence in Δn need to be considered, especially in the fitting of spectral PDL. This is clearly shown in Figure 5 where the experimental data are compared to the simulated results (dashed curves) in which only n_{eff} is polarization dependent. Using the set-up shown in Figure 3, the UV-induced spectral DGDs and PDLs in the grating are measured every 2~3 seconds with the tunable laser power of 140 mW. Due to the dynamic range of the polarization test set, the UV-induced DGD and PDL are measured up to the grating strength of -25 dB in transmission.

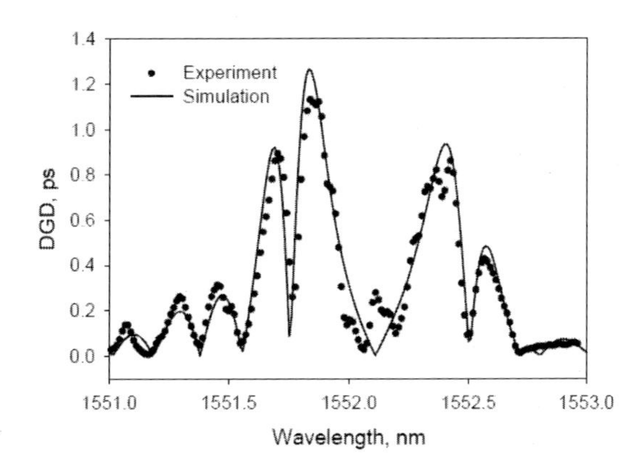

Figure 4. Measurement results of the UV-induced DGD in a 2 mm long grating. The solid curve is the simulated result.

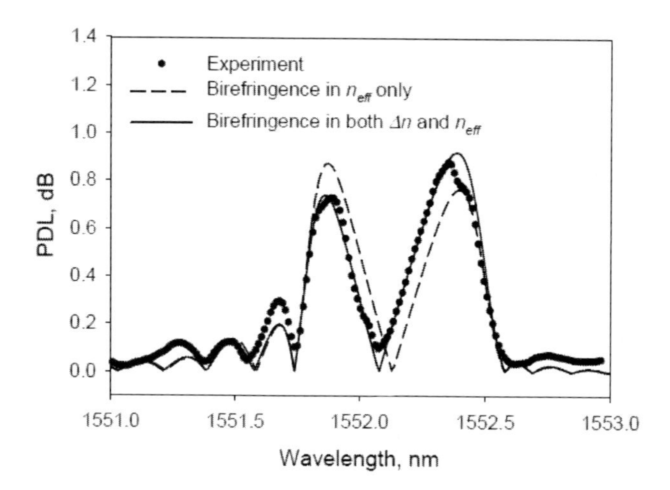

Figure 5. Measurement (dots) and simulated (solid and dashed curves) results of the UV-induced PDL in a 2 mm long grating. Solid curve: both the mode effective index and the index modulation are polarization dependent; dashed curve: the mode effective index is polarization dependent only.

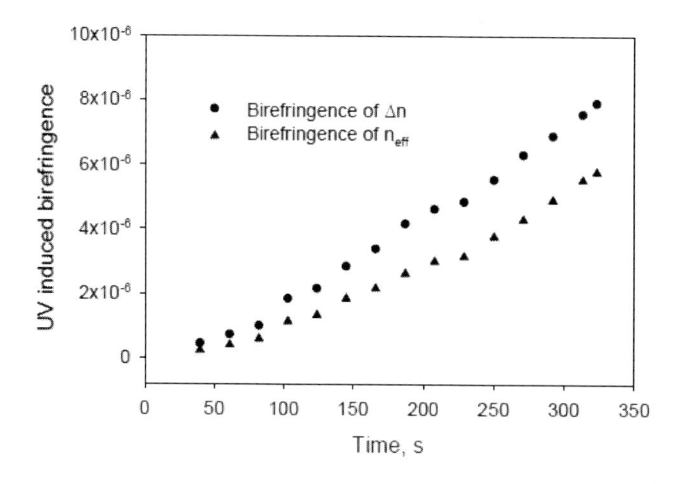

Figure 6. The growths of the birefringence in effective mode index and the birefringence in index modulation.

By fitting the spectral DGDs and PDLs, the changes in birefringence for both the n_{eff} and Δn are obtained and presented in Figure 6.

The birefringence in Δn and the birefringence in n_{eff}, shown in Figure 6 are 1.4% to 1.7% and 0.7% to 1.2% of the Δn value, respectively. The variation in induced birefringence with exposure time is not linear, as is shown in Figure 6. For various UV laser powers (from 100 to 300 mW) and various distances (from 0 to 10 mm) between the fiber and the focus, the growths of PDL and DGD are measured as well. No obvious change in birefringence is found for the same grating strength.

3.2.2. Annealing of UV-Induced Birefringence

In H_2-loaded *SMF-28* fibers, Canning et al. [19, 20] showed that the UV-induced birefringence can be annealed out at temperatures below 150 °C without significant decay of the induced index modulation. This is shown in Figure 7 where the reductions of UV-induced Δn and birefringence in H_2-loaded *SMF-28* fibers during various annealing temperatures are plotted. It can be seen clearly that for temperatures above 120 °C, the birefringence rapidly decreases to that of the fiber's intrinsic birefringence (~10^{-6}).

The grating with its annealing curve shown in Figure 7 was inscribed when the UV laser beam was focused using a lens with focal length of 150 mm. During the inscription, the temperature of the fiber section being exposed to UV laser is almost the same as room temperature (after the UV beam was turn off, no obvious Bragg wavelength shift was found).

When the UV laser was tightly focused by using a 80 mm focal length lens, the temperature of the fiber section that was exposed to UV beam can be higher than room temperature. Figure 8 shows the annealing curve of the birefringence up to 260 °C when the inscription temperature is around 140 °C during UV exposure.

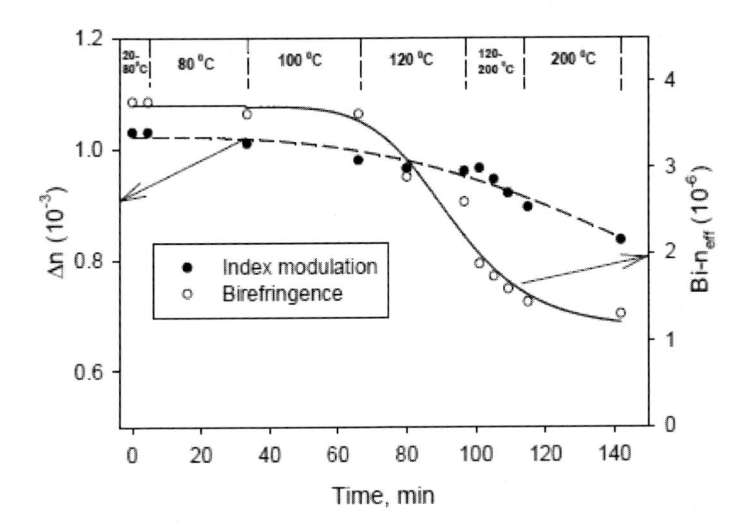

Figure 7. Annealing of UV-induced birefringence of FBGs inscribed in H_2-loaded fibers at room temperature.

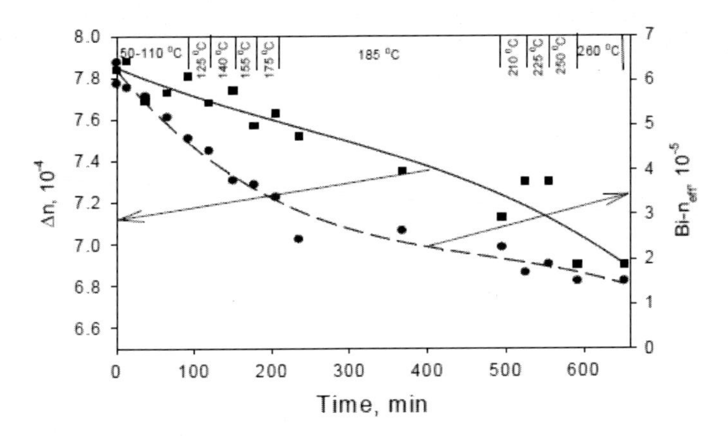

Figure 8. Annealing of UV-induced birefringence of FBGs inscribed in H_2-loaded fibers at 140 ° C.

From Figure 8 we can see that the UV-induced birefringence cannot be annealed out at low temperature. The induced birefringence is still on the order of 10^{-5} after the grating is annealed up to 260 °C.

3.3. High Peak Power IR Laser-Induced Birefringence in FBGs

Recently high peak power, regeneratively amplified Ti:Sapphire ultrafast infrared (ultrafast-IR) lasers ($\lambda = 800nm$, pulse durations from 120 fs to 2 ps) have been used to fabricate high quality FBGs in single mode fibers with or without phase masks [8, 22-27]. Depending on the pulse energy and pulse width of the ultrafast-IR laser, two types of gratings, type I-IR and type II-IR, have been observed in single mode fibers [22]. The formation of type I-IR gratings is associated with nonlinear absorption of the IR light and the grating annealing property is similar to that of type I UV gratings. Type II-IR gratings are usually fabricated with high pulse intensity and they are associated with highly localized damage in the fiber and exhibits high temperature stability [22]

3.3.1. Grating Inscription and Birefringence Monitoring

The Type I-IR and Type II-IR gratings under test are spot written in *SMF-28* fibers (both H_2-loading and non H_2-loading) by focusing the IR laser beam through a phase mask with the pitch of 1.07 μm. The focal length of the cylindrical lens is 19 mm and the focused beam spot size is approximately ~3 μm × 6.4 mm. The focused beam is scanned across the fiber core using a piezo-electric actuated stage [23]. The distance between the phase mask and the fiber for Type II-IR grating inscription is 300 μm in order to obtain high peak intensity and it is 1.3 mm for Type I-IR inscription to get a pure two-beam interference fringe [27]. The pulse durations for Type I-IR and Type II-IR grating inscriptions are measured with an autocorrelator to be 125 fs and 2.5 ps respectively. The set-up of FBG inscription and birefringence measurement is shown in Figure 3.

3.3.2. Birefringence in Type I-IR Gratings

The ultrafast-IR laser induced birefringence during the inscription of Type I-IR grating in H_2-loaded SMF-28 fibers is examined with the polarization of the inscription laser beam either normal or parallel to the fiber axis (S-polarized or P-polarized respectively). No obvious dependence of the induced birefringence on the polarization of the IR laser beam is observed. A typical PDL spectrum of a Type I-IR grating inscribed in H_2-loaded fiber is shown in Figure 9. The strength of the grating is -21 dB in transmission with an Δn of 5.2×10^{-4} at the center of the grating. Comparing Figure 9 with the simulation result shown in Figure1, we can see clearly that the major part of birefringence in the Type I-IR grating in H_2-loaded fibers is randomly distributed along the grating length, which means that the birefringence induced by the IR laser is smaller than the fiber's intrinsic birefringence. The growth of the birefringence with increasing laser induced index change is plotted in Figure 10. For comparison, the UV-induced birefringence in H_2-loaded SMF-28 fibers is plotted as well in Figure 10 with the polarization of the UV beam P-polarized, which minimizes the induced birefringence [12]. We can see that for the same induced index in H_2-loaded SMF-28 fibers, the birefringence induced by ultrafast-IR laser is much lower than that induced by the UV laser. After the Type I-IR gratings in H_2-loaded fibers are inscribed, the gratings are left at room temperature for two days. The birefringence is measured again and it is found that the birefringence induced by ultrafast-IR laser is annealed out at room temperature during the two days, which is different from the annealing of UV-induced birefringence in H_2-loaded SMF-28 fibers [19-21].

The birefringence of Type I-IR gratings induced by the ultrafast-IR laser in unloaded *SMF-28* fibers is examined and a typical spectral PDL is shown in Figure 11. It can be seen that there are clearly characteristic peaks in the spectral PDL curve. Compared to the curves in Figures 1 and 2 we can see that the laser-induced birefringence is much larger than the fiber's intrinsic birefringence and the laser-induced index is positive (similar to Figure 2(a), the PDL peak with the narrower bandwidth is on the short wavelength side). The growths of birefringence in effective index, B_i-n_{eff} with increasing index modulation, Δn, are shown in Figure 12 with the polarization of the laser beam either *P* or *S*-polarized. We can see that the IR laser with *S* -polarization induces a much higher birefringence than the *P* polarized laser beam, which is similar to the induced birefringence in H_2-loaded *SMF-28* fibers by UV laser [12]. Comparing Figure 12 with Figure 10, for the same IR laser induced Δn in Type I-IR gratings, the induced birefringence in H_2-loaded fibers is much lower than that in unloaded fibers (10^{-6} vs.10^{-5}).

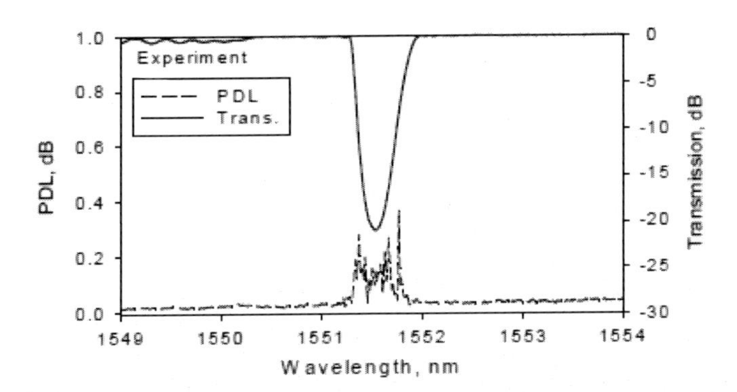

Figure 9. Spectral PDL of type I-IR grating inscribed in H_2-loaded fiber.

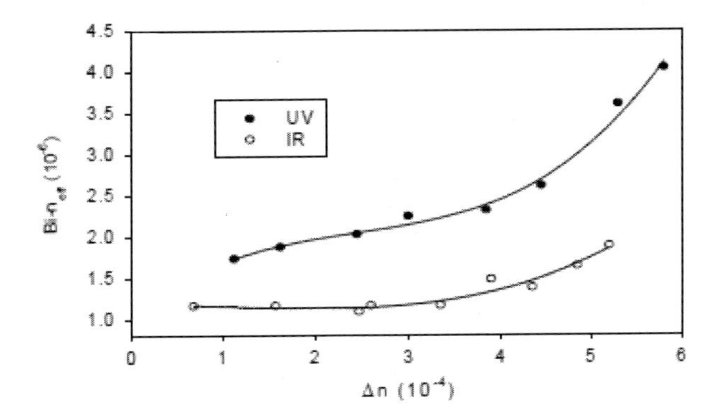

Figure 10. Comparison of the birefringence growth in H_2-loaded SMF28 fibers exposed by UV and IR lasers, respectively.

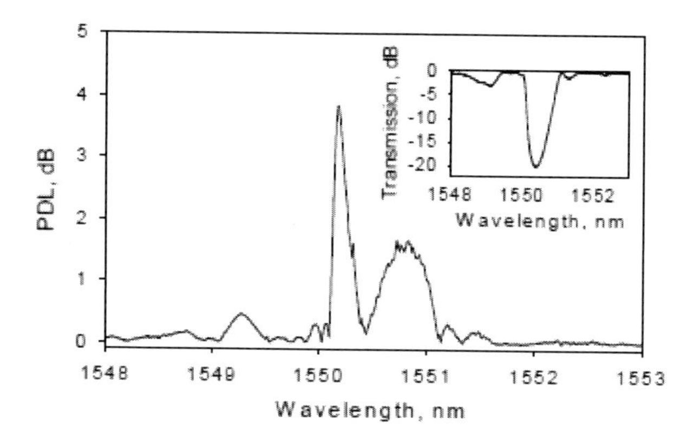

Figure 11. Spectral PDL of type I-IR grating inscribed in non H_2-loading fiber.

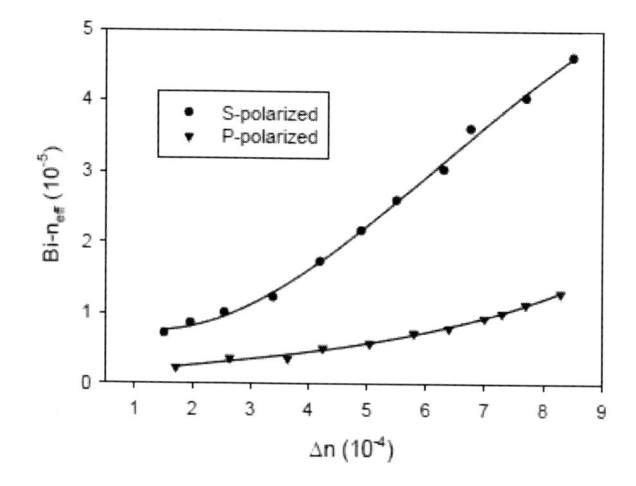

Figure 12. Growth of the IR laser induced birefringence with the polarization of the laser beam parallel and perpendicular to the fiber axis respectively.

3.3.3. Birefringence in Type II-IR Gratings

The inscription of Type II-IR gratings is usually associated with the introduction of wavelength independent scattering loss and high cladding mode coupling. The time inscription of a Type II-IR grating (Δn of up to 10^{-2}) requires far fewer laser pulses making it difficult to measure the birefringence growth during grating inscription. In H_2-loaded *SMF-28* fibers, the ultrafast-IR laser induces high birefringence and PDL for both *P* and *S*-polarized. The spectral PDL changes from sample to sample that is due to the large variation of the grating spectra between samples even for the same exposure condition.

Figure 13 shows a spectral PDL of Type II-IR grating inscribed in H_2-loaded fiber with laser beam *S*-polarized. The triangle dots present the experimental results and the solid curve is the simulation. From Figure 13 we can see that the induced Δn is negative when compared to Figure 2(b), which is different from the positive index change in Type I-IR gratings inscribed by ultrafast-IR and CW UV lasers. For the three samples of Type II-IR gratings in H_2-loaded *SMF-28* fibers, the induced index changes are all negative. In the simulation curve in Figure 13, the induced Δn and birefringence at the center of the apodized grating are

-1.7×10^{-3} and 3×10^{-4}, respectively, i.e., the birefringence is more than 15% of the Δn, which is much higher than that of the UV laser induced birefringence in H_2-loaded *SMF-28* fibers. When the IR laser beam is *P*-polarized, the induced birefringence is ~6% of the induced Δn, which is still much higher than that induced by UV laser (<2% [18]).

For non H_2-loaded fibers, four Type II-IR grating samples have been tested with two samples for each polarization of the laser beam (*S*- or *P*-polarized). It was found that, similar to the case of Type I-IR gratings, the *S*-polarized laser beam induced a higher birefringence than that of the *P*-polarized.

For the induced Δn of 10^{-3}, the birefringence induced by *P*-polarized beam is around 10^{-5} and it is around 10^{-4} for the *S*-polarized. For all the four samples in non H_2-loaded fibers, the induced refractive indices are negative, which is the same as the Type-II gratings in H_2-loaded fibers.

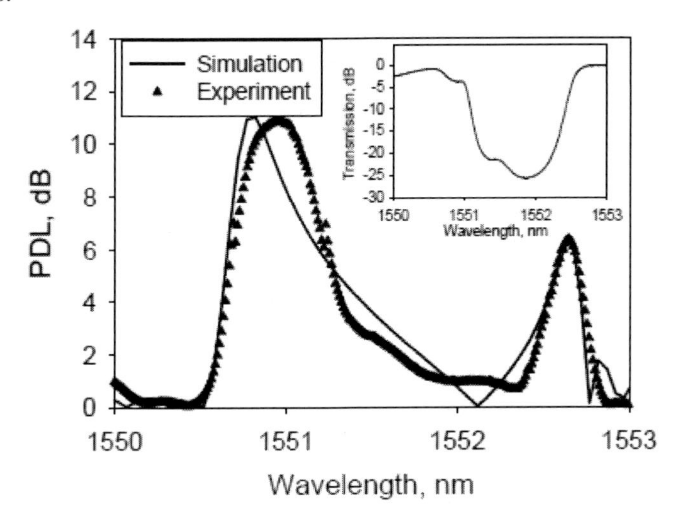

Figure 13. The spectral PDLs of Type II-IR grating inscribed in H_2-loaded fiber. Triangle dots: experiment, solid curve: simulation.

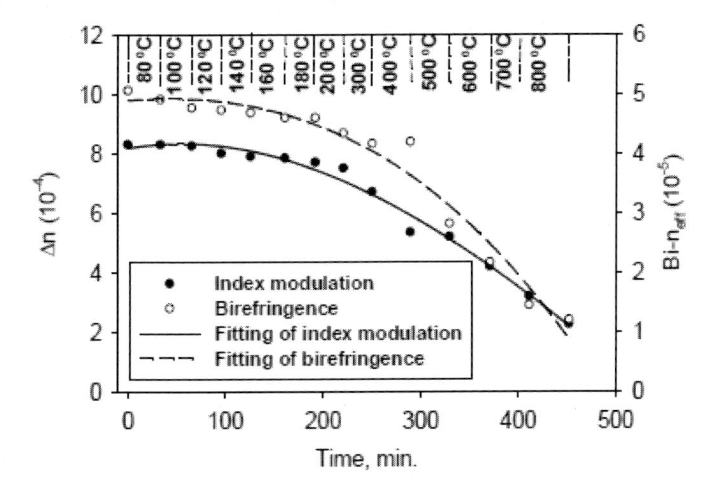

Figure 14. The annealing curves of the induced index modulation (solid dots) and birefringence (hollow dots) of Type I-IR gratings inscribed in unloaded fibers.

Negative induced index changes and generation of birefringence has been observed in bulk silica irradiated with 800 nm 200 fs pulses with intensities similar to the Type II-IR gratings presented here [28].

3.3.4. Annealing Properties of Birefringence in Type I-IR Gratings

The IR induced birefringence of Type I-IR gratings in H_2-loaded fiber anneals out at room temperature after ~48 hours with only the intrinsic birefringence of the fiber remaining and the index modulation drops by ~10%. The annealing of birefringence of Type I-IR gratings inscribed in unloaded fibers is shown in Figure 14 (the laser beam is *S*-polarized, the annealing of birefringence induced by *P*-polarized IR beam is the same). The induced birefringence has the same annealing resistance as the index modulation even for temperatures up to 800 °C, which is different from the annealing property of the induced birefringence in H_2-loaded *SMF-28* fibers both by UV and IR lasers. In Figure 14, after 450 minutes annealing at the increased temperature up to 800 °C, the induced Δn and birefringence drop by 73% and 76% respectively.

3.3.5. Annealing Properties of the Birefringence in Type II-IR Gratings

The annealing properties of Type II-IR gratings are more complicated than the Type I-IR gratings, partially due to the observation that a weak Type I-IR grating is usually inscribed initially before a Type II-IR grating is fabricated, especially in H_2-loaded fibers. Due to the different signs of induced index in the fibers (positive/negative) and different annealing properties of Type I-IR and Type II-IR gratings, the annealing property of a Type II-IR grating has a strong dependence on the grating sample, fiber (H_2-loading or non H_2-loading) and the polarization of the laser beam. For low birefringence induced by *P*-polarized laser beam in non H_2-loading fibers ($\sim 10^{-5}$), the birefringence is relatively stable during annealing at increased temperature from room temperature to 200 °C (total 3 hours) and it is still stable during a long term annealing at 200 °C (16 hours). The spectral PDLs at 50, 100, 140 and 200 °C during the annealing process are shown in Figure 15. By modelling of the spectral PDLs, the variation of the birefringence over the whole annealing period (19 hours) is on the order of the fiber's intrinsic birefringence ($\sim 10^{-6}$).

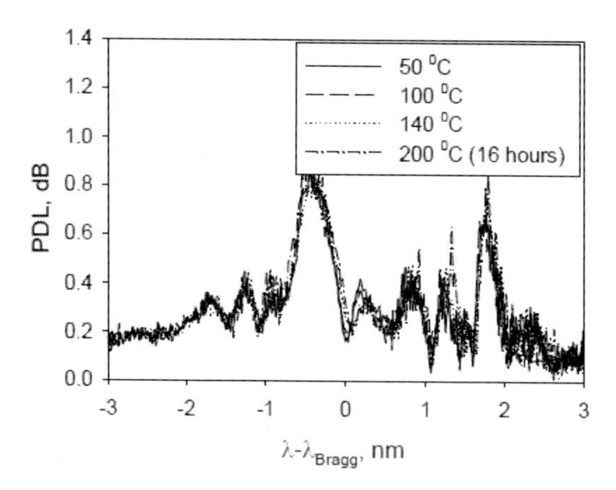

Figure 15. The type II-IR grating spectral PDLs induced by P-polarized laser beam in non H_2-loading fibers at 50, 100, 140 and 200 °C during the annealing process.

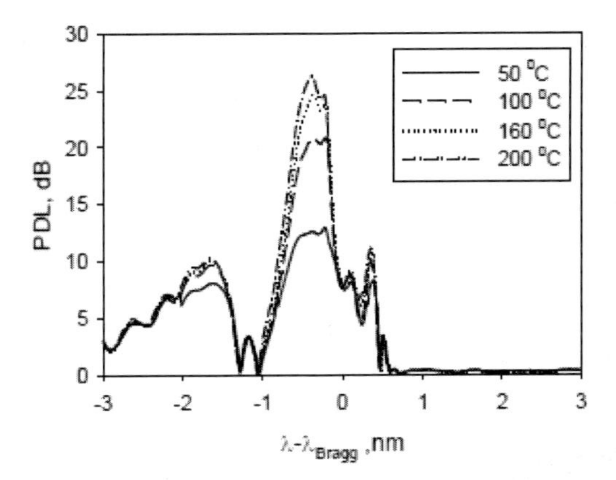

Figure 16. The type II-IR grating spectral PDLs induced by S-polarized laser beam in non H_2-loaded fibers during annealing at various increased temperatures.

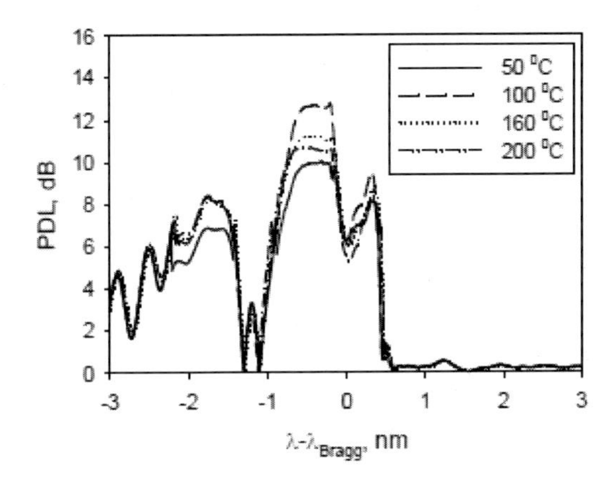

Figure 17. Another example of the changes of type II-IR grating spectral PDLs during annealing process, same as grating whose annealing properties shown in Figure 16, the type II-IR grating is written in non H_2-loaded fiber with the laser beam S-polarized.

For Type II-IR gratings inscribed by using *P*-polarized laser in H_2-loaded fibers and by using *S*-polarized laser in H_2- and non H_2-loaded fibers, the high birefringence ($>10^{-4}$) shows a much larger variation during the annealing process than the low birefringence ($\sim 10^{-5}$) induced by *P*-polarized beam in non H_2-loaded fibers. Even for the same exposure conditions (laser beam polarization, laser pulse energy, pulse width, with/without H_2-loading in fibers and exposure time etc.), different samples show different annealing behaviours. Figure 16 shows the changes of spectral PDL in non H_2-loaded fibers induced by *S*-polarized laser beam during annealing at various increased temperatures.

The spectral PDL increases with increasing annealing temperature, which is opposite to the annealing of Type I-IR gratings. By using the same laser exposure condition and the same fiber, another Type II-IR grating with the same grating strength (~23 dB in transmission) is inscribed and its spectral PDL during various annealing temperature is shown in Figure 17. Different from the annealing property shown in Figure 16, the spectral PDL increases first when the temperature increases from 50 to 100 °C, and then decreases with increasing temperature.

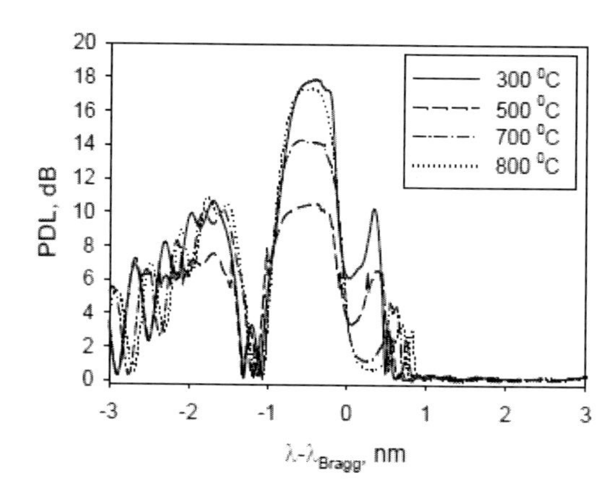

Figure 18. The spectral PDLs of type II-IR gratings at various temperatures (above 200 °C) after being annealed up to 200 °C.

The grating with its spectral PDL shown in Figure 16 was annealed with high temperature up to 800 °C. The spectral PDL still showed strong temperature dependence and no obvious decay of the birefringence was observed (see Figure 18). The variation of spectral PDL during annealing at different temperatures is due to the change of the grating strength. Through close examination of the grating spectra when the probe laser is launched into the fast and slow axes, it is found that both of grating strengths along fast and slow axes showed strong but different variations with temperature. The reason for the different variations of the grating strength with different polarizations needs to be investigated further.

4. Compensation of the Planar Waveguide Birefringence Using UV Laser Exposure

Compared with optical fibers, planar waveguides produced by plasma enhanced chemical vapor deposition (PECVD) or flame hydrolysis deposition (FHD) have a very strong intrinsic birefringence that makes the Bragg wavelength and peak reflectivity sensitive to the polarization state of the signal propagating through the waveguide. In this section we present the technique of compensating the intrinsic birefringence of planar waveguide by UV exposure. In silica-on-silicon planar technology, the ridge waveguide has both asymmetrical geometrical and material structures, and thus has larger geometrical birefringence B_g and stress birefringence B_s than that in an optical fiber. The waveguide intrinsic birefringence B_i is given by

$$B_i = B_s + B_g \tag{3}$$

where B_s is the stress-induced birefringence and B_g is the geometrical birefringence. The stress-induced birefringence is from the stress anisotropy that is induced mainly by the thermal-expansion mismatch among dissimilar materials.

In practice, some novel fabrication methods have been used to minimize waveguide birefringence, for example, trenching on the both sides of waveguide, depositing additional cladding layers, and laser processing waveguide with mid-infrared or UV, etc. [29-31]. UV irradiation is a direct and effective method to obtain polarization insensitive Bragg gratings by compensating the intrinsic birefringence B_i in the waveguide with the UV induced birefringence B_{uv}, which is produced automatically during the Bragg grating formation [32-33]. The mechanism of the UV controlled birefringence can be attributed to the stress relief of the waveguide core and the surrounding area.

It has also been suggested, that the mechanism could be a refractive index change in the waveguide core exposed by UV, which in turn changes the waveguide geometrical birefringence. In the paper [34], we addressed the relative contribution of these mechanisms and provided some clarification with respect to which process is dominant during the UV trimming of a planar waveguide's birefringence. A theoretical model is developed which reveals the relationships of the waveguide intrinsic birefringence B_i with the core refractive index, the core dimension, and the stress change of the waveguide, allowing us to separately examine the contributions of these different parameters.

In the experiments, the method to measure the intrinsic birefringence B_i and UV induced birefringence B_{uv} are established by inducing a weak Bragg grating and monitoring the polarization-dependent Bragg wavelength change under the UV irradiation. The weak Bragg grating with the modulated index Δn ~1.49 x10^{-4} are induced by using a zero-order nulled phase mask and ArF excimer laser operating under the following conditions: 40 mJ/pulse and 50 Hz. The polarization-dependent Bragg wavelength shift is given as

$$\lambda_{TM} - \lambda_{TE} = 2\Lambda(n_{TM} - n_{TE}) \tag{4}$$

where λ_{TM}, λ_{TE}, n_{TM} and n_{TE} are the Bragg wavelengths and effective indexes for TM and TE modes, respectively. The birefringence B of the waveguide and the initial birefringence B_i of the waveguide are defined as n_{TM} - n_{TE} and n_{0TM} - n_{0TE}, respectively. n_{0TM} and n_{0TE} are the initial effective indexes of the waveguide unexposed by UV for TM and TE modes, respectively. The Bragg grating is subsequently exposed to an un-modulated blanket UV irradiation by removing the phase mask. By observing the changes of two Bragg grating wavelengths for TM and TE modes, B and B_i are available. To test the change of Bragg wavelength with exposure time, the UV irradiation was interrupted periodically in order to measure the TM and TE modes of the Bragg wavelengths. As shown in Figure 19, the results were fitted with a polynomial regression, and expressed with a solid line. The waveguide birefringence B is estimated with λ_b *(TE)* - λ_b *(TM)*. The initial Bragg wavelengths λ_{b0} *(TE)*, λ_{b0} *(TM)* are then determined by extrapolating the curves to zero exposure time. It is clear that the birefringence of the waveguides can be reduced, in some cases to zero, by long UV exposures. A, B, C, D and E denote waveguides of dimensions 8.8μm×5.6μm, 7.7μm×5.6μm, 6.6μm×5.6μm, 5.7μm×5.6μm and 4.6μm ×5.8μm respectively. The birefringence in the 8.8μm×5.6μm large core size waveguide is tuned to 1×10^{-4} from -1×10^{-4} after 26 minutes UV irradiation. With the same UV irradiation condition, the birefringence in the 4.6μm×5.8μm small core size waveguide is tuned to -4×10^{-4} from -7×10^{-4}. This technique has been used successfully to fabricate polarization independent Bragg gratings in FHD waveguides with a small intrinsic birefringence of ~ 2 x10^{-4}. The conclusion is that changes of the birefringence in silica ridge waveguide structures with UV irradiation can be attributed mainly to stress changes in the waveguide core and its surrounding area, and not to the core refractive index change of 10^{-3} induced by UV irradiation. The experimental results are in agreement with this analysis.

Often high UV induced index changes and high UV exposure dosages are required in order to induce enough complementary UV-induced birefringence to compensate for the intrinsic birefringence of the waveguide.

Such extreme conditions require extremely long and intense exposures, which can result in damage to the surface of waveguides. By enhancing the photosensitivity of the planar waveguides through hydrogen loading, a larger UV induced birefringence is available in addition to a strong index change [35]; however, the induced birefringence is limited by the saturation of the index change.

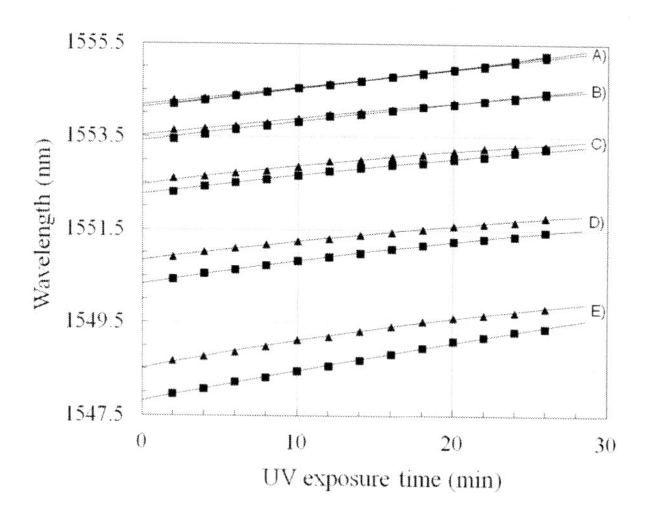

Figure 19. Experimental results of Bragg wavelength as a function of UV exposure time for different waveguide dimensions. The TE mode is denoted by squares while the TM mode is denoted by triangles. A, B, C, D and E denote waveguides of dimensions 8.8 μm X 5.6 μm, 7.7 μm X 5.6 μm, 6.6 μm X 5.6 μm, 5.7 μm X 5.6 μm and 4.6 μm X 5.8 μm respectively. The solid line is a polynomial regression through each data set.

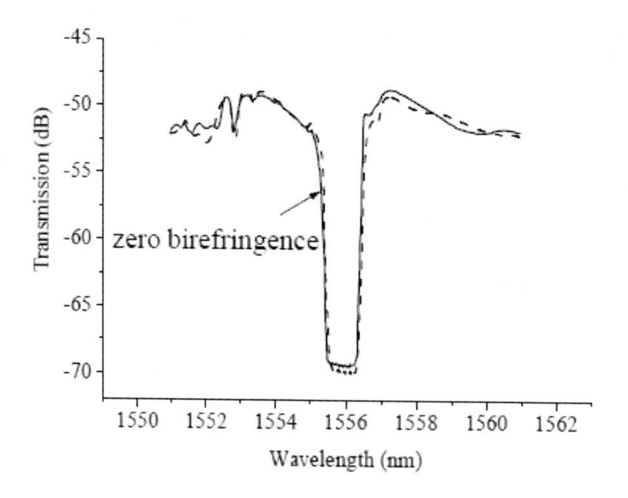

Figure 20. Transmission spectra for TM (solid line) and TE (dashed line) modes of the Bragg grating trimmed after low fluence blanket UV s-polarization exposure.

In the paper [36], a method for rapid compensation of high intrinsic birefringence in plasma-enhanced chemical vapor deposition (PECVD) based planar waveguides is presented. Compensation of the intrinsic waveguide birefringence is achieved by the induction of a large UV induced birefringence opposite that of the intrinsic birefringence. The UV induced birefringence is created with polarized UV ArF excimer laser irradiation (oriented normal to the waveguide axis) and enhanced further through hydrogen loading. The fabrication of polarization insensitive Bragg gratings (PIBG) in PECVD planar waveguides with a large intrinsic birefringence of 6.3×10^{-4} is demonstrated in Figure 20 by trimming it to zero birefringence with a low fluence blanket UV polarized exposure. After annealing for removal of the hydrogen, the polarization insensitive nature of the Bragg grating is unchanged. The big reduction in the waveguide birefringence with UV polarized irradiation is due to the change in the anisotropic densification at the core caused by UV irradiation. The birefringence of two waveguides irradiated with either UV *S*-polarization or UV *P*-polarization is different, indicating that the anisotropy at the core and its surrounding area is modified differently. The core along the direction of the TE mode is more compressed by the irradiation polarized along the same direction as the TE mode. It is clear that UV polarized irradiation with a wide beam could tune the core densification anisotropy by changing the compressive stress in the core and its surrounding area along the polarization direction of the irradiation through dichroic absorption. With the technique of the polarization independent Bragg grating by UV irradiation, the quality of planar waveguide Bragg gratings used as the telecommunication devices and sensors is improved [37].

5. The Impairments of Birefringence in FBGs and WBGs in Communication Systems

5.1. Simulation Model

The model to simulate system impairments of birefringence in FBGs and WBGs is shown in Figure 21(a) that consists of a transmitter, receiver and cascaded optical add-drop multiplexers (OADMs) with each OADM consisting a FBG (or WBG) and two optical circulators, see Figure 21(b). The FBG birefringence-induced power penalty in the network is calculated by examining the eye opening of 10 Gb/s non-return-to-zero (NRZ) signals modulated by a pseudorandom binary sequence (PRBS) with length of 2^7-1. The pulse shape of an isolated pulse in the sequence is Gaussian with 3-dB pulse duration of 60 ps.

We assume in the simulation that all gratings in Figure 21 are apodized with a 3-dB bandwidth of 60 GHz and the grating lengths are 10 mm. By using the coupled mode theory, the grating spectrum, spectral PDL, dispersion, and spectral DGD can be calculated, see Figures 22(a) and 22(b) where the index apodization profile of the FBG is Gaussian. The birefringence in mode effective index and the birefringence in index modulation used in the calculation of spectral PDL and DGD curves in Figures 22(a) and 22(b) are 5×10^{-6} and 8×10^{-6} respectively.

These two values were observed previously when H_2-loaded *SMF-28* fiber was exposed to CW frequency doubled Ar^+ laser radiation through a phase mask for a few minutes [18, 38].

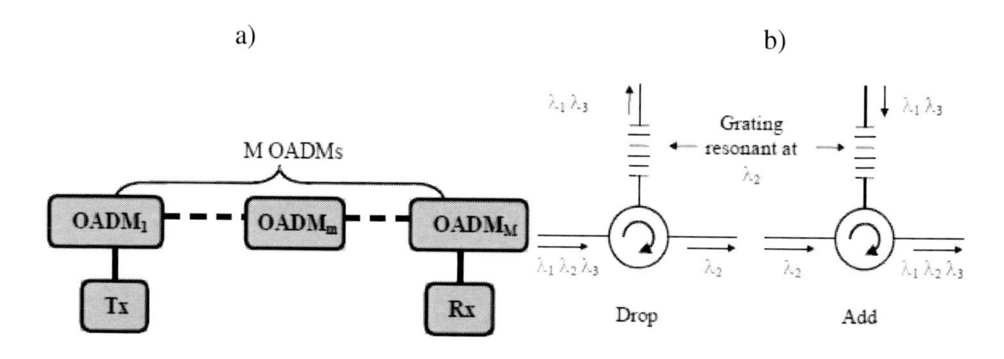

Figure 21. The simulation model: (a) the optical network consists of a transmitter, receiver and *M* FBG-based OADMs; (b) the diagram of OADM made of one FBG and one circulator.

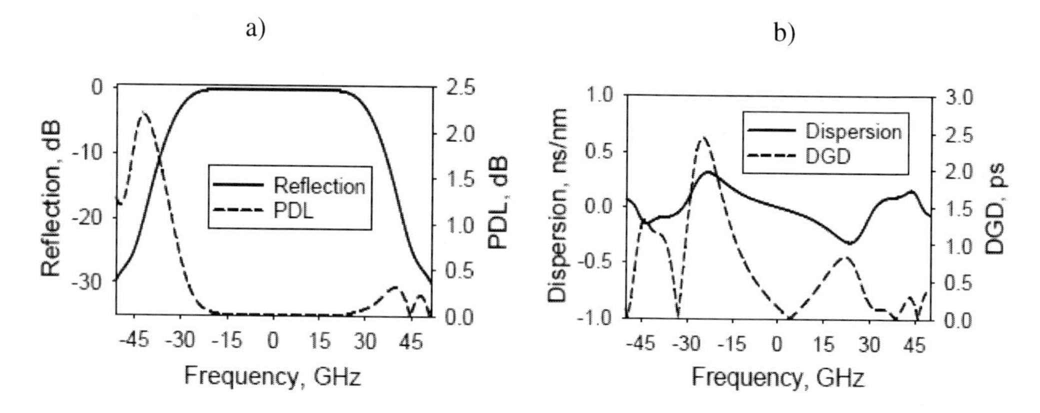

Figure 22. The properties of a FBG: (a) FBG spectrum and spectral PDL in reflection; (b) FBG dispersion and spectral DGD in reflection.

The birefringence in FBGs inscribed in non H_2-loaded *SMF-28* fiber by using ultrafast IR laser and the birefringence in planar waveguide can be an order of magnitude higher than these values [21], resulting in much larger spectral DGD and PDL. We assume that the FBG-based OADMs are linked with optical fibers and the polarization of the signal in the fiber links changes randomly with time due to variations of temperature, strain, vibration, etc. The polarization independent insertion losses from OADMs, optical fibers and connectors are not included in the simulation.

Pulse broadening due to the first order PMD can be simply expressed as $\Delta\sigma = \sigma_{out} - \sigma_{in} = \sqrt{\sigma_{in}^2 + \tau^2\gamma(1-\gamma)} - \sigma_{in}$ [39], where σ_{in} and σ_{out} are the input and output pulse widths, τ is the DGD of the system and γ $(0 \le \gamma \le 1)$ is the relative power launched to the fast (or slow) principal state of polarization.

In the presence of both PMD and PDL, especially when higher order PMD and PDL are considered as shown in Figures 22(a) and 22(b), the calculation of pulse distortion is much more complicated so that it cannot be expressed by a simple equation [40, 41].

In this work, the signal distortion induced by FBGs in the presence of birefringence is obtained by examining the evolutions of the Jones vectors at all Fourier frequencies by using the following equation,

$$\vec{E}_{out}(f_n) = T(f_n)\vec{E}_{in}(f_n) \tag{5}$$

where f_n (n=1, 2, …, N) is the discrete frequency for the Fourier transform of the signal, $\vec{E}_{in}(f_n)$ and $\vec{E}_{out}(f_n)$ are the Jones vectors (electric fields) at frequency f_n at the network's input and output respectively. The polarization state of the input signal $\vec{E}_{in}(f_n) \Big/ \left|\vec{E}_{in}(f_n)\right|$ is set to $\begin{pmatrix} 1 \\ 1 \end{pmatrix} \Big/ \sqrt{2}$. $T(f_n)$ in Equation (5) is the transmission matrix at frequency f_n of the network consisting of M OADMs and fiber links and it can be written as

$$T(f_n) = T_1(\theta_1)\cdot T_1^{FBG}(f_n)\cdot T_2(\theta_2)\cdot T_2^{FBG}(f_n)\cdots T_m(\theta_m)\cdot T_m^{FBG}(f_n)\cdots T_M(\theta_M)\cdot T_M^{FBG}(f_n)$$
$$(m = 1, 2, \ldots M) \tag{6}$$

where $T_m(\theta_m)$ represents the rotation matrix of the mth fiber link and $T_m^{FBG}(f_n)$ represents the transmission matrix of the mth FBG. They can be written as

$$T_m(\theta_m) = \begin{pmatrix} \cos(\theta_m) & \sin(\theta_m) \\ -\sin(\theta_m) & \cos(\theta_m) \end{pmatrix} \tag{7}$$

$$T_m^{FBG}(f_n) = \begin{pmatrix} r_m^{fast}(f_n) & 0 \\ 0 & r_m^{slow}(f_n) \end{pmatrix} \tag{8}$$

In Equation (7), θ_m ($0 \le \theta_m \le \pi$) is the rotation angle and it can be randomly distributed between 0 and π to represent the random polarization variation of the signal in the fiber link. $r_m^{fast}(f_n)$ and $r_m^{slow}(f_n)$ in Equation (8) are the reflection coefficients of the mth FBG when the signal is launched onto the fast and slow axes respectively. The output electrical field signal in time domain, $\vec{E}_{out}(t)$, can then be obtained through the inverse Fourier transform of $\vec{E}_{out}(f_n)$. The distorted NRZ output signal, $P(t) = \left|\vec{E}_{out}(t)\right|^2$, is compared to NRZ input signal and the power penalty of eye opening can be calculated.

5.2. Simulation Result and Discussion

5.2.1. Definition and Representation of FBG Birefringence-Induced Power Penalty

The power penalty of the cascaded Bragg grating filters in the network is examined by studying the eye-diagram and is compared to the result of a back-to-back connection. Figure 23 shows the eye diagrams for the cases of a back-to-back connection (23a), a 10 OADM cascade considering grating spectral bandwidth and dispersion but no birefringence effect (23b) and a 10 OADM cascade in the presence of grating spectral bandwidth, dispersion and birefringence (23c). The power penalty resulting from the spectral bandwidth and dispersion of FBGs in an optical network has been investigated previously [42-45]. In the presence of grating birefringence, the eye openings are reduced further and are changing with time (Figure 23c).

In order to quantitatively study the system impact of FBG birefringence, the statistical characteristics of the eye-opening diagram in Figure 23(c) needs to be examined. In this work, the power penalty induced by grating birefringence is defined as $-10\log_{10}(H/H_0)$ in the unit of dB, where H represents the eye opening in the presence of grating birefringence and H_0 is the minimum eye opening in Figure 23b.

By using this definition, the presented power penalty is induced by the birefringence only; other parameters, such as grating spectral bandwidth, dispersion, index apodization profile, and the wavelength misalignment that will be discussed later, are all included in the calculation of H_0. Due to the random variations of the polarization state of the signal, the statistical property of the power penalty is presented as a percentage value of when the power penalty is larger than a certain amount for a statistical ensemble consisting of 50,000 eye-opening values.

5.2.2. The Induced Power Penalty of Birefringence in FBGs with Various Index Apodization Profiles

The index profiles of FBG filters used in wavelength-division multiplexing (WDM) systems are usually apodized in order to promote side lobe suppression and high channel isolation. Depending on the application, various apodization profiles are often used such as Gaussian, raised cosine, Blackman, sine, sinc, tanh, super Gaussian etc. With grating lengths of 10 mm and 3-dB bandwidths of 60 GHz, the reflection spectra and dispersion of FBGs with various index apodization profiles along with the NRZ 10Gb/s signal are shown in Figures 24(a) and 24(b) respectively. In the presence of birefringence of the effective index of 1×10^{-5} and the birefringence in index modulation of 1.3×10^{-5} (i.e., the ratio between the birefringence in effective index and the birefringence in Δn as shown experimentally [18, 21]), the spectral PDL and DGD of FBGs with various index apodization profiles are shown in Figures 24(c) and 24(d), respectively, along with the 10 Gb/s NRZ signal. From Figure 24(c), while within the bandwidth of the 10 Gb/s NRZ signal (-20 to 20 GHz), the spectral PDLs in FBGs with various apodization profiles are almost zero when the birefringence in effective index is $<10^{-5}$. The spectral DGD shown in Figure 24(d) however, has a strong dependence on the index apodization profile.

Among these FBGs, the Gaussian apodization function produces the highest spectral DGD within the signal bandwidth, which is consistent with the dispersion curve associated with Gaussian apodization profile in Figure 24(b). Figures 24(b) and 24(d) also show that the FBG with the TANH index apodization profile has the lowest dispersion and spectral DGD within the signal bandwidth. From the point of view of having lower system impairment in the presence of birefringence, TANH apodization is preferred. However, this apodization function produces higher out-of-band side lobe reflection resulting in poor channel isolation which is undesirable in WDM systems, see Figure 24a.

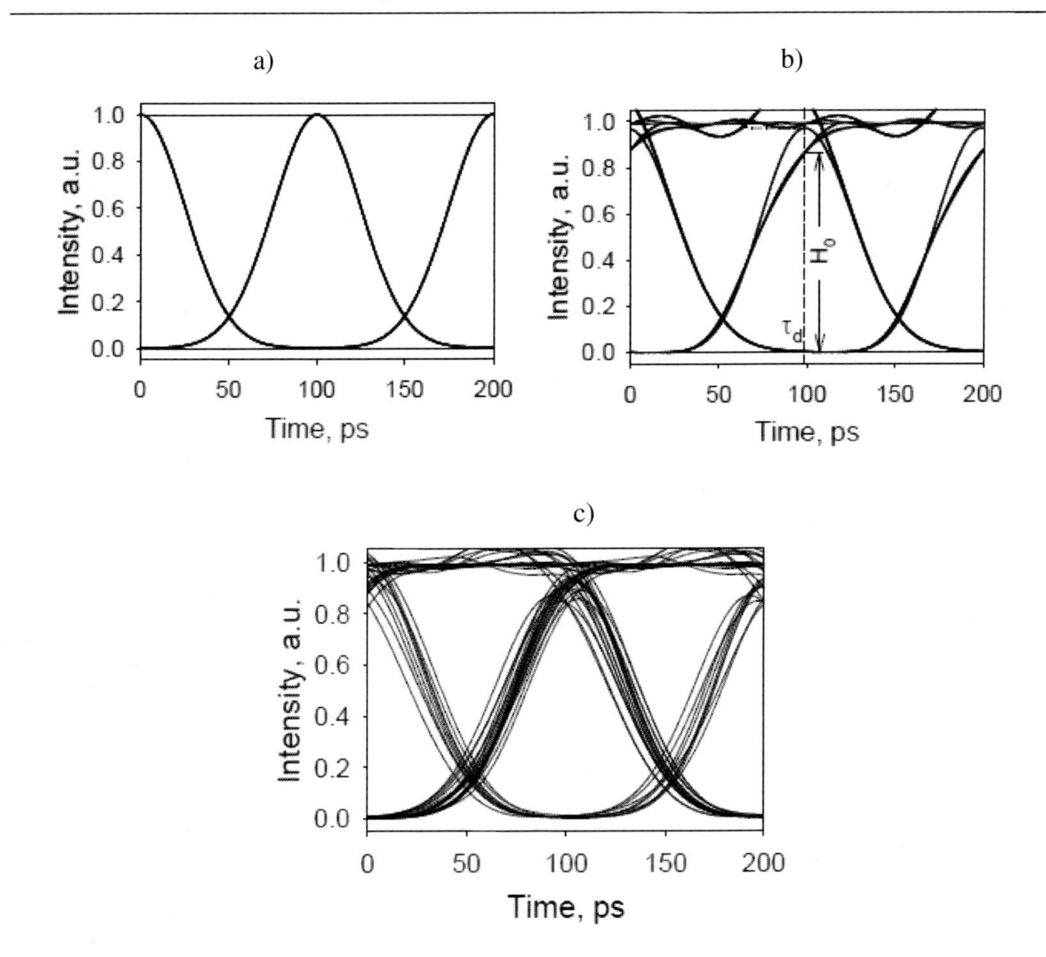

Figure 23. Eye diagrams of NRZ signal: (a) back-to-back; (b) with grating dispersion; and (c) with grating dispersion and birefringence.

It is obvious that higher spectral DGD in FBGs would produce higher system power penalty. To quantify this effect, Figure 25 presents outage probabilities for various power penalty values calculated for a network with 10 FBG-based OADMs in the presence of grating birefringence in effective index of 5×10^{-5} and birefringence in Δn of 6×10^{-5}. For the same grating birefringence values, Gaussian apodization profiles of the FBGs are more likely to induce high power penalty when compared to Raised cosine and Blackman apodization profiles. This is consistent with results shown in Figure 24(d) where within the spectral bandwidth (-20 to 20 GHz) of the 10 Gb/s NRZ signal, the spectral DGD is highest for the grating with a Gaussian apodization. The birefringence induced power penalties of FBGs with other apodization profiles, such as sine, sinc, etc., are also examined resulting in similar penalties to that for gratings with raised cosine apodization functions.

The 3-dB bandwidth of the FBGs used in above simulations is 60 GHz, which is chosen for the 10 Gb/s WDM system with channel spacing of 100 GHz. In the case where FBGs with narrower bandwidths are used for dense WDM systems, the spectral profiles of PDL and DGD shown in Figures 24c and 24d respectively will be narrower as well. As a result, the Fourier components of the signal will experience higher PDL and DGD, which induces a larger system power penalty.

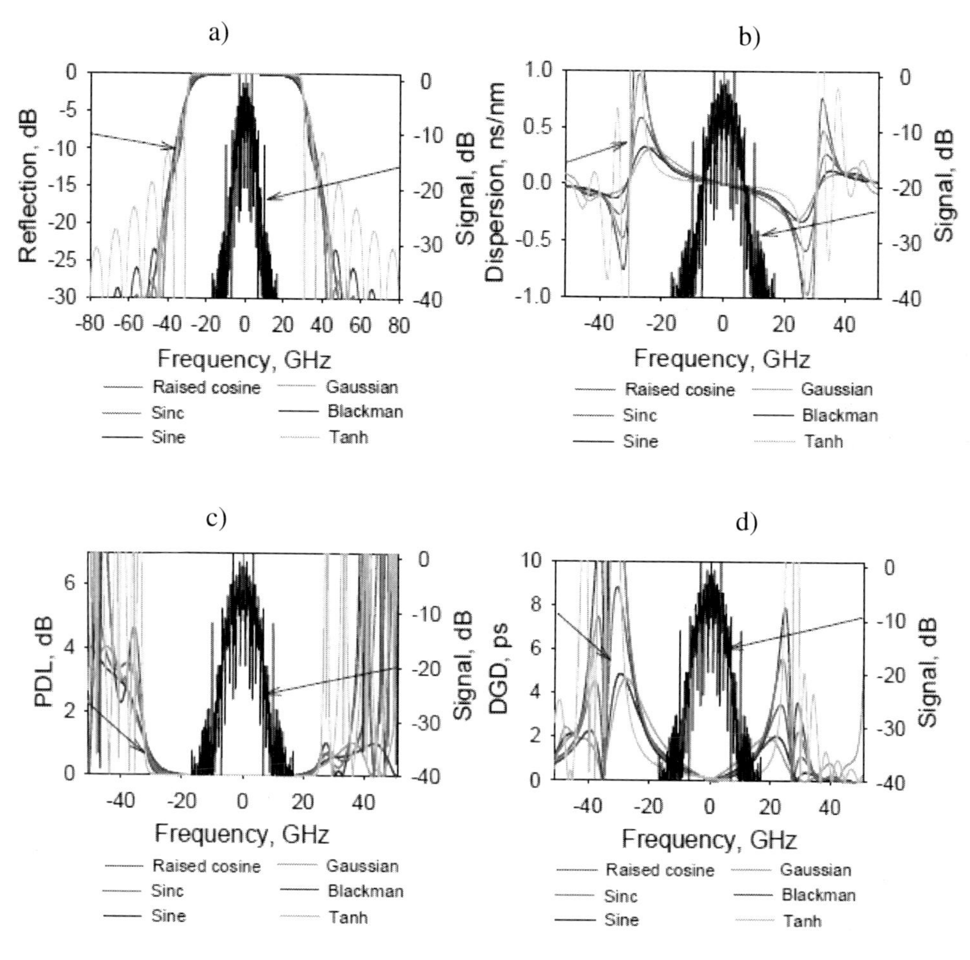

Figure 24. The responses of FBG with various index apodization profiles along with 10 Gb/s NRZ signal: (a) FBG reflection spectra; (b) FBG dispersion curves; (c) FBG spectral PDLs; and d) FBG spectral DGDs.

Similarly, for higher bit rate systems, the signal spectrum shown in Figure 24 will be broader, the frequency components then will experience higher PDL and DGD values resulting in higher power penalty. To quantify these effects, another two sets of simulations are performed. In the first statistical ensemble, the 3-dB bandwidth of the FBG filter (with Gaussian apodization) is reduced from 60 GHz to 30 GHz (appropriate for a 50 GHz channel spacing).

In the second statistical ensemble, the FBG 3-dB bandwidth is still 60 GHz but the bit rate is increased from 10 Gb/s to 20 Gb/s. Simulation results showed that in a system consisting of 10 FBGs with each of the FBG having a birefringence value of 10^{-5}, the maximum birefringence induced power penalty is 0.034 dB (based on 50,000 simulations) for the FBG 3-dB bandwidth of 60 GHz and the bit rate of 10 Gb/s.

When the 3-dB bandwidth is reduced to 30 GHz and other parameters remain the same, the outage probability for power penalty of 0.1 dB is 1.4%. When the bit rate is increased to 20 Gb/s, the outage probability for power penalty of 0.1 dB is 5.1%.

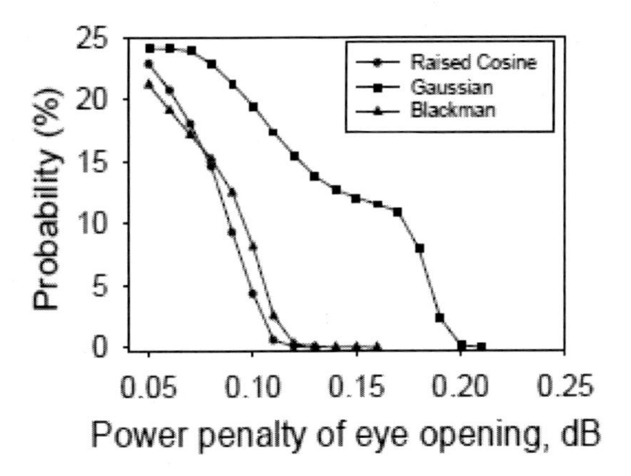

Figure 25. Outage probabilities for various power penalties for a network consisting of 10 FBG-based OADMs with various FBG apodization profiles.

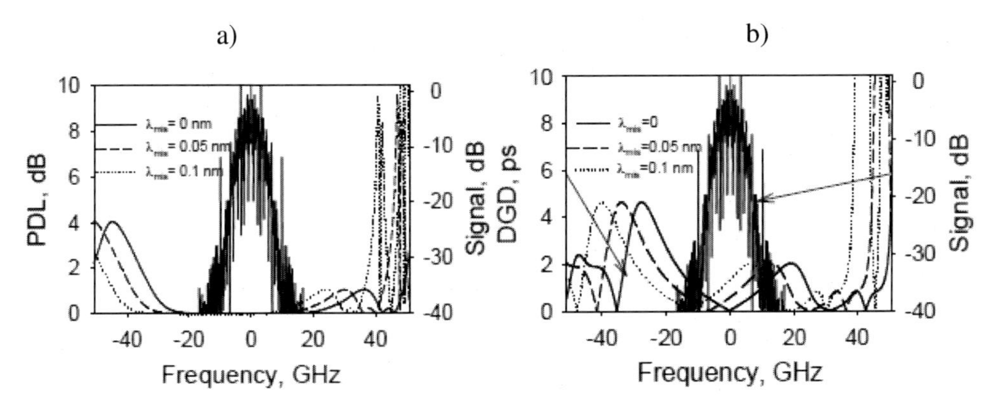

Figure 26. The spectral PDL and DGD of FBGs with wavelength misalignment of 0, 0.05 and 0.1 nm: (a) spectral PDL; (b) spectral DGD.

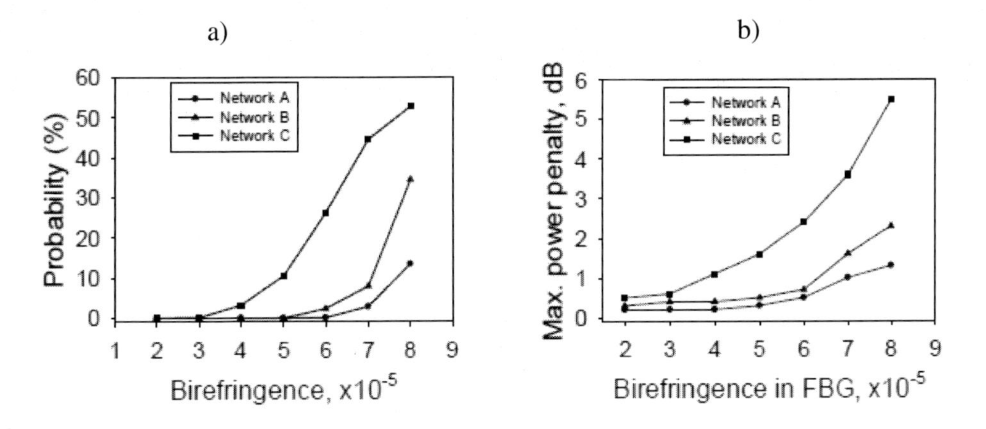

Figure 27. Birefringence induced power penalty in optical networks with various wavelength misalignment values: (a) The outage probabilities for power penalty of 0.5 dB; (b) the maximum induced power penalties based on 50,000 simulations in each statistical ensemble.

5.2.3. FBG Birefringence Induced Power Penalty in the Presence of Wavelength Misalignment

During the process of FBG fabrication, due to the inscription alignment, laser power fluctuations, and the speed accuracy of the moving stage, the Bragg wavelength of the grating can shift up to ± 0.1 nm from the designed value for a 30 dB grating. In industry, FBGs that are manufactured for network applications are typically written at a wavelength that is lower than the specified wavelength and then tension tuned to the desired value by special packaging, resulting in errors of the FBG central wavelength of ± 0.05 nm. When these gratings are used in an optical network with a laser source at a designed operating wavelength, this wavelength misalignment may introduce more signal distortion in addition to distortions arising from grating spectral band width, dispersion etc [42]. In the presence of grating birefringence, Figures 26a and 26b show the spectral PDL and spectral DGD of Gaussian apodized FBGs with 0, 0.05, and 0.1 nm wavelength misalignments along with the NRZ signal. Figure 26 shows that with the increase of wavelength misalignment, the spectral PDL and DGD at frequencies close to the center of the signal are higher when compared to the case of zero wavelength misalignment.

Obviously this increase of PDL and DGD within the signal bandwidth would introduce higher power penalty. Considering the fact that the amount of wavelength misalignment of an individual FBG is usually random within a certain range, in the simulation of this subsection the wavelength misalignments of FBGs in each network are assumed to vary from zero to a maximum value.

More specifically, the power penalties of three networks with 10 FBG-based OADMs with different wavelength misalignments are examined for various birefringence values. In the first network (network A), the Bragg resonances of the 10 FBGs are perfectly matched with the wavelength of the laser source (for the purpose of comparison).

In the second network (network B) there are two FBGs with each of following misalignments: 0.05, 0.025, 0, -0.025 and –0.05 nm. In the third network (network C) there are two FBGs with each of the following wavelength misalignments: 0.1, 0.05, 0, -0.05, and -0.1 nm.

Figure 27(a) shows the outage probabilities for power penalty of 0.5 dB in the three networks at various birefringence values and Figure 27(b) shows the maximum power penalties introduced in the three networks at various birefringence values (based on 50,000 simulation in each statistical ensemble).

Figures 27(a) and 27(b) clearly show that with wavelength misalignment in the networks, the birefringence induced power penalty is much higher when high birefringence is present in the FBGs. The power penalties induced by low birefringence ($<10^{-5}$) for various wavelength misalignments up to 0.1 nm are also examined and they are close to the result when there is no wavelength misalignment. The induced power penalty resulting from FBG birefringence is a combined effect of PMD and PDL.

In order to examine whether PMD or PDL introduces more system impairment, the birefringence induced power penalties in two systems, either with PMD only or PDL only, are studied with the birefringence value of 8×10^{-5} in each of the FBGs in network C. It should be noted however, that the treatment PMD or PDL in isolation when considering birefringence induced power penalties by FBGs in a system is not possible in reality as the two effects are coupled together [21]. Simulation results show that the outage probability for power penalty of 0.5 dB is 20% when PMD only is present and it is 27% when PDL only is present. From these simulation results and the results shown in Figure 27, we may conclude that: 1) PDL produces slightly higher system impairment than PMD in a system with a cascade of FBGs, and 2) the combination of PMD and PDL introduces more system impairment than the sum of those induced by PMD and PDL separately (the outage probability for power penalty of 0.5 dB is 53% when both PMD and PDL are considered with birefringence of 8×10^{-5}, see the result of network C in Figure 27a). The later conclusion is consistent with previous studies [6, 46].

CONCLUSION

UV and ultrafast IR lasers-induced birefringence in FBGs and WBGs is characterized experimentally. The dependence of spectral PMD and PDL on the induced birefringence in gratings is analyzed by using the coupled mode theory. It is seen that the birefringence of the index modulation is between 1.4% and 1.7%, while the birefringence of the effective index is between 0.7% and 1.2% of the index modulation value for the one-side writing technique when UV laser is used. The changes of birefringence in two types of FBGs, Type I-IR and II-IR, inscribed in *SMF-28* fibers using an ultrafast-IR laser, are studied. For Type I-IR gratings, the birefringence growth during grating inscription and its decay during grating annealing at high temperature are examined and compared to those created using a UV laser. It is shown that, in Type I-IR gratings, ultrafast IR laser-induced birefringence in H_2-loaded *SMF-28* fibers is low, can be annealed at room temperature and does not show obvious dependence on the polarization of the laser beam. The IR laser-induced birefringence in Type I-IR gratings in unloaded *SMF-28* fibers is much higher than in H_2-loaded *SMF-28* fibers and exhibits strong polarization dependence on the laser beam.

It has similar stability as the induced index for annealing temperatures up to 800 °C. The *P*-polarized ultrafast IR laser-beam–induced birefringence in Type II-IR gratings in non H_2-loaded fibers is low ($\sim10^{-5}$) and it is stable during annealing. However, the birefringence of Type II-IR gratings inscribed in H_2-loaded fibers with both *P*- and *S*-polarized laser beams and in non H_2-loaded fiber with *S*-polarized laser beam is high ($\sim10^{-4}$) and shown strong variation during annealing. The technique for compensation of the intrinsic birefringence in planar waveguide using UV laser exposure is presented.

Considering the presence of birefringence in Bragg grating based optical components in an all-optical network, the system power penalty induced by PMD and PDL is examined along with grating spectral bandwidth and dispersion by simulation. The eye diagram with spectral DGD and PDL in FBG-based OADMs is presented, and the distributions of the induced power penalties for various numbers of OADMs in the network are calculated. It is shown that when the birefringence in the Bragg grating is less than 10^{-5}, the birefringence-induced power penalty of eye opening is less than 0.1 dB with up to 30 Bragg gratings in the network. For the same birefringence values, FBGs with Gaussian apodization profile have higher spectral DGD resulting in higher induced power penalties. The presence of wavelength misalignment in optical networks significantly increases the birefringence-induced power penalty, especially when high FBG birefringence is present.

REFERENCES

[1] K. O. Hill, Y. Fujii, D. C. Johnson, and B. S. Kawasaki. (1978) Photosensitivity in optical fiber waveguides: Application to reflection filter fabrication. *Applied Physics Letters*, 32, 647–649.

[2] H. Bulow. (1998). System outage probability due to first- and second-order PMD. *IEEE Photonics Technology letters*, 10, 696-698.

[3] J. Cameron, L. Chen and X. Bao. (2000). Impact of chromatic dispersion on the system Limitation due to polarization mode dispersion. *IEEE Photonics Technology Letters*, 12, 47-49.

[4] N. Gisin, B. Huttner. (1997). Combined effects of polarization mode dispersion and polarization dependent losses in optical fibers. *Optics Communications*, 142, 119-125.

[5] A. El Amari, N. Gisin, B. Perny, et al. (1998). Statistical prediction and experimental verification of concatenation of fiber optic components with polarization dependent loss. *Journal of Lightwave Technology*, 16, 332-339.

[6] B. Huttner, C. Geiser and N. Gisin. (2000). Polarization-induced distortion in optical fiber networks with polarization-mode dispersion and polarization-dependent losses. *IEEE Journal of Selected Topics in Quantum Electronics*, 6, 317-329.

[7] K. O. Hill, G. Meltz. (1997). Fiber Bragg grating technology fundamentals and overview. *Journal of Lightwave Technology*, 15, 1263-1276.

[8] S. J. Mihailov, C. W. Smelser, D. Grobnic, R. B. Walker, P. Lu, H. Ding, and J. Unruh. (2004). Bragg gratings written in all-SiO_2 and Ge-doped core fibers with 800-nm femtosecond radiation and a phase mask. *Journal of Lightwave Technology*, 22, 94-100.

[9] P. J. Lemaire, R. M. Atkins, V. Mizrahi, and W. A. Reed. (1993). High pressure H_2 loading as a technique for achieving ultrahigh UV photosensitivity and thermal sensitivity in GeO_2 doped optical fibers. *Electronics Letters*, 29, 1191-1193.

[10] A. M. Vengsarkar, Q. Zhong, D. Inniss, W. A. Reed, P. J. Lemaire, and S. G. Kosinski. (1994). Birefringence reduction in side-written photoinduced fiber devices by a dual-exposure method. *Optics Letters*, 19, 1260-1262.

[11] H. Renner, D. Johlen and E. Brinkmeyer. (2000). Modal field deformation and transition losses in UV side-written optical fibers. *Applied Optics*, 39, 933-940.

[12] T. Erdogan and V. Mizrahi. (1994). Characterization of UV-induced birefringence in photosensitive Ge-doped silica optical fibers. *Journal of the Optical Society of America B*, 11, 2100-2105.

[13] F. Ouellette, D. Gagnon and M. Poirier. (1991). Permanent photoinduced birefringence in a Ge-doped fiber. *Applied Physics Letters*, 58, 1813-1815.

[14] K. Dossou, S. LaRochelle and M. Fontaine. (2002). Numerical analysis of the contribution of the transverse asymmetry in the photo-induced index change profile to the birefringence of optical fiber. *Journal of Lightwave Technology*, 20, 1463-1470.

[15] E. Simova, P. Berini and C. P. Grover, (1999). Characterization of chromatic dispersion and polarization sensitivity in fiber gratings. *IEEE Transactions on Instrumentation and Measurement*, 48, 939-943.

[16] Y. Zhu, E. Simova, P. Berini, and C. P. Grover. (2000). A Comparison of wavelength dependent polarization dependent loss measurements in fiber gratings. *IEEE Transactions on Instrumentation and Measurement*, 49, 1231-1239.

[17] B. L. Bachim and T. K. Gaylord. (2003). Polarization-dependent loss and birefringence in long-period fiber gratings. *Applied Optics*, 42, 6816-6823.

[18] P. Lu, D. S. Waddy, S. J. Mihailov, and H. Ding. (2005). Characterization of the growths of UV-Induced birefringence in effective mode index and index modulation in fiber Bragg gratings. *IEEE Photonics Technology Letters*, 17, 2337-2339.

[19] J. Canning, H. J. Deyerl, H. R. Sørensen and M. Kristensen. (2005). Annealing of UV-induced birefringence in hydrogen loaded germanosilicate fibres. *Proceeding of Bragg Gratings, Poling and Photosensitivity*, Star City, Sydney, Australia.

[20] J. Canning, H. J. Deyerl, H. R. Sørensen, and M. Kristensen. (2005). Ultraviolet-induced birefringence in hydrogen-loaded optical fiber. *Journal of Applied Physics*, 97, paper 053104.

[21] P. Lu, D. Grobnic and S. J. Mihailov. (2007). Characterization of the birefringence in fiber Bragg gratings fabricated with an ultrafast-infrared laser. *Journal of Lightwave Technology*, 25, 779-786.

[22] C. W. Smelser, S. J. Mihailov and D. Grobnic. (2005). Formation of type I-IR and type II-IR gratings with an ultrafast IR laser and a phase mask. *Optics Express*, 13, 5377-5386.

[23] D. Grobnic, C. W. Smelser, S. J. Mihailov, R. B. Walker, and P. Lu. (2004). Fiber Bragg gratings with suppressed cladding modes made in SMF-28 with a femtosecond IR laser and a phase mask. *IEEE Photonics Technology Letters*, 16, 1864-1866.

[24] A. Martinez, M. Dubov, I. Khrushchev, and I. Bennion. (2004). Direct writing of fibre Bragg gratings by femtosecond laser. *Electronics Letters*, 40, 1170-1172.

[25] Y. Lai, A. Martinez, I. Khrushchev, and I. Bennion. (2006). Distributed Bragg reflector fiber laser fabricated by femtosecond laser inscription. *Optics Letters*, 31, 1672-1674.

[26] A. Martinez, M. Dubov, I. Khrushchev, and I. Bennion, (2006). Photoinduced modifications in fiber gratings inscribed directly by infrared femtosecond irradiation. *IEEE Photonics Technology Letters*, 18, 2266-2268.

[27] C. W. Smelser, S. J. Mihailov, D. Grobnic, P. Lu, R. B. Walker, H. Ding, and X.Dai. (2004). Multiple-beam interference patterns in optical fiber generated with ultrafast pulses and a phase mask. *Optics Letters*, 29, 1458-1460.

[28] E. Bricchi, B. G. Klappauf and P. G. Kazansky. (2004). Form birefringence and negative index change created by femtosecond direct writing in transparent materials. *Optics Letters*, 29, 119-121.

[29] M. Huang. (2003). Thermal stress in optical waveguides. *Optics Letters*, 28, 2327-2329.

[30] H. Takahashi, Y. Hibino, Y. Ohmori, and M. Kawachi. (1993). Polarization-insensitive arrayed-waveguide wavelength multiplexer with birefringence compensation film, *IEEE Photonics Technology Letters*, 5, 707-708.

[31] M. Okuno, A. Sugita, K. Jinguji, and M. Kawachi. (1994). Birefringence control of silica waveguides on Si and its application to a polarization-beam splitter/switch. *Journal of Lightwave Technology*, 12, 625-633.

[32] S. Suzuki, Y. Inoue and Y. Ohmori. (1994). Polarization-insensitive arrayed-waveguide grating multiplexer with SiO2-on SiO2 structure. *Electronics Letters*, 30 642-643.

[33] J. Albert, F. Bilodeau, D. C. Johnson, K. O. Hill, S.J. Mihailov, D. Stryckman, T. Kitagawa and Y. Hibino. (1998). Polarization-independent strong Bragg gratings in planar lightwave circuits. *Electronics Letters*, 34, 485-486.

[34] X. Dai, S. J. Mihailov, C. L. Callender, R. B. Walker, C. Blanchetière, J. Jiang. (2005). Measurement and control of birefringence and dimension of ridge waveguides with Bragg grating and ultraviolet irradiation. *Optical Engineering*, 44, 124602.

[35] J. Canning, M. Aslund, A. Ankiewicz, M. Dainese, H. Fernando, J. K. Sahu, and L. Wosinski. (2000). Birefringence control in plasma- enhanced chemical vapor deposition planar waveguide by ultraviolet irradiation. *Applied Optics*, 39, 4296-4299.

[36] X. Dai, S. J. Mihailov, C. Blanchetière, C. L. Callender, R. B. Walker. (2005). High birefringence control and polarization insensitive Bragg grating fabricated in PECVD planar waveguide with UV polarized irradiation. *Optics Communications,* 248, 123-130.

[37] X. Dai, S. J. Mihailov, C. L. Callender, C. Blanchetiere, and R. B. Walker. (2006). Ridge-waveguide-based polarization insensitive Bragg grating refractometer. *Measurement Science and Technology*, 17, 1752-1756.

[38] P. Lu, S. J. Mihailov, D. Grobnic, and R. B. Walker. (2006). Comparison of the induced birefringence in fiber Bragg gratings fabricated with ultrafast-IR and CW-UV lasers. *Proc. of ECOC 2006*, paper Th.3.3.5.

[39] C. D. Poole, R. W. Tkach, A. R. Chraplyvy, and D. A. Fishman.(1991). Fading in lightwave systems due to polarization-mode dispersion. *IEEE Photonics Technology Letters*, 3, 68-70.

[40] L. Guo, Y. Zhoiu and Z. Fang. (2003). Pulse broadening in optical fiber with polarization mode dispersion and polarization dependent loss. *Optics Communications,* 227, 83-87.

[41] M. Wang, T. Li and S. Jian. (2003). Analytical theory of pulse broadening due to polarization mode dispersion and polarization dependent loss. *Optics Communications.* 223, 75-80.

[42] N. N. Khrais, A. F. Elrefaie, R. E. Wagner, and S. Ahmed. (1996). Performance of cascaded misaligned optical (De)multiplexers in multiwavelength optical networks. *IEEE Photonics Technology Letters*, 8, 1073-1075.

[43] B. J. Eggleton, G. Lenz, N. Litchinitser, D. B. Patterson, and R. E. Slusher. (1997). Implications of fiber Grating dispersion for WDM communication systems. *IEEE Photonics Technology Letters*, 9, 1403-1405.

[44] G. Nykolak, B. J. Eggleton, G. Lenz, and T. A. Strasser. (1998). Dispersion penalty measurements of narrow fiber Bragg gratings at 10 Gb/s. *IEEE Photonics Technology Letters*, 10, 1319-1321.

[45] M. Kuznetsov, N. M. Froberg, S. R. Henion, and K. A. Rauschenbach. (1999). Power penalty for optical signals due to dispersion slope in WDM filter cascades. *IEEE Photonics Technology Letters*, 11, 1411-1413.

[46] P. Lu, L. Chen and X. Bao. (2002). System outage probability due to the combined effect of PMD and PDL. *Journal of Lightwave Technology*. 20, 1805-1808.

In: Advances in Laser and Optics Research. Volume 10 ISBN: 978-1-62257-795-8
Editor: William T. Arkin

Chapter 2

BACTERIAL CELL INTERACTIONS WITH OPTICAL FIBER SURFACES

Natasa Mitik-Dineva,[1] Paul R. Stoddart,[2] Russell J. Crawford[1] and Elena Ivanova[1*]

[1]Faculty of Life and Social Sciences,
[2]Center for Atom Optics and Ultrafast Spectroscopy
Swinburne University of Technology,
Hawthorn, Victoria, Australia

ABSTRACT

Since 1840, investigations into the principles of fiber optics have continued to deliver new associated technologies and applications. The expansion of knowledge in the area of nano- and applied science, in particular, has predictably led to the introduction of novel techniques and instruments that exploit fiber optics. As such, the use of optical fibers as chemical or environmental/bio-sensors or as imaging devices has resulted in the need for a detailed exploration of possible interactions between the surface of optical fibers and environmental perturbations, including bacterial cells.

Although the impact of bacterial adhesion to a number of surfaces, such as glass, metal, or polymeric surfaces, has already been studied in considerable detail, there appears to be a lack of systematic investigation into the interaction of bacterial cells with optical fibers, and in many cases, their subsequent attachment. This study focuses on the influence of optical/imaging fiber surface characteristics and chemistry on bacterial attachment. Attachment profiles of medically significant and environmentally important bacteria on optical fibers, both as-received and eroded as a result of chemical etching, were evaluated. Bacterial cell characteristics such as surface charge and wettability were analyzed, together with optical fiber characteristics, such as surface chemistry, wettability, charge, surface tension and surface nano-roughness. A series of scanning electron microscopy (SEM), atomic force microscopy (AFM) and confocal laser scanning

* Paul R. Stoddart: Center for Atom Optics and Ultrafast Spectroscopy, Swinburne University of Technology, PO Box 218, Hawthorn, Victoria 3122, Australia.

microscopy (CLSM) experiments were carried out to provide an insight into bacterial attachment behaviour on the optical fiber surfaces.

The results show that the as-received fiber surface was consistently more suitable for bacterial attachment than the modified surface. Therefore this study suggests that the surface treatment of optical fibers can be an important factor in the development of optical fiber sensors.

1. Optical Fibers

Optical fibers can be defined as glass or plastic fiber structures capable of guiding light throughout their length. Modern optical fibers fabricated from high purity silica are capable of transmitting optical signals over large distances with low losses. Other advantages include the reduced need for free space optics with their associated alignment and maintenance difficulties, the large spectral bandwidth that can be exploited and the general immunity to electromagnetic interference.

From the initial introduction of the principles of fiber optics by Tyndall in the 1840's, interest in optical fibers continued to develop, leading to new technologies and applications in industries such as television, telecommunications, laser machining, dentistry and medicine. The latter discipline exploits fiber optic technology in the design of instruments such as endoscopes for minimally invasive exploratory or surgical procedures [1]. Optical fibers can also be used as environmental sensors, temperature and pressure sensors for down-hole measurements in the oil and gas industry, and for structural health monitoring in the civil engineering and aerospace industries [2]. Optical spectroscopy is one of the applications of optical fibers in biomedicine. Optical fibers are now commonly used to couple a light source to a remote measurement position and to couple the resulting signal to a spectrometer [1], thereby avoiding the need for hazardous free-space beams and tedious optical alignments. Spectroscopic imaging can be performed with an imaging optical fiber, which is composed of a number of fibers (or "pixels") fused together in a coherent arrangement i.e. each fiber maintains its relative position throughout the length of the bundle [3]. Optical fiber probes have been used to perform absorption, fluorescence and Raman spectroscopic measurements in a wide range of biomedical applications [4, 5].

Raman spectroscopy is an inelastic light scattering process, usually implemented with a laser in the visible, near infrared, or near ultraviolet range [6-8]. The light from the laser interacts with vibrational excitations in the system, resulting in characteristic shifts in the energy of the scattered light. Raman spectroscopy is capable of providing specific identification of the molecular composition of a sample (chemical/structural fingerprinting), as well as information concerning conformation and bond structure [8].

The main advantages of Raman spectroscopy over other vibrational spectroscopic techniques lie in the fact that it is relatively insensitive to water and requires no special sample preparation. Due to its characteristics, Raman spectroscopy has been used extensively in a variety of biomedical testing applications, such as the evaluation of skin composition, quantification of blood components (glucose, cholesterol and urea), estimation of protein structure and cancer and pre-cancer diagnosis [6, 7]. The technique is increasingly used to identify single bacteria [9-12]. In comparison with current tests based on bacterial cultures, the rapid identification of single bacteria by Raman scattering can help to avoid production downtime in pharmaceutical clean rooms and reduce health hazards in clinical situations and food processing. Surface-enhanced Raman scattering (SERS), initially observed by Fleishman et al. in 1974 [13], allows a significant increase in sensitivity compared to normal Raman scattering. This is achieved through an enhancement of the Raman signal by a factor of up to 10^{14}, provided the sample is in close proximity to a nanostructured metal surface (primarily gold, silver or copper). The metal also serves to quench fluorescence, thus opening the door to the development of a number of new applications. Over the years there has been increased interest in applying SERS in many fields such as forensic science, homeland security, biochemistry and medicine [14]. The spectrum obtained by SERS is the result of an analyte's molecular structure, which is useful for real-time detection of certain compounds in biofluids at sub-nanomolar concentrations. These SERS characteristics allow in-vivo measurements, highlighting its advantages over other similar analytical techniques. Attempts to use SERS as a "fingerprinting" tool for the detection of various analytes have already been reported [15-10]. The development of a number of biosensors for environmental as well as in-vivo measurements have been undertaken by Murphy et al. [16] and Stuart et al. [17], who have developed a laboratory-based SERS system for monitoring chemicals in sea-water and glucose, respectively. It has been shown that SERS is particularly sensitive to biochemicals in the immediate vicinity of the metal surface, such as flavins that are associated with the cell wall [18].

Despite its promising capabilities, SERS has been slow to reach the commercial marketplace. The main reason for the delay in more widespread use of the technique lies in the difficulty of producing uniform, reproducible substrates with high sensitivity [19]. The production of SERS substrates can be achieved by the fabrication of surfaces with precise nanometer-scale structures. [20-18] This has emerged as an important application of nanotechnology. A particular challenge involves the fabrication of SERS substrates on the tips of optical fibers, so that the technique can derive additional benefit from the advantages of optical fiber technologies [21-23].

There have been a number of noteworthy recent developments in bio-nanotechnology based on the employment of fiber optic sensors of nanometer size suitable for in-vivo monitoring of biological processes in the living cell or nano-environments [24]. Fiber optic nanosensors can be defined as nano-scale measurement devices that consist of a biologically or chemically sensitive layer [25]. The tip of the bio-sensor probe can be functionalized with biomolecules such as proteins, enzymes, antibodies or biological systems such as cells or whole organisms, thus fabricating a 'whole-cell' biosensor [26, 27]. Previously, whole cell biosensing devices measured the change in the metabolic rate of the cell, and this was interpreted as the analytical signal. More recent biosensing devices are also based on the cells' ability to respond to environmental perturbations by their expression of specific genes [26].

As an alternative, Gessner et al. have used a SERS substrate on the submicron tip of a tapered optical fiber to record the spectra of monolayers of a yeast with a spatial resolution of 200-500 nm [10]. This approach can be extended further by combining SERS with scanning near-field optical microscopy (SNOM). This approach has been used to image DNA fragments with 100 nm resolution on a silver island film SERS substrate [28]. The SERS effect can also be induced directly by applying a thin layer of silver islands to the tip of a tapered optical fiber. The probe can be placed in intimate contact with almost any type of surface because of the small size of the tip [29].

Notwithstanding recent advances in the technology, the fabrication of whole-cell biosensing devices has proven extremely challenging, mainly because the immobilization of live cells onto the fibers, or the insertion of the probe into the cell, frequently results in cellular death and impaired sensitivity [19-23]. Investigation of the cell-optical fiber surface interactions is therefore a key issue in the design of biosensing devices [19-26]. The present study aims to determine the influence of surface characteristics and chemistry on the attachment of three medically significant microorganisms on the surface of optical imaging fibers used in biomedical applications.

2. BACTERIAL ATTACHMENT

2.1. Overview

The mechanisms by which bacteria adhere to a variety of surfaces such as glass, metal, polymers, etc.) have been intensively studied over the years, yet remain far from fully understood. Bacterial characteristics such as cell shape and size, production of extracellular polymeric substances (EPS), surface protrusions such as flagella and cilia, surface tension and surface charge and wettability are all believed to significantly influence bacterial adhesion to the surfaces [30-36]. The dynamic nature of the surface characteristics of bacterial cells, when the surface molecules are changing their structural conformation as a result of physiological responses to environmental perturbations, contribute to the attachment behavior thus making an understanding of bacterial adhesion even more challenging. In addition to the variability of the cell surface characteristics, substratum surface properties play an equally important role in bacterial attachment to the surfaces [37-40]. Intensive studies have focused on the influence of physico-chemical bacterial and substratum characteristics, such as surface wettability, tension and charge [32, 36, 41, 42], while less attention has been given to investigating the effects of surface topography, roughness and porosity on the adhesive behavior of micro-organisms [43]. The overall complexity of this process is, however, the main reason why no comprehensive theory exists that can reliably predict bacterial attachment behavior.

2.2. Theoretical Considerations

The mechanisms that control bacterial adhesion have been addressed on a number of levels: theoretical approaches such as the DLVO and thermodynamic theories have revealed some of the basic physico-chemical aspects of bacterial adhesion [44, 45], such as the influence of surface charge and surface tension on the long range cell-substratum interactions and the effect of surface hydrophobicity on the short-range interactions [45-49].

In terms of the effects of surface roughness on cell-surface interactions, it has been recognized that bacterial adhesion is initiated in areas of surface irregularity that serve as microenvironments where bacteria are sheltered from unfavorable environmental factors. Bacteria attach to these sites in order to promote their survival [36, 50-52]. Although the effects of surface roughness have been studied over a wide range of physical scales [32, 41, 42, 44, 53], it has only recently been shown that surface roughness on a scale much smaller than the bacterium might be a major driver in the initial course of bacterial attachment [54-56].

One of the recently developed theories regarding the interrelation between bacterial adhesion and surface roughness is the "attachment point" theory [57]. According to this theory, organisms smaller than the scale of the surface micro-texture will attach in sufficient number and will therefore have greater adhesion strength because of the multiple attachment points on the surface when compared to micro-organisms that are of scale larger than the surface roughness [51, 52, 58, 59]. They will also be well protected from hydrodynamic forces in the micro-refuge shelters on the textured surface [43]. A number of research projects studying the relationship between surface roughness and attachment of organisms such as barnacle cyprids and algal spores have supported the applicability of the "attachment points" theory [58, 60-62].

3. Experimental Set-Up

3.1. Introduction

Several studies have already investigated the influence of substratum surface characteristics such as topography, charge, wettability and chemistry on cell-surface interactions. Most of these investigations have focused on planar surfaces of glass, different polymers or metal, whilst optical fiber surfaces have not yet been investigated. In this study, a series of experiments were designed to investigate the attachment pattern of three commonly studied medically important marine bacteria of different taxonomic affiliations during interaction with the surfaces of optical fibers.

Standard glass imaging fibers (FIGH-70-1300N, Fujikura Ltd) were used as experimental surfaces. The wettability, tension and charge of the surfaces were measured directly. The surface chemistry of the fibers was evaluated using Time-of-Flight Secondary Ion Mass Spectrometry (TOF-SIMS). The surface topography was studied using atomic force microscopy (AFM). Surface topography was modified by etching the as-received surfaces with a buffered hydrofluoric acid, as described below.

A detailed analysis of the bacterial cell surface characteristics was carried out using contact angle and surface charge measurements. The bacterial cell dimensions and surface topography were determined using AFM. Scanning electron microscopy (SEM) was used to provide a qualitative and quantitative analysis of the substratum and cell morphology. Confocal laser scanning microscopy (CLSM) enabled the visualization of EPS produced by the bacteria during attachment, and allowed the viability of the attached cells to be determined.

3.2. Methodology

3.2.1. Bacteria

The following bacterial strains were selected for this study: *Escherichia coli* K12 [63, 64-61], *Staphylococcus aureus* CIP 68.5 [58, 62] and *Pseudomonas aeruginosa* ATCC 9027 [62-

63]; and three marine bacteria: *Cobetia marina* DSM 4741T [64-65], *Pseudoalteromonas issachenkonii* KMM 3549T [66-67], and *Sulfitobacter guttiformis* DSM 11458T [68-80].

3.2.1.1. Bacterial Staining Protocols

Two dyes were used in order to simultaneously visualize viable cells and their production of extracellular polymeric substances while attaching to any of the tested surfaces. Vybrant CFDA SE Cell Tracer Kit (Invitrogen Pty Ltd) was used to color viable cells and Concanavalin A 594 (Molecular Probes Inc.) was used to label EPS.

Concanavalin A Alexa Fluor 594 Conjugate (Molecular Probes Inc.), was applied in order to visualize EPS. This dye selectively binds to α-mannopyranosyl and α-glucopyranosyl residues in EPS [65]. In neutral and alkaline solutions, the bright green dye exists primarily as a tetramer with a molecular weight of 104,000 daltons [66]. The Alexa Fluor 594 conjugate, exhibits intense fluorescence and photostability, which allows better imaging quality than spectrally similar conjugates such as fluorescein. The fluorescent emission of the Alexa Fluor 594 fluorophore is independent of pH from 4 to 10, compared to fluorescein, which is significantly affected by pH [67]. This pH insensitivity allows simultaneous use with carboxyfluorescein diacetate, succinimidyl ester (Vybrant CFDA SE Cell Tracer Kit, Invitrogen Pty Ltd) and scanning of the same field of view for both viable cells and EPS. Concanavalin A stock solution was prepared by dissolving 5 mg in 5 mL of 0.1 M sodium bicarbonate at pH 8.3 and stored at 20°C. Working solution was prepared by diluting stock solution to 1:20 using the same buffer to avoid changes in pH. It is important to mention that the overall distribution of the fluorescent signal on top of the cell/substratum surface will very much depend on the chemical composition and the distribution of the produced EPS.

The Vybrant CFDA SE (carboxyfluorescein diacetate, succinimidyl ester) Cell Tracer Kit was used to trace viable cells adsorbed on each of the surfaces. The CFDA SE is initially colorless and nonfluorescent. It passively diffuses into cells where the acetate groups are cleaved by intracellular esterases to yield highly fluorescent, amine-reactive carboxyfluorescein succinimidyl ester. The dye–protein adducts that form in labeled cells are retained by the cells and inherited by daughter cells after division. The excitation and emission wavelengths for CFDA SE are 495 nm and 517 nm, respectively. Working solutions of the dye as well as cell labeling conditions were prepared as described elsewhere [68]. In general, fiber substrates were incubated in the bacterial suspension for 11 h before an aliquot of Concanavalin A 488 was added in ratio 1:5 (cell suspension/dye). The dye was allowed 1 h to diffuse when CFDA SE dye, in the same ratio, was added to the suspension and incubated for additional 15 min at 37°C. After incubation the samples were washed with sterile nanopure H_2O (18.2 $M\Omega cm^{-1}$ Barnstead/Thermolyne NANOpure Infinity water purification system), left to dry for a few hours at room temperature (23-25°C, humidity 48%) without additional fixation to prevent the deformation of the cells and processed on the following day using confocal microscopy, as described below.

3.2.2. Optical Fibers

The experimental surfaces used for this study were glass optical fibers (FIGH-70-1300N, Fujikura Ltd). This is a standard optical imaging fiber, with approximately 10,000 picture elements (also known as "pixels") and total outer diameter of 1.3 mm. According to the manufacturer's specification, these fibers are made from silica glass cores surrounded by fluorine-doped silica cladding. This fiber surface was the base for further surface modification by chemical etching.

3.2.2.1. Initial Surface Preparation

The 100 cm long fiber, as received from the manufacturer, was initially cut into approximately 10 cm long sections which were then fixed to a glass supporting surface using crystal bond adhesive (ProSciTech Pty Ltd). Subsequently, the 10 cm long fibers were cut into 5 mm long subdivisions using an automatic dicing saw (DISCO DAD 321, Equipment Acquisition Resources Inc.) with a diamond blade (NBC-ZH 2050-J-SE, 0.09 mm). The dicing saw yields smooth, flat surfaces (facets) on the fiber end faces, as shown in the high-resolution SEM images presented in Figure 3.1.

These optical fiber sections are referred to as the 'as received' samples. Since the whole length of the fiber is protected with silicone coating, 99% acetone (Aldrich) was used to remove the coating.

All fibers were initially washed with copious amounts of sterile nanopure H_2O (18.2 $M\Omega cm^{-1}$ Barnstead/Thermolyne NANOpure Infinity water purification system), sterilized at 121°C for 15 minutes and stored under sterile conditions until just prior to use.

3.2.2.2. Surface Modification

In order to modify the optical fibers surfaces, the 5 mm long fiber sections were exposed to buffered hydrofluoric acid (BHF) by dipping one half of the sample in the etching solution for 20 min [21]. The chemical composition of the BHF was as follows: 6 parts of 40% ammonium fluoride NH_4F, 1 part of 49 % HF hydrofluoric acid and 14 parts of 36.8% HCl hydrochloric acid [21].

The ultimate effect after the acid exposure is based on the difference in the chemical structure between the fiber components. The cladding around each picture element etches at a slower rate than the silica core, thus resulting in a generally hexagonal pattern of wells on the fiber surface, with each well being approximately 2.5 μm in diameter and 2.5 μm deep. The final well size well size depends on the exposure time to the etching solution.

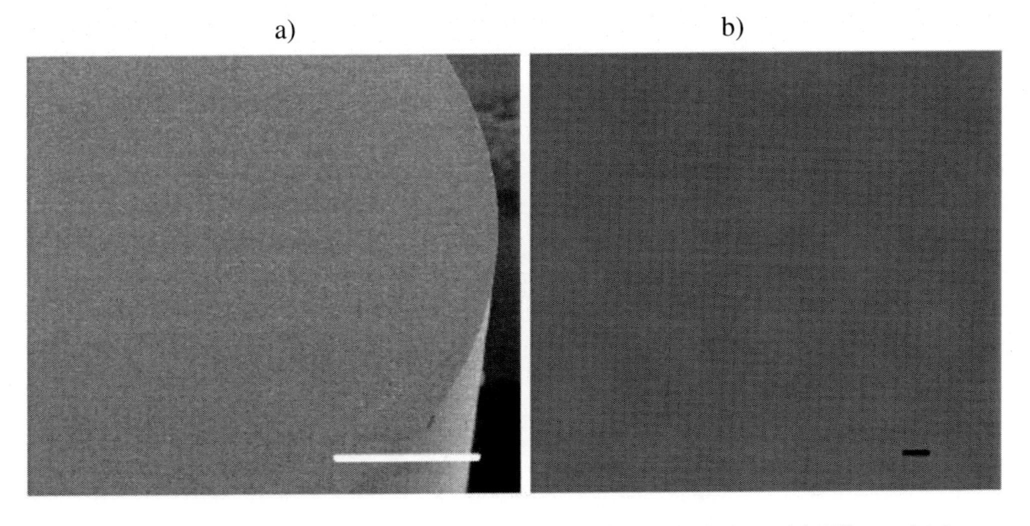

Figure 3.1. Typical SEM images of the "as-received" optical fibers. Scale bars: (a) 250 μm; (b) 1 μm.

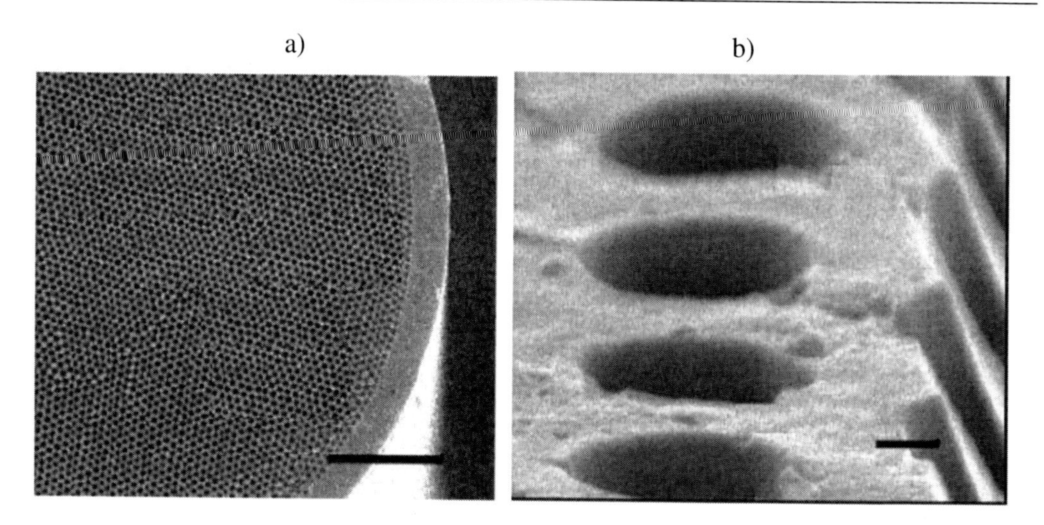

Figure 3.2. Typical SEM images of the optical fiber after exposure to the etching solution for 20 min. Scale bars: (a) 250 μm, (b) 1 μm.

In order to achieve well size equivalent to the size of a single bacterial cell (2.5 μm × 2.5 μm ± 15%), the fibers were acid treated for 20 minutes. The etched optical fiber samples are referred to as the 'modified' optical fibers. High-resolution SEM images of the modified fiber surfaces are presented in Figure 3.2.

After exposure to the etching solution, all samples were rinsed with sterile nanopure H_2O (18.2 $M\Omega cm^{-1}$ Barnstead/Thermolyne NANOpure Infinity water purification system), sterilized at 121°C for 15 minutes and stored under sterile conditions prior to inoculation.

In order to ensure that any excess BHF was washed from the fiber surface the dissipated solution was tested for acidity with Phenolphthalein and Bromothymol blue indicator dyes. The consistent color indicated an absence of BHF in the solution and on the modified fiber surfaces.

3.2.3. Cultivation Conditions

Escherichia coli, Staphylococcus aureus and *Pseudomonas aeruginosa* were routinely cultured on nutrient agar (Merck) plates and stored at -80°C in storage solution prepared from nutrient broth and 20% (v/v) glycerol.

Pseudoalteromonas issachenkonii, *Cobetia marina* and *Sulfitobacter guttiformis* were routinely cultured on marine agar 2216 (Difco) plates and stored at –80°C in marine broth (Difco) supplemented with 20% (v/v) glycerol as explained elsewhere[69]. Fresh bacterial suspensions were prepared prior to each experiment with marine (Difco) or nutrient (Merck) broth, depending on the bacterial strain. The optical density (OD) for all bacterial suspensions was adjusted to 0.2-0.3 at 600 nm on GeneQuant Pro Spectrophotometer (Amersham Biosciences) (OD_{600} = 1 corresponds to 8 × 10^8 cells/mL).

3.2.4. Sample Preparation

Prior to each experiment, bacterial cells were grown in marine/nutrient broth for 24 h, as previously described. On the day of the experiment, 2 mL of log-phase bacterial suspension was adjusted to OD_{600} 0.2-0.3 in nutrient/marine broth and kept in centrifuge tubes (Interpath

Services, Pty Ltd). Duplicate samples of both as-received and modified fibers were placed into each of the tubes and were incubated for 12 hours at room temperature (ca. 25°C). After incubation, all fibers were rinsed three times with sterilized nanopure H_2O (18.2 $M\Omega.cm^{-1}$ Barnstead/Thermolyne NANOpure Infinity water purification system), attached to glass supports and kept under sterile condition until needed.

3.2.5. Analysis of Bacterial and Fiber Surface Characteristics

3.2.5.1. Contact Angle Measurements

Bacterial and substrata surface wettability was inferred from contact angle measurements. For the purpose of this study, water was selected as the most suitable diagnostic liquid since it can be used for the measurement of contact angles on both of the substrate surfaces, as well as on lawns of bacterial cells [70].

a) Bacteria Surface Wettability

For the purpose of measuring cell surface wettability, static contact angles were measured using the sessile drop method. Bacterial cell suspensions were prepared by growing the cells overnight in nutrient/marine broth. Cells were then harvested by centrifugation and re-suspended in 0.1 M NaCl buffer (OD_{470} = 0.4). This suspension was deposited on cellulose acetate membrane filters (Sartorius, pore diameter 0.2 μm, filter diameter 47 mm). The wet filters were left at ambient temperature for approximately 30-40 minutes to air dry until a "plateau state" [71,72] was achieved.

An FTA200 instrument (First Ten Angstroms, VA), equipped with charge-coupled device (CCD) camera, was used to deposit and measure the contact angle of water droplets. After the initial deposition the drop was allowed to settle for 2 seconds without needle contact (for static contact angle measurements). Digital images were then used to determine the contact angle.

b) Substratum Surface Wettability

The surface wettability of both the as received and modified fiber surfaces was inferred from water contact angle measurements using the sessile drop method on the FTA200 instrument. The FTA200 was used to determine contact angles as previously described. Prior to measurements, all fibers were washed with sterile nanopure H_2O (18.2 $M\Omega cm^{-1}$ Barnstead/Thermolyne NANOpure Infinity water purification system) and left to dry in air at ambient temperature.

Selected fibers (as received and modified) were mounted onto a rigid support (glass microscope slide), which provided the horizontal surface orientation required for deposition of 1 μL water droplet onto the fiber surface. Due to the limited fiber surface area (1.33 mm^2) each fiber was used for only a single measurement. The experiments were repeated 10 times and the final contact angles given below represent an average of those measurements.

3.2.5.2. Surface Free Energy

The optical fiber surface tension was calculated by means of the Young equation:

$$\gamma_{l/v} \cos\theta = \gamma_{s/v} - \gamma_{s/l}$$

where: $\gamma_{l/v}$ is the surface tension or surface free energy of liquid to air; $\gamma_{s/v}$ is the surface tension or surface free energy of solid to air; $\gamma_{s/l}$ is the surface tension or surface free energy of interface between solid and liquid.

According to Van Oss et al. [73], the surface free energy (γ_i) is the sum of the Lifshitz-van der Waals apolar component (γ^{LW}) and the Lewis acid-base polar component (γ^{AB}), where:

$$\gamma_i = \gamma_i^{LW} + \gamma_i^{AB} \quad (1)$$

The Lifshitz-van der Waals apolar component (γ^{LW}) can be formulated as follows [74]:

$$\gamma_{ij}^{LW} = [(\gamma_i^{LW})^{1/2} - (\gamma_j^{LW})^{1/2}]^2 \quad (2)$$

$$\gamma_{ij}^{LW} = \gamma_i^{LW} + \gamma_j^{LW} - 2(\gamma_i^{LW}\,\gamma_j^{LW})^{1/2} \quad (3)$$

Whereas the Lewis acid-base polar (γ^{AB}) component of the equation can be divided into an electron donor (γ_i^-) and an electron acceptor (γ_i^+) component [75], where:

$$\gamma_i^{AB} = 2(\gamma_i^+\gamma_i^-)^{1/2} \quad (4)$$

In that case the definition of the Lewis acid-base interactions between two substances in the condensed state will be as follows[75]:

$$\gamma_{s/l}^{AB} = 2(\gamma_s^+\gamma_l^-)^{1/2} + 2(\gamma_l^+\gamma_s^-)^{1/2} \quad (5)$$

Therefore, (γ_S^{LW}) can be evaluated from contact angles between solid and liquid. According to van Oss [75], the Young equation now becomes:

$$\gamma_l^{LW}\cos\theta = \gamma_s^{LW} - \gamma_{s/l}^{LW} \quad (6)$$

and can be expressed as follows:

$$1 + \cos\theta = 2(\gamma_s^{LW} / \gamma_l^{LW})^{1/2} \quad (7)$$

in which case the complete Young Equation is:

$$\gamma_{l/v}(1+\cos\theta) = 2\,[(\gamma_s^{LW}\,\gamma_l^{LW})^{1/2} + (\gamma_s^+\,\gamma_l^-)^{1/2} + (\gamma_s^-\,\gamma_l^+)^{1/2}] \quad (8)$$

The optical fiber surface tension was calculated from the contact angle measurements of three diagnostic liquids with well-known surface tension properties (Table 3.1) such as water (γ^{LW} = 21.8 and $\gamma^+ = \gamma^-$ = 25.5 mJ.m^{-2}), formamide (γ^{LW} = 39.0, γ^+ = 2.28 mJ.m^{-2}, and γ^- = 39.6 mJ.m^{-2}) as polar fluids, and diiodomethane (γ^{LW} = 50.8 and $\gamma^+ = \gamma^-$ = 0 mJ.m^{-2}) as apolar [76-78].

Table 3.1. Surface tensions and its parameters (mJ/m^2) for solvents used in the measurement of contact angles

Solvent	Surface tensions (γ), mJ/m^2				
	γ_{lv}	γ_{lv}^{LW}	γ_{lv}^{AB}	γ_{lv}^{+}	γ_{lv}^{-}
Water	72.8	21.8	51.0	25.5	25.5
Formamide	58	39	19	2.28	39.6
Diiodomethane	50.8	50.8	~0		

*Contact angle of water, formamide and diidomethane (θ_W, θ_F and θ_D respectively);
**Lifshitz/van der Waals component (γ^{LW}), acid/base component (γ^{AB}), electron acceptor (γ^{+}) and electron donor (γ^{-}).

3.2.5.3. Surface Charge Measurements

The bacterial cell surface charge was inferred from zeta potential measurements. Zeta potentials provide an indication of the overall net surface charge and can be obtained by measuring the electrophoretic mobility (EPM) [5, 6, 79]. For the purpose of measuring cell surface charge it is important to always have pure cell cultures, as the presence of subpopulations with different surface properties in single-strain cultures can lead to erroneous conclusions [80]. Even then it is to be remembered that no zeta (ζ) potential value can be assigned to bacteria at the strain and species level and that even different isolates of the same strain can express different ζ potentials [81]. For the purpose of this study the EPM was measured as a function of ionic strength by microelectrophoresis using a zeta potential analyser (ZetaPALS, Brookhaven Instruments Corp). The data were processed with the accompanying software, which employs the Smoluchowski equation.

Prior to measurements, cells were cultivated for 24 hours in nutrient or marine broth depending on the strain and were latter harvested by centrifugation at 5000 rpm for 5 minutes [82]. Harvested cell pellets were re-suspended in 10 mM potassium chloride (KCl), washed, then re-centrifuged. This step was repeated four times to eliminate residual extracellular polysaccharides that, as previously mentioned, may influence the surface electric potential. After the final wash, the cell pellets were re-suspended in 10 mM KCl solution to OD_{600nm} = 1, as suggested by De Kerchove *et al.* [82]. This cell solution was then diluted 1000 times in 5 mL 10 mM KCl, pH 7.5, for use in the EPM measurements. Measurements were conducted at an electric field of 2.5 V.cm^{-1} and frequency of 2 Hz [83]. All measurements were done in triplicates and for each sample the final EPM represents the average of five successive ZetaPALS readings, each of which consisted of 14 cycles per run.

3.2.5.4. AFM Characterization of the Surface

A scanning probe microscope (SPM) (Solver P7LS, NT-MDT) was used to image the fiber surface morphology whilst also providing a quantitative analysis of the surface roughness. The analysis was performed in the semi-contact mode, which reduces the interaction between the tip and the sample and thus allows the destructive action of lateral forces that exist in contact mode to be avoided. The carbon “whisker” type silicon cantilevers (NSC05, NT-MDT) with a spring constant of 11 N/m, tip radius of curvature of 10 nm, aspect ratio of 10:1 and resonance frequency of 150 kHz were used to obtain good topographic resolution. Scanning was performed perpendicular to the axis of the cantilever at a typical rate of 1 Hz.

Image processing of the raw topographical data was performed with first order leveling in the horizontal plane and the topography and surface profile of the samples were obtained simultaneously. In this way the surface features of the samples were measured with a resolution of a fraction of nanometer and the surface roughness of the investigated areas (shown in Figure 1 and 2) could be statistically analyzed using the standard instrument software (LS7-SPM v.8.58).

3.2.5.5. TOF-SOMS Analysis of the Fiber Surface Chemistry

The Time-of-Flight Secondary Ion Mass Spectrometry (TOF-SIMS) uses a pulsed primary ion beam, typically liquid metal ions such as Ga^+ and Cs^+ to bombard and ionize species from a sample surface. The resulting secondary ions that emit from the surface are then electrostatically accelerated into a mass spectrometer, where they are mass analyzed by measuring their time-of-flight from the sample surface to the detector. TOF-SIMS can provide mass spectroscopy for surface chemical characterization, images to visualize the distribution of individual chemical species on the surface and depth profiles for thin film characterization and can be used for surface analysis of inorganic, organic materials and biological cells, applied to conductors, insulators and semiconductors.

The primary requirement from this investigation was to enhance our understanding of the fiber surface chemistry and thereafter determine any difference in the surface composition between the two fiber surfaces that may have influenced the cells' behavior. All measurements were performed using a ToF-SIMS IV instrument (ION-TOF GmbH, Munster, Germany) with a reflection analyzer and a pulsed electron flood source for charge neutralization. Both positive and negative spectra were acquired from a 100 μm × 100 μm area. Samples were exposed to the atmosphere for less than 5 min during mounting in the TOF-SIMS instrument.

All experiments were performed using a cycle time of 100 μs. A monoisotopic [69] Ga^+ primary ion source was operated at 25 keV in the "burst alignment" mode, which gives very high spatial resolution at the expense of mass resolution and positive and negative spectra were acquired with a mass resolution typically greater than 6000 at $m/z = 27$, sufficient to identify most of the fragments. The spectra acquired were analyzed using the accompanying software.

3.2.5.6. Scanning Electron Microscopy (SEM)

Scanning electron microscopy (SEM) was employed to provide a more in-depth understanding of bacterial adhesion on all micro-nano structured surfaces used throughout the experiments. A FeSEM (ZEISS SUPRA 40VP) was used to obtain the high-resolution images of the substratum surface, bacterial morphology as well as the adhesion pattern. Primary beam energies of 3 to 15 kV were used, which allowed features on the sample surface or within a few microns of the surface to be observed.

Prior to imaging all samples were mounted on pin type aluminium SEM mounts with double-sided conducting carbon tape and then coated in a Dynavac CS300 coating unit with carbon and gold to reduce charging effects on the specimen surface. The thickness of the coating was not measured but was assumed to be in the order of few nm. The working distance (WD) varied between 6-7 mm, and images were mostly captured on 500×, 1000×

and 1000× magnification. Control images of all surfaces before bacterial inoculation and both with and without broth were also taken.

To allow quantification of the number of adsorbed bacteria, cell numbers from at least ten representative images/areas were transformed into a number of bacteria per unit area using Image-Pro software [84]. The average densities have estimated errors of approximately 10-15% due to local variability in the coverage.

3.2.5.7. Confocal Laser Scanning Microscopy (CLSM)

The confocal laser scanning microscope (CLSM, Olympus Fluoview FV1000 Spectroscopic Confocal System) which includes an inverted microscope system (Olympus IX81 with 20×, 40× (oil), 100× (oil) UIS objectives) that operates using multiple Ar, He and Ne laser lines (458, 488, 515, 543, 633 nm) was used.

The system was equipped with a transmitted light differential interference contract attachment and a CCD camera (Cool View FDI). Fluorescence was scanned at 560Em/590Ex nm and 492Em/517Ex nm for Concanavalin A and Vybrant respectively.

3.3. Results

3.3.1. Cell Surface Wettability and Charge

Bacterial cell surface wettability is presented in Table 3.2. Cell surface wettability varied amongst the bacteria studied, most likely reflecting the different chemical composition of surface-expressed polymeric substances. The classification of hydrophilic or hydrophobic cell surfaces remains rather unclear as different values have been suggested to define the borderline denoting cell hydrophobicity [36, 76, 77]. In this study, water contact angle (θ) values of 60° were considered as the borderline between hydrophilic and hydrophobic [85,86], which led to a conclusion that the cell surface of three bacteria, *E. coli, Sulfitobacter guttiformis* and *P. aeruginosa,* out of six tested strains studied were slightly hydrophilic. In contrast, the cell surface of *Coberia marina* and *Staphylococcus aureus* was of a highly hydrophobic nature.

The values obtained are consistent with data already reported for *E. coli* [36, 86] and *Pseudomonas aeruginosa* [36]. It appeared, however, that the surface of the strain *Staphylococcus aureus* CIP 68.5 was hydrophobic with water contact values reaching 72°, which is significantly higher than for *Staphylococcus aureus* 835 previously reported by Vermeltfoort et *al.* [87] Since the difference in measured contact angle was considerable, the surface wettability of two more *Staphylococcus aureus* strains was also measured. Water contact angles for *Staphylococcus aureus* ATCC 25923 and *Staphylococcus aureus* ATCC 12600^T were found to be 50° and 27°, respectively, thus suggesting that the surface wettability is most likely a strain specific characteristic. The hydrohobic nature of *Staphylococcus aureus* cells might be due to the presence of highly negatively charged and hydrophobic teichoic and lipoteichoic acids which are the main constitutes of *Staphylococcus aureus* cell wall [88]. Cell wall teichoic acids are composed of a linear chain of approximately 40 1,3-phosphodiester-linked ribitol phosphate residues linked to O6 of the N-acetylmuramyl residues of peptidoglycan [89]. Lipoteichoic acids are composed of a single, unbranched 1,3-linked poly (glycerophosphate) chain, in which units may be partly substituted with positively charged D-alanine ester [89].

In the latter case *S. aureus* strains exhibited significantly less hydrophobic characteristics; as reported by Vermeltfoort *et al.* [87].

Cell surface wettability is a key parameter for the thermodynamic theory, where it would be expected that the bacteria with hydrophilic (hydrophobic) characteristics would show preference for attachment to hydrophilic (hydrophobic) substrata [44] In that case, *Staphylococcus aureus, Cobetia marina* and *Sulfitobacter guttiformis* would be expected to have a stronger propensity for attaching to hydrophilic surfaces, whilst *E. coli, Pseudomonas aeruginosa* and *Pseudoalteromonas issachenkonii* would be expected to have a stronger propensity for hydrophobic surfaces [44, 48].

The electrokinetic microbial surface properties are another bacterial cell surface characteristic used for predicting bacterial attachment tendencies. The bacterial surface charge was inferred via measurement of the electrophoretic mobility of the cells and converted into zeta potential using Smoluchowski's approximation, as described above. The zeta potential of all six strains were measured in 10mM KCl to maintain the same ionic strength of the solution including initial re-suspension [81 82].

Bacterial cell surface charges are presented in Table 3.3. The results obtained for *E. coli* [36, 90], *Pseudomonas aeruginosa* [36, 91] and *Staphylococcus aureus* [91] were in reasonable agreement with data already reported.

It follows from the data presented in Table 3.3 that the least negatively charged strain was *Pseudomonas aeruginosa* (ζ = -14.4 ± 0.7 mV) and the most negatively charged strain was *Sulfitobacter guttiformis* (ζ = -43.2 ± 0.2 mV). Accounting for the suggested inverse correlation between cell surface charge and bacterial adhesion [36] and the electrostatic repulsion between negatively charged bacteria and negatively charged surfaces under commonly encountered pH conditions [92] it would be expected that *Sulfitobacter guttiformis* would exhibit the weakest and *Pseudomonas aeruginosa* the strongest attachment preferences.

Nevertheless, it has been reported that more electronegative cells can still exhibit hydrophobic characteristics, mainly due to the constant dynamic motion of the outer surface proteins that can result in varying levels of molecular polarity or charge over the cell surface [36, 93, 94].

Table 3.2. Surface wettability of bacterial cells

Strain	Water contact angle θ (°)
Escherichia coli	33 ± 4
Pseudomonas aeruginosa	43 ± 8
Staphylococcus aureus	72 ± 8
Cobetia marina	75 ± 9
Pseudoalteromonas issachenkonii	52 ± 3
Sulfitobacter guttiformis	56 ± 4

Table 3.3. Electrophoretic mobility and zeta potential for the studied strains

Species	Electrophoretic Mobility ($\mu s^{-1} V cm^{-1}$)	Zeta Potential (mV)
Escherichia coli	-3.1 ± 0.6	-38.4 ± 0.3
Pseudomonas aeruginosa	-1.1 ± 0.1	-14.4 ± 0.7
Staphylococcus aureus	-2.8 ± 0.8	-35.2 ±1.0
Cobetia marina	-2.5 ± 0.6	-32.5 ± 0.5
Pseudoalteromonas issachenkonii	-2.9 ± 0.2	-35.3 ± 0.2
Sulfitobacter guttiformis	-3.4 ± 0.5	-43.2 ± 0.2

3.3.2. Substratum Surface Chemistry, Topography and Wettability

In order to better understand the possible mechanisms of attachment between bacteria and optical fibers, it was considered essential to obtain a detailed analysis of some of the physico-chemical surface characteristics of the 'as received' and modified fiber surfaces.

The chemical composition of both fiber surfaces was analyzed using ToF-SIMS. The results obtained showed a similarity between both surfaces. As evident from the images presented in Figure 3.3 (a) and (b) and the positive and negative spectra presented in Figure 3.4, the most abundant component on both surfaces was Si, followed by SiC_3H_9, $SiCH_3$, CH_3 and Na, representing 70% of the elemental composition for both surfaces. The only difference was the lesser representation of Ge on the modified surface. This observation was expected since the precise effect of the etching solution is to removal of some of the germanium and fluorine ions from the fiber surface. Higher concentrations of F were found on the modified sample, indicating that residual F remained on these surfaces from the etching solution. Both surfaces showed an appreciable presence of carbon, most likely due to surface air contamination.

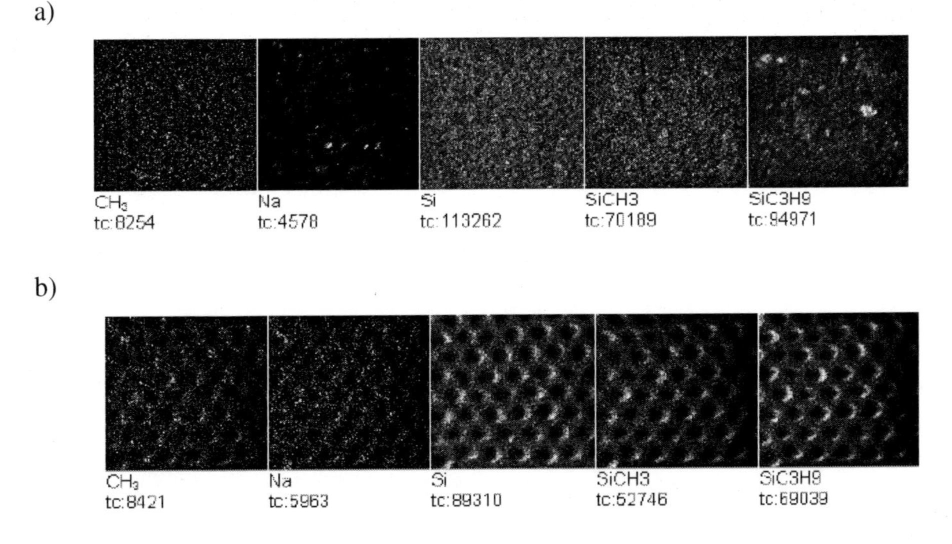

Figure 3.3. ToF-SIMS scans from the (a) as-received and (b) modified fiber surface.

Since the ToF-SIMS data did not reveal substantial differences in the chemical composition between the two surfaces, AFM analyses were conducted to characterize the surface topography. The typical AFM images presented in Figure 3.5 show a topographical profile of the two surfaces.

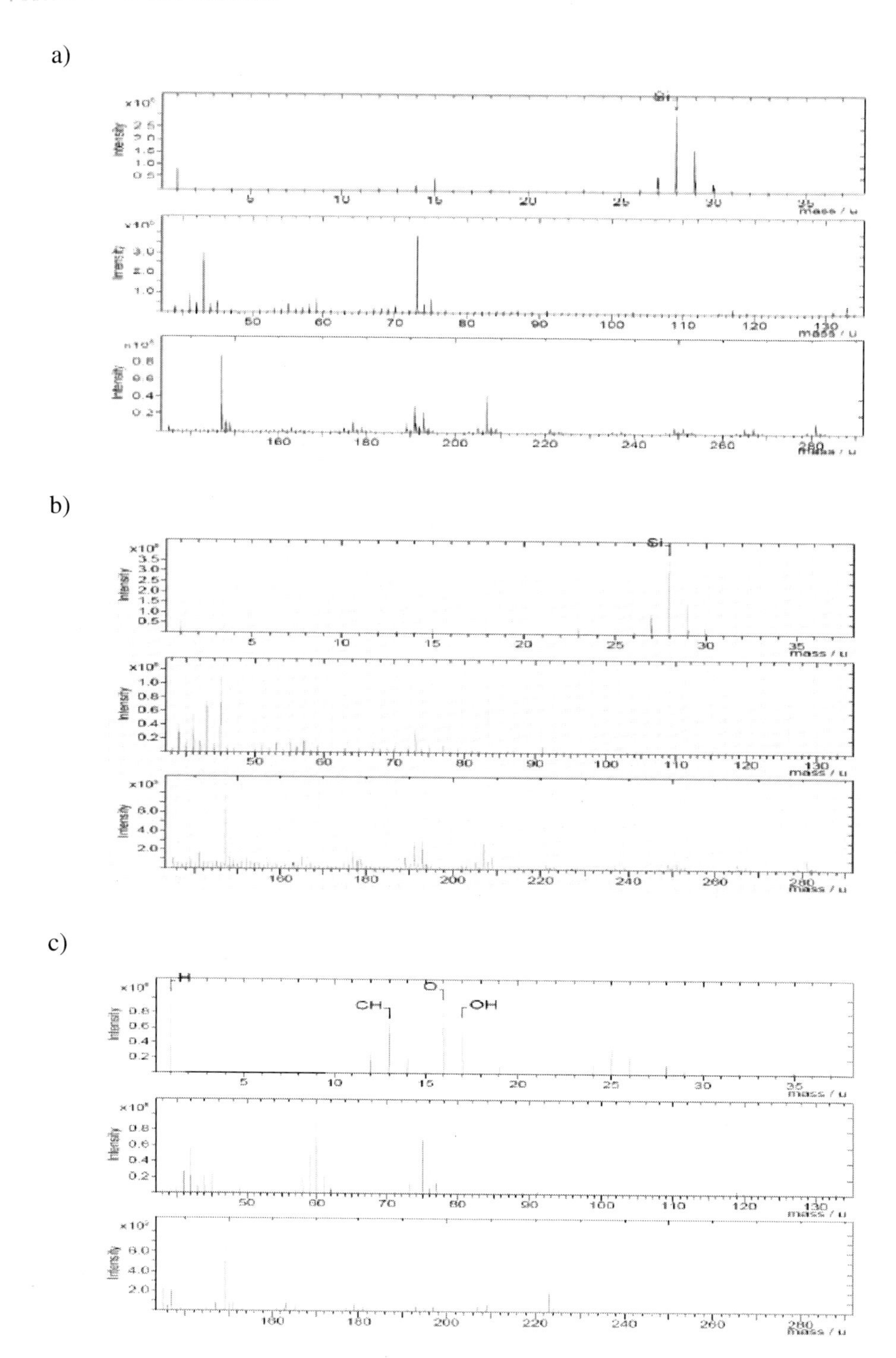

Figure 3.4. (Continued on next page).

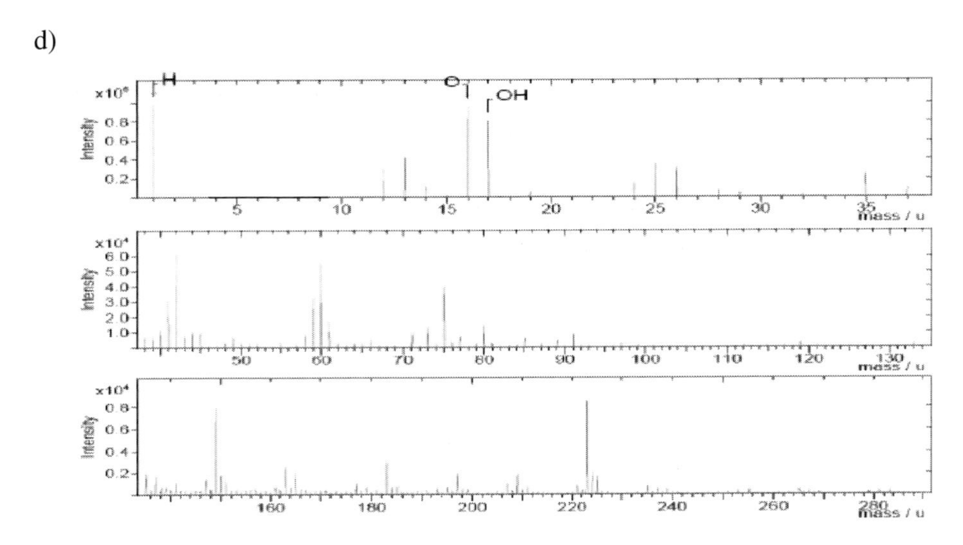

Figure 3.4. Positive (a, b) and negative (c, d) spectra collected from the as-received (a, c) and the modified (b, d) fiber surface.

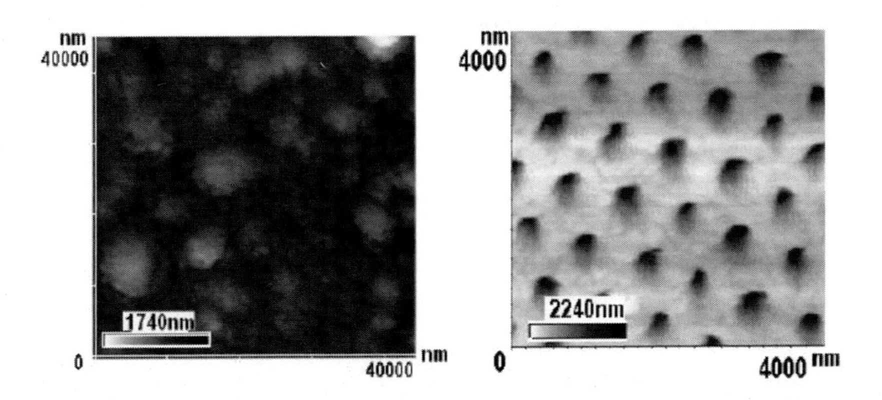

Figure 3.5. Surface topography of the 'as-received' and the modified fiber surfaces.

Table 3.4. Roughness parameters from the as-received and the modified fiber surface as inferred from AFM

Fiber surface*	R_a (nm)	R_q (nm)	R_{max} (nm)
'As received' fiber surface	180 ± 9	235 ± 9	1740 ± 84
Modified fiber surface	1563 ± 77	3428 ± 173	2400 ± 96

*Scanned area is approximately 40 μm x 40 μm.

The observed differences in the surface topography were confirmed by a statistical analysis of the surface height, leading to the roughness parameters presented in Table 3.4.

Obtained results indicate that the fiber surface containing the micro-structured honeycomb pattern of wells as a result of exposure to the etching solution is significantly rougher than the as-received fiber. The modified fiber is more than 10 times rougher that the 'as received' surface according to the R_q measure. Further analyses of the fiber surfaces involved the quantification of the surface wettability and tension before and after modification. Three diagnostic fluids, water, formamide and diiodomethane, were used for this purpose. Measured contact angles on both the 'as received' and the modified fiber surface are presented in Table 3.1, whereas the surface tensions (and their parameters (mJ/m^2)) of selected diagnostic fluids are presented in Table 3.2 [78].

3.3.3. Bacterial Attachment Patterns on the As-Received and Modified Fiber Surfaces

3.3.3.1. Overview

A detailed visualization of the substratum surface morphology before and after bacterial cultivation, together with the bacterial attachment pattern, was obtained via SEM analysis (Figure 3.7). An analysis of the images indicated that the presence of growth medium did not modify either of the fiber surfaces in a way that might influence bacterial adhesive behavior.

Table 3.5. Surface wettability and surface tensions of the as-received and the modified fiber surfaces

Fiber surface	Contact angle*, θ (°)			Surface tensions**, γ (mJ/m^2)			
	Water	Diiodomethane	Formamide	γ^{LW}	γ^{AB}	γ^{+}	γ^{-}
As-received	107.2	129.6	102.6	1.67	7.51	3.34	4.22
Modified	106.9	102.3	101.3	7.89	1.77	0.18	4.38

*Contact angle of water, formamide and diiodomethane (θ_W, θ_F and θ_D respectively);
**Lifshitz/van der Waals component (γ^{LW}), acid/base component (γ^{AB}), electron acceptor (γ^{+}) and electron donor (γ^{-}).

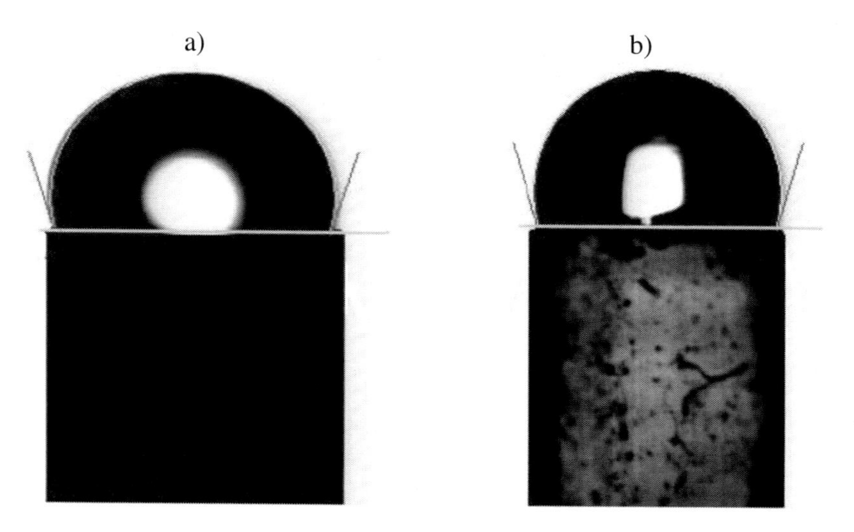

Figure 3.6. Typical images representing measured water contact angles on (a) the 'as received' and (b) the modified fiber surface.

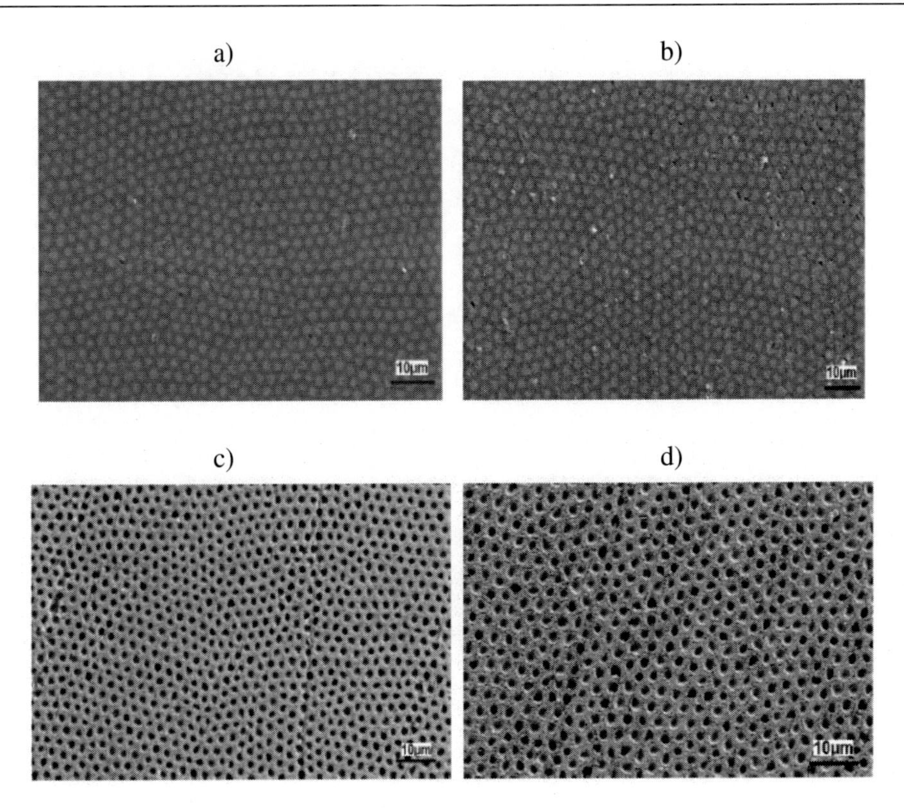

Figure 3.7. Typical SEM images of the 'as-received' fiber surfaces (a) without and (b) with marine broth and the modified fiber surface (c) without and (d) with marine broth. The scale bar on all images is 10 μm.

3.3.3.2. Bacterial Attachment Pattern

High resolution SEM images representing the attachment behavior of the six bacterial strains on the 'as-received' fiber surfaces are given in Figure 3.8. As evident in the images, the fiber surface has a different appearance as a result of the presence of the extra-cellular products (EPS) secreted by the cells. Granular deposits of variable size can be seen, particularly in the case of *E. coli, Pseudomonas aeruginosa, Pseudoalteromonas issachenkonii* and *Sulfitobacter guttiformis* attachment. It is likely that these deposits serve as primers, facilitating the modification of the fiber surfaces to assist in the bacterial adhesion process. The secretion of EPS by these strains during colonization of other surfaces has already been reported [54, 56, 95]. In contrast, *C. marina* and *Staphylococcus aureus* cells, while also being successful colonizers of the 'as-received' fiber surface, did not produce EPS to the same extent as *E. coli*, *Pseudomonas aeruginosa, Pseudoalteromonas issachenkonii* or *Sulfitobacter guttiformis.* The difference between the observed bacterial attachment patterns was seen via the number of bacteria attached to the fiber surfaces. The data presented in Table 3.6 clearly indicates that *Cobetia marina* and *Pseudoalteromonas issachenkonii* were the two most successful colonizers of the surface, with 55,000 and 53,000 cells attached per square mm, respectively. A remarkably different bacterial response was observed on the modified fiber surfaces. All six of the studied bacterial strains failed to attach to the modified fiber surface. Despite varying quantities of granular EPS detected on the modified fiber surface (around and inside the wells), no bacterial cells were observed to adhere to this surface.

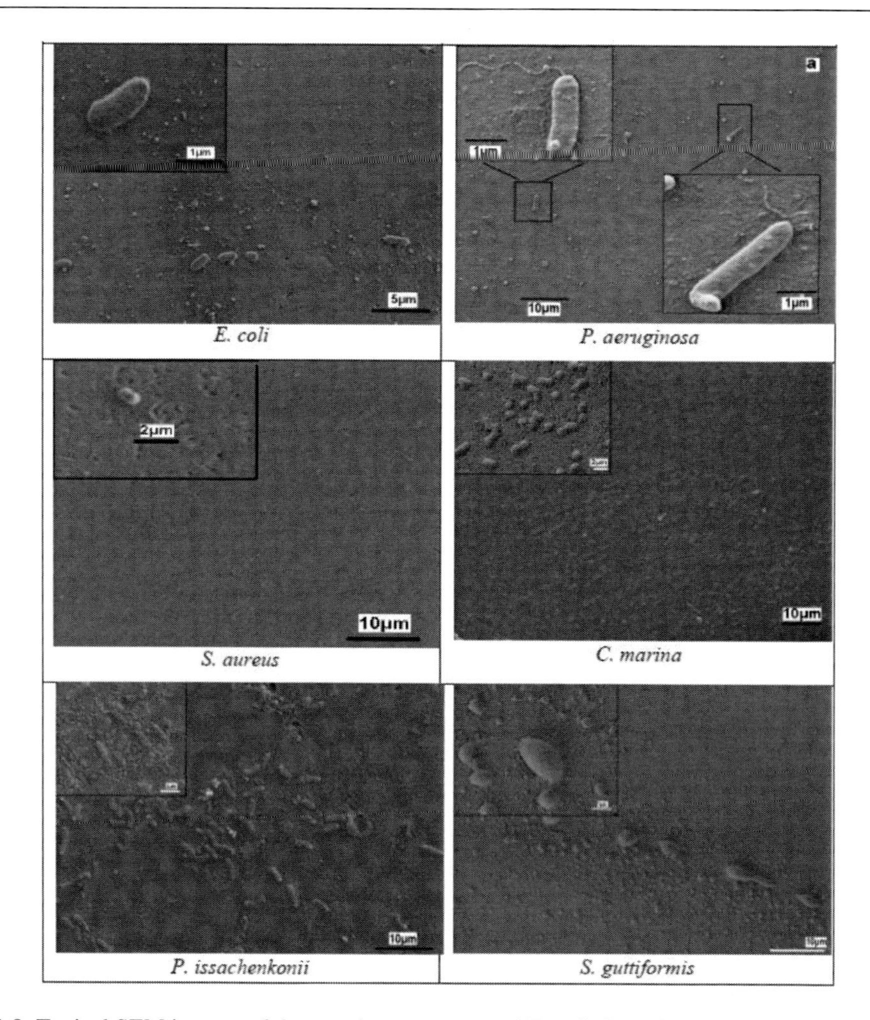

Figure 3.8. Typical SEM images of the attachment pattern of E. coli, Pseudomonas aeruginosa, Staphylococcus aureus, Cobetia marina, Pseudoalteromonas issachenkonii and Sulfitobacter guttiformis on the as-recieved fiber surface.

Visualization of the extra-cellular products produced by the cells while interacting with both surfaces was achieved by labeling α-mannopyranosyl and α-glucopyranosyl residues commonly found in bacterial EPS [65] with the fluorescent dye Concanavalin A. CLSM was subsequently used to image the EPS distribution. The images presented in Figure 3.9 confirmed the presence of EPS produced by *E. coli, Pseudomonas aeruginosa, Pseudoalteromonas issachenkonii* and *Sulfitobacter guttiformis* after 12 h incubation on the as-received fiber surface.

Granular EPS deposits produced by *E. coli* (a), *Pseudomonas aeruginosa* (b) and *Psedoalteromonas issachenkonii* (d) were mostly located on and around fiber cores, whilst granular EPS of different size produced by *Sulfitobacter guttiformis* (c) were randomly deposited over the fiber cores and the surrounding cladding. No EPS produced by *Cobetia marina* and *Staphylococcus aureus* were observed using CLSM, confirming the data obtained using SEM.

It is noteworthy that despite appreciable amounts of EPS being produced, no viable cells could be detected on these surfaces. Figures 3.9 (a) – (c) highlight that the bacterial cells can be clearly seen on the surfaces, however, cell viability was not confirmed by the fluorescent labeling process.

The EPS produced by the six tested strains whilst interacting with the modified fiber surfaces was clearly detectable (Figure 3.10). However, the amount, size and area of the EPS deposition varied, depending on the strain.

Table 3.6. Number of bacterial cells attached to the as-received fiber surfaces

Strain	Attached cells (mm^2)
E. coli	2,980
Pseudomonas aeruginosa	1,150
Staphylococcus aureus	1,380
Cobetia marina	55,460
Pseudoalteromonas issachenonii	53,300
Sulfitobacter guttiformis	3,600

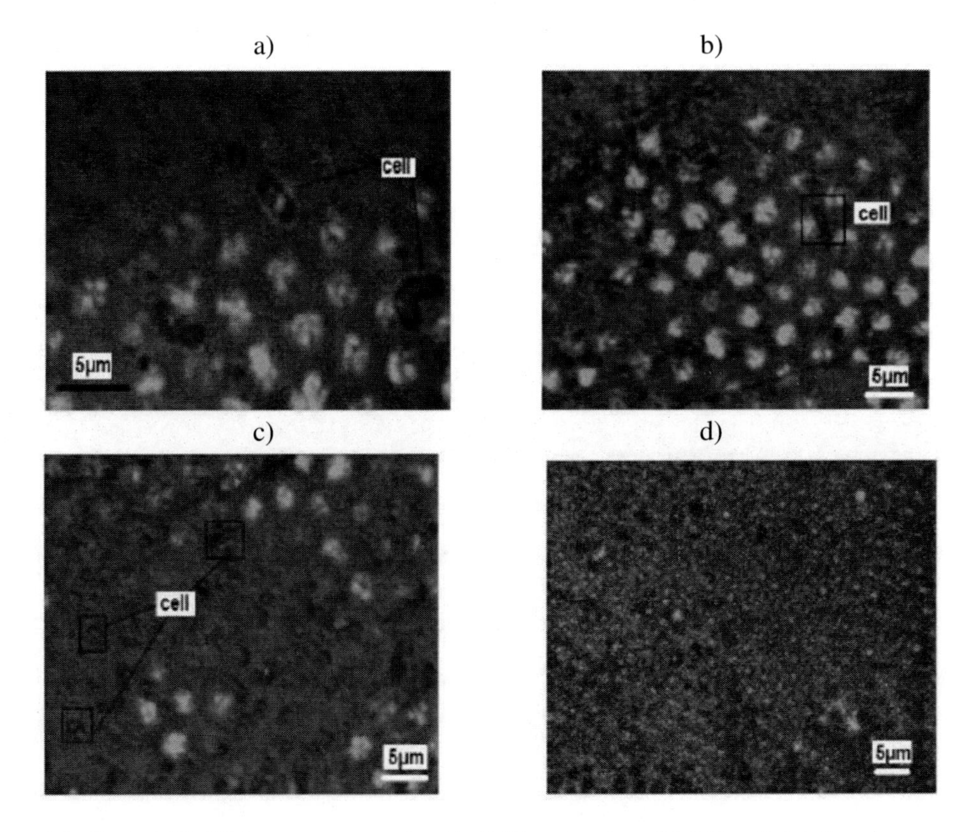

Figure 3.9. Typical CLSM images representing the EPS production of (a) *E. coli*, (b) *Pseudomonas aeruginosa*, (c) *Sulfitobacter guttiformis* and (d) *Pseudoalteromonas issachenkonii* after 12h incubation on the as-received fiber surface.

For instance, EPS synthesized by *Sulfitobacter guttiformis* (f) were deposited inside and around the wells i.e. on the fiber cores and the surrounding cladding, whereas EPS produced by *E. coli* (a), *Pseudomonas aeruginosa* (b) and *Pseudoalteromonas issachnekonii* (e) were predominantly located inside the wells (on the fiber cores). As expected *Staphylococcus aureus* (c) and *C. marina* (d) produced minimal amounts of EPS.

3.4. Discussion

Even though the interrelationship between bacterial attachment patterns and cell surface characteristics (such as wettability and charge) has been studied intensively over the past decades, the somewhat contradictory results do not allow the formulation of a reliable correlation.

Some reported data have suggested that there is a strong linear dependence between bacterial adhesion and bacterial surface wettability [96], whilst others have provided evidence suggesting that bacterial adhesion is inversely correlated with bacterial surface contact angle [36]. The latter statement is in favor of a hypothesis that bacterial survival strategies include an attachment process that is dependent on the presence, chemical composition and structure of surface exo-cellular properties [97-99]. A correlation between the substratum surface wettability and the extent of bacterial adhesion has also proven elusive due to contradictory results.

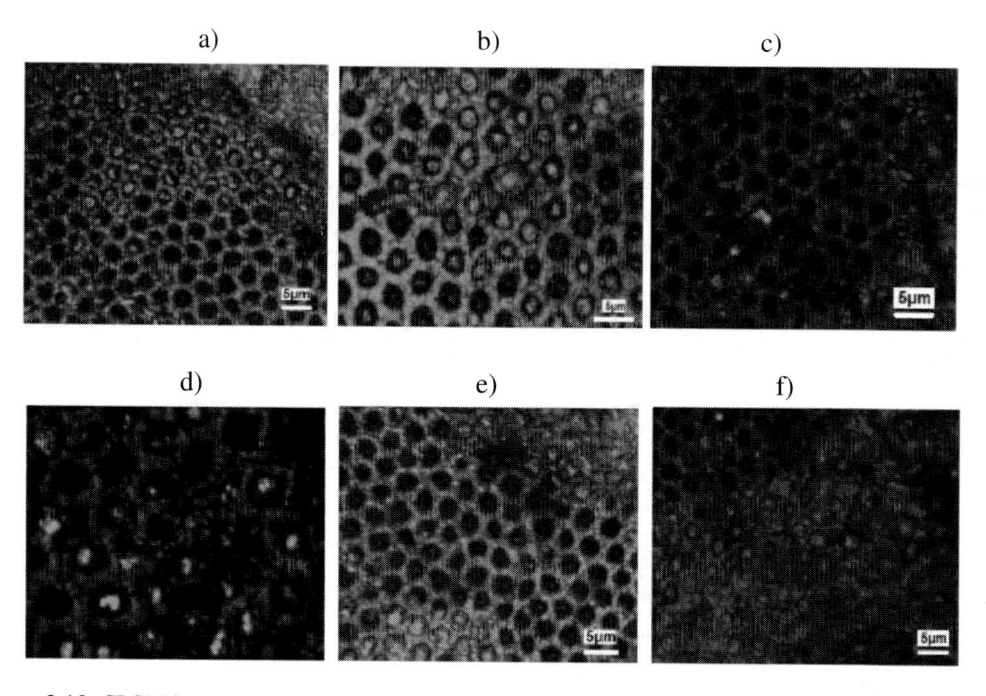

Figure 3.10. CLSM image representing the EPS production of (a) *E. coli*, (b) *Pseudomonas aeruginosa*, (c) *Staphylococcus aureus*, (d) *Cobetia. marina*, (e) *Pseudoalteromonas issachenkonii* and (f) *Sulfitobacter guttiformis* after 12 h of incubation on the modified fiber surface.

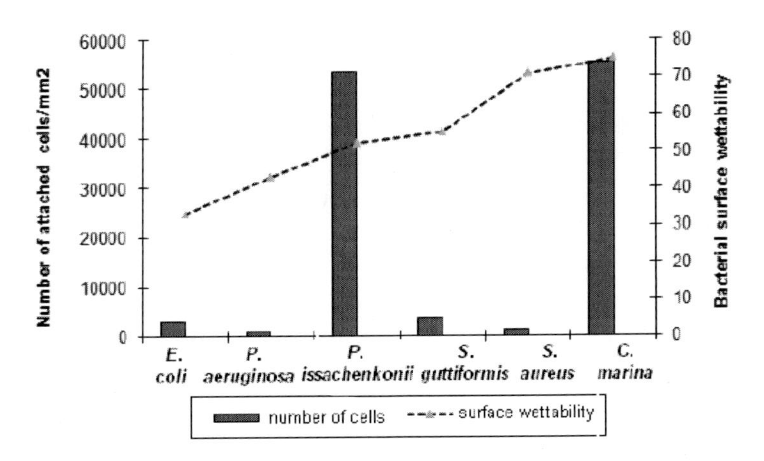

Figure 3.11. Evaluation of bacterial attachment patterns on the as-received fiber surfaces: number of the attached cells versus bacterial surface wettability.

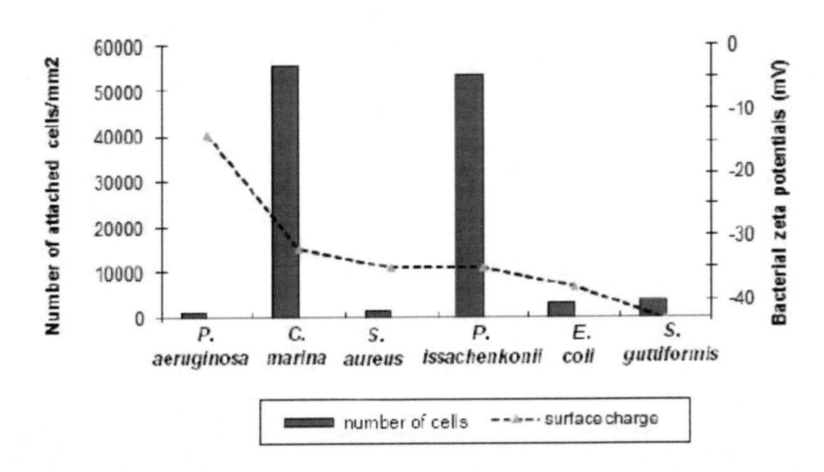

Figure 3.12. Evaluation of bacterial attachment patterns on the as-received fiber surfaces: number of the attached cells versus bacterial surface charge.

For example, even though cells possessing hydrophobic characteristics are believed to adhere to hydrophobic substrata to a greater degree than hydrophilic substrata, and vice versa, the attachment pattern can be different as a result of the constant dynamic motion of the outer-surface proteins [44]. To the extent that the data obtained in this study do not conform to a simple reductionist interpretation, the results confirm that the interpretation of bacterial attachment behavior is very complex.

For instance, the two strains with highly hydrophobic cell surfaces, *Cobetia marina* and *Staphylococcus aureus,* displayed very different attachment tendencies. *Cobetia marina* cells were successful colonizers of the as-received fiber surface, thus complying with van Loosdrecht's theory [96], while *Staphylococcus aureus* cells did not attach to the same surfaces to any significant extent (Figure 3.11).

From an electrostatic perspective, *Sulfitobacter guttiformis* cells, having the lowest negative charge, would be expected to have the strongest propensity for attachment and *Pseudomonas aeruginosa* the weakest propensity for attachment towards the fiber surfaces.

However, *Sulfitobacter guttiformis* cells attached in greater number compared to *Pseudomonas aeruginosa* cell (data presented in Figure 3.12), yet the most successful colonizers were *Cobetia marina* and *Pseudoalteromonas isachenkonii.* This observation is supported by data from earlier studies where it was reported that bacterial cells can exhibit variable surface characteristics due to the presence of outer surface EPS, their constant dynamic motion and the presence/absence of molecular areas of variable polarity or charge [36, 94].

Consequently, the results obtained in our study indicated that there was no correlation between cell surface wettability and charge with the level of bacterial attachment to the as-received fiber surfaces. Hence neither of the cell surface characteristics studied here could provide a reasonable explanation for the different bacterial responses towards the two surfaces. Substratum physico-chemical characteristics such as wettability, charge, chemistry and topography have been intensively studied over several decades in an attempt to predict trends in bacterial attachment behavior [39, 46, 48, 49, 100]. Physico-chemical analysis of both the as-received and modified fiber surfaces reported herein suggested that almost all of the surface modifications occurring as a result of exposure to the etching solution have an insignificant effect on the surface properties. For example, water contact angles and surface tension on both surfaces were almost identical. Analysis of the surface chemistry of the as-received and modified fibers by ToF-SIMS also did not reveal appreciable differences between the surfaces (except for a slight discrepancy in the presence of Ge and F on the modified fiber surfaces).

The changes in the fiber surface topography after chemical etching are believed to be significant enough to influence the different bacterial response to the as-received and modified fiber surfaces. A significant increase in all the roughness parameters was found on the chemically eroded fiber surfaces. Existing understanding of the effects of surface roughness on bacterial adhesion suggests that bacteria prefer microscopic surface irregularities as the starting point for their attachment, as these provide shelter from unfavorable environmental influences [41, 101]. However, recent work on this field, suggests that this might not always be the case [102]. In particular, these studies have demonstrated that nano-scale surface roughness may have a stimulating effect on bacterial adhesion.

Moreover, differences in the surface roughness of just a few nanometers appear capable of exerting a strong influence on the cellular response towards certain surfaces [54, 56]. The results presented here show that a variable number of bacterial cells were able to successfully colonize the nano-scale rough 'as-received' fiber surface, whereas no cells were detected on the microscopically rough modified fiber surfaces. It is notable that the same adhesion tendency was observed for all of the tested strains, regardless of their different taxonomic affiliation and their cell surface characteristics. On the other hand, the results do not appear to support the "attachment point" theory, as the presence of cavities of comparable size to the bacterial cells did not serve to promote adhesion.

Conclusion

Bacterial attachment profiles have been measured on optical fibers surfaces by comparing the bacterial cell surface properties (surface wettability, surface charge, and EPS production) and fiber surface properties (surface chemistry, surface tension, wettability and surface topography). The results highlighted the difficulty of interpreting bacterial attachment behavior based solely on their physico-chemical surface characteristics, and that strain-specific attachment profiles are consistently maintained throughout the attachment process.

Although a limited number of bacterial strains (representing different phylogenetic lineages) were studied, all of the selected strains showed a consistent preference for attachment to nano-smooth fiber surfaces. Contrary to the predictions of the "attachment point" theory, the presence of microscopic cavities on the modified surface did not serve to promote bacterial attachment. An analysis of the attachment profiles has led to a conclusion that the majority of the studied strains employed a similar strategy for attachment to the as-received fiber by producing appreciable quantities of EPS.

In summary, the as-received optical fibers were found to be appropriate for use in the development of SERS substrates or chemical or environmental sensors, since they possess surface characteristics that allow them to effectively undergo bacterial attachment.

However, the surfaces of modified optical fibers were not found to be amenable to bacterial attachment. The data produced in this study will be of use for the construction of chemical sensors, whole-cell biosensors, SERS probes or other optical fiber instrumentation.

REFERENCES

[1] Gannot, I.; Ben-David, M., Optical fibers and waveguides for medical applications. In: *Biomedical Photonics Handbook*, Vo-Dinh, T., Ed. CRC Press, Boca Raton: 2003.

[2] Udd, E., *Fiber Optic Smart Structures*. Wiley: New York, 1995.

[3] Dubaj, V.; Mazzolini, A.; Wood, A.; Harris, M., Optic fibre bundle contact imaging probe employing a laser scanning confocal microscope. *Journal of Microscopy* 2002, 207, 108-117.

[4] Marazuela, M. D.; Moreno-Bondi, M. C., Fiber-optic biosensors – an overview. *Analytical and Bioanalytical chemistry* 2002, 372, 664–682.

[5] Potyrailo, R. A.; Hobbs, S. E.; Hieftje, G. M., Optical waveguide sensors in analytical chemistry: today's instrumentation, applications and trends for future development. *Journal of Analytical Chemistry* 1998, 362, 349–373.

[6] Petry, R.; Schmitt, M.; Popp, J., Raman Spectroscopy - a prospective tool in the life sciences. *ChemPhysChem* 2003, 4, (1), 14-30.

[7] Carey, P. R., Raman spectroscopy, the sleeping giant in structural biology, awakes. *Journal of Biological Chemistry* 1999, 274, (38), 26625-26628.

[8] Keller, M. D.; Kanter, E. M.; Mahadevan-Jansen, A., Raman spectroscopy for cancer diagnosis. *Spectroscopy* 2006, 21, (11), 33-41.

[9] Schuster, K. C.; Reese, I.; Urlaub, E.; Gapes, J. R.; Lendl, B., Multidimensional information on the chemical composition of single bacterial cells by confocal Raman microspectroscopy. *Analytical Chemistry* 2000, 72, (22), 5529-5534.

[10] Gessner, R.; Rosch, P.; Kiefer, W.; Popp, J., Raman spectroscopy investigation of biological materials by use of etched and silver coated glass fiber tips. *Biopolymers* 2002, 67, (4-5), 327-330.

[11] Rosch, P.; Harz, M.; Schmitt, M.; Peschke, K. D.; Ronneberger, O.; Burkhardt, H., Chemotaxonomic identification of single bacteria by micro-Raman spectroscopy: Application to clean-room-relevant biological contaminations. *Applied and Environmental Microbiology* 2005, 71, (3), 1626-1637.

[12] Xie, C.; Mace, J.; Dinno, M. A.; Li, Y. Q.; Tang, W.; Newton, R. J., Identification of single bacterial cells in aqueous solution using confocal laser tweezers Raman spectroscopy. *Analytical Chemistry* 2005, 77, (14), 4390-4397.

[13] Fleischman, M.; Hendra, P. J.; McQuillan, A. J., Raman spectra of pyridine adsorbed at a silver electrode. *Chemical Physics Letters* 1974, 26, (2), 163-166.

[14] Haynes, C. L.; Yonzon, C. R.; Zhang, X.; Van Duyne, R. P., Surface-enhanced Raman sensors: early history and the development of sensors for quantitative biowarfare agent and glucose detection. *Journal of Raman Spectroscopy* 2005, 36, 471.

[15] Crow, P.; Molckovsky, A.; Uff, J.; Stone, N.; Wilson, B.; Wongkeesong, L. M., *J. Urol.* 2004, 170, 68.

[16] Murphy, T.; Lucht, S.; Schmidt, H.; Kronfeldt, H. D., Surface-enhanced Raman scattering (SERS) system for continuous measurements of chemicals in sea-water. *Journal of Raman Spectroscopy* 2000, 31, (10), 943-948.

[17] Stuart, D. A.; Yuen, J. M.; Shah, N.; Lyandres, O.; Yonzon, C. R.; Glucksberg, M. R.; Walsh, J. T.; Van Duyne, R. P., In: vivo glucose measurement by surface-enhanced Raman spectroscopy. *Analytical Chemistry* 2006, 78, 7211-7215.

[18] Zeiri, L.; Efrima, S., Surface-enhanced Raman spectroscopy of bacteria: the effect of excitation wavelength and chemical modification of the colloidal milieu. *Journal of Raman Spectroscopy* 2005, 36, (6-7), 667-675.

[19] Vo-Dinh, T.; Stokes, D. L., In: *Handbook of Vibrational Spectroscopy*, Chalmers, J. M.; Griffiths, P. R., Eds. Wiley: New York, 2002; Vol. 2, p 1302.

[20] Vo-Dinh, T., Surface-enhanced Raman spectroscopy using metallic nanostructures. *Trends in Analytical Chemistry* 1998, 17, (8-9), 557.

[21] White, D. J.; Stoddart, P. R., Nanostructured optical fiber with surface-enhanced Raman scattering functionality. *Optics Letters* 2005, 30, (6), 598-600.

[22] White, D. J.; Mazzolini, A. P.; Stoddart, P. R., Fabrication of a range of SERS substrates on nanostructured multicore optical fibres. *Journal of Raman Spectroscopy* 2007, 38, (4), 377-382.

[23] Polwart, E.; Keir, R. L.; Davidson, C. M.; Smith, W. E.; Sadler, D. A. A. S., Novel SERS-active optical fibers prepared by the immobilization of silver colloidal particles. *Applied Spectroscopy* 2000, 54, 522.

[24] Stoddart, P. R.; Brack, N., Physical techniques for cell surface probing and manipulation. In: *Nanoscale Structure and Properties of Microbial Cell Surfaces*, Ivanova, E. P., Ed. Nova Science: 2007; pp 1-158.

[25] Vo-Dinh, T.; Kasili, P., Fiber-optic nanosensors for single-cell monitoring. *Analytical and Bioanalytical Chemistry* 2005, 382, (4), 918-925.

[26] Biran, I.; Rissin, D. M.; Ron, E. Z.; Walt, D. R., Optical imaging fiber-based live bacterial cell array biosensor. *Analytical Biochemistry* 2003, 315, (1), 106-113.

[27] D'Souza, S. F., Microbial biosensors. *Biosensors and Bioelectronics* 2001, 16, 337-353.

[28] Deckert, V.; Zeisel, D.; Zenobi, R.; Vo-Dinh, T., Near-field surface enhanced Raman imaging of dye-labeled DNA with 100-nm resolution. *Analytical Chemistry* 1998, 70, (13), 2646-2650.

[29] Stokes, D. L.; Chi, Z. H.; Vo-Dinh, T., Surface enhanced Raman scattering inducing nanoprobe for spectrochemical analysis. *Applied Spectroscopy* 2004, 58, (3), 292-298.

[30] Eboigbodin, E. K.; Newton, A. R. J.; Routh, F. A.; Biggs, A. C., Role of non-adsorbing polymers in bacterial aggregation. *Langmuir* 2005, 21, 12315-12319.

[31] Wong, H.-C.; Chung, Y.-C.; Yu, J.-A., Attachment and inactivation of *Vibrio parahaemolyticus* on stainless steel and glass surface. *Food Microbiology* 2002, 19, (4), 341-350.

[32] Sharon, N., Carbohydrates as future anti-adhesion drugs for infectious diseases. *Biochimica et Biophysica Acta (BBA) - General Subjects* 2006, 1760, (4), 527-537.

[33] Stoodley, P.; Sauer, K.; Davies, D. G.; Costerton, J. W., Biofilms as complex differentiated communities. *Annual Review of Microbiology* 2002, 56, (1), 187-209.

[34] Bell, C. H.; Arora, B. S.; Camesano, T. A., Adhesion of *Pseudomonas putida* KT2442 Is mediated by surface polymers at the nano- and microscale. *Environmental Engineering Science* 2005, 22, (5), 629-641.

[35] Fletcher, M., *Bacterial attachment in aquatic environments: a diversity of surface and adhesion strategies*. Wiley-Liss: New York, 1996; p 1-24.

[36] Li, B.; Logan, B. E., Bacterial adhesion to glass and metal-oxide surfaces. *Colloids and Surfaces B: Biointerfaces* 2004, 36, (2), 81-90.

[37] Pringle, J. H.; Fletcher, M., Influence of substratum hydration and adsorbed macromolecules on bacterial attachment to surfaces. *Applied and Environmental Microbiology* 1986, 51, (6), 1321-1325.

[38] Simoni, S. F.; Bosma, T. N. P.; Harms, H.; Zehnder, A. J. B., Bivalent cations increase both the subpopulation of adhering bacteria and their adhesion efficiency in sand columns. *Environmental Science and Technology* 2000, 34, (6), 1011-1017.

[39] Teixeira, P.; Oliveira, R., Influence of surface characteristics on the adhesion of *Alcaligenes denitrificans* to polymeric substrates. *Journal of Adhesion Science and Technology* 1999, 13, 1243-1362.

[40] Fletcher, M.; Loeb, G. I., Influence of substratum characteristics on the attachment of a marine *Pseudomonas* to solid surfaces. *Applied and Environmental Microbiology* 1979, 37, (1), 67-72.

[41] Whitehead, A. K.; Verran, J., The effect of surface topography on the retention of microorganisms. *Food and Bioproducts Processing* 2006, 84, (C4), 253-259.

[42] Emerson, R. J.; Bergstrom, T. S.; Liu, Y.; Soto, E. R.; Brown, C. A.; McGimpsey, W. G.; Camesano, T. A., Microscale correlation between surface chemistry, texture, and the adhesive strength of *Staphylococcus epidermidis*. *Langmuir* 2006, 22, (26), 11311-11321.

[43] Scardino, A. J.; Harvey, E.; De Nys, R., Testing attachment point theory: diatom attachment on microtextured polyimide biomimics. *Biofouling* 2006, 22, (1), 55-60.

[44] Bruinsma, G. M.; Rustema-Abbing, M.; van der Mei, H. C.; Busscher, H. J., Effects of cell surface damage on surface properties and adhesion of *Pseudomonas aeruginosa*. *Journal of Microbiological Methods* 2001, 45, (2), 95-101.

[45] Cao, T.; Tang, H.; Liang, X.; Wang, A.; Auner, G. W.; Salley, S. O.; Ng, S. K. Y., Nanoscale investigation on adhesion of *Escherichia coli* to surface modified silicone using atomic force microscopy. *Biotechnology and Bioengineering* 2006, 94, (1), 167-176.

[46] Pereira, M. A.; Alves, M. M.; Azeredo, J.; Mota, M.; Oliveira, R., Influence of physico-chemical properties of porous microcarriers on the adhesion of an anaerobic consortium. *Journal of Industrial Microbiology and Biotechnology* 2000, V24, (3), 181-186.

[47] Castellanos, T.; Ascencio, F.; Bashan, Y., Cell-surface hydrophobicity and cell-surface charge of *Azospirillum* spp. *FEMS Microbiology, Ecology* 1997, 24, (2), 159-172.

[48] Bos, R.; van der Mei, H. C.; Busscher, H. J., Physico-chemistry of initial microbial adhesive interactions - its mechanisms and methods for study. *FEMS Microbiology Reviews* 1999, 23, 179-230.

[49] Busscher, H. J.; van der Mei, H. C., Physico-chemical interactions in initial microbial adhesion and relevance for biofilm formation. *Advances in Dental Research* 1997, 11, (1), 24-32.

[50] Jones, J. F.; Velegol, D., Laser trap studies of end-on E. coli adhesion to glass. *Colloids and Surfaces B: Biointerfaces* 2006, 50, (1), 66-71.

[51] Chae, M. S.; Schraft, H.; Truelstrup Hansen, L.; Mackereth, R., Effects of physicochemical surface characteristics of *Listeria monocytogenes* strains on attachment to glass. *Food Microbiology* 2006, 23, (3), 250-259.

[52] Shellenberger, K.; Logan, B. E., Effect of molecular scale roughness of glass beads on colloidal and bacterial deposition. *Environmental Science and Technology* 2002, 36, (2), 184-189.

[53] Li, B.; Logan, B. E., The impact of ultraviolet light on bacterial adhesion to glass and metal oxide-coated surface. *Colloids and Surfaces B: Biointerfaces* 2005, 41, (2-3), 153-161.

[54] Mitik-Dineva, N.; Wang, J.; Mocanasu, C. R.; Stoddart, P. R.; Crawford, R. J.; Ivanova, E. P., Impact of nano-topography on bacterial attachment. *Biotechnology Journal* 2008, 3, 536-544.

[55] Mitik-Dineva, N.; Wang, J.; Stoddart, R. P.; Crawford, R. J.; Ivanova, P. E. In: *Nano-structured surfaces control bacterial attachment* ICONN, Melbourne, Australia, 2008; Melbourne, Australia, 2008.

[56] Ivanova P., E.; Mitik-Dineva, N.; Wang, J.; Pham, K. D.; Wright, P. J.; Nicolau, D. V.; Mocanasu, C. R.; Crawford, R. J., *Sulfitobacter guttiformis* attachment on poly(tert-butylmethacrylate) polymeric surfaces. *Micron* in press.

[57] Howell, D.; Behrends, B., A review of surface roughness in antifouling coatings illustrating the importance of cutoff length. *Biofouling* 2006, 22, (6), 401-410.

[58] Callow, M. E.; Jennings, A. R.; Brennan, A. B.; Seegert, C. E.; Gibson, A.; Wilson, L.; Feinberg, A.; Baney, R.; Callow, J. A., Microtopographic cues for settlement of zoospores of the green fouling alga *Enteromorpha*. *Biofouling* 2002, 18, (3), 229-236.

[59] Verran, J.; Boyd, R. D., The relationship between substratum surface roughness and microbiological and organic soiling: A review. *Biofouling* 2001, 17, 59-71.

[60] Hoipkemeier-Wilson, L.; Schumacher, J. F.; Carman, M. L.; Gibson, A. L.; Feinberg, A. W.; Callow, M. E.; Finlay, J. A.; Callow, J. A.; Brennan, A. B., Antifouling potential of lubricious, micro-engineered, PDMS elastomers against zoospores of the green fouling alga *Ulva* (Enteromorpha). *Biofouling* 2004, 20, (1), 53-63.

[61] Petronis, S.; Berntsson, K.; Gold, J.; Gatenholm, P., Design and microstructuring of PDMS surfaces for improved marine biofouling resistance. *Journal of Biomaterials Science, Polymer Edition* 2000, 11, 1051-1072.

[62] Berntsson, K. M.; Jonsson, P. R.; Lejhall, M.; Gatenholm, P., Analysis of behavioural rejection of micro-textured surfaces and implications for recruitment by the barnacle *Balanus improvisus*. *Journal of Experimental Marine Biology and Ecology* 2000, 251, (1), 59-83.

[63] Garrity, G. M., *Bergey's manual of systematic bacteriology*. 2nd ed.; Williams and Wilkins: 1984; Vol. 2, p 2816.

[64] Castellani, A.; Chalmers, A. J., *Manual of topical medicine 3rd ed., Type genus: Bacteroides*. New York, 1919.

[65] Goldstein, I., J., Hollerman, C. E., Smith, E, E., Protein-carbohydrate interaction. II. Inhibition studies on the interaction on concanavalin A with polysaccharides. *Biochemistry* 1964, 4, (5), 876-883.

[66] Sumner, J., B., Howell, S, F., The role of divalent metals in the reversible inactivation of jack bean hemagglutinin. *J. Biol. Chem.* 1936, 115, 583.

[67] MolecularProbes Alexa Fluor Dyes: Simply the Best and Brightest: Fluorescent Dyes and Conjugates. www.probes.com (May),

[68] Invitrogen, Vybrant CFDA SE Cell Tracer Kit. In: www.probes. invitrogen.com: 2006.

[69] Ivanova, E. P.; Pham, D. K.; Wright, P. J.; Nicolau, D. V., Detection of coccoid forms of *Sulfitobacter mediterraneus* using atomic force microscopy. *FEMS Microbiology Letters* 2002, 214, (2), 177.

[70] Dong, H.; Onstotta, T. C.; Kob, C.-H. A.; Hollingsworth, A. D.; Brown, D. G.; Mailloux, B. J., Theoretical prediction of collision efficiency between adhesion-deficient bacteria and sediment grain surface. *Colloids and Surfaces B: Biointerfaces* 2002, 24, (3-4), 229.

[71] Korenevsky, A.; Beveridge, T. J., The surface physicochemistry and adhesiveness of *Shewanella* are affected by their surface polysaccharides. *Microbiology* 2007, 153, (6), 1872-1883.

[72] Bakker, D. P.; Busscher, H. J.; van der Mei, H. C., Bacterial deposition in a parallel plate and a stagnation point flow chamber: microbial adhesion mechanisms depend on the mass transport conditions. *Microbiology* 2002, 148, (2), 597-603.

[73] Etzler, F. M., Surface free energy of solids: A comparison of models. In *Contact angle, Wettability and Adhesion*, 1st ed.; Mittal, K. L., Ed. Brill Academic: 2006; Vol. 4, pp 215-236.

[74] Girifalco, L. A.; Good, R. J., A theory for the estimation of surface and interfacial energies. I. Derivation and application to interfacial tension. *Journal of Physical Chemistry* 1957, 61, (7), 904-909.

[75] Van Oss, C. J., *Interfacial forces in aqueous media*. Marcel Dekker: New York, 1994.

[76] Ong, Y.; Razatos, A.; Georgiou, G.; Sharma, M. M., Adhesion forces between E. coli bacteria and biomaterial surfaces. *Langmuir* 1999, 15, 2719-2725.

[77] Brant, A. J.; Childress, E. A., Assessing short-range membrane–colloid interactions using surface energetics. *Journal of membrane science* 2002, 203, (1-2), 257.

[78] Van Oss, C. J.; Good, R. J.; Chaudhury, M. K., Additive and nonadditive surface tension components and the interpretation of contact angles. *Langmuir* 1988, 4, 884-891.

[79] Sanders, R. S.; Chow, R. S.; Masliyah, J. H., Deposition of bitumen and asphaltene-stabilized emulsions in an impinging jet cell *Journal of Colloid and Interface Science* 1995, 174, (1), 230-245.

[80] Van der Mei, H. C.; Busscher, H. J., Electrophoretic mobility distributions of single-strain microbial populations. *Applied and Environmental Microbiology* 2001, 67, (2), 491-494.

[81] Busscher, H. J.; Norde, W., Limiting values for bacterial ζ potentials. *Journal of Biomedical Materials Research* 2000, 50, 463-464.

[82] De Kerchove, A. J.; Elimelech, M., Relevance of electrokinetic theory for "soft" particles to bacterial cells: implication for bacterial adhesion. *Langmuir* 2005, 21, 6462-6472.

[83] Eboigbodin, E. K.; Newton, A. R. J.; Routh, F. A.; Biggs, A. C., Bacterial quorum sensing and cell surface electrokinetic properties. *Applied Microbiology and Biotechnology* 2006, 73, 669-675.

[84] Waar, K.; van der Mei, H. C.; Harmsen, H. J. M.; Degener, J. E.; Busscher, H. J., *Enterococcus faecalis* surface proteins determine its adhesion mechanism to bile drain materials. *Microbiology* 2002, 148, (6), 1863-1870.

[85] Vogler, E. A., Structure and reactivity of water at biomaterial surfaces. *Advances in Colloid and Interface Science* 1998, 74, (1-3), 69-117.

[86] Burks, G. A.; Velegol, S. B.; Paramonova, E.; Lindenmuth, B. E.; Feick, J. D.; Logan, B. E., Macroscopic and nanoscale measurements of the adhesion of bacteria with varying outer layer surface composition. *Langmuir* 2003, 19, (6), 2366-2371.

[87] Vermeltfoort, P. B. J.; van Kooten, T. G.; Bruinsma, G. M.; Hooymans, A. M. M.; van der Mei, H. C.; Busscher, H. J., Bacterial transmission from contact lenses to porcine corneas: An ex vivo study. *Investigative Ophthalmology and Visual Science* 2005, 46, (6), 2042-2046.

[88] Gross, M.; Cramton, S. E.; Gotz, F.; Peschel, A., Key role of teichoic acid net charge in *Staphylococcus aureus* colonization of artificial surfaces. In 2001; Vol. 69, pp 3423-3426.

[89] Canepari, P.; Boaretti, M.; Lleó, M. M.; Satta, G., Lipoteichoic acid as a new target for activity of antibiotics: mode of action of daptomycin (LY146032). *Antimicrobial Agents and Chemotherapy* 1990, 34, (6), 1220–1226.

[90] Soni, K. A.; Balasubramanian, A. K.; Beskok, A.; Pillai, S. D., Zeta potential of selected bacteria in drinking water when dead, starved, or exposed to minimal and rich culture media. *Current Microbiology* 2007.

[91] Gottenbos, B.; Grijpma, D. W.; Van der Mei, C. H.; Feijen, J.; Busscher, H. J., Antimicrobial effects of positively charged surfaces on adhering Gram-positive and Gram-negative bacteria. *Journal of Antimicrobial Chemotherapy* 2001, 48, 7-13.

[92] Jucker, B. A.; Harms, H.; Zehnder, A. J., Adhesion of the positively charged bacterium *Stenotrophomonas (Xanthomonas) maltophilia* 70401 to glass and Teflon. In 1996; Vol. 178???, pp 5472-5479.

[93] Van Loosdrecht, M. C.; Lyklema, J.; Norde, W.; Schraa, G.; Zehnder, A. J., Electrophoretic mobility and hydrophobicity as a measure to predict the initial steps of bacterial adhesion. *Applied and Environmental Microbiology* 1987, 53, (8), 1898-1901.

[94] Vadillo-Rodriguez, V.; Busscher, H. J.; Norde, W.; Vries, J.; van der Mei, H. C., Atomic force microscopy corroboration of bond aging for adhesion of Streptococcus thermophilus to solid substrata. *Journal of Colloid and Interface Science* 2004, 278, 251-254.

[95] Lam, J. S.; Graham, L. L.; Lightfoot, J.; Dasgupta, T.; Beveridge, T. J., Ultrastructural examination of the lipopolysaccharides of Pseudomonas aeruginosa strains and their isogenic rough mutants by freeze-substitution. *Journal of Bacteriology* 1992, 174, (22), 7159-7167.

[96] Van Loosdrecht, M. C. M.; Norde, W.; Lyklema, J.; Zehnder, A. J. B., Hydrophobic and electrostatic parameters in bacterial adhesion. *Aquatic Sciences - Research Across Boundaries* 1990, 52, (1), 103-114.

[97] Danese, P. N.; Pratt, L. A.; Kolter, R., Exopolysaccharide production is required for development of *Escherichia coli* K-12 biofilm architecture. *Journal of Bacteriology* 2000, 182, (12), 3593-3596.

[98] Auerbach, I. D.; Sorensen, C.; Hansma, H. G.; Holden, P. A., Physical morphology and surface properties of unsaturated *Pseudomonas putida* biofilms. *Journal of Bacteriology* 2000, 182, (13), 3809-3815.

[99] Jain, A.; Gupta, Y.; Agrawal, R.; Khare, P.; Jain, S. K., Biofilms—A microbial life perspective: a critical review *Critical Reviews in Therapeutic Drug Carrier Systems* 2007, 24, (5), 393-443.

[100] Bos, R.; van der Mei, C. H.; Gold, J.; Busscher, H. J., Retention of bacteria on a substratum surface with micro-patterned hydrophobicity. *FEMS Microbiology Letters* 2000, 189, (2), 311-315.

[101] Riedewald, F., Bacterial adhesion to surfaces: The influence of surface roughness. *PDA Journal of Pharmaceutical Science and Technology* 2006, 60, (3), 164-171.

[102] Mitik-Dineva, N.; Wang, J.; Truong, V. K; Stoddart, P. R.; Malherbe, F.; Crawford, R. J.; Ivanova, E. P., Escherichia coli, Pseudomonas aeruginosa and Staphylococcus aureus Attachment Patterns on Glass Surfaces with Nano-scale Roughness *Current Microbiology* 2008.

In: Advances in Laser and Optics Research. Volume 10 ISBN: 978-1-62257-795-8
Editor: William T. Arkin

Chapter 3

PASSIVELY MODE-LOCKED FIBER LASERS WITH NONLINEAR OPTICAL LOOP MIRRORS

Nak-Hyun Seong[1*] and Dug Young Kim[2†]

[1]School of Chemical Sciences, University of Illinois at Urbana-Champaign,Urbana, US
[2]Center for Nano Optical Imaging Systems (CNOIS), GIST,
Department of Information and Communication, Gwang-ju, South Korea

ABSTRACT

We present the operational principles of two recently introduced nonlinear optical loop mirrors (NOLMs): dispersion imbalanced NOLM (DI-NOLM) and an absorption imbalanced NOLM (AI-NOLM). These NOLMs with 3-dB fiber couplers can break the symmetry of nonlinear phase shift in a fiber loop either with two different dispersion fibers or an asymmetrically located attenuator in a fiber loop mirror. Mode-locked laser pulses were obtained utilizing these two types of NOLMs and the pulse properties of them were investigated. It is demonstrated that these lasers show different pulse operation modes where the pulsed output can be characterized as stable mode-locked pulses, broad square pulses, harmonically mode-locked pulses, bound solitons, etc. depending on the parameters of a laser system. Stable mode-locking condition for a figure-eight fiber laser with a DI-NOLM is relatively hard to obtain due to the sensitive changes of the optical dispersion parameter of a fiber with respect to parameters and external conditions such as wavelength, temperature, bending, etc. Meanwhile, it is shown that the station mode-locked output pulses can be obtained easily with an AI-NOLM because of its simple and robust operating condition.

* Nak-Hyun Seong: School of Chemical Sciences, University of Illinois at Urbana-Champaign,Urbana, IL 61801, US. E-mail address: nhseong@uiuc.edu.

† Center for Nano Optical Imaging Systems (CNOIS), GIST, Department of Info. and Comm.,1 Oryoung-dong, Buk-gu, Gwang-ju, 500-712, South Korea. E-mail address: dykim@gist.ac.kr.

1. INTRODUCTION

Compact laser sources for generating ultrashort optical pulses are studied a lot lately in high-speed optical measurements and testing, optical devices, material studies, and bio and medical applications. In turn, the realization of practical fiber optic devices developed during the last decade has become possible due to the help of successful developments for ultrafast optical pulses. All-optical demultiplexing, optical gating or switching, and optical logic gates are studied in these days based on ultrafast optical pluses. Temporal optical soliton generation and its propagation, nonlinear dynamics associated with high intensity optical pulses in optical fibers are other fascinating topics related to optical short pulse generation. Optical solitons are the result of the balance between the group velocity dispersion and self-phase modulation. Optical solitons has attracted a lot of attention for last two decades not only because of its physical properties but also for its potential applications for distortionless propagation of pulses in the optical communication systems. With the help of noticeable advances in fiber optic components and technologies developed lately in optical communications there has been much improvements in fiber lasers. Among many approaches for ultrafast optical pulse generation, mode-locked fiber lasers are considered to be one of the most compact, reliable and powerful pulse sources in many applications.

A mode-locked fiber laser can easily generate ultrashort optical pulse and also be simply constructed by the help of a fusion splicer without the problem of component alignment, which is frequently very problematic in a bulk laser system. Many higher order nonlinear optical phenomena are involved in most of fiber laser cavities are because of the small core areas of optical fibers and long cavity lengths. As we can easily bend or wind an optical fiber on a spool, we can make the cavity length of a fiber laser much longer than a bulk laser up to a few tens or hundreds of meters. This long cavity is very advantageous in generating high energy pulses when a laser is mode-locked, where pulse energy is inversely proportionally to the length of a laser cavity when the average power of a laser is same.

Because of capability in generating ultrafast optical pulses, passive mode-locking is normally preferred in a fiber laser system. All-optical switching elements with intensity dependent throughput characteristics are essential to attain passive mode-locking of a laser. Many kinds of all-optical switching elements such as a nonlinear optical loop mirror (NOLM), a nonlinear amplifying loop mirror (NALM), and a nonlinear polarization rotator have been invented and studied intensively [1]–[4]. These switching elements are also used for time-domain filtering of optical pulses, all-optical demultiplexing, pedestal suppression of noisy pulses in optical communications. The first passively mode-locked all fiber laser [5] was reported by Duling with an NALM, where a 3-dB optical coupler was used. An NALM needs an optical gain medium such as an Erbium-doped fiber amplifier (EDFA) or a semiconductor optical amplifier (SOA) to break the symmetry of an optical loop mirror [5], [7]. Anisotropy in nonlinear phase shift was achieved in NALM by placing an EDFA at an asymmetric position inside the NALM. As the 3-dB optical coupler was used in the NALM, the efficiency of a fiber laser was much improved. However, a fiber laser using an NALM is normally unstable due to the active switching element and nonlinear optical interactions caused by two counter propagating optical pulses in an optical gain medium.

A passively mode-locked figure-eight fiber laser with an NOLM was introduced later, where an uneven optical coupler was used instead of an optical amplifier [6]. Stability problem associated the gain medium used in an NALM is eliminate in this case. Here, optical intensity difference between two counter propagating waves in an NOLM induces a self-switching due to the difference of the nonlinear phase shifts of the two counter propagating waves.

On the other hand, as a self-switching element, an NOLM with an asymmetric optical coupler shows its own problems such as large optical power loss, background ASE noises, and lasing with some continuous wave components because of asymmetric fiber coupler used in an NOLM.

Recently, dispersion imbalanced NOLM (DI-NOLM) and an absorption imbalanced NOLM (AI-NOLM) by using a symmetric 3-dB fiber coupler were reported to effectively remove continuous wave components and ASE noise. These NOLMs with 3-dB fiber couplers can break the symmetry of nonlinear phase shift in a fiber loop by using two different dispersion fibers or an asymmetrically located attenuator in a loop mirror. A figure-eight fiber laser with a DI-NOLM [8, 9, 10] was proposed, and it showed different pulse operation modes whose pulse output can be characterized with broad square pulses, harmonically mode-locked pulses, bound solitons, and stable mode-locked pulses.

However, a successful mode-locking operation with a DI-NOLM needs very complicated calculation and arrangement in the dispersion parameters of optical fibers used in it for a given center wavelength. The major advantage of AI-NOLM is that it can operate at any given wavelengths regardless of dispersion characteristics of fibers used in it.

A successful demonstration of mode-locking with AI-NOLM was reported lately[11, 12]. In this chapter, we describe and explain the basic operational principles of NOLM. Various properties of a mode-locked figure-eight fiber laser with DI-NOLM or AI-NOLM is presented as well.

2. FIBER SAGNAC INTERFEROMETER

When a sangnac loop is constructed by an uneven coupler, the transmitted electric field follows the difference of the nonlinear phase shifts between two counter propagating waves [1].

Figure 1 shows the configuration of a sagnac loop with the coupling ratio α of a center coupler. If the input electric field E_{in} is launched into the loop, the electric fields of E_1 and E_2 are expressed by

$$E_1 = \sqrt{\alpha} E_{in}, \tag{2-1}$$

$$E_2 = i\sqrt{1-\alpha} E_{in}. \tag{2-2}$$

Those fields pass through the loop of a length L, then the fields are written as

$$E_1^{'} = \sqrt{\alpha} E_{in} \exp(i\alpha\phi), \tag{2-3}$$

$$E_2^{'} = i\sqrt{1-\alpha}E_{in}\exp[i(1-\alpha)\phi], \tag{2-4}$$

where ϕ is defined by

$$\phi = \frac{2\pi}{\lambda}n_2|E_{in}|^2 L = \gamma P_0 L, \tag{2-5}$$

and n_2 is nonlinear refractive index, and γ is a nonlinear coefficient. The output electric field is given by

$$E_{out} = \sqrt{\alpha}E_1^{'} - \sqrt{1-\alpha}E_2^{'} = \alpha E_{in}\exp(i\alpha\phi) - (1-\alpha)E_{in}\exp[i(1-\alpha)\phi]. \tag{2-6}$$

The transmissivity T is written by

$$T = |E_{out}|^2 / |E_{in}|^2 = 1 - 2\alpha(1-\alpha)\{1+\cos[(1-2\alpha)\phi]\}. \tag{2-7}$$

If the coupling ratio α is 0.5, T becomes zero, thus the launched light from the input port is totally reflected by the saganc loop. Therefore, the sagnac loop is used as the device of a perfect mirror. However, if the coupling ratio α is off 0.5, then the two counter propagating lights passing through the loop have different nonlinear phase shifts due to the asymmetric power. Figure 2 shows the transmissivity for different α values. The peaks represent that the nonlinear phase difference between two counter propagating lights is $(2n + 1)\pi$ and the deeps express the nonlinear phase difference is $2n\pi$, where n is integer. Thus the input power should be adjusted to increase the efficiency as the nonlinear phase difference becomes π, but the output for the low input power leakages as shown in the circle region of figure 2 due to the difference of the nonlinear phase shifts. This switching device uses the transmissivity curvature for input power versus output power, and it can compress the pulse width by this switching characteristics.

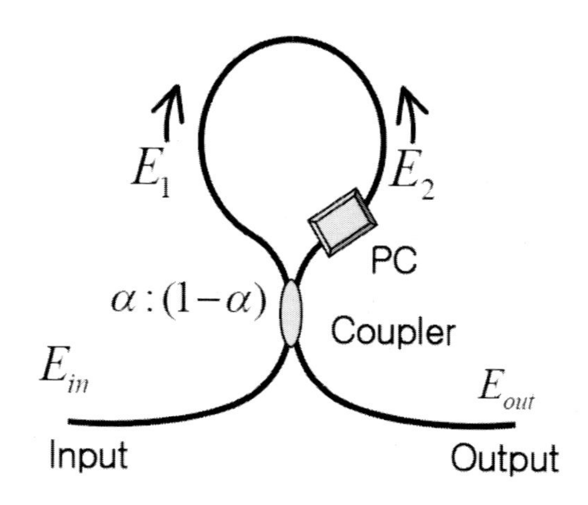

Figure 2.1. The sagnac loop with coupling ratio α is configured. PC is a polarization controller.

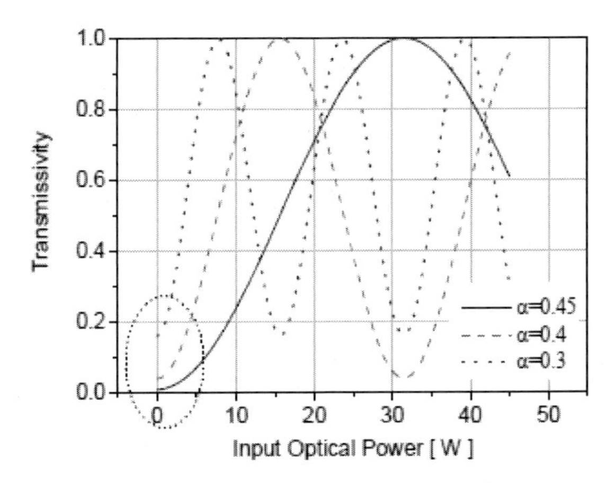

Figure 2.2. Switching characteristics of a nonlinear optical loop mirror.

In above analysis, an NOLM did not consider the polarization dependency. However, an optical fiber has birefringence since an optical fiber is not perfectly circular and the refractive index of an optical fiber is asymmetry by residual or externally induced stress. A birefringence causes the maximum different path length when two lights are passed through the same length of an optical fiber but launched into orthogonal principle axes. The path length difference can be applied as an optical filter, especially comb filter. In WDM system, this comb filter is very useful as a WDM filter. Thus a sagnac loop should consider the polarization dependency to analyze the performance of a WDM filter [13]–[16].

A linear birefringence can be represented by the combination of two circular polarized bases and also a circular birefringence can be expressed by the combination of two linear polarized bases.

Any arbitrary polarization state can be expressed by a rotatable phase variation, which can consist of a polarization rotator and a wave retarder. The Jones matrixes of a wave retarder and a polarization rotator can be written as [13]

$$L = \begin{bmatrix} \exp(i\delta/2) & 0 \\ 0 & \exp(-i\delta/2) \end{bmatrix}, \quad R = \begin{bmatrix} \cos(b/2) & -\sin(b/2) \\ \sin(b/2) & \cos(b/2) \end{bmatrix}, \tag{2-8}$$

where δ is the linear birefringence and defined by

$$\delta = \frac{2\pi}{\lambda} L_{length} \Delta n , \tag{2-9}$$

and b is the circular birefringence, which is similar to the linear birefringence. Figure 2.3 shows two counter propagating lights passing through a sagnac loop. The sagnac loop is assumed as the combination of a polarization rotator and a wave retarder, and these elements locate at one part of the loop as shown in figure 2.3. This combination can produce any arbitrary polarization state for any input polarization sate. Also a polarization controller can be expressed by this combination. The Jones matrix along the sagnac loop for a clockwise propagating light can be expressed by

$$A_{CW} = RL = \begin{bmatrix} \cos(b/2)\exp(i\delta/2) & -\sin(b/2)\exp(-i\delta/2) \\ \sin(b/2)\exp(i\delta/2) & \cos(b/2)\exp(-i\delta/2) \end{bmatrix}, \tag{2-10}$$

while the Jones matrix for a counter propagating light can be represented as

$$A_{CCW} = L^T R^T = \begin{bmatrix} \cos(b/2)\exp(-i\delta/2) & \sin(b/2)\exp(-i\delta/2) \\ -\sin(b/2)\exp(i\delta/2) & \cos(b/2)\exp(i\delta/2) \end{bmatrix}. \tag{2-11}$$

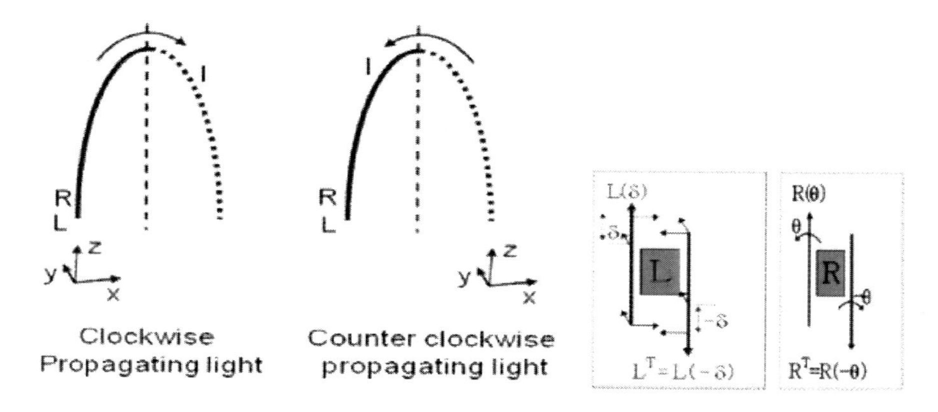

Figure 2.3. The decomposed sagnac loop and the transposes of a wave retarder and a polarization rotator.

The transmitted electric field by using 50/50 center coupler is given by

$$E_t = \frac{1}{2}[IA_{CW} - A_{CCW}I]E_{in}, \tag{2-12}$$

where the coordinate inversion matrix I is defined by

$$I = \begin{bmatrix} -1 & 0 \\ 0 & 1 \end{bmatrix}. \tag{2-13}$$

The I expresses the coordinate inversion since the coordinate x becomes $-x$ after half position as shown in figure 2.3. The clockwise light passes through a wave retarder first and a polarization rotator next, and after half position the light transforms the coordinate. Thus it is given as IRL.

However, the counter clockwise light firstly transforms the coordinate as shown in figure 2.3, and the light passes through a transposed wave retarder next and a transposed polarization rotator later. Thus the transmitted electric field is written as

$$E_t = -i\cos(b/2)\sin(\delta/2)\begin{bmatrix} 1 & 0 \\ 0 & 1 \end{bmatrix}E_{in}. \tag{2-14}$$

Therefore, the transmissivity as the input electric field E_{in} of arbitrary polarization state is launched into the sagnac loop is given by

$$T = \cos^2(b/2)\sin^2(\delta/2). \tag{2-15}$$

Equation (2-15) represents that the transmissivity does not depend on the input state of polarization. When the any arbitrary input polarization state is launched into a saganc loop, the transmissivity is decided only by the circular birefringence and the linear birefringence inside the loop, even though it uses a 3 dB center coupler. If the circular birefringence b is to be $\pi \pm \pi$ by twisting or spinning the principal axes of the linear birefringence, equation (2-15) becomes

$$T = \sin^2(\delta/2). \tag{2-16}$$

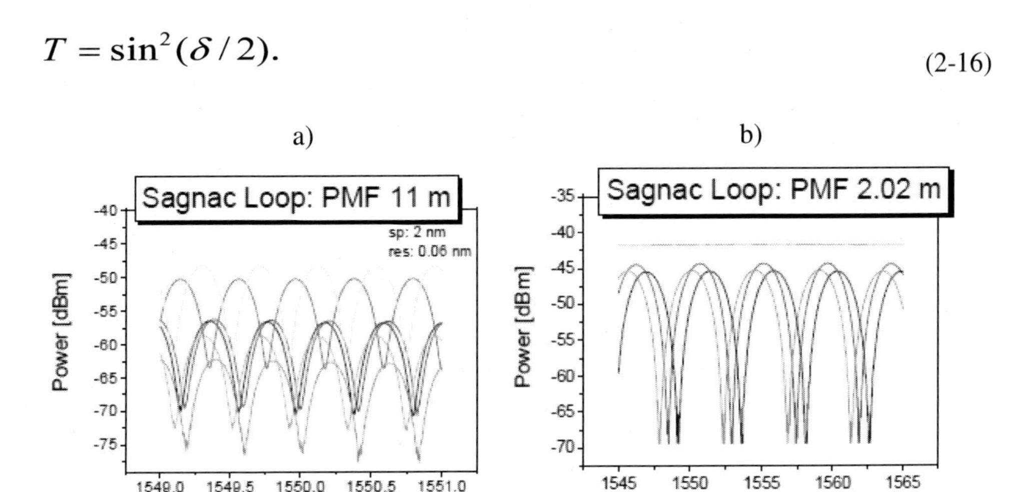

Figure 2.4. Sagnac loop response of polarization maintained fibers.

Since δ depends on the operation wavelength, the sagnac loop can be a WDM filter with wavelength period as

$$\Delta\lambda = \frac{\lambda^2}{\Delta n L_{length}}, \tag{2-17}$$

where

$$\Delta\lambda = -\frac{\lambda^2}{c}\Delta f, \quad \Delta f = \frac{c}{\Delta n L_{length}}, \tag{2-18}$$

and Δn is the refractive index difference of two principle axes. To demonstrate the WDM filter, the high birefringence fiber, especially polarization maintaining fiber (PMF) is used to construct the sagnac loop. Figure 2.4 shows the characteristics of the WDM filter which the peak period is 0.41 nm when the length of the PMF is 11 m. As the length of the PMF is reduced to 2.02 m, the peak period is 4.45 nm. Therefore the Δn is 10^{-4}. In experiment the transmitted spectrum does not change by adjusting the PC outside the loop, but the spectrum is tuned by adjusting the PC inside the loop as shown in figure 2.4.

The spectrum changes by adjusting the PC are due to the variation of a circular birefringence. Equation (2-15) represents these variations.

3. Mode-Locked Fiber Laser

A passive mode-locking usually operates at the fundamental repetition rate, and this method can generate an ultrashort pulse and a high energy pulse due to a long cavity length. A passive mode-locking has a saturable absorber to shorten an optical pulse. When the width of an optical pulse is ultrashort, the pulse has extremely high peak power due to confining the intensity as the average power is the same.

Since a passive mode-locking method does not use an external force, the construction scheme of this method is very simple. In a fiber laser, it just splices the optical fibers with a fusion splicer to construct a passive mode-locked fiber laser.

The resonance frequency in a laser cavity is called a mode frequency or an axial mode spacing frequency. A cavity round trip length, L is associated with the fundamental resonance frequency, ω_m. The relation between L and ω_m follows $\omega_m = 2\pi c/L$.

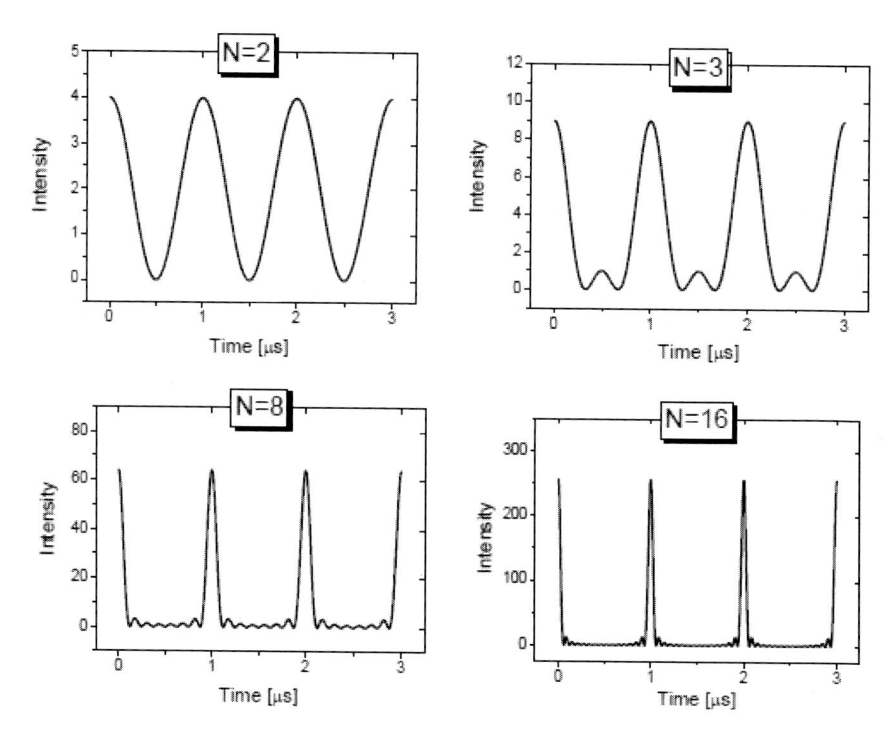

Figure 3.1. The mode locked intensity of the total electric field is calculated for different mode number, N.

The frequencies in a laser cavity exists on the same mode spacing from low frequency to the infinite frequency. If there are N axial modes and the same amplitude of electric field, then the total electric field can be written as

$$E(t) = \sum_{n=0}^{N-1} \exp i(\omega_0 + n\omega_m)t = \frac{\exp(iN\omega_m t) - 1}{\exp(i\omega_m t) - 1} \exp(i\omega_0 t) \quad . \tag{3-1}$$

For mode-locking, the phase difference between neighbor modes should be constant. Then, the intensity of the total electric field is written as

$$I(t) = |E(t)|^2 = \frac{1 - \cos(N\omega_m t)}{1 - \cos(\omega_m t)} = \frac{\sin^2(N\omega_m t/2)}{\sin^2(\omega_m t/2)} \quad . \tag{3-2}$$

Figure 3.1 shows the intensity profile of the mode-locked electric field for the different mode number, N. When the electric field has two modes, the intensity profile is in the form of sine wave. However, the intensity profile becomes pulse shape for N=3. As the number of mode is increased, the intensity profile is to be sharp. In a fiber laser, many modes are contributed to a mode-locked pulse due to a long cavity length, which is inverse of mode spacing frequency, thus a fiber laser can generate an ultrashort optical pulse.

Figure 3.2 shows the relation between time and frequency for a mode-locked pulse. The mode spacing f_m in spectrum domain is inverse of repetition rate T_R such as $f_m = 1/T_R$, and the spectral width δf of a mode-locked pulse in spectrum domain is also inverse of pulse width $\tau \approx 1/\delta f$. An optical component which has intensity dependent transmission characteristics such as an NOLM or a saturable absorber are key components for mode locking. These devices work as an optical switch to suppress the temporal width of a laser pulse. Pulse shortening is obtained from a nonlinear transmission effect where low intensity light is absorbed more compared to high intensity light in these devices.

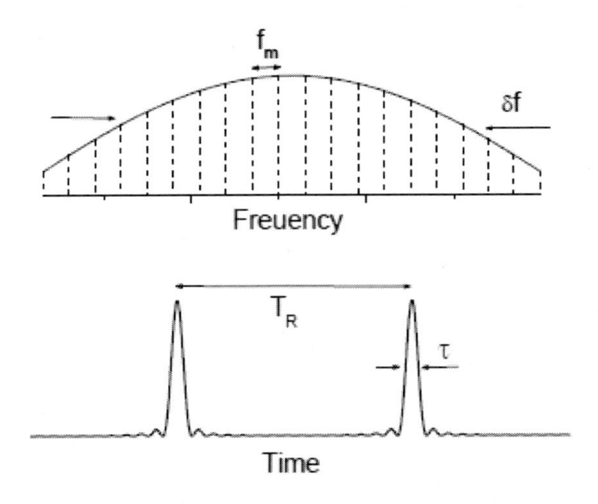

Figure 3.2. Time and frequency are inversely related for a mode locked pulse.

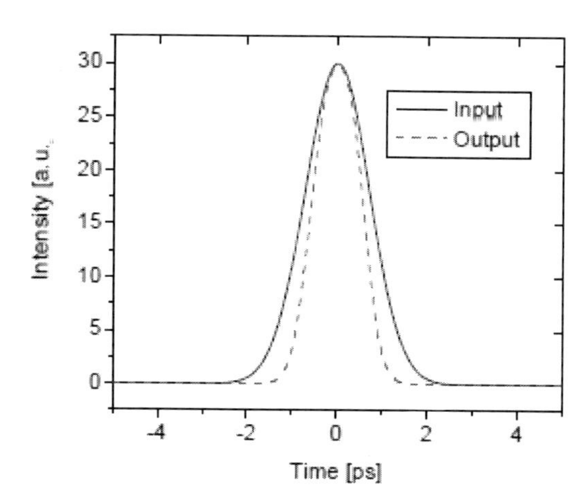

Figure 3.3. Saturable absorber characteristics of an NOLM when gaussian parameters are given as P_0=30 W, T_0=1 ps and α is 0.45.

The optical properties of an NOLM can be represented by equation (2-7) for the suppressing process of an optical pulse width. If we assume that the input pulse width to an NOLM is 1 ps and the peak power is 30, then the output can be calculated and is shown in figure 3.3. The compressed pulse shape depends on the peak power of the pulse and the coupling ratio of an NOLM. Two polarization controllers (PCs) are normally used in a fiber laser with an NOLM. One polarizer is placed inside an NOLM and the other one is placed outside. By adjusting the two PCs in a fiber laser properly, a figure-eight fiber laser with an NOLM produces mode locked laser pulses. In practice, the mode-locking normally obtained just by adjusting a PC which is placed outside of an NOLM. Firstly, we set the PC inside of an NOLM such that the linear loss of an NOLM becomes independent with the polarization state of input light. After this condition is obtained, the PC outside of an NOLM is adjusted to maximize the output power of a laser.

If the PC outside of an NOLM is in a state for the maximum output power, and the PC inside of an NOLM is properly set for mode-locking, then the fiber laser can be self-started with a pulsed operation mode just by increasing the current of a pump laser without moving any mechanical adjustment.

4. Dispersion Imbalanced NOLM

The invention of an EDFA has stimulated the development of practical optical communication systems, but amplified spontaneous emission (ASE) noise in an EDFA is a major problem which degrades the performance of an optical communication system. The effective suppression or the removal of the ASE noise is quite important for a high quality optical communication system. ASE noise reduction methods used in optical communication can be applied to fiber laser systems since ASE noise is also quite significant artifact in a fiber laser system. There are many methods to remove ASE noise. An optical band pass filter (BPF) or a saturable absorber (SA) are normally used for this purpose. Lately Wong [8] has proposed an effective method to suppress ASE noise by using a DI-NOLM.

The asymmetry of a DI-NOLM is achieved by using two fibers with different dispersion values. Compared to an NOLM or an NALM, a DI-NOLM has a unique property that no continuous wave component of light of any intensity can pass through the switching element when the polarization state of a DI-NOLM is properly set. A DI-NOLM consists of two fibers of different dispersion values, and dispersion acts only on optical pulses not on continuous light. For low power light, it just accumulates the linear phase shift by dispersion as the light passes through a optical fiber. If two different fibers different dispersion values in a DI-NOLM, after passing through two different fibers, the difference of the linear phase shifts of two counter propagating lights is zero, because the linear phase shift is the superposition of two phase shifts for the different fibers. There is no active element inside the loop mirror so that the switching function of the device is not affected by the interaction of two counter propagating lights. The next important improvement in a nonlinear optical loop mirror is accomplished by using lumped dispersive elements with a DI-NOLM [17], where the total dispersion is managed by putting an oppositive dispersion fiber outside the DI-NOLM. Thus the incoming pulse is first broadened by the dispersive fiber element before it enters the DI-NOLM. Figure 4.1 shows the setup of a DI-NOLM with different optical fibers.

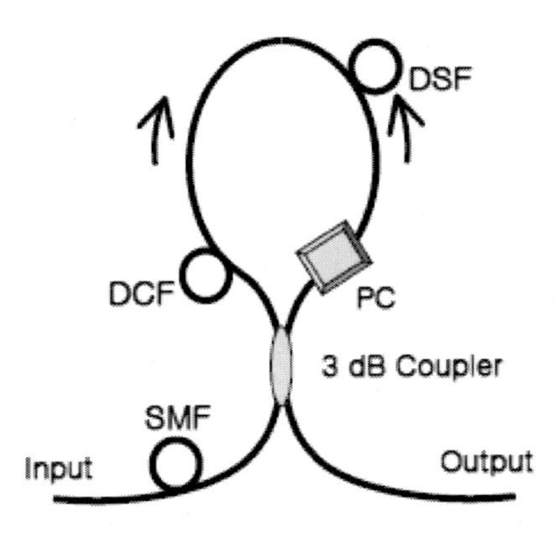

Figure 4.1. Configuration of a DI-NOLM with lumped elements.

One is a dispersion compensating fiber (DCF) whose dispersion coefficient is negative at 1550 nm wavelength, and the other one is a dispersion shifted fiber (DSF) whose dispersion coefficient is positive at 1550 nm wavelength.

As a DI-NOLM is based on an asymmetric pulse broadening effect, the continuous wave component of any incoming light is reflected while the high intensity part of an optical pulse is transmitted. Therefore, a figure-eight fiber laser by using a DI-NOLM has many potential advantages over the conventional figure-eight fiber laser. In general, the output of a figure-eight fiber laser has several different operation modes that produce a broad square pulse, an unstable bunch of spike pulse, harmonic mode-locked multi-pulses, and a single pulse [18] depending on its pump power and the states of polarization controllers in the cavity. However, it is reported that only a broad square pulses are observed in a fiber laser with a DI-NOLM [9]. As various applications need different types of optical pulses, it is important to know the conditions for each operation mode of the laser output.

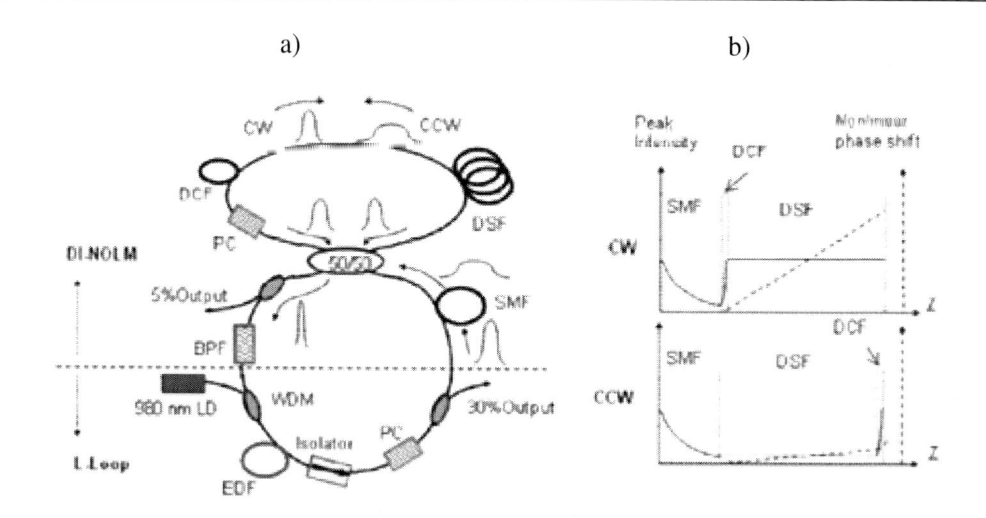

Figure 4.2. (a) Experimental configuration of figure-eight fiber laser by using a DI-NOLM with lumped dispersion elements. LL: linear loop. (b) Peak intensity and accumulated peak nonlinear phase shift as a function of propagation distance Z for counterclockwise(CCW) and clockwise(CW) propagation directions inside the DI-NOLM.

The detailed structure of the figure-eight fiber laser is shown in figure 4.2(a). The upper loop is a DI-NOLM as a self-switching element and the lower loop is an unidirectional linear loop with an EDF as a gain medium. The upper and the lower fiber loops are connected with a 50/50 directional coupler. The lower fiber loop consists with a PC, a gain medium, a polarizing optical isolator, a 30 % output coupler, and a 20 m long SMF with D = +16.98 [ps/km-nm] at 1550 nm wavelength. The gain medium of the laser is a 10 m long commercially available EDF made by SAMSUNG. Its absorption coefficient at 1550 nm wavelength is 4.5 dB/m. The EDF is pumped by a 91 mW pump LD at 980 nm wavelength. The polarization dependent isolator is put between the EDF and the 30 % output coupler to block any reflected light coming into the gain medium. A 5 % output coupler is put just after the DI-NOLM to monitor the switching characteristics of the DI-NOLM, and a 3 nm optical band pass filter is placed after it. The DI-NOLM is composed of a PC, a 100 m long DSF with D = -0.22 [ps/km-nm] at 1550 nm wavelength and a 3.4 m long DCF with D = -100 [ps/km-nm]. Figurc 4.2(b) also shows how a pulse is self-switched and shortened as it passes through a DI-NOLM. An incoming optical pulse is first broadened by the 20 m long SMF whose chromatic dispersion is +0.34 ps/nm just before it enters the DI-NOLM. The optical pulse is split into two equal parts by a 50/50 directional coupler. The clockwise propagating (CW) pulse is immediately compressed to its original pulse width by the 3.4 m long DCF whose chromatic dispersion is -0.34 ps/nm and then, passes through the 100 m long DSF to have a large peak nonlinear phase shift. The counter-clockwise propagating (CCW) pulse has small nonlinear phase shift as it propagates through the 100 m long DSF first and is compressed by the 3.4 m long DCF later. When the difference of the peak nonlinear phase shift between the CW and the CCW pulses becomes π, the peak of the pulse is completely transmitted through the DI-NOLM while the low intensity part of the pulse is reflected back by the DI-NOLM.

The noiselike square pulse [19]–[21] of the laser can be easily achieved by adjusting the the PC inside the laser cavity. Figure 4.3 shows that the pulse width of the square pulse measured with a fast photodetector. The full-width half maximum is about 3.26 ns when the pump LD power is 91 mW. It is measured with a 20 GHz fast detector and a sampling oscilloscope. It was observed that the pulse width is increased as the pump power is increased. The repetition rate of the pulse train is 1.37 MHz.

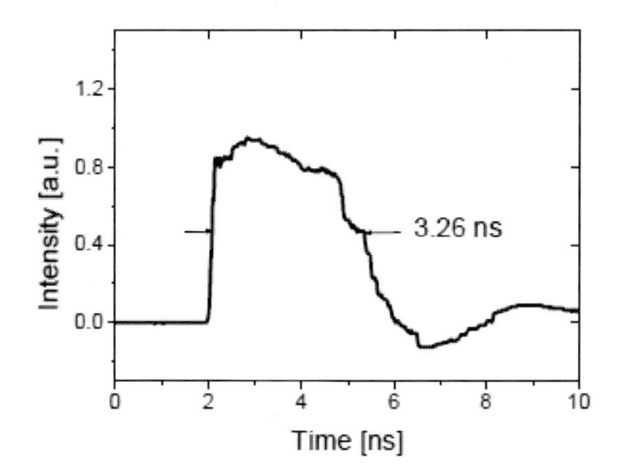

Figure 4.3. Temporal behavior of a noiselike pulse measured by a fast detector.

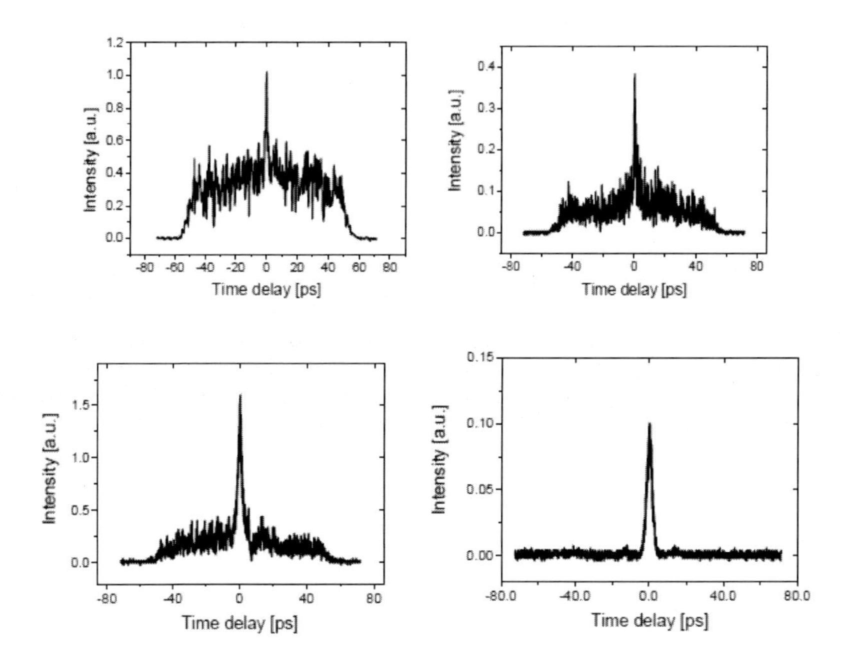

Figure 4.4. Autocorrelation traces are measured for the different operation mode.

Even though no continuous wave can pass through the DI-NOLM, there exist many sharp spike pulses inside the broad square envelope as shown in figure 4.4. If this square pulse does not consist of many sharp pulse, then this quasi-continuous pulse cannot be transmitted through the DI-NOLM. These short pulses undergo complicated evolution within the broad envelope when these pass through the DI-NOLM and the overall structure of the pulse looks square due to averaging effect. Figure 4.4 is the SHG autocorrelation trace of a square noiselike pulse measured with a background free non-collinear type autocorrelator. This autocorrelator is made by two fast rotating mirrors, which have very rapid scanning rate.

Therefore, it can measure the substructure of the square pulse. It shows the coherent spike of the laser output and the SHG intensity ratio between the background and the peak of the coherent spike is about 1:2. Adjusting the PC inside the cavity, it can vary the intensity ratio between the background and the coherent spike of the autocorrelation trace, and also generate the mode locked pulse.

5. Attenuation-Imbalanced NOLM

An AI-NOLM consists of an NOLM with a 50/50 coupler and an attenuator placed asymmetrically in a fiber Sagnac loop mirror. The basic operation principle of an AI-NOLM was first introduced by Smith et al. [11] for pulse shaping and pedestal suppression. As it uses a 50/50 coupler, it can reject ASE noise or background continuous wave more effectively than a conventional NOLM. Another advantage of an AI-NOLM is that there is no active element inside the loop mirror so that the switching function of the device is not affected by the interaction of two counter propagating waves. An AI-NOLM is very simple in its structure as only an optical power-attenuating element is needed in addition to a fiber Sagnac interferometer. Meanwhile, a DI-NOLM needs two or three different fibers with specific dispersion characteristics at a given wavelength and complicated calculation of dispersion parameters for all fibers used in a fiber loop mirror [8]. Unlike a DI-NOLM, an AI-NOLM works for any wavelengths regardless of its dispersion property. However, it has inherent power loss due to an attenuator placed inside.

The basic principle of an AI-NOLM is explained in figure 5.1. Input optical field is split into two counter propagating waves in a fiber Sagnac loop by a 50/50 fiber coupler. The optical power of a counter clockwise propagating wave is reduced immediately after the fiber coupler by an optical attenuator before it is propagated through a fiber loop. Meanwhile, the optical power of a clockwise propagating wave is reduced after it is propagated through the whole fiber loop. When P_0 is the input optical power to the fiber with a length L and effective area A_{eff}, the nonlinear phase shift of light after propagating the fiber can be written as [11]

$$\phi_{NL} = k_0 n_2 L P_0 / A_{eff}, \quad (5\text{-}1)$$

where n_2 is the nonlinear refractive index coefficient and k_0 is the free space propagation constant of light. The power difference created among the two counter propagating waves in the fiber Sagnac loop leads to a difference in nonlinear phase shift among two waves given by

$$\delta\phi = (1 - T_L)\phi_{NL} / 2, \quad (5\text{-}2)$$

where T_L is the transmissivity of the attenuator inside the fiber Sagnac loop. The transmissivity of this AI-NOLM can be written as

$$T = \frac{T_L}{2}\left\{1 - \cos\left[\frac{(1-T_L)\phi_{NL}}{2}\right]\right\}. \quad (5\text{-}3)$$

If there is no attenuation i.e. $T_L = 1$, then the transmissivity T of the AI-NOLM is zero. However, the transmissivity T can be changed by tuning the transmissivity of the attenuator T_L. Figure 5.2 shows the calculated switching characteristics of an AI-NOLM for three different T_L values when $k_0 n_2 L/A_{eff} = 1$ [1/W]. A peak transmissivity is obtained wherever the nonlinear phase difference of the two counter propagating light $(1\text{-}T_L)\phi_{NL}/2$ is $(2n + 1)\pi$, and it is minimum wherever the phase difference is $2n\pi$. The maximum transmissivity of the AI-NOLM is proportional to the transmissivity of the attenuator T_L. However, larger T_L requires larger input power to have maximum transmissivity. As the minimum transmissivity is always zero for an AINOLM regardless of T_L value as shown in figure 5.2, there is no background ASE noise transmitted trough an AI-NOLM. This is a major advantage of an AI-NOLM over a conventional NOLM for making a fiber laser out of it.

A new self-starting figure-eight fiber laser based on an AI-NOLM was reported [12]. Low power optical input is rejected by an AI-NOLM much more effectively than a conventional NOLM. As there is no active medium inside an AI-NOLM, this laser potentially has much advantage over a conventional figure-eight fiber laser system. The detailed configuration structure of the figure-eight fiber laser is shown in figure 5.3. The upper loop in an AI-NOLM is a self-switching element and the lower loop is a unidirectional linear loop with a gain medium. The upper and the lower fiber loops are connected with a 3 dB directional coupler. The lower fiber loop consists of a polarization controller (PC), an Erbium doped fiber as a gain medium, a polarizing optical isolator, a 10% optical coupler (Monitor-2), a 50% output coupler, and a 20 m long single mode fiber (SMF) whose chromatic dispersion coefficient D is +16.98 ps/km-nm at 1550 nm wavelength.

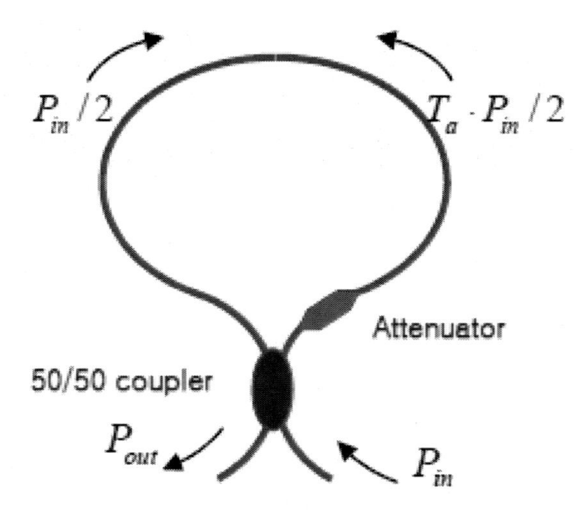

Figure 5.1. Configuration of an AI-NOLM.

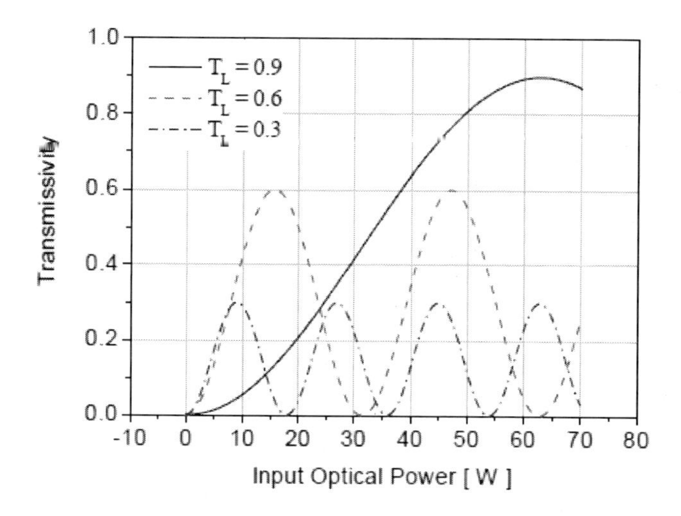

Figure 5.2. Switching characteristics of an AI-NOLM.

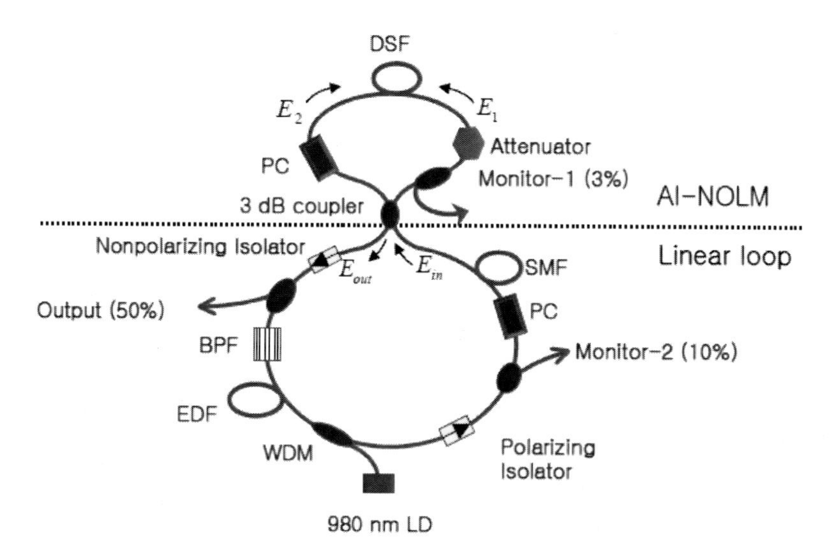

Figure 5.3. Setup of a figure-eight fiber laser with an AI-NOLM.

The gain medium of the laser is a 10 m long commercially available EDF made by SAMSUNG. Its absorption coefficient at 1550 nm wavelength is 4.5 dB/m. The EDF is pumped by a 100 mW pump LD at 980 nm wavelength. The polarizing optical isolator is put between the EDF and the 10% coupler to block the reflected light from the AI-NOLM and any externally reflected light coming into the gain medium. A 50% output port is put just after the AI-NOLM to obtain a minimum pulse width. A very large portion of laser power is coupled out with a 50% output coupler to have enough power for pulse width measurement using an autocorrelator. An optical band pass filter whose 3 dB bandwidth is 3 nm is placed after the output coupler in order to obtain stable operation of the fiber laser. The AI-NOLM is composed of a PC, a 3% optical directional coupler (Monitor-1), an attenuator, and a 100 m long dispersion shifted fiber (DSF) with D = -0.22 ps/km-nm at 1550 nm wavelength. The fiber laser is designed to have a pulse self-switched and shortened as it is passed through the laser cavity. An incoming optical pulse to the AI-NOLM is first broadened by the 20 m long SMF whose total accumulated chromatic dispersion coefficient D is +0.34 ps/nm just before it enters the AI-NOLM.

The optical pulse is split into two equal parts by a 50/50 directional coupler. The counter clockwise (CCW) propagating pulse is immediately attenuated, and then passed through a 100 m long DSF to have a small nonlinear phase shift, whereas, the clockwise (CW) propagating pulse has a large nonlinear phase shift as it is propagated through the 100 m long DSF first and attenuated later. When the difference of the peak nonlinear phase shift between the CW and the CCW propagating pulses becomes π, the peak of the pulse is completely transmitted through the AI-NOLM while the low intensity part of the pulse is reflected back by the AI-NOLM.

A variable optical attenuator inside the AI-NOLM is used by bending loss of an optical fiber. When the bending diameter of an optical fiber is very small, the amount of attenuation can be easily controlled by changing the bending diameter. For example, the bending loss of an SMF-28 fiber for a single turn around a circle of 3 mm diameter is about 40 dB. A single turn of a dispersion shifted fiber (DSF) around a circle of 1.5 cm diameter is used as the optical attenuator inside the AI-NOLM in the fiber laser. Figure 5.4 shows the attenuation spectrum of the optical attenuator, which is measured by the 3% power coupler (Monitor-1) inside the AI-NOLM. An ASE signal from an EDFA is used as an input light source. The optical power difference due to the bending of 1.5 cm diameter circle is about 1.66 dB at 1550 nm wavelength. This small loss difference by an optical attenuator inside the AI-NOLM is good enough to drive the nonlinear phase shift of π for this case.

The repetition rate of a fiber laser with an AI-NOLM can be increased by decreasing the length of an AI-NOLM. When the peak optical power is fixed, decreasing the length of an AI-NOLM requires an increase of the optical loss of the attenuator in an AI-NOLM to obtain the same nonlinear phase shift according to figure 5.2 or equation (5-3). In this case, the insertion loss for the AI-NOLM can be large due to the power loss by the attenuator. Thus, the optical loss of an attenuator in an AI-NOLM should not be very large.

Various operation modes of a figure-eight fiber laser [5, 10] such as a regularly spaced mode-locked pulse, a noise-like square pulse, and multiple pulse operation modes are observed by tuning the state of the PC inside the AI-NOLM. In each operation mode, the laser generates stable pulses for a few hours when no external perturbation is applied in a laboratory environment; the amplitudes and positions of output pulses stand still when they are observed by an autocorrelator or a sampling scope. The pulse width of the regularly spaced mode-locked pulse train is measured by a background free autocorrelator with a SHG crystal. Figure 5.5 shows that the measured autocorrelation pulse width of the input pulse to the AI-NOLM is 2.08 ps, while the autocorrelation pulse width of the output pulse from the AI-NOLM is 457 fs. If it is assumed to be a sech^2 pulse, the actual pulse width for the output pulse is 296 fs. The repetition rate of the pulse is about 1.5 MHz, and the average output power out of the 50% output coupler in figure 5.3 is 2.45 dBm. The optical spectra of the input and output pulses are obtained by an optical spectrum analyzer. By controlling the PC inside the AI-NOLM, a continuous wave component such as ASE noise can be completely rejected by the AI-NOLM. This has been verified by observing the optical spectrum just before and after the AI-NOLM. Figure 5.6 (a) is the optical spectrum of input optical pulse corresponding to the temporal pulse shape in figure 5.5 (a).

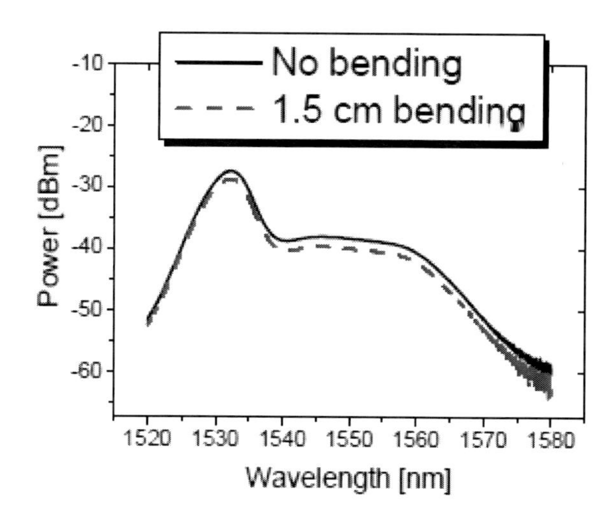

Figure 5.4. Measured attenuation loss spectrum of an optical attenuator in an AI-NOLM.

It is collected from the 10% monitor port in figure 5.3. Its shape is mostly determined by the optical band pass filter; the center wavelength is 1550.1 nm and the 3 dB bandwidth is about 2.6 nm.

It has a small hump at about 1532 nm wavelength due to the ASE peak of an EDFA. The output optical spectrum from the AI-NOLM corresponding to the autocorrelation trace of figure 5.5 (b) is shown in figure 5.6 (b). The 3 dB bandwidth of the output pulse is about 16.8 nm. It is taken just after the pulse pass through the AI-NOLM. Note that there is no ASE peak of an EDFA at 1532 nm wavelength. It is because no continuous wave can propagate through the AI-NOLM. This is the major advantage of using an AI-NOLM as a switching element in a figure-eight fiber laser over a conventional NOLM or a NALM.

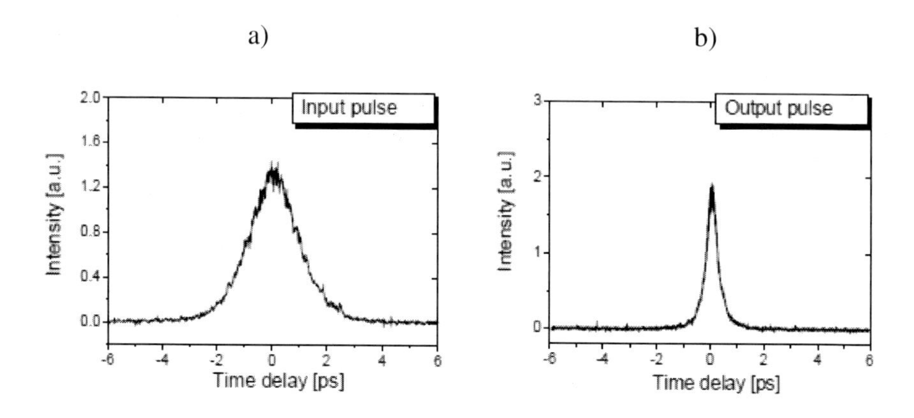

Figure 5.5. Measured autocorrelation traces for (a) input and (b) output pulses.

Equations (5-1) and (5-2) indicate that the difference of nonlinear phase shift between the two counter propagating waves in an AI-NOLM is proportional to attenuation ($1-T_L$) and the peak optical power P_0. Assuming that the peak nonlinear phase shift difference for an optimum switching inside the AI-NOLM is the same, the peak optical power P_0 is inversely proportional to the attenuation ($1-T_L$). If the optical average power supplied by the EDFA inside the laser cavity is constant, the pulse width becomes inversely proportional to the peak optical power P_0. This means that the pulse width of this laser is proportional to the attenuation ($1-T_L$) inside the AI-NOLM. This concept is verified by measuring pulse widths for three different attenuation values inside an AI-NOLM.

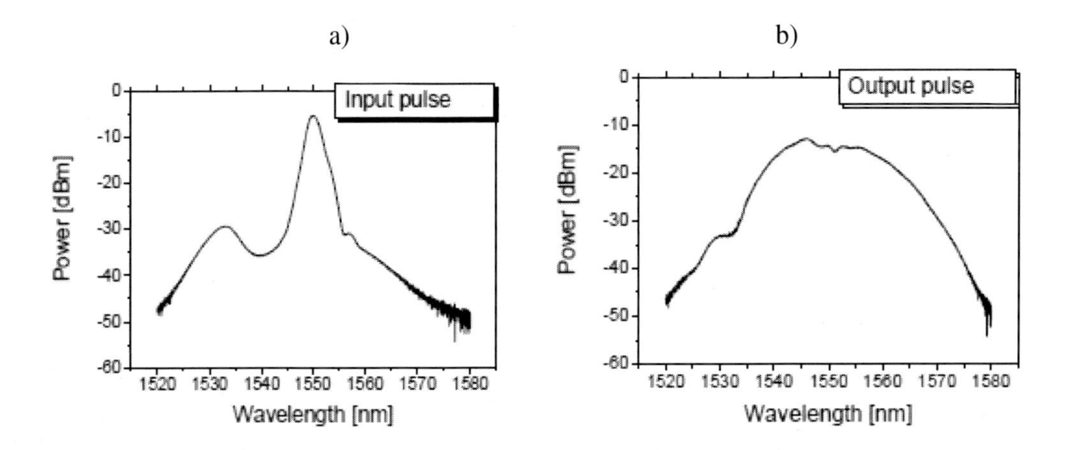

Figure 5.6. Measured optical spectra for (a) input and (b) output pulses.

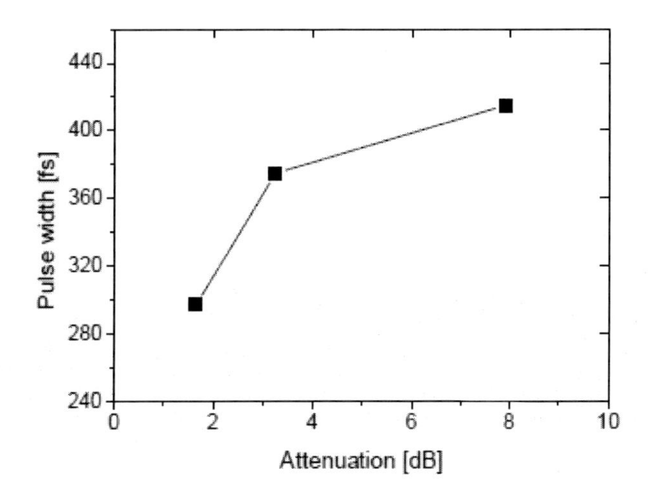

Figure 5.7. Variation of laser pulse width as a function of optical attenuation in an AI-NOLM.

Figure 5.7 shows that the output pulse width of the fiber laser can be varied from 296 fs to 415 fs simply by changing the transmissivity T_L of an optical attenuator inside an AI-NOLM.

The attenuation values for pulse widths of 296 fs, 374 fs and 415 fs are 1.66 dB, 3.24 dB, and 7.9 dB, respectively. The corresponding peak powers of the input pulse which gives the maximum transmission is calculated to be 17.7 W, 25.5 W and 74.7 W, respectively, obtained from equations (5-1) and (5-2). The effective area of the DSF used in AI-NOLM is assumed to be 50 μm^2.

CONCLUSIONS

The basic operational principles of DI-NOLM and AI-NOLM are explained with proper mathematical formulae. Then, we have demonstrated the successful operation of mode-locked fiber lasers based base on DI-NOLM and AI-NOLM. We have verified that these lasers can operate in three different operational modes: single pulse mode-locked pulse, square noiselike pulse, and harmonic multi-pulse modes depending upon the polarization state of the laser cavity. Pulse width shorting as well as spectrum broadening between input and output pulses for both DI-NOLM and AI-NOLM were explained and measured data are shown which shows the detailed switching effect of these devices. We have shown that these devices can reject the ASE noise inside a figure-eight fiber laser much more effectively than a conventional NOLM. Compared to a recently reported figure-eight fiber laser with a DI-NOLM, the structure and the operational constraints of AI-NOLM are much simpler, because it does not need complex calculation or arrangement of fiber dispersion inside a laser cavity.

ACKNOWLEDGMENT

This work was supported by the Creative Research Initiatives Program of Korea Science and Engineering Foundation (KOSEF)/Ministry of Education Science and Technology (MEST).

REFERENCES

[1] N. J. Doran and D. Wood, "Nonlinear-optical loop mirror," *Opt. Lett.* 13, 56 (1988).
[2] K. Smith, N. J. Doran and P. G. J. Wigley, "Pulse shaping, compression, and pedestal suppression employing a nonlinear-optical loop mirror," *Opt. Lett.* 15, 1294 (1990).
[3] M. E. Fermann, F. Haberl, M. Hofer, and H. Hochreiter, "Nonlinear amplifying loop mirror," *Opt. Lett.* 15, 752 (1990).
[4] D. B. Mortimore, "Fiber Loop Reflectors," *J. Light. Tech.* 6, 1217 (1988).
[5] I. N. Duling III, "Subpicosecond all-fibre erbium laser," *Electron. Lett.* 27, 544 (1991).
[6] S. Wu, J. Strait, R. L. Fork, and T. F. Morse, "High power passively mode-locked Er-doped fiber laser with a nonlinear optical loop mirror," *Opt. Lett.* 18, 1444 (1993).
[7] D. J. Richardson, R. I. Laming, D. N. Payne, M. W. Phillips, and V. J. Matsas, "320 fs soliton generation with passively mode-locked erbium fibre laser," *Electron. Lett.* 27, 730 (1991).
[8] W. S. Wong, S. Namiki, M. Margalit, H. A. Haus, and E. P. Ippen, "Self-switching of optical pulses in dispersion-imbalanced nonlinear loop mirrors," *Opt. Lett.* 22, 1150 (1997).
[9] D. S. Lim, H. K. Lee, M. Y. Jeon, J. T. Ahn, K. H. Kim, H. Y. Kim, and El-Hang Lee, "Passively mode-locked fiber laser with dispersion-imbalanced nonlinear loop mirrors," in *Proc. CLEO*, 419 (1998).

[10] N. H. Seong, D. Y. Kim, "A new figure-eight fiber laser based on a dispersion-imbalanced nonlinear optical loop mirror with lumped dispersive elements," *IEEE Photon. Techn. Lett.* 14, 459 (2002).

[11] K. Smith, N. J. Doran, P. G. J. Wigley, "Pulse shaping, compression, and pedestal suppression employing a nonlinear-optical loop mirror,"*Opt. Lett.* 15, 1294 (1990).

[12] N. H. Seong, D. Y. Kim, S. P. Veetil, "Mode-locked fiber laser based on an attenuation-imbalanced nonlinear optical loop mirror," *Opt. Comm.* 280, 438 (2007).

[13] X. Fang and R. O. Claus, "Polarization-independent all-fiber wavelength-division multiplexer based on a saganc interferometer," *Opt. Lett.* 20, 2146 (1995).

[14] H. Y. Rhy, B. Y. Kim and Hai-W. Lee, "Self-Switching with a Nonlinear Birefringent Loop Mirror," *IEEE J. Quantum Electron.* 36, 89 (2000).

[15] C. S. Kim, Y. G. Han, R. M. Sova, U. C. Paek, Y. Chung, and J. U. Kang, "Optical Fiber Modal Birefringence Measurement Based on Lyot-Sagnac Interferometer," *IEEE Photon. Technol. Lett.* 15, 269 (2003).

[16] E. A. Kuzin, J. M. E. Ayala, B. I. Escamilla, and J. W. Haus, "Measurements of beat length in short low-birefringence fibers," *Opt. Lett.* 26, 1134 (2001).

[17] M. Matsumoto and T. Ohishi, "Dispersion-imbalanced nonlinear optical loop mirror with lumped dispersive elements," *Electron. Lett.* 34, 1140 (1998).

[18] B. Grudinin, D. J. Richardson and D. N. Payne, "Energy quantisation in figure eight fiber laser," *Electron. Lett.* 28, 67 (1992).

[19] M. Horowitz and Y. Silberberg, "Nonlinear filtering by use of intensity-dependent polarization rotation in birefringent fibers," *Opt. Lett.* 22, 1760 (1997).

[20] M. Horowitz and Y. Silberberg, "Control of Noiselike Pulse Generation in Erbium-Doped Fiber Lasers," *IEEE Photon. Technol. Lett.* 10, 1389 (1998).

[21] M. Horowitz, Y. Barad and Y. Silberberg, "Noiselike pulses with a broadband spectrum generated from an erbium-doped fiber laser," *Opt. Lett.* 22, 799 (1997).

In: Advances in Laser and Optics Research. Volume 10 ISBN: 978-1-62257-795-8
Editor: William T. Arkin

Chapter 4

OPTICAL BREAKDOWN IN GASES INDUCED BY HIGH-POWER IR CO_2 LASER PULSES

J. J. Camacho,[1*] ***L. Díaz,***[2] ***M. Santos,***[2] ***L. J. Juan***[1] ***and J. M. L. Poyato***[1†]

[1]Departamento de Química-Física Aplicada, Facultad de Ciencias, Universidad Autónoma de Madrid, Cantoblanco, Madrid, Spain
[2]Instituto de Estructura de la Materia, CFMAC, CSIC, Serrano, Madrid, Spain

ABSTRACT

This chapter reviews some fundamentals of laser-induced breakdown spectroscopy (LIBS) and describes some experimental studies developed in our laboratory on gases such as nitrogen, oxygen and air. LIBS of these gases at different pressures, in the spectral range ultraviolet-visible-near infrared (UV-Vis-NIR), was excited by using a high-power transverse excitation atmospheric (TEA) CO_2 laser (λ=9.621 and 10.591 μm; τ(FWHM)=64 ns; and different laser power densities). The spectra of the generated plasmas are dominated by emission of strong atomic, ionic species and molecular bands. Excitation temperatures were estimated by means of atomic and ionic lines. Electron number densities were deduced from the Stark broadening of several ionic lines. The characteristics of the spectral emission intensities from different species have been investigated as functions of the gas pressure and laser irradiance. Optical breakdown threshold intensities in different gases have been measured experimentally. The physical processes leading to laser-induced breakdown of the gases have been analyzed. Plasma characteristics in LIBS of air were examined in detail on the emission lines of N^+, O^+ and C by means of time-resolved optical-emission spectroscopy (OES) technique. The results show a faster decay of continuum and ionic spectral species than of neutral atomic and

* Corresponding author: J. J. Camacho. Departamento de Química-Física Aplicada. Facultad de Ciencias. Universidad Autónoma de Madrid. Cantoblanco. 28049-Madrid. Spain. E-mails: j.j.camacho@uam.es; luisol@cfmac.csic.es; magna@iem.cfmac. csic.es; jml.poyato @uam.es.

† L. J. Juan, J. M. L. Poyato: Departamento de Química-Física Aplicada. Facultad de Ciencias. Universidad Autónoma de Madrid. Cantoblanco. 28049-Madrid. Spain. L. Díaz, M. Santos: Instituto de Estructura de la Materia, CFMAC, CSIC, Serrano 121. 28006-Madrid, Spain.

molecular ones. The velocity and kinetic energy distributions for different species were obtained from time-of-flight (TOF) OES measurements. Excitation temperature and electron density in the laser-induced plasma were estimated from the analysis of spectral data at various times from the laser pulse incidence. Temporal evolution of electron density has been used for the estimation of the three-body recombination rate constant.

1. INTRODUCTION

A remarkable demonstration of the interaction of radiation with matter is the plasma production which occurs when a powerful laser is focused in a gas. Strong pulses of infrared (IR) laser light can cause breakdown and plasma formation in gases which are usually transparent at these wavelengths. By this means gases are transformed into opaque, highly conducting plasmas in times of the order of nanoseconds. If the radiant energy in the focus exceeds the threshold value for breakdown, as happens with high-power lasers (ruby, Nd:YAG, CO_2, excimer, etc), the gas becomes highly ionized and the plasma thus produced will practically absorb the beam. In atmospheric air, for example, laser-beam-induced breakdown is characterized by a brilliant flash of bluish-white light at the lens focus accompanied by a distinctive cracking noise. This transformation from neutral gas into hot plasma takes place in three quite distinct stages: initiation, formative growth, and plasma development accompanied by shock wave generation and propagation in the surrounding gas. A fourth and final stage, extinction, follows.

The formation of laser-induced breakdown (LIB) plasma in a gas has been investigated since its discovery by Maker *et al* [1] resulting in several studies that have been summarized by different authors [2-16]. Several hundred papers describing various aspects of the phenomenon have been published, and a healthy controversy appears to be in existence concerning the mechanisms by which gases can become almost perfect conductors under the influence of short-duration pulses of long-wavelength light alone. The LIB plasma develops a shock wave into the ambient medium and the gas is heated to high temperatures [17]. It is evident that after breakdown, the plasma becomes very opaque and an abrupt shutoff of the laser transmitted light occurs. Due to the many experimental parameters involved in the LIB, an exhaustive investigation of the plasma would involve the processing of an impressive number of records. To investigate LIB of gases several diagnostic techniques have been employed in the last years. Optical emission spectroscopy (OES) is a powerful tool to get information on the LIB species. Because of the transient features of the plume created by LIB, OES technique with time and space resolution is especially appropriate to obtain information about the behaviour of the created species in space and time as well as the dynamics of the plasma evolution. Although OES gives only partial information about the plasma particles, this diagnostic technique helped us to draw a picture of the plasma in terms of the emitting chemical species, to evaluate their possible mechanisms of excitation and formation and to study the role of gas-phase reactions in the plasma expansion process.

The objectives of this work are: (i) to show some fundamentals of laser-induced breakdown spectroscopy (LIBS) and, (ii) to review of our recent results on LIBS analysis of chemical species in gases plasmas induced by high-power IR CO_2 laser, adding some new results. This chapter describes the experimental results obtained from UV-Visible-near IR spectra of LIB plasmas generated by carbon dioxide laser pulses in gases such as N_2, O_2 and air. The major parts of this work are already published by us in different journals [18-21]. We consider here only research on the plasma induced when a high-intensity laser radiation is focused in a gas. The emission observed in the laser-induced plasma region is due to electronic relaxation of excited atoms, ions and molecular bands of different diatomic molecules. Excitation temperatures and electron number densities were calculated by means of different atomic and ionic lines. Estimates of vibrational and rotational temperatures of some molecules electronically excited species are also reported. The characteristics of the spectral emission intensities from different species have been investigated as functions of the gas pressure and laser irradiance. Optical breakdown threshold intensities in the different studied gases have been determined.

For air we discuss the dynamics of the plume expansion and formation of different atomic, ionic and molecular species for different delay times with respect to the beginning of the laser pulse. The velocity and kinetic distributions for the different species were obtained from the time-of-flight (TOF) measurements using time-resolved OES. Possible mechanisms for the production of these distributions are discussed. Line intensities from different atomic and ionic species were used for determining electron temperature and Stark-broadened profiles of some lines were employed to calculate electron density. The temporal behavior of electron number density has been employed for the estimation of three-body recombination rate constant and recombination time. The present paper is aimed at discussing thermo-chemical processes produced by a high-power IR CO_2 pulsed laser on several gases and at evaluating LIB changes which are of fundamental importance in establishing the mechanisms responsible for the plasma emission.

1.1. Laser-Induced Breakdown Spectroscopy (LIBS)

Excellent textbooks and reviews about the fundamentals of laser-induced breakdown spectroscopy (LIBS) and examples of various processes are readily available today [13-16]. LIBS, also sometimes called laser-induced plasma spectroscopy, is a technique of atomic-molecular emission spectroscopy which utilizes a highly-power laser pulse as the excitation source. LIBS can analyze any matter regardless of its physical state, being it solid, liquid or gas. Because all elements emit light when excited to sufficiently high energy, LIBS can detect different species (atomic, ionic and molecular) and limited only by the power of the laser as well as the sensitivity and wavelength range of the spectrograph/detector. Basically LIBS makes use of OES and is to this extent very similar to arc/spark emission spectroscopy. LIBS operates by focusing the laser onto a small volume of the sample; when the laser is triggered it breaks a very small amount of gas which instantaneously generates a plasma plume with temperatures of about 10000–30000 K. At these temperatures, the gas dissociates (breakdown) into excited ionic and atomic species. At the early time, the plasma emits a continuum of radiation which does not contain any known information about the species present in the plume and within a very small timeframe the plasma expands at supersonic velocities and cools. At this point the characteristic atomic/ionic and molecular emission lines of the species can be observed. The delay between the emission of the continuum and characteristic radiation is of the order of 1 μs, this is one of the reasons for temporally gating the detector. LIBS is technically similar and complementary to a number of other laser-based techniques (Raman, laser-induced fluorescence etc).

In fact devices are now being manufactured which combine these techniques in a single instrument, allowing the atomic, molecular and structural characterization of a sample as well as giving a deeper insight into physical properties.

A typical LIBS system consists of a pulsed laser and a spectrometer with a wide spectral range and a high sensitivity, fast response rate and time gated detector. The principal advantages of LIBS over the conventional analytical spectroscopic techniques are its simplicity and the sampling speed.

1.2. Laser Parameters

The variables that can influence the LIBS measurements are mainly the laser properties i.e. wavelength, energy, pulse duration, focusing spot size, shot-to-shot energy fluctuations, ambient conditions, physical properties of the sample and the detection window (delay time and gate width). How these parameters affect the precision and accuracy of LIBS are addressed below. In LIBS a high-power laser is used to breakdown a gaseous sample in the form of plasma.

The primary energy related parameters influencing the laser-gas interaction are the laser peak power P_W (or radiant pulse energy per time, in W) and the laser peak intensity I_W (power density or irradiance; energy per unit area and time, W×cm^{-2}) given by

$$P_W = E_W / \tau^{Las}_{FWHM}, z \qquad (1.1)$$

$$I_W = P_W / \pi r^2 \qquad (1.2)$$

where E_W (in J) is the pulse energy, τ^{Las}_{FWHM} (in s) is the laser pulse duration of the full width at half maximum (FWHM) and πr^2 is the focal spot area (cm^2). The fluence Φ_W (in J×cm^{-2}) on the focused spot area, the photon flux density F_{ph} (photon×cm^{-2}×s^{-1}), electric field F_E (V× cm^{-1}) and pressure radiation p_R (in Pa) are given by

$$\Phi_W = E_W / \pi r^2 \qquad (1.3)$$

$$F_{ph} = I_W \lambda / hc \qquad (1.4)$$

$$F_E = \sqrt{I_W / c\varepsilon_0} \qquad (1.5)$$

$$p_R = 2 I_W / c \qquad (1.6)$$

where λ is the laser wavelength, h is the Planck constant, c is the speed of light, and ε_0 is the electric constant. In Equation (1.6) we suppose that the laser radiation is totally reflected and therefore the pressure radiation can be doubled. The laser peak intensity I_W, fluence, photon flux, electric field and pressure radiation are inversely proportional to the focused spot area. For LIBS, the peak intensity I_W (and the other properties Φ_W, F_{ph}, F_E and p_R) that can be delivered to the sample is more important than the absolute value of the laser power. For the formation of plasma, the laser irradiance needs to exceed the threshold value, typically of the order of several GW×cm^{-2} for a nanosecond laser pulse.

If the laser energy is very close to the breakdown threshold, the pulse-to-pulse fluctuations can cause the plasma condition to be irreproducible, which results in poor measurement precision. The intensities of the emission lines are proportional to the laser energy while the laser plasma is in the optical thin region.

When the laser energy increases further, it produces very dense and hot plasma that can absorb laser energy. This will lead to an increase in the continuum emission and a decrease in the signal intensity. Besides, the laser pulse duration and the shot-to-shot fluctuations can also affect the signal reproducibility and hence LIBS precision.

1.3. Focal Properties

The laser power density at the focal volume is inversely proportional to the focused spot size. For a laser beam with a Gaussian profile, the focused bean waist w_0 is given by

$$w_0 = \frac{\lambda f}{\pi w_s} \tag{1.7}$$

where f is the focal length of the lens and w_s is the radius of the unfocused beam. The higher laser power density at the focal point can be achieved by reducing the focused beam waist using a shorter focal length lens. On the other hand, the angular spread in laser light generated by the diffraction of plane waves passing through a circular aperture consists of a central, bright circular spot (the Airy disk) surrounded by a series of bright rings. The beam divergence angle θ, measured to edges of Airy disk, is given by $\theta=2.44\times\lambda/d$, where λ is the laser wavelength and d is the diameter of the circular aperture. It can be shown that a laser beam, with beam divergence θ_i, incident on a lens of focal length f, whose diameter is several times larger than the width of the incident beam, is focused to a diffraction-limited spot of diameter approximately equal to $f\times\theta_i$. If the focal region of the laser beam is assumed to be cylindrical is shape, the spot size in terms of length l, can be approximated as

$$l = (\sqrt{2} - 1)\theta_i f^2 / d \tag{1.8}$$

2. Optical Breakdown in Gases

Optical breakdown in gases leads to the generation of free electrons and ions, electrons in gases are either bound to a particular molecule or are quasifree when they have sufficient kinetic energy to move without being captured by local molecular energy potentials. Thus, transitions bound and quasi-free states are the equivalent of ionization of molecules in gases. The optical breakdown process describes in greater detail by Raizer [2-3, 23], starts when a laser beam with sufficient power density is focused down, and a sufficient radiation flux density is achieved, leading to a discharge/spark. This discharge is somewhat similar to the discharge induced by a sufficient electric field between the electrodes of a spark plug in an internal combustion engine.

The temperature and pressure of the gas in the region of this discharge will be increased significantly as the energy of the laser is absorbed to cause this so called laser induced optical breakdown. The energy deposition into a gas by a focused laser beam can be described by four progressive steps: (i) initial release of electrons by multi-photon effect; (ii) ionization of the gas in the focal region by the cascade release of electrons; (iii) absorption and reflection of laser energy by the gaseous plasma, rapid expansion of the plasma and detonation wave formation; and (iv) the propagation of the detonation wave into the surrounding gas and relaxation of the focal plasma region.

2.1. LIB Plasma

Plasma is a local assembly of atoms, molecules, ions and free electrons, overall electrically neutral, in which the charged species often act collectively. The LIB plasma is initiated by a single laser pulse. If we consider the temporal evolution of LIB plasma, at early times the ionization grade is high. As electron-ion recombination proceeds, neutral atoms and molecules form.

A recombination occurs when a free electron is captured into an ionic or atomic energy level and gives up its excess kinetic energy in the form of a photon. LIB plasmas are characterized by a variety of parameters, the most basic being the degree of ionization. A weakly ionized plasma is one in which the ratio of electrons to other species is less than 10%. At the other extreme, high ionized plasmas may have atoms stripped of many of their electrons, resulting in very high electron to atom/ion ratios. LIB plasmas typically, for low power laser intensities, fall in the category of weak ionized plasmas. At high laser power densities, LIB plasmas correspond to strong ionized plasmas.

2.2. Initiation Mechanism: Multiphoton Ionization (MPI) and Electron Impact Ionization (EII)

Plasma is initiated by electron generation and electron density growth. The conventional LIB plasma can be initiated in two methods: multiphoton ionization (MPI) and electron impact ionization (EII) both followed by electron cascade. EII is sometimes denominated as cascade ionization process or avalanche ionization due to inverse bremsstrahlung (IB) heating of electrons. MPI involves the simultaneous absorption of a number of photons *n*, required to equal the ionization potential $I_P(A)$ of an atom or molecule *A*

$$nh\nu + A \rightarrow A^{+} + e + I_P(A); \; nh\nu \geq I_P(A), \quad (2.1)$$

where n is the number of photons needed to strip off an electron, which corresponds to the integer part of the quantity:

$$n = \frac{I_P + \varepsilon_{osc}}{h\nu} + 1 \quad (2.2)$$

here ε_{osc} is the oscillation energy of a free electron in the alternating electric field. Within the classical microwave breakdown theory [22], a free electron oscillates in the alternating electric field E of the laser electromagnetic wave with frequency ω and wavelength λ (μm), and its oscillation energy

$$\varepsilon_{osc}[eV] = \frac{e^2E^2}{4m\omega^2} = \frac{e^2}{4m\pi c^3} I_W \lambda^2 = 4.67 \times 10^{-14} I_W \lambda^2 \quad (2.3)$$

remains constant. In Equation (2.3) e is the electron charge and I_w is the laser intensity [irradiance, power density or flux density in Wxcm^{-2}; Equation 1.2]. The probability of MPI W_{MPI}, by absorbing simultaneously n laser photons to strip off an electron, is expressed by the classical formula [23]

$$W_{MPI}[s^{-1}] \cong \omega\ n^{3/2} \left(1.36 \frac{\varepsilon_{osc}}{I_P}\right)^n = 1.88 \times 10^{15} \lambda^{2n-1} n^{3/2} \left\{\frac{6.35 \times 10^{-14} I_W}{I_P}\right\}^n, \quad (2.4)$$

where I_P is in eV. Besides, the probability of simultaneous absorption of photons decreases with the number of photons n necessary to cause ionization. Therefore, the MPI rate is proportional to I_W^n and inversely proportional to I_P^n.

EII process consists on the absorption of light photon by free or quasifree electrons, producing electrons with enough kinetic energy e^* to ionize one atom or molecule

$$e + nh\nu + A \rightarrow e^* + A \rightarrow 2e + A^+. \quad (2.5)$$

Two conditions must co-exist for EII to initiate: (i) an initial electron must reside in the focal volume; and (ii) the initial electron must acquire energy which exceeds the ionization energy of the material in the focus. These free or quasifree electrons can be produced by the effect of cosmic ray ionization (natural ionization), by means of MPI, or by a breakdown induced in some impurity. The equilibrium number of free electrons and ions per cm^3 in the atmosphere at the earth's surface is about 500 [24]. The electron cascade ionization process requires the presence of either free electrons or excited atoms or molecules in the focal region for their initiation, but the possibility of a free electron or an excited atom produced by natural causes being present there during the laser flash can be discounted, as Tozer pointed out [25]. They occur naturally in the earth's atmosphere, produced at a rate of about 10 cm^{-3}xs^{-1} at the earth's surface by the presence of local radioactivity and the passage of ultraviolet radiation and cosmic rays. Most electrons become rapidly attached to electronegative atoms and molecules to form negative ions. The mean lifetime of a free electron in the atmosphere is about 10^{-7} s, so that the aggregate life of free electrons is about 10^{-6} s when liberated at the rate of ≈ 10 $cm^{-3}xs^{-1}$. The probability of finding a free electron in the focal region ≈ 10^{-4} cm^3 during a laser flash ≈ 10^{-8} s is thus entirely negligible, as is the chance for finding an excited atom. One concludes that the laser light itself produces the initiatory electrons. These electrons in the focal volume gain sufficient energy, from the laser field through IB collision with neutrals, to ionize atoms, molecules or ions by inelastic electron-particle collision resulting in two electrons of lower energy being available to start the process again

$$e^*[\varepsilon \geq IP(A)] + A \rightarrow A^+ + 2e;\ e^*[\varepsilon \geq IP(A^+)] + A^+ \rightarrow A^{2+} + 2e. \tag{2.6}$$

Thus a third species (atom-molecule/ion) is necessary for conserving momentum and energy during optical absorption. The recurring sequences of IB absorption events and subsequent EII lead to a rapid growth in the number of free electrons, if the laser intensity is sufficient to overcome the losses of free electrons through diffusion out of the focal volume and through recombination. The MPI mechanism dominates electron generation only for low exciting wavelengths. Therefore initial EII becomes a problem at a higher wavelength because neither cascade nor MPI can furnish sufficient number of electrons. At higher laser intensities, electric field of the laser is able to pull an outer shell electron out of its orbit. After the initial electron ejection the LIB plasma is commonly maintained by the absorption of optical energy and the EII. Electrons in the laser field will gain energy through electron-neutral IB collisions and will lose energy by elastic and inelastic collisions with the neutral species through excitation of rotational and vibrational degree of freedom of molecules and excitation of electronic states. While some electrons will be lost by attachment, new electrons will be produced by ionizing collisions. At high laser intensity, few electrons will be generated with energy larger than the ionization energy. The wavelength-resolved emission spectra from the laser plasma are not expected to vary due to the plasma origin. However plasma origin may be relevant, if the enhancement is observed between UV, visible and IR excitation wavelengths.

Once that LIB plasma is formed, its growth is governed by the continuity rate equation for the electron density [26] due to the combined effect of EII and MPI

$$\frac{dn_e}{dt} = \nu_i n_e + W_n I_W^n N - \nu_a n_e - \nu_R n_e^2 + D_e \nabla^2 n_e \ , \tag{2.7}$$

where ν_i is the impact ionization rate, W_n is the multiphoton ionization rate coefficient, I_w is the intensity of the laser beam, n is the number of photons required for MPI, N is the number of atoms/molecules per unit volume, ν_a is the attachment rate, ν_R is the recombination rate and D_e is the electron diffusion coefficient. The term dn_e/dt is the net rate of change in electron concentration at a point in the focal volume at a time t after the release of initiatory electrons. On the right side of the equation (2.7), the first term is the electron generation due to impact ionization. The second term on the right is MPI rate. The third, fourth and fifth terms are sink terms which represent electron attachment, recombination and diffusion, respectively. Impact ionization is defined by multiplying the number of electrons per unit volume to the impact ionization rate ν_i. The impact ionization rate refers to the rate at which electrons are generated as a result of ionizing collisions. At high laser intensity, a few new electrons can be generated and gain energy larger than their ionization energy which leads to the generation of new electrons by impact ionization, thereby leading to the cascade growth. Recombination losses are usually not important in the breakdown forming stage.

2.3. Electron Attachment, Recombination and Diffusion

Electron attachment is the rate of electron attachment ν_a multiplied by the number of electrons per unit volume. The LIB plasma typically loose electrons to the neutral species via the attachment mechanism in the form of three-body attachment or two-body dissociative attachment. Three-body attachment is: $e + AB + X \rightarrow AB^- + X$, where X appears to be a facilitator that allows the electrons to be gained by AB even through X remains unchanged throughout the process. Two-body dissociative attachment is: $e + AB \rightarrow A^- + B$. In this mechanism the electrons must exhibit a threshold electron energy that is equal to the difference between the dissociative energy of AB and the attachment energy of A, which results in the separation of A and B.

Electron recombination is the rate of electron recombination ν_R multiplied to the number of electrons per unit volume. When the electron density is high, such as during the last stage of cascade breakdown, the LIB plasma can lose electrons to ions through electron-ion recombination. Similar to the electron attachment, three-body recombination and two-body recombination occurs as: $e + AB^+ + X \rightarrow AB + X$, $e + AB^+ \rightarrow A + B$. The electron-ion recombination rate has been studied theoretically for a three-body recombination by Gurevich and Pitaevskii [27]

$$\nu_R = 8.8 \times 10^{-27} \frac{n_e^2}{T_e^{3.5}} \; [s^{-1}] \quad , \qquad (2.8)$$

where ne is the electron density in cm-3 and Te is the electron temperature in eV. The electron diffusion term is expressed as $D_e \nabla^2 n_e$ [Equation (2.7)]. This loss mechanism, more important for a small diameter laser beam, is the diffusion of electrons out of the focal volume. Morgan [6] referred to the combined effect of diffusion and cascade ionization as the responsible for top-hat intensity profile. By imposing an electron skin at the edge of the intensity profile, they found that the electron density grows exponentially as

$$\langle v_e \rangle = \frac{2.408 D_e}{r^2} \quad , \qquad (2.9)$$

where $\langle v_e \rangle$ is the average electron velocity, D_e is electron diffusion coefficient and r is the radius of the beam. The equation (2.9) is intended to be an upper boundary for diffusion losses only because laser beams typically have a radial distribution closer to the Gaussian rather than top-hat distribution.

In LIB plasmas, the decrease of electron density n_e is mainly due to recombination between electrons and ions in the plasma. These processes correspond to the so-called radiative recombination and three-body recombination processes in which a third body may be either a heavy particle or an electron. The electron number density n_e (cm^{-3}) in the laser induced plasma is governed by the kinetic balance equation [28, 29]

$$\frac{\mathrm{d}n_e}{\mathrm{d}t} = k_{\mathrm{ion}} n_e N_i - k_{\mathrm{rec}} n_e^3, \tag{2.10}$$

where N_i indicates the concentration of heavy particles (neutrals and ions) and k_{ion} (cm^3 s^{-1}) and k_{rec} (cm^6 s^{-1}) denote the rate constants of ionization ($e + A \rightarrow A^+ + 2e$) and tree-body electron-ion recombination ($2e + A^+ \rightarrow A^* + e$), respectively. The excess of energy in three-body electron-ion recombination is deposited as kinetic energy to a free electron, which participates in the recombination process as a third body partner. The three-body electron-ion recombination energy can be converted into radiation in the process of radiative electron-ion recombination ($e + A^+ \rightarrow A^* \rightarrow A + h\nu$). The cross section of this process is relatively low and it can be competitive with three-body electron-ion recombination only when the plasma density is low. If $\mathrm{d}n_e/\mathrm{d}t=0$ an equilibrium condition can be established; if $\mathrm{d}n_e/\mathrm{d}t\neq0$, then the ionization ($\mathrm{d}n_e/\mathrm{d}t>0$) or the three-body recombination ($\mathrm{d}n_e/\mathrm{d}t<0$) prevails and departure from equilibrium occurs [28]. The second derivative of $Y=\mathrm{d}n_e/\mathrm{d}t$ with respect to the electron number density is given by

$$\frac{\mathrm{d}^2 Y}{\mathrm{d}n_e^2} = -6k_{\mathrm{rec}} n_e. \tag{2.11}$$

The recombination time can be determined by the value of the rate constant of the recombination process as $t_{rec}=1/(n_e^2.k_{rec})$ [29].

In summary, the process of plasma initiation essentially consists of the formation of free or quasi-free electrons by interplay of MPI and EII. Therefore, two mechanisms MPI and EII can initiate a conventional LIB plasma formation. After the LIB plasma formation the temporal growth is governed by the equation (2.7). The recombination of these two source terms (MPI and EII) and three sink terms (electron attachment, electron recombination and electron diffusion) controls the development of the conventional LIB plasma. The decrease of n_e is mainly due to the so-called radiative recombination and three-body recombination processes in which a third body may be either a heavy particle or an electron. These mechanisms that directly affect the temporal development of the LIB plasma, determine the necessary spectroscopic techniques required to spectrally resolve elemental species inside the LIB plasma.

2.4. Optical Breakdown Threshold Intensities

The minimum power density required to form a plasma is called the breakdown threshold; different types of laser, sample, and environmental conditions will have different breakdown thresholds. Breakdown thresholds of solids and liquids are usually much lower than those for gases. The principal method of investigation has been to measure the beam intensity required for electron liberation and the minimum intensity needed to produce breakdown as a function of the radiation wavelength and pressure of a variety of gases. Precise measurements of the intensities of laser radiation required to release initiatory electrons or to lead to breakdown are made only with the greatest difficulty. The difficulties arise on account of the imprecise definition of the extent of the focal region and inaccurate knowledge of the spatial-temporal characteristics of the beam intensity within the focal region, which, in turn, lead to uncertainties in the absolute value of the instantaneous radiation intensity. The parameters which characterize a focused laser beam are its polarization, wavelength, line width, duration, divergence and the temporal and spatial distribution of intensity.

For a given pulse these are functions of the laser and focusing system governed by the mode structure within the laser cavity, by the aberration functions of the lens or focusing mirror, and by the beam diameter at the lens or mirror. In specifying the electron liberation or breakdown threshold intensities all these factors should ideally be specified, but, regrettably, in the literature there is often inadequate detail and essential features of experimental procedures are frequently omitted. In consequence many published data are of little value, serving merely to indicate orders of magnitude and broad trends only rather than absolute values in well-defined conditions. For these reasons data published by various workers are often contradictory, and reliable interpretation is sometimes difficult to make.

For gases to breakdown, a certain concentration of electrons has to be reached before the end of the laser pulse. Laser-induced breakdown is frequently defined [5, 30] as an electron density multiplication during the laser pulse by a factor of 10^{13} corresponding to 43 electron generations.

In fact, multiplying the natural electron density by 10^{13} leads to $n_e \approx 10^{16}$ cm^{-3} which is the electron density of plasmas at atmospheric pressure for which electron-ion IB dominates electron-neutral IB. With respect to electron-neutral IB, the electron-ion IB has a much higher efficiency as a result of the long range Coulomb interaction, and a plasma with an electron density $n_e = 10^{16}$ cm^{-3} is quasi instantaneously completely ionized.

The condition for optical breakdown is taken to occur when the number density of the induced electrons equals the critical density for the laser wavelength. The critical plasma density $n_e^{crit}\left[\text{cm}^{-3}\right] = m\,\omega^2/4\pi e^2 \cong 1.1\times 10^{21}/\lambda^2[\mu\text{m}]$ ($n_e^{crit} \cong 10^{19}\ cm^{-3}$ for CO_2 laser) is the density where the electron plasma frequency equals to the laser frequency. When the electron density exceeds the critical density the sample is not transparent any more. Energetic electrons produce excited species through impact excitation, dissociation and ionization of gas molecules. According to the microwave theory [22], electrons gain energy from the laser radiation field by elastic collisions with neutral atoms at the rate: $(d\varepsilon/dt)_{gain} = (e^2 F_E^2/m)\cdot[\nu_c^2/(\omega^2+\nu_c^2)]$, where F_E and ω are the root-mean-square electric field and angular frequency of the radiation and ν_c is the electron-neutral collision frequency.

Several models have been developed to describe the optical breakdown and to compute the breakdown threshold. Chan *et al.* [30] proposed a model based on the energy balance of electrons neglecting their energy distribution. According to this work, breakdown occurred if the laser heating of electrons by IB induces a gain of electron energy that overcomes the energy losses. Thus, one requires a laser power density (power threshold density)

$$I_{\text{las}} \geq \frac{m\ c\ I_{\text{P}}}{4\pi e^2 \ln 2} \frac{\omega^2 + \nu_{\text{c}}^2}{\nu_{\text{c}}} \left[\frac{43}{\tau_{\text{las}}} \ln 2 + \frac{D_e}{\Lambda^2} + \frac{2m\ \langle \varepsilon \rangle\ \ln 2\ \nu_{\text{c}}}{M\ I_{\text{P}}} + (\alpha + \frac{\beta}{\Lambda^2})\nu_{\text{c}} \right], \tag{2.12}$$

where m and e are the mass and charge of electrons, c is the light velocity, and I_{P} and M are the ionization potential and the atomic mass of the gas. The terms ω, ν_{c}, τ_{las}, D_{e}, Λ and $<\varepsilon>$ are the laser frequency, effective electron-neutral collision frequency, laser pulse duration, diffusion coefficient, diffusion length and average electron energy, respectively. The terms α (dimensionless) and β (length^2) are two parameters which depend on the atomic structure of the gas. The terms inside the brackets represent various loss terms. The first term in the brackets stands for the generation of 43 electrons necessary for breakdown. The second, third, and fourth terms take into account the electron energy loss due to diffusion out of the focal volume, elastic and inelastic collisions, respectively. The loss due to electron attachment is very low and therefore not considered in Equation (2.12). The losses due to elastic and inelastic collisions are proportional to ν_{c} that increases linearly with the gas pressure p. At low pressure, collisional loss can be neglected and electron heating by IB varies linearly with p according to $\nu_{\text{c}}^2 << \omega$. Thus, the threshold decreases with p in the low-pressure range. When increasing p to sufficiently high values, the collisional losses overcome the terms of electron generation and diffusion loss. If $\nu_{\text{c}}^2 << \omega$ still holds, both gain and loss terms are proportional to p and the threshold is pressure independent. In the high pressure range ($\nu_{\text{c}}^2 >> \omega$) the gain by IB diminishes as p^{-1} and the threshold increases. Thus, it exists an optimum pressure for which the optical breakdown threshold is minimum. According to Equation (2.12), the breakdown power threshold density in general is directly proportional to the ionization potential of the gas. Moreover, the breakdown threshold passes through a minimum at the pressure when laser angular frequency ω is equal to the effective electron-neutral collision frequency ν_{c} as indicated by the term outside the brackets. Depending on their relative magnitudes, breakdown may be termed to be limited, diffusion limited or attachment limited [30].

Time-limited breakdown occurs if the first term in the brackets dominates, that is when the laser pulse duration τ_{las} is so short that the growth rate of electron density required to induce a visible breakdown exceeds any losses. Thus, the threshold power density varies inversely with pulse duration and the breakdown is determined by the product of the intensity times the pulse duration. Diffusion-limited processes occur when the second term in the brackets dominates, that is for gas breakdown to take place in a small focal volume at low gas pressure. The breakdown power density threshold in this case decreases as Λ^{-2} with the focal size and it also decreases as p^{-2} with the gas pressure in the range of pressure so that $\omega >> \nu_{\text{c}}$. The third and fourth terms are the attachment and elastic collision loses. They are relatively unimportant and are dependent on the type and masses of the gas. For inert gases, the attachment loss can be completely neglected. The last term is the energy loss due to inelastic collision and it should be important for molecular gases because the large number of excited states they possess. Because the attachment rate and collision frequency are assumed to be proportional to the gas pressure p, these three terms are independent of the gas pressure.

At low pressure and in particular for small waist, the electron diffusion out of the focal volume is the dominating loss term. The electron diffusion length Λ can be estimated assuming a focal volume of cylindrical shape with radius

$$r = \frac{f\Theta}{2}, \tag{2.13}$$

and length $l = \left(\sqrt{2}-1\right)\frac{f^2\Theta}{d}$ (Equation 1.8), where f is the focal length of the focussing lens, Θ the angle of laser beam divergence, and d the laser beam diameter incident on the lens. For a Gaussian laser beam, one has

$$\frac{1}{\Lambda^2} = \left(\frac{2.405}{r}\right)^2 + \left(\frac{\pi}{l}\right)^2, \tag{2.14}$$

For a large numerical opening ($\geq f/5$), Equation (2.14) is simplified to $\Lambda = r/2.405$. Using this expression with r from Equation (2.13) and computing the electron diffusion coefficient as

$$D_e = \frac{2\langle\varepsilon\rangle}{3m\nu_{\text{eff}}}, \tag{2.15}$$

the energy loss due to electron diffusion is evaluated. With respect to the loss due to elastic collisions, the energy loss by diffusion can be neglected if

$$\nu_{\text{eff}} >> \frac{4.81\sqrt{I_P M}}{\sqrt{3m}f\Theta}. \tag{2.16}$$

A large dispersion of breakdown threshold values exists in literature. It is attributed to the large number of parameters on which the optical breakdown depends. Several mechanisms have been found to reduce the threshold of optical breakdown. Smith and Haught [31] observed threshold lowering by Penning effect during ruby laser breakdown in a high-pressure Ar atmosphere when adding 1% Ne. The phenomenon was due to Ar ionization by collisions with excited Ne atoms which were produced by a resonant excitation process. However, the threshold lowering was at maximum of about 50% [31]. For CO_2 laser radiation, a resonant excitation process can be excluded because of the small photon energy and the Penning effect does not contribute to optical breakdown threshold lowering in the far IR spectral range. For laser radiation of sufficiently high photon energy, the presence of impurities with low ionization energy led to the threshold lowering [32] that was attributed to multiphoton ionization.

However, this effect was not observed for CO_2 laser radiation. Gas impurities with the lowest ionization potential like hydrocarbon radicals require at least the simultaneous absorption of more than 50 photons that is a process of vanishing probability. Contrarily, molecular species such as hydrocarbon or other radicals brake the ionization avalanche. They have many vibrational and rotational excitation levels which cause electron energy loss by inelastic collisions [see Equation (2.12)].

Several authors [33, 34] reported threshold lowering when initiating the breakdown by ablation of a solid target. A threshold reduction by a factor of 10^2 was observed for CO_2 laser radiation [33]. The threshold lowering was explained by shock wave generation as an effect of strong material ablation. The shock wave heats up the surrounding gas which is instantaneously transformed in a strongly ionized plasma. The optical breakdown from solid material ablation has been shown to be a multistage plasma initiation process that is characterized by three thresholds [34]: (i) the material ablation threshold I_{vap}; (ii) the breakdown threshold of the evaporated material I_{vap}^*; and (iii) the breakdown threshold of the surrounding gas I_{gas}^*. It is noted that I_{vap}^* and I_{gas}^* are the thresholds of preionized vapour and the gas, respectively. For the case of CO_2 laser ablation, the initial ionization stage of the ablated material vapour is $n_e/n_{vap}<10^{-5}$-10^{-4}[34], n_{vap} being the ablated material vapor density.

As a consequence of preionization, the number of electrons generations necessary for complete ionization is strongly reduced (<<43). Thus, the electron generation term can be neglected in Equation (2.12). The diffusion loss can be also neglected according to the relative large volume preionized by the shock wave. In the case of rare gases, the loss by inelastic collisions is much smaller than that due to elastic collisions and the avalanche ionization is determined by the balance between IB heating of electrons and losses by elastic collisions. The breakdown threshold (in Wxcm^{-2}) given by Equation (2.12) is simplified to [34]

$$I_{las} \geq I_{EC}^* = 1.8 \times 10^7 \frac{I_P}{M}, \tag{2.17}$$

where the ionization potential is in eV and the atomic mass of the gas is in atomic mass units. The index EC stands for elastic collisions to recall that only this loss term has been taken into account. Barchukov et al. [33] proposed a threshold criterion similar to Equation (2.17) with a three times larger numerical constant. The difference is due to the average electron energy which was supposed to be equal to the ionization potential by Barchukov et al. [33] while $<\varepsilon>=(1/3)IP$ was taken for Equation (2.17).

2.5. Laser-Plasma Interaction

The interaction between the laser radiation and free electrons of the plasma is described by the Drude model considering the electron motion in the laser field as a harmonic oscillator. For collision frequencies $\omega_P^2 >> \nu_c^2$, where ω_P is the plasma frequency. The dielectric constant is given by $\varepsilon \approx 1 - \omega_P^2/\omega^2$, where ω is the laser frequency. Optical breakdown in gases at atmospheric pressure leads to an electron density equal to the critical density (for CO_2 laser radiation $n_e^{crit} \cong 10^{19}$ cm^{-3}). In the region where the critical density is reached, the plasma frequency is equal to the laser frequency and $n = \sqrt{\varepsilon} = 0$. The plasma is thus completely reflecting in the corresponding zone. Only a few authors have investigated the laser beam reflection by the plasma. In fact, the studies show that breakdown plasmas reaching the critical density absorb most of the incident laser energy [35-37]. The fraction of reflected radiation is small because of strong absorption in the zone adjacent to the plasma sheet of critical density.

Donaldson *et al.* [36] showed that 80% of incident laser energy was absorbed in a zone of weak thickness where the electronic density varied from $0.83n_e^{crit}$ to n_e^{crit} when generating breakdown with a (Nd:YAG) laser of τ_{las}=35 ps and $I_W=10^{14}$ Wcm^{-2}. Offenberger and Burnett [35] measured the reflected and transmitted power of TEA-CO_2 laser pulses during breakdown in hydrogen. The reflected power was always below 2% of the incident laser power. The major absorption mechanism of CO_2 laser radiation during breakdown ignition is electron-neutral IB. Once strongly ionized plasma is formed, the electron impact ionization or electron-ion IB dominates as a result of the long range Coulomb interaction between charged particles.

Several experimental and theoretical studies have been performed to investigate the IB effect. Among numerous expressions for the determination of the IB absorption coefficient [2, 5, 36] the formula

$$\alpha_{IB} = 3.69 \times 10^8 \frac{n_e^2 \sum_i f_i Z_i^2}{\sqrt{T_e}\,\nu^3}\left(1 - e^{-h\omega/2\pi k_B T}\right) \quad [\text{cm}^{-1}], \tag{2.18}$$

proposed by Spitzer [38] was used by many authors to estimate the laser energy absorption by the plasma. Here, T_e and n_e are in K and cm^{-3}, respectively. The factors f_i are the fractional abundances of ions, Z_i the corresponding ion charge. In the case of CO_2 laser radiation, one has $\left(1 - e^{-h\omega/2\pi k_B T}\right) \approx h\omega/2\pi k_B T$ and Equation (2.18) simplifies for a singly ionized plasma to

$$\alpha_{IB} = 1.8 \times 10^{-35} \frac{n_e^2}{T_e^{3/2}} \quad [\text{cm}^{-1}], \tag{2.19}$$

where T_e is in eV. Equation (2.19) shows that the efficiency of IB absorption decreases with increasing electron temperature. At high plasma temperatures, other absorption mechanisms dominate. They have been made in evidence during studies of laser-plasma interaction related to thermonuclear fusion using power densities several orders of magnitude higher than breakdown thresholds.

2.6. Absorption Wave Propagation

After breakdown ignition, the strongly absorbing plasma will propagate in the direction opposite to the laser beam. The absorption wave formation has been observed in many experiments using CO_2 laser sources [33, 35, 39]. The theoretical analyses of optical breakdown and absorption wave propagation performed by Raizer [2] using a hydrodynamic model have been widely accepted and became a standard theory in the field. According to this model, the laser-induced absorption waves propagate by the following stepwise mechanisms: (i) A small plasma zone is heated up by the laser beam. It reaches the critical density and strongly absorbs the laser radiation. (ii) The electron density in the adjacent zones increases. (iii) The adjacent preionized zone that is irradiated by the laser beam is heated up and becomes absorbent. Thus, the strongly absorbing plasma zone propagates in the direction opposite to the laser beam. Three different propagation modes are distinguished. The breakdown wave is characterized by the following propagation mechanism: (i) breakdown occurs initially in the region of the highest laser power density and later in the zones of lower power density. The expansion of high pressure plasma compresses the surrounding gas and drives a shock wave. Thus, the breakdown propagates in the direction opposite to the laser beam. The plasma also tends to expand back along the beam path toward the laser, a phenomenon known as moving breakdown. The velocity of the breakdown wave is given by

$$v_{bw} = \frac{w_0}{t_b \, \mathrm{tg}\varphi}, \tag{2.20}$$

where w_0 and φ are the minimum radius and opening angle of the focused laser beam, respectively. For example, for our typical experimental conditions with the CO_2 laser, taking w_0=0.05 cm, t_b=100 ns and tgφ=0.2, the breakdown wave propagates with v_{bw}=2.5□10^6 cm/s;

(ii) Propagation through the detonation wave mechanisms occurs when rapid heating of the gas in the region of strong absorption induces a spherical shock wave. The latter propagates into the surrounding gas that is heated and preionized. The part of the preionized gas that is further illuminated by the laser beam absorbs the laser radiation and becomes opaque. Thus, the absorption zone follows the shock wave. The propagation velocity of the detonation wave is given by

$$v_{bw} = \left[2\left(\gamma^2 - 1\right)\frac{I_W}{\rho_0}\right]^{1/3}, \tag{2.21}$$

where γ and ρ_0 are the adiabatic constant of the gas and the specific mass, respectively. The specific energy that is injected into the gas is

$$\varepsilon_{bw} = \frac{\gamma}{\left(\gamma^2 - 1\right)\left(\gamma + 1\right)} v_{bw}^2. \tag{2.22}$$

It is noted that detonation wave propagation velocity and injected specific energy are independent of the atomic structure of the gas. The gas influences the detonation wave propagation only through its specific mass and adiabatic constant. Thus, a change of gas nature is equivalent to a pressure variation if γ is unchanged. Consequently, the detonation wave propagation velocity in Ar is equal to that in Xe at three times lower pressure. For Xe at atmospheric pressure and $I_W=10^8$ W×cm^{-2}, the detonation wave propagates with a velocity of 6x10^5 cmxs^{-1} and heats up the gas to a temperature of 28 eV. The temperature is obtained by assuming an ideal gas, for which the specific energy is related to the temperature by $\varepsilon_{bw}=(3/2) k_B T N_A/M$. Here, k_B, T, N_A and M are Boltzmann's constant, plasma temperature, Avogadro's constant and mass, respectively.

(iii) For $I_W>10^{10}$ Wcm^{-2}, the plasma is heated up to $T<10^2$ eV. According to the high temperature, the plasma strongly radiates in the UV and soft X-ray spectral range ionizing thus the surrounding gas. Once preionized, the gas in the zone illuminated by the laser beam absorbs the laser radiation and a laser sustained radiation wave propagates in the direction opposite to the laser beam. The dominating propagation mechanism of the absorption wave depends on the experimental conditions. Breakdown waves are formed in the case of very small opening angle of the focused laser beam whereas radiation waves occur at very high laser power density. For moderate power density and sufficiently wide opening angle, the optical breakdown propagates as a detonation wave. A portion of the laser pulse energy is absorbed by the expanding plasma generating three different types of waves: (i) laser-supported combustion (LSC) waves; (ii) laser-supported detonation (LSD) waves; and (iii) laser-supported radiation (LSR) waves [40]. They differ in their predictions of the opacity and energy transfer properties of the plasma to the surrounding gas. At low-power laser regime (I_W<1 MW/cm^2), LSC waves are produced, which comprise of a precursor shock, that is separated from the absorption zone and the plasma. The shock wave results in an increase in the ambient gas density, temperature and pressure, whereas the shock edges remain transparent to the laser light. At medium-power laser regime (1 MW/cm^2<I_W<4 GW/cm^2), the precursor shock is sufficiently strong and the shocked gas is hot enough to begin absorbing the laser radiation without requiring additional heating by energy from the plasma. The laser absorption zone follows directly behind the shock wave and moves at the same velocity. In this case a LSD wave is produced and has been modelled by several Raizer [2-3, 23]. The propagation of the LSD wave is controlled by the absorption of the laser energy. At high-power laser regime (I_W >4 GW/cm^2), the plasma is so hot that, prior to the arrival of the shock wave, the gas it heated to temperatures at which laser absorption begins. Laser radiation is initiated without any density change and the pressure profile results mainly from the strong local heating of the gas rather than a propagating shock wave. The LSR wave velocity increases much more rapidly with irradiance than those of the LSC and LSD waves.

3. LIB Plasma Analysis

In contrast to conventional spectroscopy, where one is mainly concerned with the structure of an isolated atom and molecule, the radiation from the plasma also depends on the properties of the plasma in the intermediate environment of the atomic or molecular radiator. This dependence is a consequence of the long-range Coulomb potential effects which dominate the interactions of ions and electrons with each other and with existing neutral particles. These interactions are reflected in the characteristic radiations in several ways. They can control population densities of the discrete atomic states, spectral shift and broadening by Stark effect, decrease of ionization potentials of the atomic species, cause continuum radiation emissions and emission of normally forbidden lines. Generally, the radiation emitted from self-luminous plasma can be divided into bound-bound (b-b), bound-free (b-f), and free-free (f-f) transitions.

3.1. Local Thermodynamic Equilibrium (LTE)

Plasma description starts by trying to characterize properties of the assembly of atoms, molecules, ions and electrons rather than individual species. If thermodynamic equilibrium exits, then plasma properties can be described through the concept of temperature. Thermodynamic equilibrium is rarely complete, so physicists have settled for a useful approximation, local thermodynamic equilibrium (LTE). In LTE model it is assumed that the distribution of population densities of the electrons is determined exclusively through collisional processes and that they have sufficient rate constants so that the distribution responds instantaneously to any change in the plasma conditions.

In such circumstances each process is accompanied by its inverse and these pairs of processes occur at equal rates by the principle of detailed balance. Thus, the distribution of population densities of the electrons energy levels is the same as it would be in a system in complete thermodynamic equilibrium. The population distribution is determined by the statistical mechanical law of equipartition among energy levels and does not require knowledge of atomic cross sections for its calculation. Thus, although the plasma density and temperature may vary in space and time, the distribution of population densities at any instant and point in space depends entirely on local values of density, temperature, and chemical composition of plasma. If the free electrons are distributed among the energy levels available for them, their velocities have a Maxwellian distribution

$$dn_{\mathrm{v}} = n_e 4\pi \left(\frac{m}{2\pi k_B T_e} \right)^{3/2} \exp\left(-\frac{m\,\mathrm{v}^2}{2k_B T_e} \right) \mathrm{v}^2 d\mathrm{v} , \tag{3.1}$$

where n_e is the electron density, m is the electron mass, k_B is the Boltzmann constant, T_e is the electron temperature and v is the electron velocity. For the bound levels the distributions of population densities of neutrals and ions are given by the Boltzmann (3.2) and Saha (3.3) equations

$$\frac{N_j}{N_i} = \frac{g_j}{g_i} \exp\left(-\frac{(E_j - E_i)}{k_B T_e} \right), \tag{3.2}$$

$$\frac{N_{z+1,k} n_e}{N_{z,k}} = \frac{g_{z+1,k}}{g_{z,k}} 2 \left(\frac{2\pi\, m\, k_B\, T_e}{h^2} \right)^{3/2} \exp\left(-\frac{Ip_{z,k}}{k_B T_e} \right), \tag{3.3}$$

where N_i, N_j, $N_{z+1,k}$ and $N_{z,k}$ are the population densities of various levels designated by their quantum numbers j (upper), i (lower) and k (the last for the ground level) and ionic charge z and z+1. The term $g_{z,i}$ is the statistical weight of the designated level, E_j and E_i are the energy of the levels j and i and $Ip_{z,k}$ is the ionization potential of the ion of charge z in its ground level k. Equations (3.1)-(3.3) describe the state of the electrons in an LTE plasma.

For complete LTE of the populations of all levels, including the ground state, a necessary condition is that electron collisional rates for a given transition exceed the corresponding radiative rates by about an order of magnitude [41]. This condition gives a criterion [42] for the critical electron density of the level with energy $\Delta E = E_j - E_i$

$$n_e^{\mathrm{crit}} \geq \frac{5}{8\sqrt{\pi}} \left(\frac{\alpha}{a_0} \right)^3 z^7 \left(\frac{\Delta E}{z^2 E_H} \right)^3 \sqrt{\left(\frac{k_B T_e}{z^2 E_H} \right)} \cong 1.6 \times 10^{12} T_e^{1/2} (\Delta E)^3, \tag{3.4}$$

where α is fine-structure parameter, a_0 is Bohr radius, and E_H is the hydrogen ionization potential. In the numerical relationship of Equation (3.4), n_e^{crit} is given in cm^{-3}, T_e in K and ΔE (energy difference between the two neighboring states) in eV. Many plasmas of particular interest do not come close to complete LTE, but can be considered to be only in partial thermodynamic equilibrium in the sense that the population of sufficiently highly excited levels are related to the next ion's ground state population by Saha-Boltzmann relations, respective to the total population in all fine-structure levels of the ground state configuration [41]

For any atom or ion with simple Rydberg level structure, various criteria were advanced for the minimum principal quantum number n_{crit} for the lowest level, often called thermal or collision limit, for which partial thermodynamic equilibrium remains valid to within 10%. One criterion with quite general validity is given by Griem [42]:

$$n_{crit} \approx \left[\frac{10z^7}{2\sqrt{\pi n_e}} \left(\frac{\alpha}{a_0} \right)^3 \right]^{2/17} \left(\frac{k_B T_e}{z^2 E_H} \right)^{1/17} . \tag{3.5}$$

3.2. Line Radiation

Line radiation from plasma occurs for electron transitions between the discrete or bound energy levels in atoms, molecules or ions. In an optically thin plasma of length l along the line of sight [43], the integrated emission intensity I_{ji} of a spectral line arising from a transition between bound levels j and i is given by

$$I_{ji} = \frac{A_{ji} h\nu_{ji}}{4\pi} \int N_j ds = h\nu_{ji} A_{ji} N_j l , \tag{3.6}$$

where N_j is the population density of the upper level j, $h\nu_{ji}$ is the photon energy (energy difference between levels j and i), and A_{ji} is the spontaneous transition probability or Einstein A coefficient.

The integration is taken over a depth of plasma viewed by the detector, and the intensity of radiation is measured in units of power per unit area per unit solid angle. Transition probabilities can be sometimes expressed via the oscillator strength f_{ji}. This is defined as the ratio of the number of classical oscillators to the number of lower state atoms required to give the same line-integrated absorption [44]. Its relationship to the Einstein coefficient is

$$f_{ji} = \frac{4\pi\varepsilon_0}{e^2} \frac{mc^3}{8\pi^2 \nu_{ji}} \frac{g_j}{g_i} A_{ji} . \tag{3.7}$$

The usefulness of f_{ji} is that it is dimensionless, describing just the relative strength of the transition. The detailed values of A_{ji} , g_i, and g_j can be obtained from reference compilations or from electronic databases, i.e by NIST [45].

3.3. Continuum Radiation

The origins of continuum radiation are both bound-free and free-free transitions. Free-free emission or IB radiation is due to the interaction of free electrons with positively charged ions. In free-bound emission (recombination radiation), a free electron is captured by an ion in a bound level. The energy of the photon given off is the difference between original energy of the electron and its new energy in whatever level of whatever atom it ends up in. Since this difference can have any value, the result of many free-bound transitions is a continuous spectrum. Transitions between two free energy levels can occur in plasmas increasing the energy exchanges of charged particles.

Classically, this takes place because a moving charge radiates when it is accelerated or retarded. For most cases of practical importance, these free-free transitions are classified as bremsstrahlung or cyclotron spectra. In bremsstrahlung, the acceleration of charged particle takes place via the Coulomb field of charged particles. In cyclotron radiation, the acceleration is due to the gyration of charged particles in a magnetic field. The total continuum radiation at any particular frequency $I(\nu)$ is the sum of the contributions from all such processes having components at the specified frequency. Thus

$$I(\nu)d\nu = \frac{1}{4\pi}\int n_e \sum_i N_i \left[\gamma(i, T_e, \nu) + \sum_p \alpha(i, j, T_e, \nu) \right] h\nu \; ds \; d\nu \quad , \tag{3.8}$$

where $\gamma(i, T_e, \nu)$ is the atomic probability of a photon of frequency ν being produced in the field of an atom or ion (specified by i) by an electron of mean kinetic temperature T_e making free-free transition; $\alpha(i, j, T_e, \nu)$ is the corresponding probability where the electron makes a free-bound transition into a level j. As before, the integration is taken over the plasma depth s.

3.4. Line Broadening; Determination of Electron Number Density from Stark Broadening of Spectral Lines

The shape of the spectral lines in the LIB has been studied since the first observation of the laser-induced breakdown in early 1960s. It plays an important role for the spectrochemical analysis and quantification of the plasma parameters. The observed spectral lines are always broadened, partly due to the finite resolution of the spectrometers and partly to intrinsic physical causes.

In addition, the center of the spectral lines may be shifted from its nominal central wavelength. The principal physical causes of spectral line broadening are the Doppler, resonance pressure, and Stark broadening. There are several reasons for this broadening and shift. These reasons may be divided into two broad categories: broadening due to local conditions and broadening due to extended conditions.

Broadening due to local conditions is due to effects which hold in a small region around the emitting element, usually small enough to assure LTE. Broadening due to extended conditions may result from changes to the spectral distribution of the radiation as it traverses its path to the observer. It also may result from the combining of radiation from a number of regions which are far from each other.

3.4.1. Natural Broadening

The uncertainty relates the lifetime of an excited state (due to the spontaneous radiative decay) with the uncertainty of its energy. This broadening effect results in an unshifted Lorentzian profile. The FWHM of natural broadening for a transition with a natural lifetime of τ_{ji} is: $\Delta\lambda^{N}{}_{FWHM}=\lambda^2/\pi c\tau_{ji}$. The natural lifetime τ_{ji} is dependent on the probability of spontaneous decay: $\tau_{ji}=1/A_{ji}$. Natural broadening is usually very small compared with other causes of broadening.

3.4.2. Doppler

The Doppler broadening is due to the thermal motion of the emitting atoms, molecules or ions. The atoms in a gas which are emitting radiation will have a distribution of velocities. Each photon emitted will be "red" or "blue" shifted by the Doppler effect depending on the velocity of the atom relative to the observer. The higher the temperature of the gas, the wider the distribution of velocities in the gas. Since the spectral line is a combination of all of the emitted radiation, the higher the temperature of the gas, the broader will be the spectral line emitted from that gas. This broadening effect is described by a Gaussian and there is no associated shift. For a Maxwellian velocity distribution the line shape is Gaussian, and the FWHM may be estimated as (in Å):

$$\Delta\lambda^{D}_{FWHM} = 7.16\times 10^{-7}\cdot\lambda\cdot\sqrt{T/M}\,, \tag{3.9}$$

being λ the wavelength in Å, T the temperature of the emitters in K, and M the atomic mass in amu.

3.4.3. Pressure Broadening

The presence of nearby particles will affect the radiation emitted by an individual particle. There are two limiting cases by which this occurs: (i) Impact pressure broadening: The collision of other particles with the emitting particle interrupts the emission process. The duration of the collision is much shorter than the lifetime of the emission process. This effect depends on both the density and the temperature of the gas. The broadening effect is described by a Lorentzian profile and there may be an associated shift. (ii) Quasistatic pressure broadening: The presence of other particles shifts the energy levels in the emitting particle, thereby altering the frequency of the emitted radiation. The duration of the influence is much longer than the lifetime of the emission process. This effect depends on the density of the gas, but is rather insensitive to temperature. The form of the line profile is determined by the functional form of the perturbing force with respect to distance from the perturbing particle. There may also be a shift in the line center. Pressure broadening may also be classified by the nature of the perturbing force as follows: (i) *Linear Stark broadening* occurs via the linear which results from the interaction of an emitter with an electric field, which causes a shift in energy which is linear in the field strength (~E and ~$1/r^2$); (ii) *Resonance broadening* occurs when the perturbing particle is of the same type as the emitting particle, which introduces the possibility of an energy exchange process (~E and ~$1/r^3$); (iii) *Quadratic Stark broadening* occurs via the quadratic Stark effect which results from the interaction of an emitter with an electric field, which causes a shift in energy which is quadratic in the field strength (~E and ~$1/r^4$); (iv) *Van der Waals broadening* occurs when the emitting particle is being perturbed by Van der Waals forces. For the quasistatic case, a Van der Waals profile is often useful in describing the profile. The energy shift as a function of distance is given in the wings by e.g. the Lennard-Jones potential (~E and ~$1/r^6$).

3.4.4. Stark Broadening

Stark broadening of spectral lines in the plasma occurs when an emitting species at a distance r from an ion or electron is perturbed by the electric field. This interaction is described by the Stark effect. The linear Stark effect exists for hydrogen and for all other atoms. Stark broadening from collisions of charged species is the primary mechanism influencing the emission spectra in LIBS.

Stark broadening of well-isolated lines in the plasma can be used to determine the electron number density n_e(cm^{-3}). In the case of a non-H-like line, an estimation of the Stark width (FWHM) and line shift of the Stark broadened lines is given as [41-44]:

$$\Delta\lambda^{\mathrm{Stark}}_{\mathrm{FWHM}} = 2W\left(\frac{n_e}{10^{16}}\right) + 3.5A\left(\frac{n_e}{10^{16}}\right)^{1/4}\left(1 - BN_D^{-1/3}\right) W\left(\frac{n_e}{10^{16}}\right), \tag{3.10}$$

$$\Delta\lambda^{\mathrm{Shift}} = D\left(\frac{n_e}{10^{16}}\right) \pm 2A\left(\frac{n_e}{10^{16}}\right)^{1/4}\left(1 - BN_D^{-1/3}\right) W\left(\frac{n_e}{10^{16}}\right), \tag{3.11}$$

where W is the electron impact parameter or half-width, A is the ion impact parameter both in Å, B is a coefficient equal to 1.2 or 0.75 for ionic or neutral lines, respectively, D (in Å) is the electron shift parameter and N_D is the number of particles in the Debye sphere $N_D=1.72\mathrm{x}10^9 T^{3/2} n_e^{-1/2}$. The electron and the ion impact parameters are functions of temperature.

The first term on the right side of Equation (3.10) refers to the broadening due to the electron contribution, whereas the second one is the ion broadening. The minus sign in Equation (3.11) applies to the high-temperature range of those few lines that have a negative value of D/W at low temperatures. Since for LIB conditions Stark broadening is predominantly by electron impact, the ion correction factor can safely be neglected, and Equation (3.10) becomes

$$\Delta\lambda^{\mathrm{Stark}}_{\mathrm{FWHM}} \approx 2W\left(\frac{n_e}{10^{16}}\right). \tag{3.12}$$

The coefficients W are independent of n_e and slowly varying functions of electron temperature. A comprehensive list of width and shift parameters W, A and D is given by Griem [42].

In the quasi-static approximation, the interaction between slowly moving ions and radiating species can be approximated by a perturbation which remains nearly constant over the whole time that the species is radiating. Hydrogen and hydrogen-like ions exhibits linear Stark effect. The FWHM (in Å) of a hydrogen or H-like ion spectral line, in the quasi-static approximation, is given by [41, 42]

$$\Delta\lambda_{\mathrm{FWHM}}^{\mathrm{Stark}} = 8.16\times10^{-19}\left(1-0.7N_D^{-1/3}\right)\lambda_0^2(n_2^2-n_1^2)(Z_p^{1/3}-Z_e)n_e^{2/3}, \tag{3.13}$$

where λ_0 is the wavelength line centre, n_2 and n_1 are the principal quantum numbers of the upper and lower states, respectively, Z_p and Z_e are the nuclear charge on the perturbing ion and the emitting species (atom or ion) and n_e is the electron number density in cm^{-3}. Although the line shapes do depend on the electron contribution, the FWHM are generally insensitive. Equation (3.13) represents a very good estimate of the Stark broadening in those hydrogenic lines that do not have a strong undisplaced Stark component as for example L_β, L_δ, H_β (Balmer) and H_δ transitions. On the other hand, the FWHM of hydrogenic lines with strong central Stark components are dominated by interaction of the electrons with the emitting hydrogenic species such as L_α and H_α transitions. Such lines have a Lorentzian line shape and FWHM for L_α transition in the impact approximation is given by

$$\Delta\lambda_{\mathrm{FWHM}}^{\mathrm{Stark}} \approx 1.62\times10^{-17}\frac{n_e}{\sqrt{T}}\left(13.76-\log\frac{n_e^{1/2}}{T}\right), \tag{3.14}$$

where $\Delta\lambda_{\mathrm{FWHM}}^{\mathrm{Stark}}$ is in Å, T is in K and n_e is in cm^{-3}. It is seen from Equations (3.13) and (3.14) that the ion broadening, in the quasi-static approximation, varies as $n_e^{2/3}$ and is independent of the temperature whereas the collisional broadening varies approximately as n_e and it is very much temperature dependent. It is to be noted the electron densities determined from Equations (3.13) and (3.14) are only crude estimations and one must compute the entire line profile to extract the total line width for an accurate estimation of n_e.

3.5. Determination of Excitation, Vibrational and Rotational Temperatures

The excitation temperature T_{exc} can be calculated according to the Boltzmann equation under the assumption of LTE (Section 3.1). The significance of this temperature depends on the degree of equilibrium within the plasma. For plasma in LTE, any point can be described by its local values of temperature, density, and chemical composition. By considering two lines λ_{ji} and λ_{nm} of the same species, characterized by different values of the upper energy level ($E_j \neq E_n$), the relative intensity ratio can be used to calculate the plasma excitation temperature

$$T_{\mathrm{exc}} = \frac{E_j - E_n}{k_B \ln\left[\frac{I_{nm}\cdot\lambda_{nm}\cdot g_j\cdot A_{ji}}{I_{ji}\cdot\lambda_{ji}\cdot g_n\cdot A_{nm}}\right]}. \tag{3.15}$$

When selecting a line pair, it is advisable to choose two lines as close as possible in wavelength and as far apart as possible in excitation energy. This is to limit the effect of varying the spectral response of the detection system. The use of several lines instead of just one pair leads to greater precision of the plasma excitation temperature estimation. In fact, though the precision of the intensity values can be improved by increasing the signal intensity, the transition probabilities A_{ji} reported in the literature exhibit significance degree of uncertainty (5-50%).

The excitation temperature can be calculated from the relative intensities of a series of lines from different excitation states of the same atomic or ionic species from the slope of the Bolztmann plot $\ln[I_{ji}\cdot\lambda_{ji}/g_j\cdot A_{ji}]$ versus E_j/k_B

$$\ln\left[\frac{I_{ji}\cdot\lambda_{ji}}{g_j\cdot A_{ji}}\right]=C-\frac{E_j}{k_B\cdot T_{exc}}, \tag{3.16}$$

where I_{ji} is the emissivity (W m^{-3} sr^{-1}) of the emitted $j\rightarrow i$ spectral line, λ_{ji} is the wavelength, $g_j=2J_j+1$ is the statistical weight, A_{ji} is the Einstein transition probability of spontaneous emission, E_j/k_B is the normalized energy of the upper electronic level and $C=\ln(hcN_j/4\pi Q(T))$ ($Q(T)$ is the partition function). The values of the λ_{ji}, g_j, A_{ji} and E_i for selected atomic or ionic lines can be obtained from the NIST Atomic Spectral Database [45]. A set of selected spectral lines can be chosen based on their relative strengths, accuracies and transition probabilities.

The emission spectra of the diatomic species reveal a relatively complex structure which is due to the combination of the electronic transitions from the different rotational and vibrational states [46-48]. The emission intensities of the molecular bands can be analyzed in order to calculate the molecular vibrational temperature T_{vib}. For a plasma in LTE, the intensity of an individual vibrational v'-v" band $I_{v'-v"}$ is given by

$$\ln\left(\frac{I_{v'-v"}\cdot\lambda^4_{v'-v"}}{q_{v'-v"}}\right)=A-\frac{G(v')hc}{k_B\cdot T_{vib}}, \tag{3.17}$$

where A is a constant, $\lambda_{v'-v"}$ is the wavelength corresponding to the band head, $q_{v'-v"}=\left|\int_0^\infty \Psi_{v'}(R)\Psi_{v"}(R)dR\right|^2$ is the Franck-Condon factor and $G(v')hc/k_B$ is the normalized energy of the upper vibrational level. A line fit to $\ln\left(I_{v'-v"}\cdot\lambda^4_{v'-v"}/q_{v'-v"}\right)$ as a function of the upper normalized electronic-vibrational energies has a slope equal to $-1/T_{vib}$.

On the other hand, the emission intensities of the rotational lines of a vibrational band can be analyzed in order to estimate the effective rotational temperature T_{rot}.

In this case it is necessary to consider the Hund´s coupling case for the both electronic states implied in the transition.

From the assignment of the rotational spectrum it is possible to estimate the effective rotational temperature by considering the J value for the maximum of the band $T_{rot}=(2B_v h c/k_B)(J_{max}+1/2)^2$, being B_v the rotational constant for v' vibrational level and J_{max} the total angular momentum at the maximum.

Another method for estimating the vibrational and rotational temperatures is based on a simulation program of the spectra. Software developed in our laboratory [49] calculated the spectra of a diatomic molecule by summing the intensity of all rovibrational levels and convoluting the results with the instrumental line shape of the optical system. The emission intensity $I_{v',J'-v",J"}$ of a molecular line can be approximated by

$$I_{v'J'-v''J''} \approx \frac{64\pi^4 \tilde{\nu}^4_{v'J'-v''J''}}{3(2J'+1)} N_{v'J'} \overline{R}_e^{\,2} q_{v',v''} S_{J'J''} \quad , \qquad (3.18)$$

where $\tilde{\nu}_{v'J'-v''J''}$ is the wavenumber of the transition, $2J'+1$ is the rotational degeneracy of the upper state, $N_{v',J'}$ is the population in the initial (upper) state, $\overline{R}_e$ is the average electronic transition moment, $q_{v',v''}$ is the Franck-Condon factor and $S_{J',J''}$ is the Hönl-London factor [50]. Spectrum simulations are based on comparison of experimental and calculated spectra for different rotational and vibrational population distributions which depend on temperature.

3.6. Ionization Degree of the Plasmas: Saha Equation

In plasma there is a continuous transition from gases with neutral atoms to a state with ionized atoms, which is determined by an ionization equation. The transition between gas and plasma is essentially a chemical equilibrium, which shifts from the gas to the plasma side with increasing temperature. Let us consider the first three different ionization equilibria of an atom A:

$A \leftrightarrow A^+ + e + IP(A\text{-}I)$,

$A^+ \leftrightarrow A^{2+} + e + IP(A\text{-}II)$,

$A^{2+} \leftrightarrow A^{3+} + e + IP(A\text{-}III)$.

For each ionization equilibrium, considering the atoms and ions in their ground electronic state, the LTE between ionization and recombination reactions at temperature T is described by the Saha equation (see Equation 3.3)

$$\frac{n_e \cdot N_i}{N_0} = \frac{g_e \cdot g_i}{g_0} \frac{(2\pi\ m\ k_B T)^{3/2}}{h^3} e^{-I_P/k_B T} \quad , \qquad (3.19)$$

where $n_e = N_i$ are the electron and ion densities in the different ionization equilibria in the second member of ionization equilibria.From this equation, ionization degree $n_e{\cdot}N_i/N_0$ can be estimated.

4. Experimental Details

LIBS is a plasma based method that uses instrumentation similar to that used by other spectroscopic methods (atomic emission spectroscopy, laser-induced fluorescence etc). A typical LIBS apparatus utilizes a pulsed laser that generates the powerful optical pulses used to form the plasma. Principles of laser operation in general and the operation of specific lasers are described in detail in numerous books. The discussion here will be limited to the fundamentals of the operation of the transversely excited atmospheric (TEA) carbon dioxide laser used in this work. The CO_2 laser is a near-infrared gas laser capable of very high power and with the highest efficiency of all gas lasers (≈10-20%) and for cw operation the highest output power. Although CO_2 lasers have found many applications including surgical procedure, their popular image is as powerful devices for cutting, drilling, welding or as weapons for military applications. The linear CO_2 molecule has three normal modes of vibration, labelled v_1 (symmetry stretch), v_2 (bending vibration) and v_3 (asymmetric stretch). The fundamental vibration wavenumbers are 1354, 673 and 2396 cm^{-1}, respectively. The vibrational state of the molecule is described by the number of vibrational quanta in these modes. The bending vibrational mode is twofold degenerate and can have a vibrational angular momentum along the CO_2 axis. The number of quanta of this vibrational angular momentum is stated as an upper index to the vibrational v_2 quanta. The upper laser level (00^01) denotes the ground vibrational state for the mode v_1, the ground vibrational state for the mode v_2 which is doubly degenerate, and the first excited vibrational state for the mode v_3. The active medium is a gas discharge in a mixture of He, N_2 and CO_2. By electron impact in the discharge excited vibrational levels in the electronic ground states of N_2 and CO_2 are populated. The vibrational levels v = 1 in the N_2 molecule and (v_1, v_2, v_3) = (00^01) in the CO_2 molecule are near-resonant and energy transfer from the N_2 molecule to the CO_2 molecule becomes very efficient. This populates the (00^01) level in CO_2 preferentially, creates inversion between the (00^01) and the (02^00) levels, and allows laser oscillations on many rotational transitions between these two vibrational states in the wavelength range 9.6-10.6 μm. The main laser transitions in CO_2 occur between the excited states of the mode $v_3(00^01)$ and the symmetric stretching mode $v_1(10^00)$ (10.6 μm) or the bending mode $v_2(01^10)$ (9.6 μm). A single line can be selected by a Littrow-grating, forming one of the resonator end mirrors. Helium atoms do not take part directly in the excitation of CO_2 molecules but do play an important role in heat-transfer from the gas mixture to the tube walls, as well as facilitating the depopulation of the lower vibrational levels in CO_2, contributing in this way to the maintenance of the population inversion. The power of CO_2 lasers depends on their configuration. The laser used in these experiments was a transversely excited atmospheric (TEA) CO_2 laser in which the gas-flow is transverse to the laser cavity's axis. The pressure in the tube is close to atmospheric pressure. The CO_2:N_2:He mixture is exchange in a continuous way, enhancing the output power of the laser. The laser can achieve a power of 50 MW. The optical materials used in lasers emitting radiation in the infrared range are obviously different than those used in the visible range. For example, materials such as germanium (Ge) or gallium arsenide (GaAs) are completely opaque in the visible range, while being transparent in the infrared range. Some materials, such as zinc selenide (ZnSe), are transparent in both spectral ranges. Typical materials transparent in the IR range are: NaCl or CsI. Metal mirrors (copper, molybdenum) are used in the IR range, owing to their small absorption (and large reflectivity) as well as their large heat capacity which enables removal of heat from the active medium. A schematic diagram of the experimental configuration used for time-resolved TEA-CO_2 pulsed laser gas breakdown diagnostics is shown in Figure 1. The experiments were carried out with a transverse excitation atmospheric (TEA) CO_2 laser (Lumonics model K-103) operating on an 8:8:84 mixture of CO_2:N_2:He, respectively. The laser is equipped with frontal Ge multimode optics (35 % reflectivity) and a rear diffraction grating with 135 lines mm^{-1} blazed at 10.6 μm. The laser pulse repetition rate was usually 1 Hz. The divergence of the emitted laser beam is 3 mrad. The laser delivered up to 3.16 J at a wavelength of 10.591

μm, leading to an estimated power of 49.5 MW (Equation 1.1), intensity (power density or irradiance) of 6.31 GW×cm^{-2} (Equation 1.2), fluence of 403 J×cm^{-2} (Equation 1.3), photon flux of 3.1x10^{29} photonxcm^{-2}xs^{-1} (Equation 1.4), electric field of 1.54 MV×cm^{-1} (Equation 1.5) and radiation pressure of 421 kPa (Equation 1.6) on the focal position. The focused-spot area (7.85×10^{-3} cm^2) of the laser beam was measured with a pyroelectric array detector (Delta Development Mark IV). The temporal shape of the TEA-CO_2 laser pulse, monitored with a photon drag detector (Rofin Sinar 7415), consisted in a prominent spike of a FWHM of around 64 ns carrying ~90% of the laser energy, followed by a long lasting tail of lower energy and about 3 μs duration. The primary laser beam was angularly defined and attenuated by a diaphragm of 17.5 mm diameter before entering to the gas cell. A beam splitter was used to redirect 10% of the laser pulse energy on a pyroelectric detector (Lumonics 20D) or on a photon-drag detector (Rofin Sinar 7415) for energy and temporal shape monitoring and triggering, respectively, through a digital oscilloscope (Tektronix TDS 540). The laser-pulse energy was varied with the aid of several calibrated CaF_2 attenuating plates. The shot-to-shot fluctuation of the laser energy was approximately 5%. In time-resolved gas breakdown, the pulsed laser light was focused by a NaCl lens of 24 cm focal lens onto the surface of a 0.7 mmx0.7 mm stainless steel mesh in gas at atmospheric pressure. This allows us to fix the focal position for LIB at any fluence inducing strong gas breakdown plasma.

No lines from metals were found in the spectra,meaning that the metal mesh was practically never ablated. The high purity gases (~99.99 %) were placed in a medium-vacuum cell equipped with a NaCl window for the laser beam and two quartz windows for optical access. The gas is initially at ambient temperature (298 K). The cell was evacuated with the aid of a rotary pump, to a base pressure of 4 Pa that was measured by a mechanical gauge. Optical emission from the plume was imaged by a collecting optical system onto the entrance slit of different spectrometers.

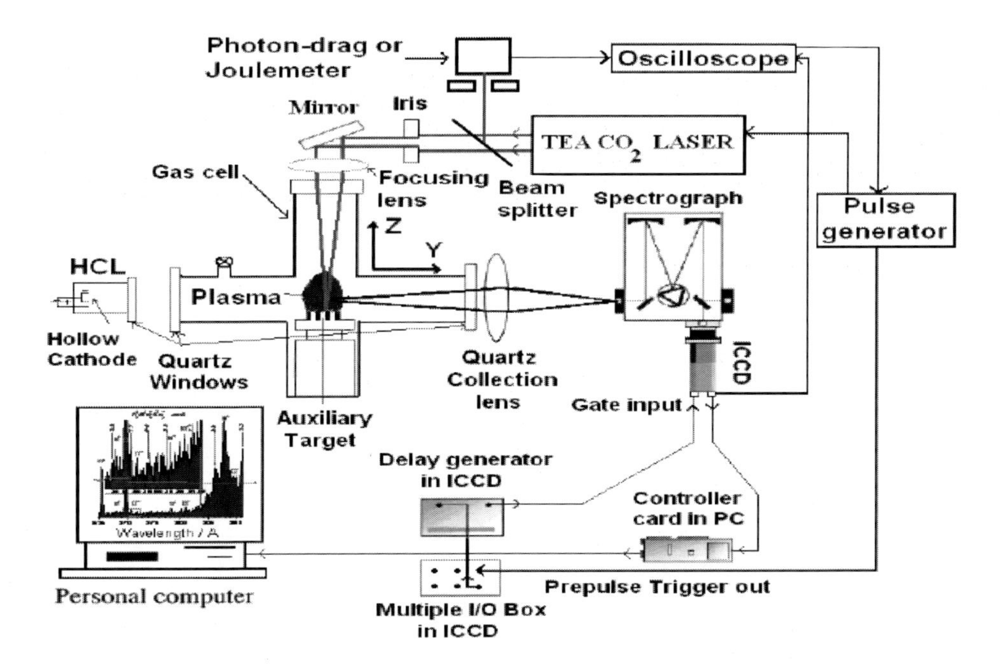

Figure 1. Schematic diagram of the experimental set-up used for time-resolved TEA-CO_2 pulsed laser gas breakdown diagnostics.

The light emitted from the laser-induced plasma was optically imaged 1:1, at right angles to the normal to the focal volume, by a quartz lens (focal length 4 cm, f-number = f/2.3) onto the entrance slit of the spectrometer. The distance between gas plasma axis and entrance slit was typically y=16 cm. Optical emission accompanying the laser-induced gas plasma was viewed in a *XZ* parallel plane to the front face of the metal mesh for different distances *z* along the plasma Y axis. Two spectrometers were used: 1/8 m Oriel spectrometer (10 and 25 μm slits) with two different gratings (1200 and 2400 grooves mm-1) in the spectral region 2000-11000 Å at a resolution of ~1.3 Å in first-order (1200 grooves mm-1 grating), and an ISA Jobin Yvon Spex (Model HR320) 0.32 m equipped with a plane holographic grating (2400 grooves mm^{-1}) in the spectral region 2000-7500 Å at a resolution of ~0.12 Å in first-order.

Figure 2. (Continued on next page).

Figure 2. Four images of the LIB of nitrogen at atmospheric pressure (λ=9.621 μm and power density of 4.5 GW×cm^{-2}), oxygen at 49.0 kPa (λ=10.591 μm and power density of 1.93 GW×cm^{-2}), air at atmospheric pressure (λ=9.621 μm and intensity of 5.86 GW×cm^{-2}) and He at atmospheric pressure (λ=10.591 μm and power density of 6.31 GW×cm^{-2}) at different times of the experiment induced by a TEA-CO_2 laser pulse. Laser beam direction is from right to left.

Two detectors were attached to the exit focal plane of the spectrographs and used to detect the optical emissions from the laser-induced plasma: an Andor DU420-OE (open electrode) CCD camera (1024x256 matrix of 26×26 μm^2 individual pixels) with thermoelectric cooling working at –30 ºC; A 1024×1024 matrix of 13x13 μm^2 individual pixels ICCD (Andor iStar DH-734), with thermoelectric cooling working at –20 ºC. The low noise level of the CCD allows long integration times and therefore the detection of very low emission intensities.

The spectral window in high-resolution experiments was about 12 nm. The intensity response of the detection systems was calibrated with a standard (Osram No.6438, 6.6-A, 200-W) halogen lamp and Hg/Ar pencil lamp. Several (Cu/Ne, Fe/Ne and Cr/Ar) hollow cathode lamps (HCL) were used for the spectral wavelength calibration of the spectrometers.

In time-resolved measurements, for synchronization, the CO_2 laser was operated at the internal trigger mode and ICCD detector was operated in external and gate modes.

The external trigger signal generated by the laser was fed through the scope and delay generator into the back of the ICCD detector head. The total insertion delay (45 ± 2 ns) is the total length of time taken for the external trigger pulse to travel through the digital delay generator and gater so that the ICCD will switch on.

The time jitter between the laser and the fast ICCD detector gate was about ± 2 ns. The delay time t_d is the time interval between the arrival of the laser pulse on the metal mesh and the activation of the ICCD detector. The gate width time t_w is the time interval during which the plasma emission is monitored by the ICCD. Both parameters were adjusted by the digital delay generator of the ICCD detector.

The CO_2 laser pulse picked up with the photon drag detector triggered a pulse generator (Stanford DG 535) through the scope and this pulse was used as external trigger in the ICCD camera.

The laser pulse and the gate monitor output were displayed in a digital oscilloscope. In this way, by using the output of the photon drag detector, the oscilloscope, the delay pulse generator and the gate monitor output of the ICCD camera, the gate width time t_w and the delay time t_d could be adjusted without insertion time.

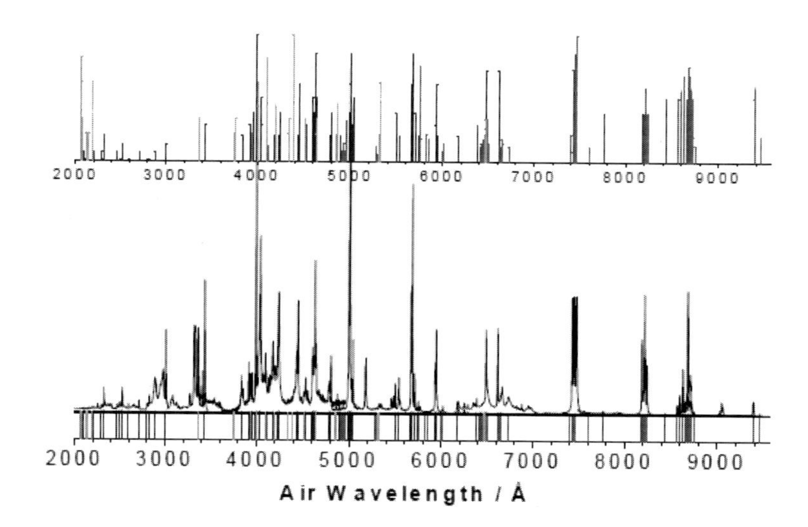

Figure 3. An overview of the LIB emission spectrum of nitrogen at a pressure of 1.2×10^5 Pa, excited by the 9P(28) line at 1039.36 cm^{-1} of the TEA CO_2 laser, compared with atomic lines of N (red), N^+ (blue) and N^{2+}(green).

5. Results and Discussion

When a gas is irradiated by CO_2 laser lines of sufficient power, a visible breakdown occurs. The onset of LIB in air is a sudden dramatic event involving the production of more than ~ 10^{16} electron-ion pairs and the emission of radiation characteristic of the gas-plasma. Figure 2 shows a series of images of the LIB nitrogen (up to the left), oxygen (up to the right), air (below to the left) and helium (below to the right) plasma at different times of the experiment. Although the laser-induced plasma appears spatially uniform to the naked eye, it is indeed elongated along the direction of the incoming carbon dioxide laser beam. For a laser power density around 4.5 GW$\times$cm^{-2}, the laser-induced plasma may be ≈ 6 cm long and a few centimeters in diameter. A number of plasma lobes can be distinguished as well as diffuse, luminous cloud surrounding the central plasma core.

The observations of the LIB geometry during the experiments indicate that the actual plasma region is not spherical, but elongated in the direction of the laser beam propagation. There is an expansion back toward the laser that essentially fills the converging cone of the CO_2 laser radiation. The growth of the plasma in the direction opposite to the laser beam had lead to the model of a radiation-supported detonation wave (Section 2.6). A shock wave propagates from the focal region (a point at the centre of the cell) into the gas and absorption of energy from the laser beam drives the shock wave, causing it to spread. The structure of the LIB plasma is complex, and indeed there may be several distinct plasma regions produced along the laser beam axis. This multiple collinear plasmas in gases at pressures around the atmospheric level are observable by the naked eye. The CO_2 laser pulse remains in the focal volume after the plasma formation for some significant fraction of its duration and the plasma formed can be heated to very high temperatures and pressures by IB absorption. Since plasmas absorb radiation much more strongly than ordinary mater, plasmas can block transmission of incoming laser light to a significant degree; a phenomenon known as plasma shielding [51]. The high temperatures and pressures produced by plasma absorption can lead to thermal expansion of the plasma at high velocities, producing an audible acoustic signature, shock waves, and cavitation effects. The plasma also tends to expand back along the beam path toward the laser, a phenomenon known as moving breakdown. The shock wave heats up the surrounding gas which is instantaneously transformed in strongly ionized plasma.

5.1. LIBS of Nitrogen

In this section we present our results on the large-scale plasma produced in nitrogen gas at room temperature and pressures ranging from 4×10^3 to 1.2×10^5 Pa by high-power TEA-CO_2 LIB plasma [18]. The time-integrated spectrum of the generated plasma is dominated by emission of strong N^+ and N and very weak N^{2+} atomic lines and molecular features of $N_2^+(B^2\Sigma_u^+\text{-}X^2\Sigma_g^+)$, $N_2^+(D^2\Pi_g\text{-}A^2\Pi_u)$, $N_2(C^3\Pi_u\text{-}B^3\Pi_g)$ and very weak $N_2(B^3\Pi_g\text{-}A^3\Sigma_u^+)$. Figure 3 displays an overview of the optical emission spectrum of LIDB in nitrogen (2000–9565 Å) compared with atomic lines of N, N^+ and N^{2+} [45].

Strong atomic N^+ lines dominate the spectrum but, atomic N lines (about 3 times weaker) and very weak N^{2+} lines (about 10^2-10^3 times weaker) also are present. The assignments of the atomic N (mainly in the 5700–9565 Å spectral region), N^+ (2000–6000 Å) and N^{2+}(2000–5500 Å) individual lines are indicated by stick labels. In the upper part of Figure 3 we indicate in a column graph, the relative intensities of atomic observed N, N^+ and N^{2+} lines tabled in NIST Atomic Spectral Database [45]. There is a good agreement between line intensities tabulated in NIST and the measured intensities observed here for N, N^+ and N^{2+}. The spectrum of Figure 3 has been obtained with six successive exposures on the CCD detector using a 1/8 m Oriel spectrometer (1200 grooves/mm grating). In addition to identified atomic lines, molecular bands associated to N_2^+ and N_2 diatomic molecules are observed. The analysis of the molecular emission has already been used for a long time to get information on the structure and symmetry of excited states [46-48].

Studies of the electronic spectra of N_2 and N_2^+ in a number of discharge tubes, such as electrodeless microwave discharges and conventional *ac* and *dc* discharges, are well known. In many electrical discharges, the most prominent electronic transitions of N_2 are the first positive $B^3\Pi_g\text{-}A^3\Sigma^+_u$ system (between 480 and 2530 nm) and the second positive $C^3\Pi_u\text{-}B^3\Pi_g$ system (between 270 and 550 nm) [52]. For the electronic states implied in the electronic transitions of N_2(C-B and B-A) and N_2^+(B-X and D-A) the vibrational quanta $\Delta G(v+½)$ in the upper and lower electronic states have similar magnitudes and therefore the vibrational transitions with $\Delta v=v'\text{-}v''$ constant (sequences) appear quite close. The known part of the C-B second positive system of molecular nitrogen consist of the Δv = 4, 3, 2, 1, 0, -1, -2, -3, -4, -5, -6 and -7 triple-headed band sequence, all degraded to the violet. A sharp cutting-off of the rotational in v'=4 vibrational level of the C state is observed, which Herzberg [53] attributed to a predissociation. Pannetier *et al* [54] observed the 5-5 band of the C-B system with band-head at 3259.2 Å.

Also Tanaka and Jursa [55] studied this band system with high intensity in the aurora afterglow observed for weak red-degraded triplet bands originate from v'=5 ($C^3\Pi_u$). Perturbations of various types in the $C^3\Pi_u$ state of molecular nitrogen were also observed. Moreover, the most prominent electronic transitions of N_2^+ are the first negative $B^2\Sigma_u^+\text{-}X^2\Sigma_g^+$

system (between 280 and 590 nm) and the $A^2\Pi_u$-$X^2\Sigma_g^+$ Meinel system (550 and 1770 nm) [52]. The A-X Meinel bands for nitrogen cation were first identified in the aurora borealis [56]. The analysis of these band systems of N_2 and N_2^+ was accomplished in the early work of many authors and played an important part in the development of our understanding of the spectra of diatomic molecules.

In order to assign the molecular features of the LIDB in nitrogen, its spectrum is compared to that of the *dc* electric glow discharge at low pressure (~5 Torr). Typical spectra recorded with the oriel spectrometer (25 μm slit and grating of 1200 grooves/mm) after CO_2 laser excitation and in the cathode glow discharge of N_2 are given in Figure 4. A global analysis of the latter allows one to distinguish the second positive C-B system (between 270 and 530 nm) and the first positive B-A system (between 570 and 970 nm) of N_2 and very weak emissions corresponding to the first negative B-X system of N_2^+ ions. Moreover, spectroscopic measurements performed on the dc electric glow in N_2 spectrum showed that although numerous molecular bands appear, nitrogen atomic lines are not present. Besides, the second positive system of N_2 and the first negative system of N_2^+ spectra are frequently observed simultaneously in plasma containing nitrogen.

In the glow discharge in N_2 at 5977.4 Å, the 0-0 band of the $c_4{}^1\Pi_u$-$a''{}^1\Sigma_g^+$ Ledbetter Rydberg series [57] of nitrogen is observed overlapped with the 8-4 band sequence of the first positive B-A system. The LIB emission spectrum of N_2 (Figure 4) shows six red-degraded heads in the region 225-275 nm which were readily assigned to the Δv=0 (v=0, 1, ...6) band sequence of the $D^2\Pi_g$-$A^2\Pi_u$ Janin-d'Incan system [52] of N_2^+. In the spectral range between 2700-5300 Å, the second positive system of N_2(C-B) and the first negative system of N_2^+(B-X) spectra are observed simultaneously. Table 1 lists the different molecular species that have been observed in the LIB spectrum of nitrogen.

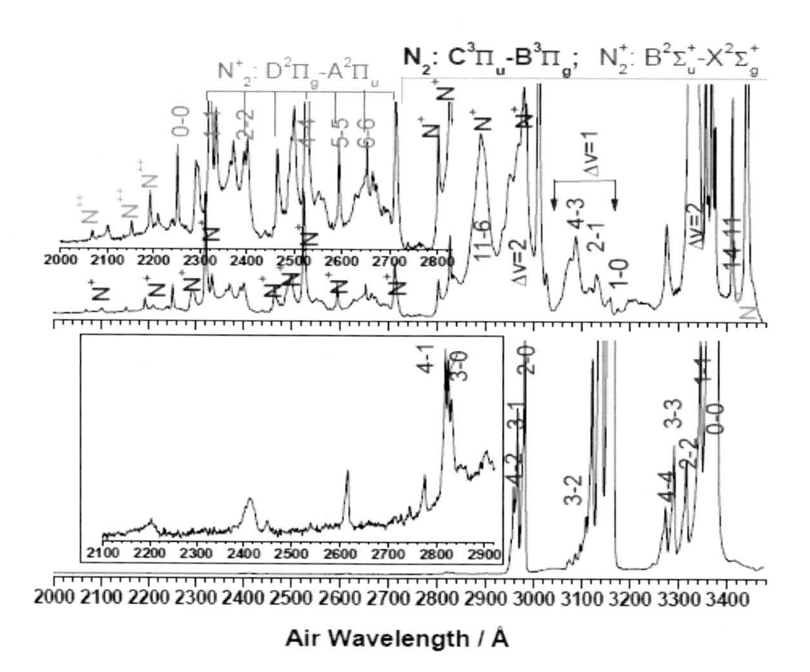

Figure 4. (Continued on next page).

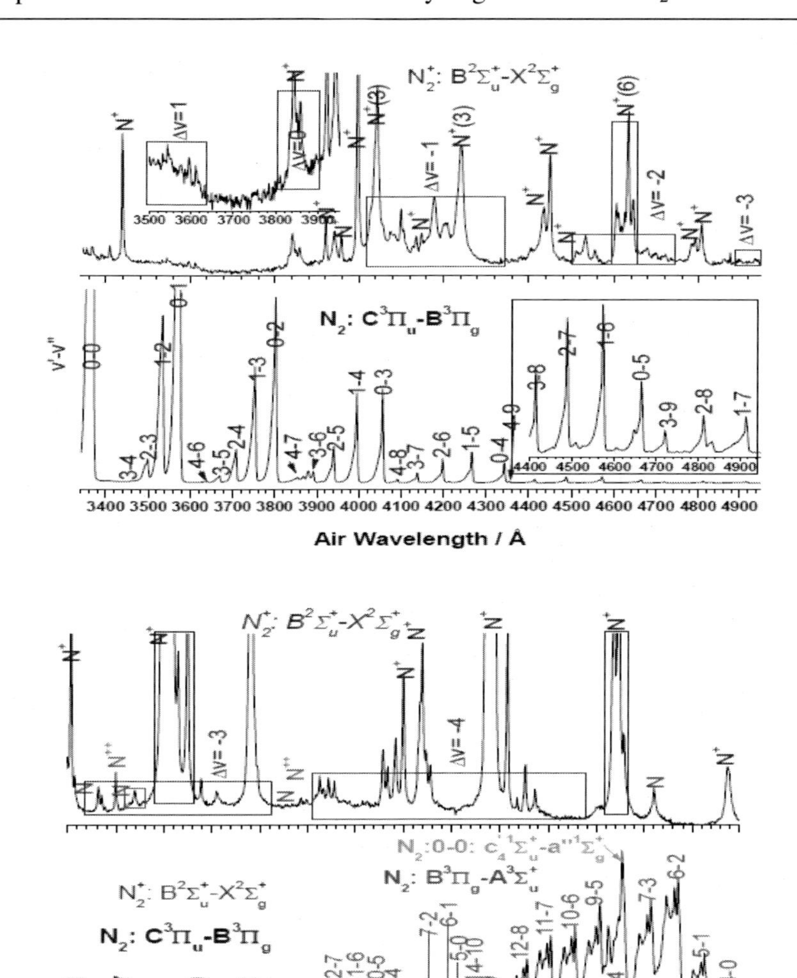

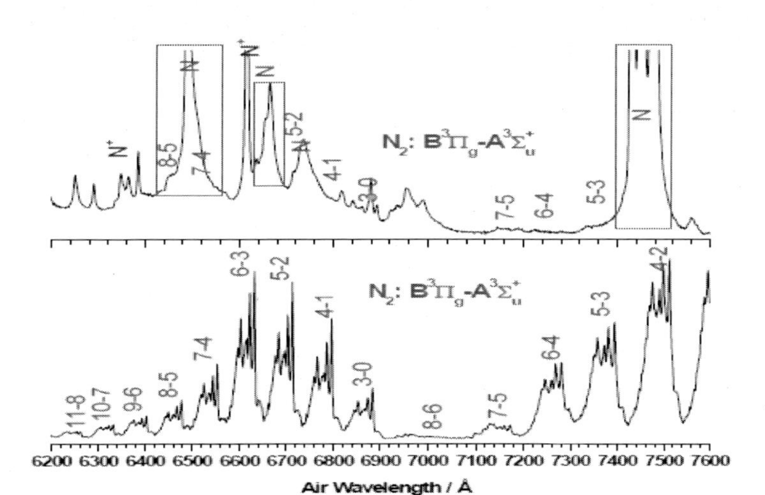

Figure 4. (Continued on next page).

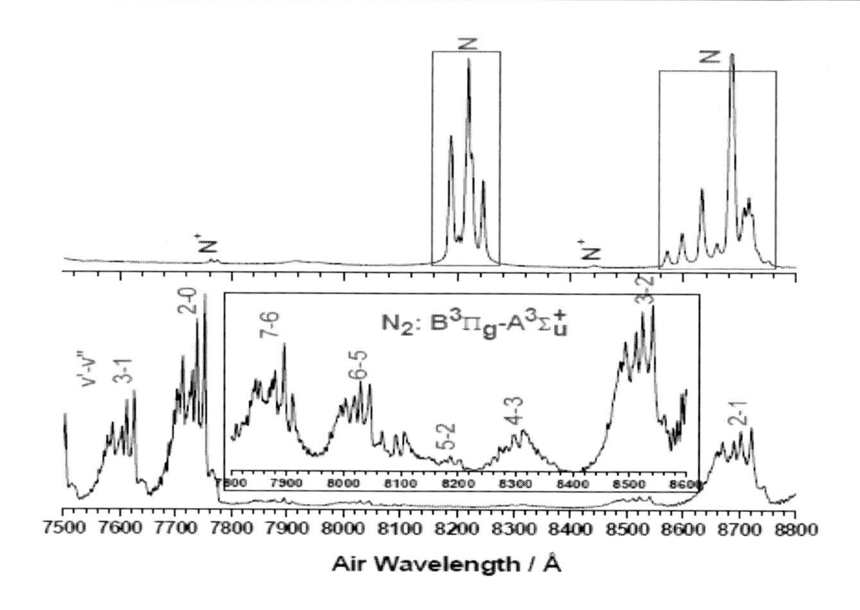

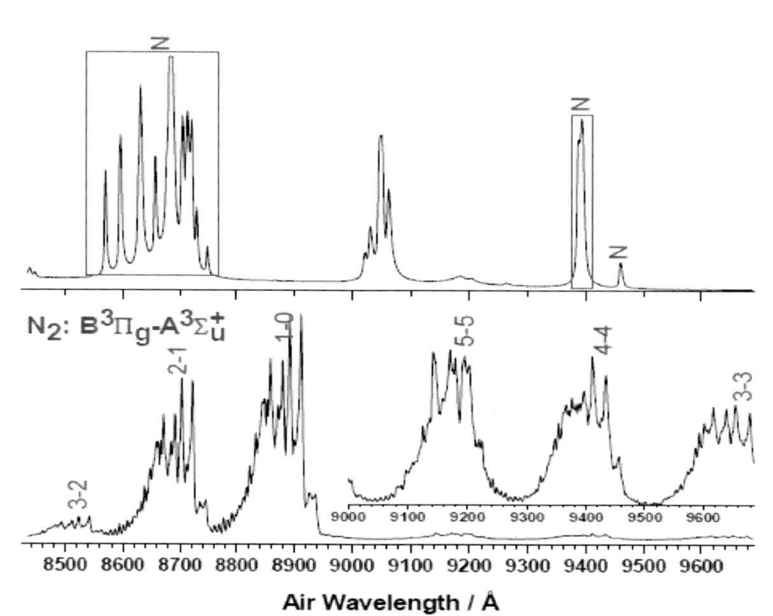

Figure 4. An overview of the low-resolution emission spectra observed in the 2000-9690 Å region. Upper panel: LIDB in nitrogen at a pressure of 1.2×10^5 Pa, excited by the 9P(28) line at 1039.36 cm^{-1} of the CO_2 laser; Lower panel: dc electric glow discharge spectrum of nitrogen at low pressure.

In order to investigate the different electronic bands of N_2 and N_2^+ both LIDB and high-voltage dc electric glow discharge spectra were recorded with a resolution of ~0.12 Å by a ISA Jobin Yvon Spex 0.32 m spectrometer. The high-resolution of LIB spectrum allowed us to resolve partially the vibrational bands of the second positive C-B system of N_2 and the first negative B-X system of N_2^+. Both spectra have been obtained with forty successive exposures on the CCD camera in the spectral region 2000-7500 Å. In the high-resolution spectra, no new processes were detected but allow us to identify unequivocally the band structure of the different transitions. As examples, Figure 5(a)-(f) shows a comparison between two spectra, the lower one obtained for the high-voltage dc electric discharge, and the upper one recorded in the LIB experiment. We indicate with italic the position of the band-head (v'-v") of first negative system of N_2^+(B-X) while in regular typeface the bands of the second positive system of N_2(C-B). The upper panel of Figure 5(a) shows the LIB emission spectrum of nitrogen in the spectral region 2925-3175 Å of nitrogen. Assignment of the emission band heads is shown in table 1 and indicated also on the spectra. The two sets of three blue degraded band heads (2925-2980 Å) and (3105-3165 Å) are readily assigned to the Δv=2 and Δv=1 sequences of the second positive system of N_2(C-B), respectively. The series of

Table 1. List of the most intense observed molecular bands in the LIB in nitrogen, corresponding electronic transition and wavelength in Å (air) for the major band heads

Molecule	Name system	Observed band system	Major band heads (Å) (v', v")
N_2^+	Janin-d'Incan	$D^2\Pi_g - A^2\Pi_u$	2343 (0, 0); 2398 (1, 1); 2456 (2, 2); 2516 (3,3); 2579 (4, 4); 2645 (5, 5); 2714 (6, 6)
Molecule	Name system	Observed band system	Major band heads (Å) (v', v")
N_2^+	First negative	$B^2\Sigma_u^+ - X^2\Sigma_g^+$	2861.7 (11, 6); 3033.0 (11, 7); 3065.1 (15, 10); 3076.4 (4, 1); 3078.2 (3, 0); 3291.6 (5, 3); 3293.4 (4, 2); 3298.7 (3, 1); 3308.0 (2, 0); 3319.9 (8, 6); 3349.6 (18, 13); 3381.5 (10, 8); 3419.6 (14, 11); 3447.3 (23, 16); 3460.8 (17, 13); 3493.4 (12, 10); 3532.3 (5, 3); 3538.3 (4, 3); 3548.9 (3, 2); 3563.9 (2, 1); 3582.1 (1, 0); 3588.6 (16, 13); 3612.4 (10, 9); 3806.8 (5, 5); 3818.1 (4, 4); 3835.4 (3, 3); 3857.9 (2, 2); 3884.3 (1, 1); 3914.4 (0, 0); 4110.9 (6, 7); 4121.3 (5,6); 4140.5 (4, 5); 4199.1 (2, 3); 4236.5 (1, 2); 4278.1 (0, 1); 4459.3 (7, 9); 4466.6 (6, 8); 4490.3 (5, 7); 4515.9 (4, 6); 4554.1 (3, 5); 4599.7 (2, 4); 4651.8 (1, 3); 4709.2 (0, 2); 4864.4 (7, 10); 4881.7 (6, 9); 4913.2 (5, 8); 4957.9 (4, 7); 5012.7 (3, 6); 5076.6 (2, 5); 5148.8 (1, 4); 5228.3 (0,3); 5485.5 (4, 8); 5564.1 (3, 7); 5653.1 (2, 6); 5754.4 (1, 5); 5864.7 (0, 4)
N_2	Second positive	$C^3\Pi_u - B^3\Pi_g$	2953.2 (4, 2); 2962.0 (3, 1); 2976.8 (2, 0); 3116.7 (3, 2); 3136.0 (2, 1); 3159.3 (1, 0); 3268.1 (4, 4); 3285.3 (3, 3); 3311.9 (2, 2); 3338.9 (1, 1); 3371.3 (0, 0); 3500.5 (2, 3); 3536.7 (1, 2); 3576.9 (0, 1); 3641.7 (4, 6); 3671.9 (3, 5); 3710.5 (2, 4); 3755.4 (1, 3); 3804.9 (0, 2); 3857.9 (4, 7); 3894.6 (3, 6); 3943.0 (2, 5); 3998.4 (1, 4); 4059.4 (0, 3); 4094.8 (4, 8); 4141.8 (3, 7); 4200.5 (2, 6); 4269.7 (1, 5); 4343.6 (0, 4); 4355.0 (4, 9); 4416.7 (3, 8); 4490.2 (2, 7); 4574.3 (1, 6); 4667.3 (0, 5); 4723.5 (3, 9); 4814.7 (2, 8); 4916.8 (1, 7); 5031.5 (0, 6); 5066.0 (3, 10); 5179.3 (2, 9); 5309.3 (1, 8)

bands between 3020-3095 Å belong to any of the bands 11-7, 4-1 and 3-0 of the first negative system of N_2^+(B-X). The lower panel, corresponding to the dc electric glow discharge of nitrogen at low pressure, shows practically the same bands with different intensity distributions and spectral widths. In the spectrum of the lower panel of Figure 5(b) (nitrogen electric glow discharge) we easily identified five bands of the Δv=0 sequence and the 1-0 band of the second positive system of N_2(C-B). The main intensity is observed for the dominant transition N_2(C, *v*'=0)→(B, v"=0) which corresponds to the most intense nitrogen laser line at 3371 Å. However, this simple picture changes drastically in the LIB emission spectrum of nitrogen excited by the CO_2 laser, being now the 0-0 band very weak. A large number of additional strong bands mainly in the region 3260-3410 Å are detected in the LIDB spectrum and can be attributed to emissions from the first negative system of N_2^+ (especially for Δv=2). The B-X system of N_2^+ has been observed over a wide range of vibrational levels. The highest values so far observed are *v'*=29 for the excited B state and *v"*=23 for the X state. The main bands of this system lie in the -2≤Δv≤2 with v'<5 sequences and all of these bands are blue degraded. However under certain circumstances many more bands, some red degraded, are observed, most being tail bands of these sequences. The observed bands with v'≤7 and also those with 8≤v'≤11 and Δv<-1 are shaded to the violet. All the observed bands with v'≥12 and those with *v*'=10 and 11, Δv≥0 are shaded to the red although some bands appear headless. These bands are so-called tail bands taking place a reversal in the successions of the bands in the sequence. Moreover, in the bands of the first negative system of N_2^+ have been observed numerous rotational line displacements and intensities anomalies arising from perturbations in the B state. The perturbing state is $A^2\Pi_u$. In the Deslandres table listing the observed band heads, there is a pronounced gap in on arm of the Condon locus, and a less marked one in the other [52]. Franck-Condon factors indicate that the missing bands should be as intense as many of the observed. The few bands observed in the region of the gap are 8-6 and 9-8, which are headless and the 10-9 and 10-8 which have no definite heads. The missing bands coincide with the strong bands of either first negative system of N_2^+(B-X) or the second positive system of N_2(C-B), and are difficult to detect. In the spectrum of the lower panel of Figure 5(c) corresponding to the nitrogen electric glow discharge we easily identified several bands of the Δv=-1 sequence and two weak 4-6 and 3-5 bands of the second positive system of N_2(C-B). In this spectral region (3425-3675 Å) the most intense bands are due to the transitions N_2(C, v'=0)→(B, v"=1) and N_2(C, v'=1)→(B, v"=2), also present in the LIB emission spectrum of nitrogen in the upper panel of Figure 5(c). As in the previous cases, the LIB emission spectrum excited by the CO_2 laser changes drastically regarding the emission spectrum of nitrogen electric glow discharge. In it a large number of additional strong bands corresponding mainly to the Δv=+1 sequence B-X band system of N_2^+ are detected which are partially overlapped by the weak bands of the C-B system of N_2. For low *v*' the 1-0, 2-1, 3-2 ... bands of the B-X system of N_2^+ are degraded to shorter wavelengths and for high v' values the bands are degraded to longer wavelengths (tail bands).

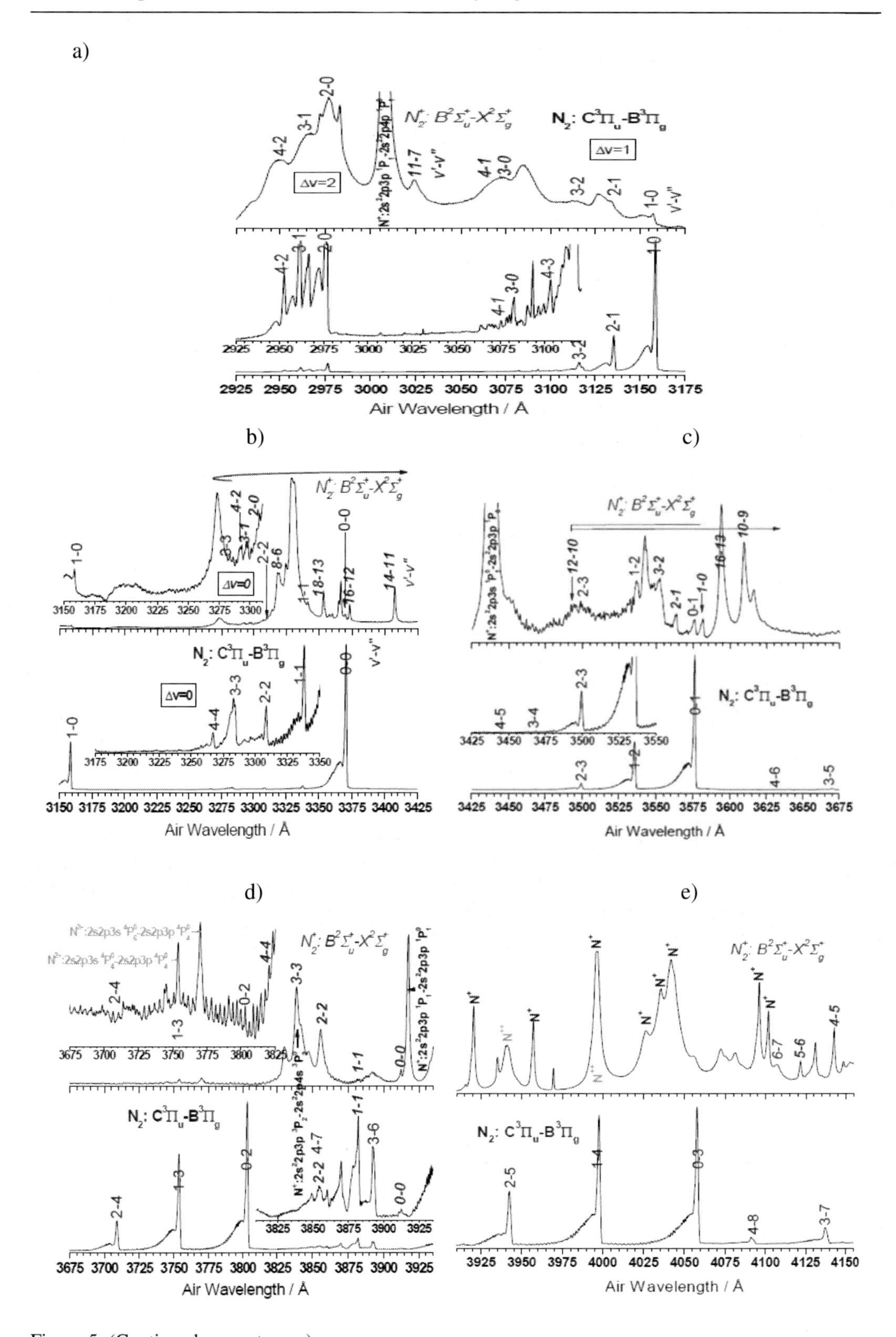

Figure 5. (Continued on next page).

f)

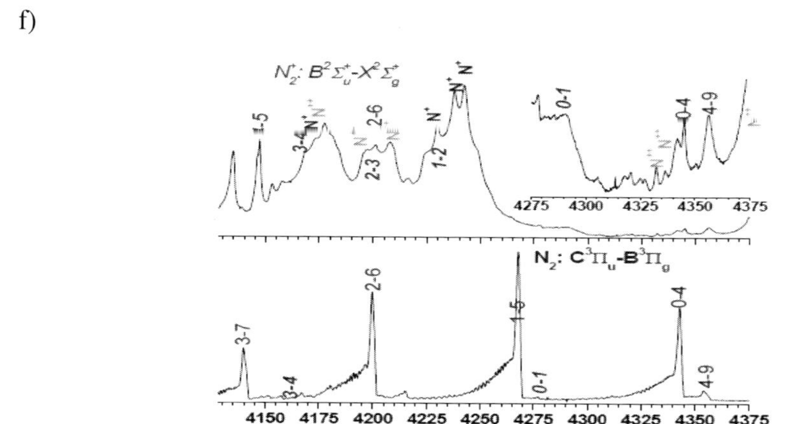

Figure 5 (a)-(f). Comparison between high-voltage dc electric discharge (lower panels) and LIDB (upper panels) emission spectra for nitrogen in several spectral regions.

In the nitrogen electric glow discharge spectrum [lower panel of Figure 6(d)] we identified 3-6, 4-7, 0-2, 1-3 and 2-4 bands of the second positive system of N_2(C-B) and 0-0, 1-1 and 2-2 bands of the first negative system of N_2^+. In this spectral region (3675-3935 Å) the most intense band is the 0-2 band of N_2(C-B). The LIB emission spectrum in this region hardly shows bands of the second positive system of N_2(C-B). However, this emission spectrum exhibits a large number of bands corresponding mainly to the Δv=0 sequence B-X band system of N_2^+. In the spectra of the lower panel of figs. 5(d)-(f) (nitrogen electric glow discharge) we identified several bands of the second positive system of N_2(C-B), indicated on the spectra, and some very weak bands of the first negative system of N_2^+. As in the previous cases, the spectra change drastically in the LIDB emission of nitrogen excited by the CO_2 laser. A large number of N^+ and N^{2+} atomic lines and additional strong bands of the first negative system of N_2^+ are now present.

To understand the different processes involved in the analyzed emission, Rydberg-Klein-Rees (RKR) potential energy curves for some bound electronic states of N_2 and N_2^+ have been calculated. The potential energy curves for the $X^1\Sigma_g^+$, $A^3\Sigma_u^+$, $B^3\Pi_g$, $C^3\Pi_u$ states of N_2 and $X^2\Sigma_g^+$, $A^2\Pi_u$, $B^2\Sigma_u^+$, $C^2\Sigma_u^+$ and $D^2\Pi_g$ states of N_2^+ were obtained from the experimental information reported by Huber and Herzberg [58] and Laher and Gilmore [59].

Figure 6 shows the calculated RKR potentials and associated transitions for electronic states of N_2 and N_2^+ which can be relevant to interpret the results of the present work. Many perturbations are known in molecular states of nitrogen although a depth explanation of these features in terms of mixing of electronic states is not yet available. A useful graphical summary of many potential energy curves has been reported by Gilmore [60].

Perturbations are often accompanied by complex intensity irregularities as happens in the first negative system of N_2^+. Whereas the vibrational and rotational constants run quite normal for the $X^2\Sigma_g^+$ state of N_2^+, this is not at all the case for the $B^2\Sigma_u^+$ electronic state. Both the B_v and $G(v)$ curves versus the vibrational quantum number v have unusual shapes. This can be interpreted as caused by a strong mutual vibrational perturbation between the $B^2\Sigma_u^+$ and $C^2\Sigma_u^+$ states of the same species of N_2^+ (see Figure 6).

As this perturbation is homogeneous (i.e., $\Delta\Lambda$=0) the shifts in both levels will be nearly independent of J producing that the $B^2\Sigma_u^+$ potential curve to flatten out in the middle of its energy range. The minimum of the $C^2\Sigma_u^+$ potential energy curve should be moved to smaller internuclear distances. There are also observed numerous rotational perturbations in the B-X system of N_2^+, caused by an interaction between the $A^2\Pi_u$ and the $B^2\Sigma_u^+$ states. The excitation temperature T_{exc} was calculated from the relative intensities of some N^+ ionic lines (3400–4800 Å spectral region) according to the Boltzmann equation (3.16).

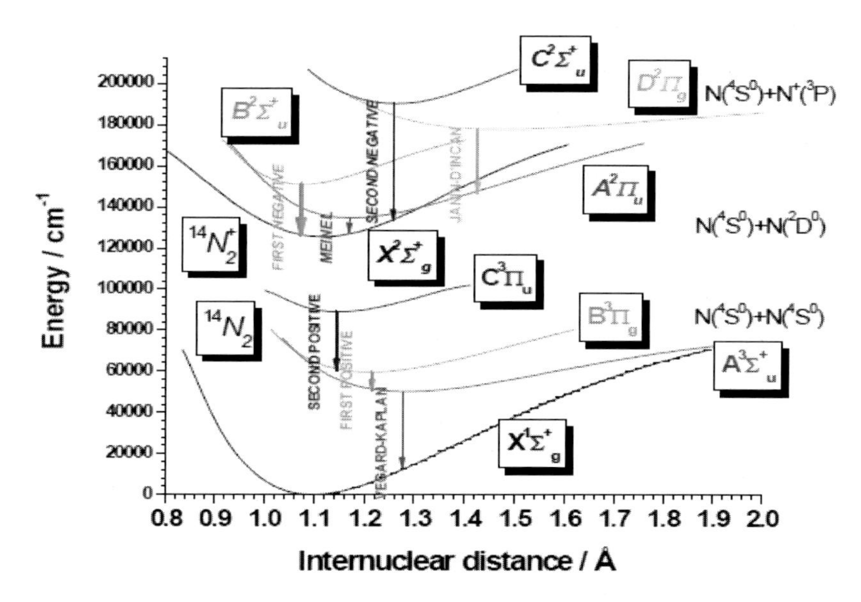

Figure 6. RKR potential energy curves for some bound electronic states of N_2 and N_2^+.

The estimated excitation temperature was T_{exc}= 21000±1300 K. However, if the excitation temperature is determined using only the relative intensities of N atomic lines (7300–8800 Å spectral region) a value of T_{exc}=7900±1300 K is obtained. This behaviour is observed by other authors [61] and may be interpreted to result from the different emissivity distributions of neutral atoms and ion lines. The emissivity of the ion lines is produced, on the average, near the inner region with higher temperature. On the contrary, the emissivity of the neutral atom lines comes, on the average, from the low temperature region close to the plasma front, where the neutral atom density is higher. The intensity measurements correspond to the integration of the local emissivity values along the line-of-sight, integrated in turn in the perpendicular directions. As a consequence, the neutral atom Boltzmann plot provides a temperature value which is a certain average of the low-temperature values in the plasma (7900 K), whereas the temperature obtained from the ion Boltzmann plot (21000 K) averages the values existing in the high-temperature region. On the other hand, we have carried out simulations of the Δv=+1 sequence of B-X band of N_2^+ for different vibrational temperatures finding that a value around 20000 K reasonably reproduces the experimental spectrum. Also, if we consider a temperature of 7900 K the ionization degree obtained by means of the Saha equation is of 0.00064. Such a low ionization degree does not justify the observed emission spectra of N^+ and N_2^+. Keeping in mind these results, the temperature obtained from relative intensity of N^+ atomic lines (21000±1300 K) was chosen as the first approximation for the excitation temperature.

5.2. LIBS of Oxygen

In this section we present our recent results on LIBS in oxygen gas and pressures ranging from 4.6 to 75 kPa was studied using a high-power transverse excitation atmospheric CO_2 laser (λ=9.621 and 10.591 μm; τ_{FWHM}=64 ns; power densities ranging from 0.87 to 6.31 GW×cm^{-2}) [20].

For the present experiments the measured focused-spot area was 7.85×10^{-3} cm^2. This value is higher than the calculated area (2.2×10^{-4} cm^2) obtained from the beam waist (Equation 1.7). This fact is due to the non-Gaussian profile of the CO_2 laser beam. Moreover the CO_2 laser beam passes through a circular aperture of diameter 17.5 mm. For this diaphragm the calculated divergence angle for the laser beams at 9.621 and 10.591 μm are 1.3 and 1.5 mrad, respectively. Thus, considering the total beam divergence (~4.4 mrad), the calculated diameter of the focused TEA-CO_2 laser (beam waist) is 1.06 mm, which is very similar to the measured value (~1 mm). If the focal region of the laser beam is assumed to be cylindrical in shape, the spot size in terms of length l (Equation 1.8) of the focused TEA-CO_2 laser is 6.0 mm, which is similar to the measured value (~7 mm). For the different pulse laser energies measured in LIB of oxygen, the calculated laser peak power (Equation 1.1), intensity (Equation 1.2), fluence (Equation 1.3), photon flux (Equation 1.4), electric field (Equation 1.5) and pressure radiation (Equation 1.6) are given in Table 2.

Table 2. Laser parameters for the TEA CO_2 LIB experiments of oxygen

Laser λ(μm)	Energy E_W (mJ)	Power P_W (MW)	Intensity I_W (GW cm^{-2})	Fluence Φ_W (J cm^{-2})	Photon Flux, F_{ph} (ph. cm^{-2} s^{-1})	Electric Field F_E (MV cm^{-1})
9.621	2685	42.1	5.36	342	2.60×10^{29}	1.50
9.621	2256	35.4	4.50	287	2.18×10^{29}	1.37
9.621	1732	27.1	3.46	220	1.67×10^{29}	1.20
9.621	1209	19.0	2.41	154	1.17×10^{29}	1.01
9.621	503	7.88	1.00	64.0	4.86×10^{28}	0.65
10.591	3161	49.5	6.31	402	3.36×10^{29}	1.63
10.591	2145	33.6	4.28	273	2.28×10^{29}	1.34
10.591	1481	23.2	2.96	189	1.58×10^{29}	1.11
10.591	968	15.2	1.93	123	1.03×10^{29}	0.90
10.591	624	9.78	1.25	79.5	6.64×10^{28}	0.72
10.591	436	6.83	0.87	55.5	4.64×10^{28}	0.60

Figures 7(a-f) display an overview of the low-resolution LIB emission spectrum (2320-9690 Å) in oxygen at a pressure of 53.2 kPa, excited by the 10P(20) line of the CO_2 laser, and assignment of the atomic lines of O, O^+, O^{2+}, N and N^+ [45]. Strong atomic O lines dominate the spectrum but, ionic O^+ lines (about 8 times weaker) and weak O^{2+} lines (about 150 times weaker) also are present. Some atomic and ionic nitrogen lines were also present, as well as, the first negative band system 330-400 nm corresponding to the transition $B^2\Sigma_u^+ - X^2\Sigma_g^+$ in

N_2^+. In the acquisition of the spectrum of the Figure 7(d-f) a cut-off filter was used to suppress the second order intense UV oxygen atomic lines. This cut-off filter produces a decrease of the intensity with regard to the spectra of the Figure 7(a-c). In order to get more insight into LIB of oxygen and to obtain an unambiguous assignment of the emission lines, we have scanned the corresponding wavelength regions with higher resolution (~0.10 Å in first-order), which was sufficient to distinguish clearly between nearly all observed lines.

The spectra have been obtained with fifty successive exposures on the ICCD camera in the spectral region 1900-7500 Å. As example, Figure 8(a-f) shows several spectra recorded in the LIB experiment. These LIB emission spectra were recorded under the following experimental conditions: oxygen pressure of 48.8 kPa, excitation line 10P(20) at 10.591 μm and CO_2 laser power density 4.28 GW×cm^{-2}. Figures 8(a-f) display some details of the large features found in Figs. 7(a-f). No new features were observed in these high-resolution spectra. In these Figures, multiplet transitions between different J levels for O^+ and O^{2+} are observed.

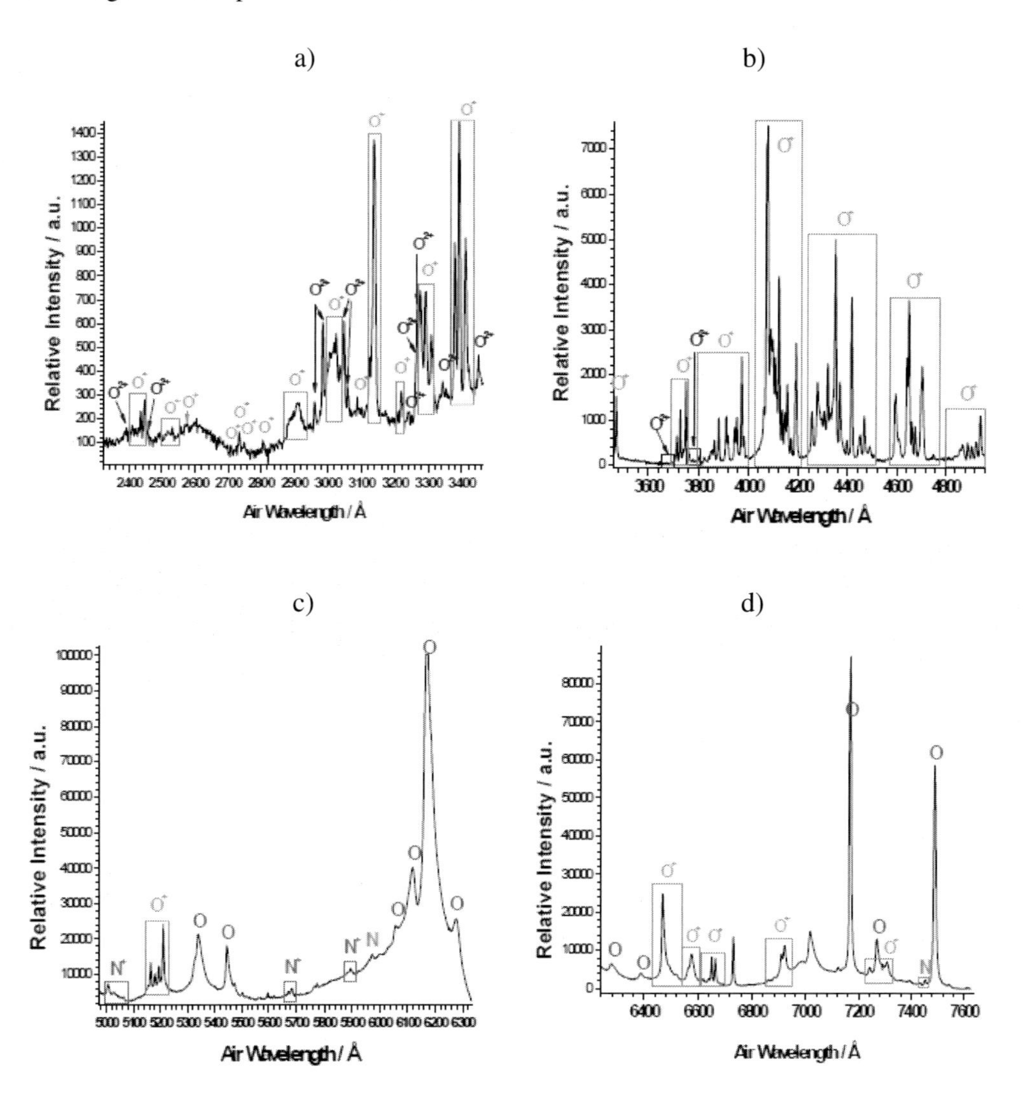

Figure 7. (Continued on next page).

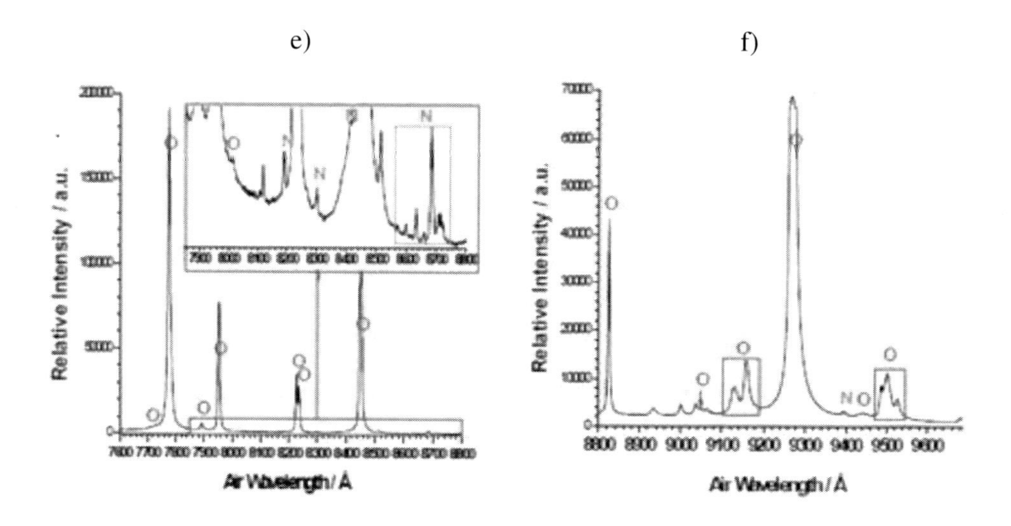

Figure 7(a-f). Low-resolution LIB emission spectrum observed in the 2320-9690 Å region in oxygen at a pressure of 53.2 kPa, excited by the 10P(20) line of the TEA-CO_2 laser at 10.591 μm and a power density of 0.87 GW×cm^{-2}, and assignment of the atomic lines of O, O^+, O^{2+}, N and N^+.

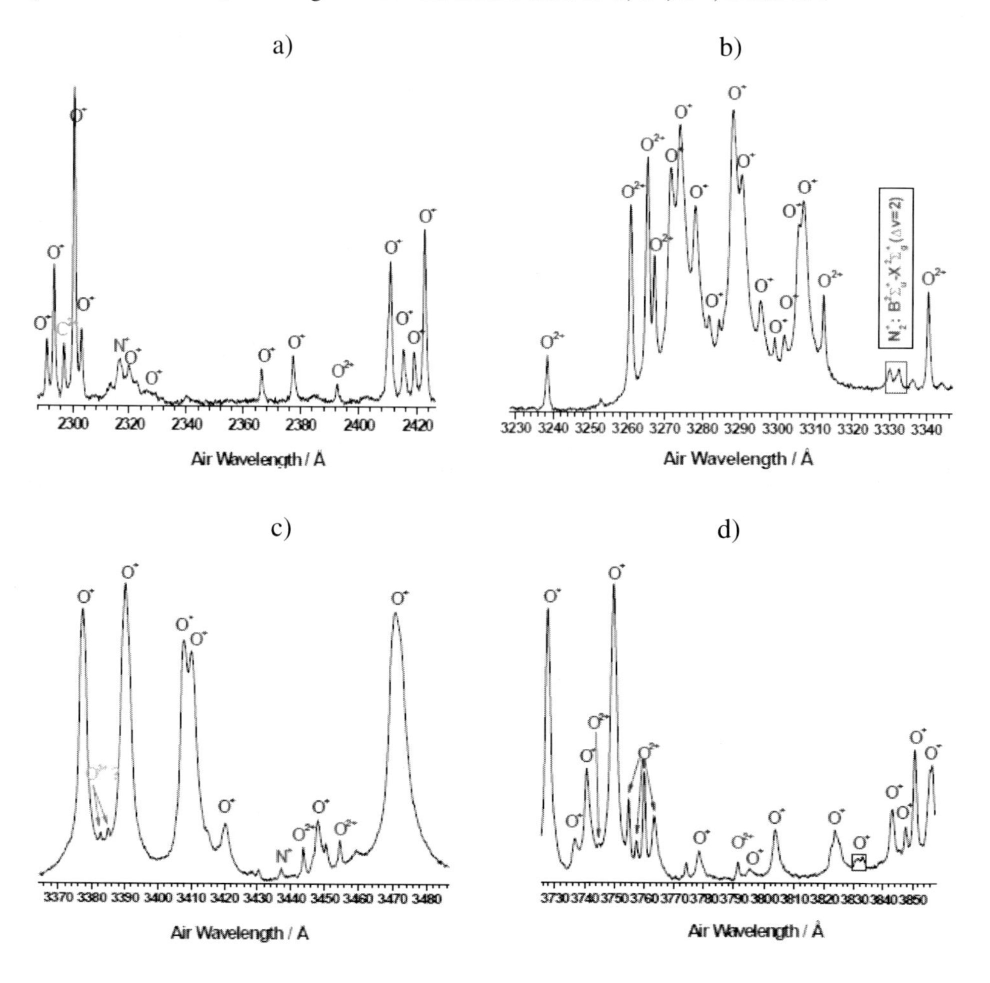

Figure 8. (Continued on next page).

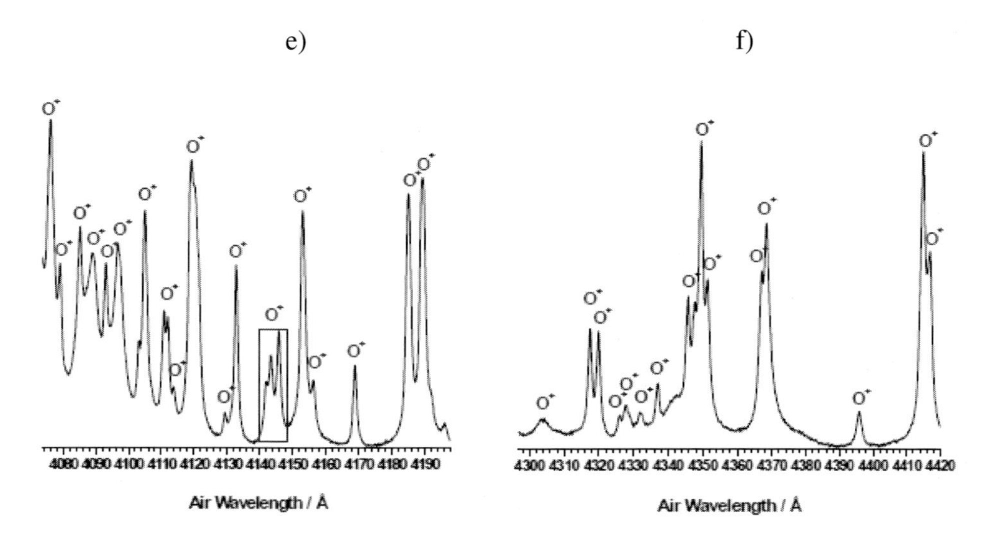

Figure 8 (a-f). High-resolution LIB emission spectrum observed in several spectral regions in oxygen at a pressure of 48.8 kPa, excited by the CO_2 laser at 10P(20) line (10.591 μm) with a power density of 4.28 GW×cm^{-2}, and assignment of some atomic lines of O^+, O^{2+}, and N^+.

In some cases these multiplet structures are not completely resolved due to Stark broadening of ionic lines. The spectral features clearly show the complexity of the relaxation process and bring out the possibility of cascading processes.

The excitation temperature was calculated from the relative intensities of several O^+ (3270–3310 Å spectral region) and O^{2+} (2900–3350 Å spectral region) atomic lines and the slope of the Boltzmann plot (Equation 3.16). The values of the λ_{ki}, g_k, A_{ki} and E_k for O^+ and O^{2+} selected atomic lines were obtained from the NIST Atomic Spectral Database [45].

The excitation temperatures were determined under the following experimental conditions: oxygen pressure of 48.8 kPa, excitation line 10P(20) at 10.591 μm and CO_2 laser power density 4.28 GW×cm^{-2}. The obtained excitation temperatures, in the case of O^+ and O^{2+}, were 23000 ± 3000 K and 31500 ± 1600 K, respectively (Figure 9).

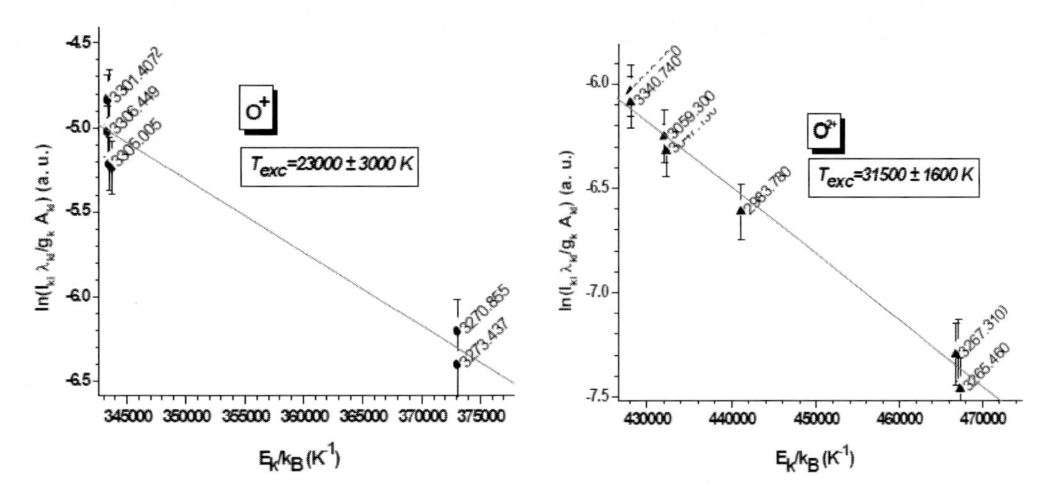

Figure 9. Linear Boltzmann plots for several O^+ and O^{2+} transition lines used to calculate plasma temperature, T_{exc}. Plots also show linear fit to the data.

This behavior can be due to the different quenching rate coefficients between each species. Also this fact may be interpreted to result from the different emissivity distributions of single ionized and double ionized oxygen lines. When LIB is produced in oxygen under high intensity laser radiation, some molecules can obtain an energy that exceeds the binding energy. Also some of their electrons become so energetic that the atoms and molecules ionize. Taking into account our experimental spectral observations, at these high temperatures oxygen becomes a mixture mainly of primary O_2, O, O^+, O^{2+} and electrons.

The transition between a gas and plasma is essentially a chemical equilibrium which shifts from the gas to plasma side with increasing temperature. Let us consider the first two different ionization equilibria of oxygen:

$$\text{O (2s22p4 } {}^3P_2\text{)} \leftrightarrow \text{O+(2s24p3 } {}^4S^0_{3/2}\text{)+e+IP(O - I)},$$

$$\text{O+(2s22p3 } {}^4S^0_{3/2}\text{)} \leftrightarrow \text{O2+(2s22p2 } {}^3P_0\text{)+e+IP(O - II)},$$

where the first two ionization potentials for oxygen are $I_P(\text{O}-\text{I}) = 13.618\text{eV}$, and $I_P(\text{O}-\text{II}) = 35.121\text{eV}$ [62]. Taking into account the consideration of section 3.6, we can obtain the ionization degree. Figure 10 shows the ionization degree $N_i/(N_0+N_i)$ of O and O^+, plotted as a function of the gas temperature T at a constant total pressure $P=(N_0+n_e+N_i)k_BT$ of 53.2 kPa. The graph shows that oxygen is already fully ionized at thermal energies well below the first ionization-energy of 13.618 eV (equivalent to 158000 K).

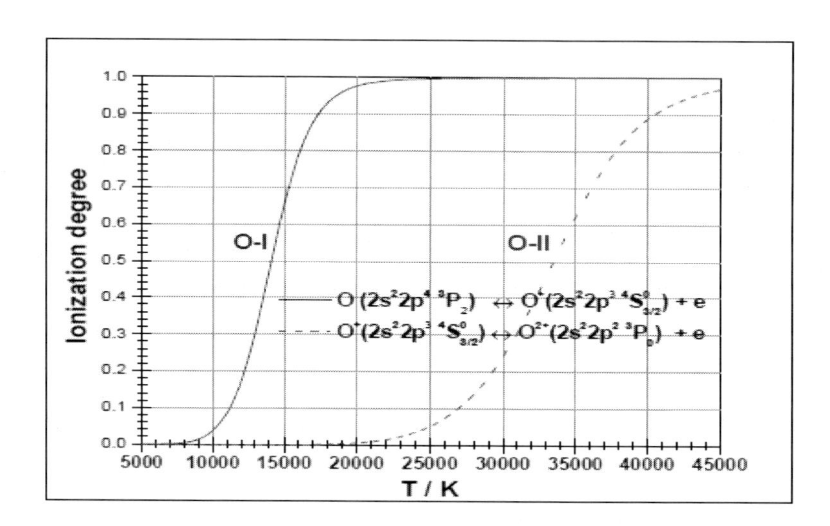

Figure 10. Temperature dependence of the ionization degree $N_i/(N_0+N_i)$ of atomic oxygen O and oxygen singly ionized O^+ at a constant pressure of 53.2 kPa.

If we consider a temperature of 23000 K, the ionization degrees of O and O^+ obtained by means of the Saha equation are 0.994 and 0.022, respectively. For T=31500 K, the ionization degrees of O and O^+ obtained by means of the Saha equation are 0.999 and 0.34, respectively. These so high values of the ionization degrees justify the observed emission spectra. The electron number density was obtained by considering the discussion reported in section 3.4.

The Doppler line widths for some lines of O^+ for different temperatures are shown in Figure 11. In our experiments, for O^+ lines, the Doppler line widths are 0.07-0.12 Å at 23000 K. The choice of plasma emission for n_e measurements is made to ensure that the O^+ spectral lines are sensitive enough to Stark effect and do not suffer from interference by other species. The estimation of electron density n_e has been carried out by measuring the broadening of the spectral profiles of isolated lines of O^+ (2738, 3386, 3809, 4075, and 4418 Å) from the high-resolution spectra. The electron impact parameters for the different O^+ lines were approximated to a first-order exponential decay. The electron number densities of the laser-induced plasma were determined from the high-resolution emission spectra in oxygen at a pressure of 48.8 kPa, excited by the CO_2 laser at 10.591 μm with a power density of 4.28 GW×cm^{-2}. A Lorentz function was used to fit the spectra.

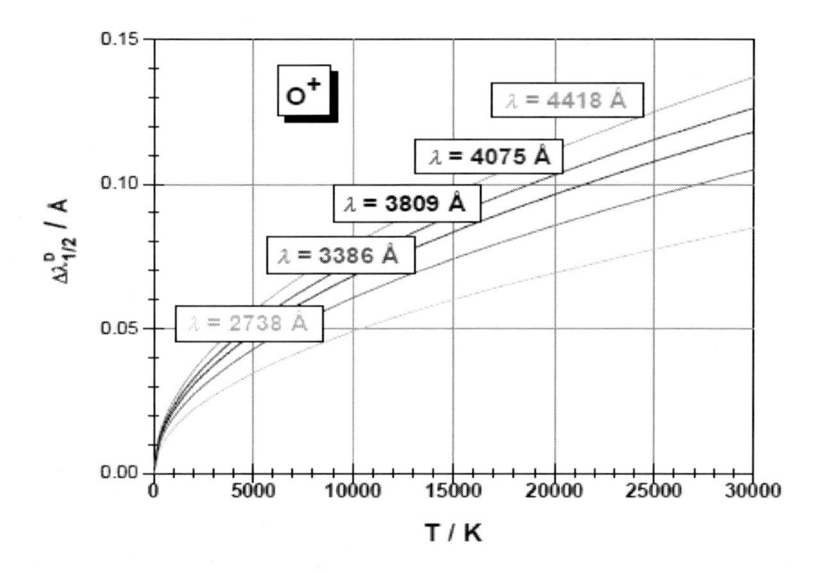

Figure 11. Doppler line widhs for some lines of O^+.

In order to extract the Stark broadening from the total experimentally measured line broadening, we have to previously deconvolute the different effects that contribute to the broadening of the spectral line: The instrumental, Doppler and Stark broadenings. Electron densities in the range (3.5-16.5)×10^{16} cm^{-3}, with an estimated uncertainty of 10%, were determined from the Stark broadening [Equation (3.12)] data of several singly ionized oxygen lines. At the evaluated temperature of 23000 ± 3000 K, Equation (3.4) yields n_e≈(0.54-2.3)×10^{16} cm^{-3}. These electron densities are lower to deduced values from the Stark broadening (3.5-16.5)×10^{16} cm^{-3}, approximately one order of magnitude. Based on these calculations, the validity of the LTE assumption is supported.

The interactions between the incoming laser radiation and the gas sample depend upon numerous variables related to the laser and the gas. These variables include laser wavelength, energy, spatial and temporal profile of the laser beam, and the thermal properties of the sample.

The incident beam is partially reflected and partially absorbed by the bulk to a degree that depends on the nature of the gas and the temperature it reaches under laser irradiation. LIBS spectra of oxygen obtained by laser irradiation at the different wavelengths are compared in

Figure 12. These high-resolution LIB emission spectra in oxygen were obtained to a pressure of 48.8 kPa, excited by two TEA-CO_2 laser wavelengths at 10.591 μm (I_W=6.31 GW×cm^{-2}) and 9.621 μm (I_W=5.36 GW×cm^{-2}). The spectral range was chosen in order to detect both single and double ionized oxygen species. Also this spectral region has been selected to show differences in signal intensity and background emission in detail. The first remark that we should make is that the background continuum emission after the same optimization was performed on the data acquisition window is much stronger for spectrum produced by the 10.591 μm laser line.

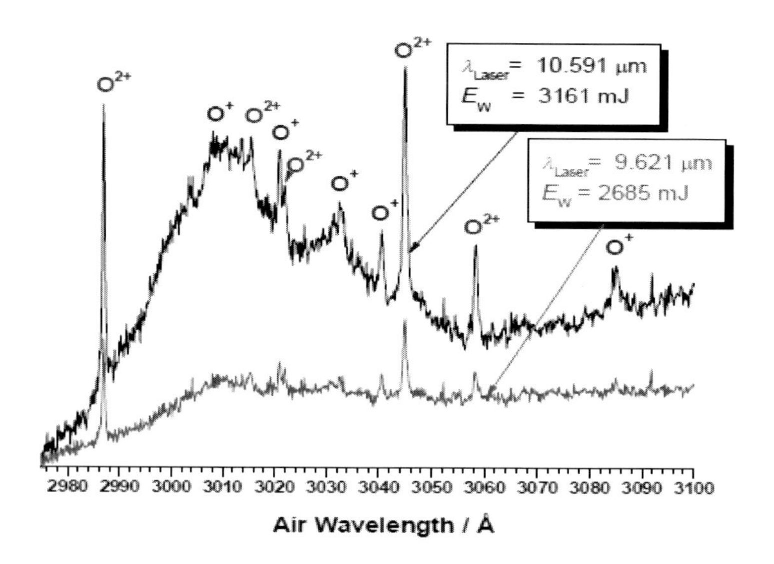

Figure 12. High-resolution LIB emission spectra in oxygen at a pressure of 48.8 kPa, excited by two TEA-CO_2 laser wavelengths at 10.591 μm (E_W=3161 mJ) and 9.621 μm (E_W=2685 mJ).

This is due to the higher laser power density (I_W=6.31 GW×cm^{-2}) and the higher absorption in the plasma caused by the IB, whose cross-section is proportional to $\lambda^3[1-\exp(-hc/\lambda k_B T)]^{-1}$ or approximately λ^2 (T is the electron temperature during the electron avalanche or cascade growth of ionization and λ is the laser excitation wavelength). The spectral lines of O^+ and O^{2+}(2983.78, 3017.63, 3023.45, 3043.02, 3047.13, 3059.30 Å) were clearly observed. It is clear that the ionic spectral lines for both O^+ and O^{2+} were enhanced by a factor of 4 when the LIBS is induced by the TEA-CO_2 laser at 10.591 μm. Besides, the irradiation at this wavelength favors the formation of the doubly ionized species (~ 30%), as it is obtained from the ratio of the intensities of the O^{2+} and O^+ lines.

Moreover, plasma electron densities were determined from Stark broadening of the O^{2+} double ionized line at 2983.78 Å. The measured electron densities for the LIB emission spectra in oxygen were (3.5±0.2)×10^{16} cm^{-3} and (3.2±0.2)×10^{16} cm^{-3} for excitation at 10.591 μm and 9.621 μm, respectively. Values of the electron impact half-width W for O^{2+} were taken from the reported values given by Sreckovic *et al.* [63]. It is noted that the emission intensity for O^+ and O^{2+} shows different picture than the electron density possibly due to the effect of the laser wavelength.

To see the effect laser irradiance the measurements were also carried out at different laser fluences. Optical emission spectra of the oxygen plasma plume at a pressure of 48.8 kPa as a function of the laser intensity are shown in Figs. 13(a) and 13(b).

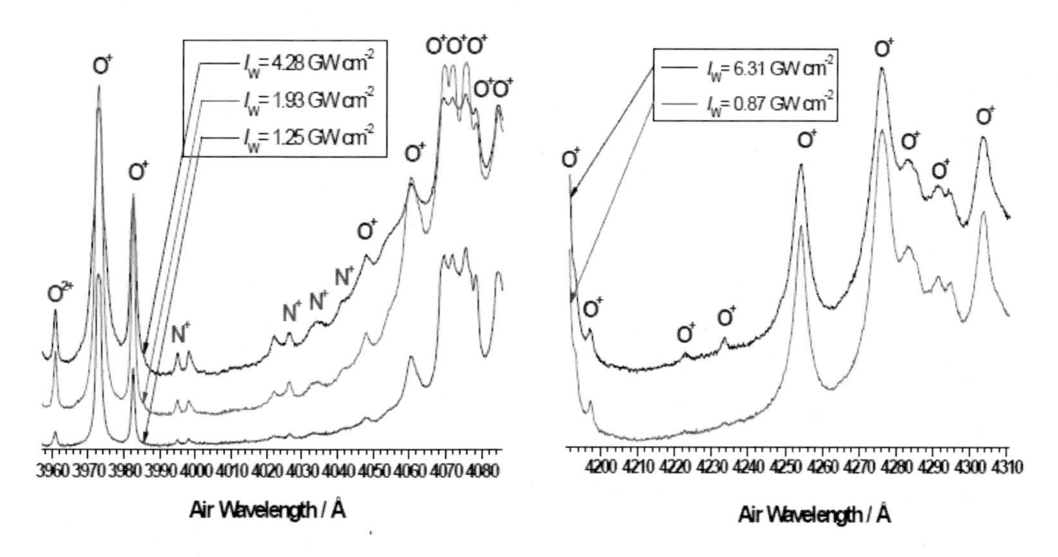

Figure 13 (a-b). High-resolution LIB in oxygen (at a pressure of 48.8 kPa) emission spectra observed in the (a) 3957-4085 Å and (b) 4195-4310 Å regions, excited by two TEA-CO_2 laser wavelengths at 10.591 μm as a function of the laser power density.

These spectra were recorded after the incidence of only one pulse of the TEA-CO_2 laser. The data were measured at a delay of 20 ms. An increase of atomic/ionic emission intensity and of the doubly ionized O^{2+} formation respect to O^+ with increasing the laser irradiance was observed. Also the background increases with the laser power.

At higher laser power densities (6.31-4.28 GW cm^{-2}), the spectral lines are more broadened than at lower power densities as a result of the high pressure associated with the plasma. It is assumed that at higher laser fluence the LIB plasma is more energetic and more ionized.

The emission characteristics of the laser-induced plasma are influenced by the composition of the gas atmosphere. The pressure of the gas is one of the controlling parameters of the plasma characteristics, as well as the factors related to the laser energy absorption. Also the presence of air gas (vacuum conditions) during the LIB process has consequences on the expansion dynamics. An interesting observation was the effect of the oxygen pressure. Nanosecond TEA CO_2-laser produced plasma emission has been characterized as a function of oxygen pressure. Experiments were performed in the pressure interval from 8 to 50 kPa and at pulse energy of 3161 mJ.

Figure 14 shows LIB emission spectra at various oxygen pressures, excited by the 10.591 μm line at a power density of 6.31 GWxcm^{-2}. As can be seen in Figure 14, the intensities of different spectral lines of O^+ and O^{2+} increase with decreasing pressure, reach a maximum at about 12.5 kPa, and then decrease for lower pressures. Characteristic emission lines from O^+ and O^{2+} elements exhibited significant enhancement in signal intensity at a few kPa oxygen pressure as compared to high pressures below atmospheric pressure.

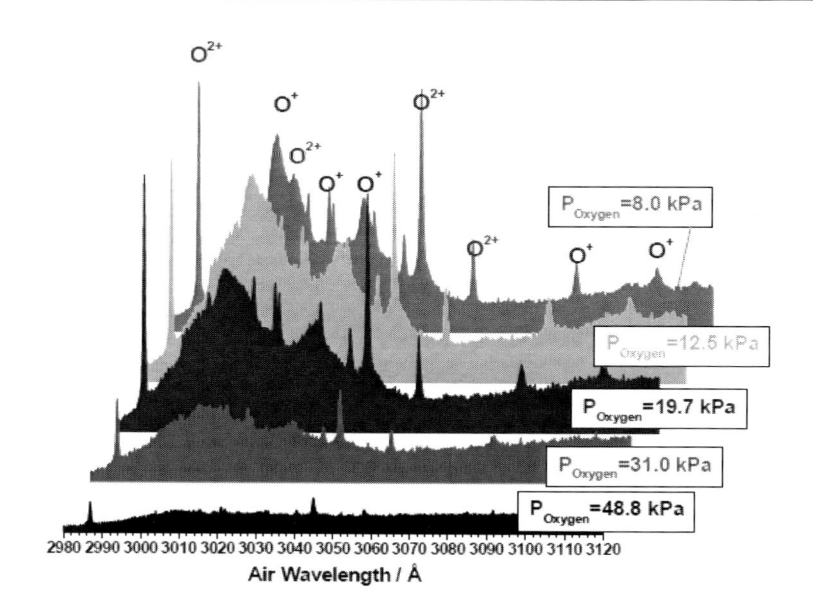

Figure 14. High-resolution LIB emission spectra at various oxygen pressures, excited by the TEA-CO_2 laser (10.591 μm) at a power density of 6.31 GWxcm^{-2}.

However, the ratio of the intensities among the O^{2+} and O^+ lines remains the same at all pressures. The measurements indicate enhancement due to a longer lifetime of the plasma expanding to a larger size at lower oxygen pressures.

Further reduction in oxygen pressure down to ~ 12.5 kPa resulted in a decrease in signal intensity, as a result of a reduction of collisional excitation of the emission lines which occurs when the plasma plume expands into the oxygen atmosphere.

5.3. LIBS of Air

A spectroscopic study of ambient air plasma, initially at room temperature and pressures ranging from 32 to 101 kPa, produced by TEA-CO_2 laser (λ=9.621 and 10.591 μm; $\tau_{FWHM}\approx$64 ns; power densities ranging from 0.29 to 6.31 GW×cm^{-2}) has been carried out in an attempt to clarify the processes involved in laser-induced breakdown (LIB) air plasma.

To understand the detailed aspects of laser-beam interaction with air and recombination processes following the breakdown, OES studies of the emission spectra from the plasma offer the most convenient method. The strong emission observed in the plasma region is mainly due to electronic relaxation of excited N, O and ionic fragments N^+. The medium-weak emission is due to excited species O^+, N^{2+}, O^{2+}, C, C^+, C^{2+}, H, Ar and molecular band systems of $N_2^+(B^2\Sigma_u^+-X^2\Sigma_g^+)$, $N_2(C^3\Pi_u-B^3\Pi_g)$, $N_2^+(D^2\Pi_g-A^2\Pi_u)$ and $OH(A^2\Sigma^+-X^2\Pi)$. Figure 15(a-f) displays an overview of the low-resolution LIB emission spectrum in air at atmospheric pressure, excited by the 10P(20) line of the CO_2 laser with an intensity of 2.2 GW×cm^{-2}, and assignment of the atomic lines of N, O, C, C^+, H, Ar, N^+, O^+, N^{2+}, O^{2+}, C^{2+} tabulated in NIST Atomic Spectral Database [45] and molecular bands of $N_2^+(B^2\Sigma_u^+-X^2\Sigma_g^+)$, $N_2^+(D^2\Pi_g-A^2\Pi_u)$, $N_2(C^3\Pi_u-B^3\Pi_g)$ and $OH(A^2\Sigma^+-X^2\Pi)$. Strong atomic N^+, N and O lines dominate the spectrum but, atomic lines of C, C^+, H, Ar, O^+, N^{2+}, O^{2+}, C^{2+} also are present.

LIB spectrum of air was compared with the LIB spectra obtained in our laboratory for nitrogen [18] and oxygen [20].

The spectra have been obtained at higher resolution with fifty successive exposures on the ICCD camera in the spectral region 2000-7500 Å. As examples, Figure 16(a-l) shows several spectra recorded in the air LIB experiment.

These LIB emission spectra were recorded under the following experimental conditions: air pressure of ~101 kPa, excitation line 9P(28) at 9.621 μm and laser irradiance of 5.36 GW×cm^{-2}. Note that spectra of Figs. 15(a-f) and 16(a-l) were excited by two different laser wavelengths. No new features were observed in these high-resolution spectra.

In these Figures, multiplet transitions between different *J* levels for N^+ and O^+ are observed. In some cases these multiplet structures are not completely resolved due to Stark broadening of atomic/ionic lines.

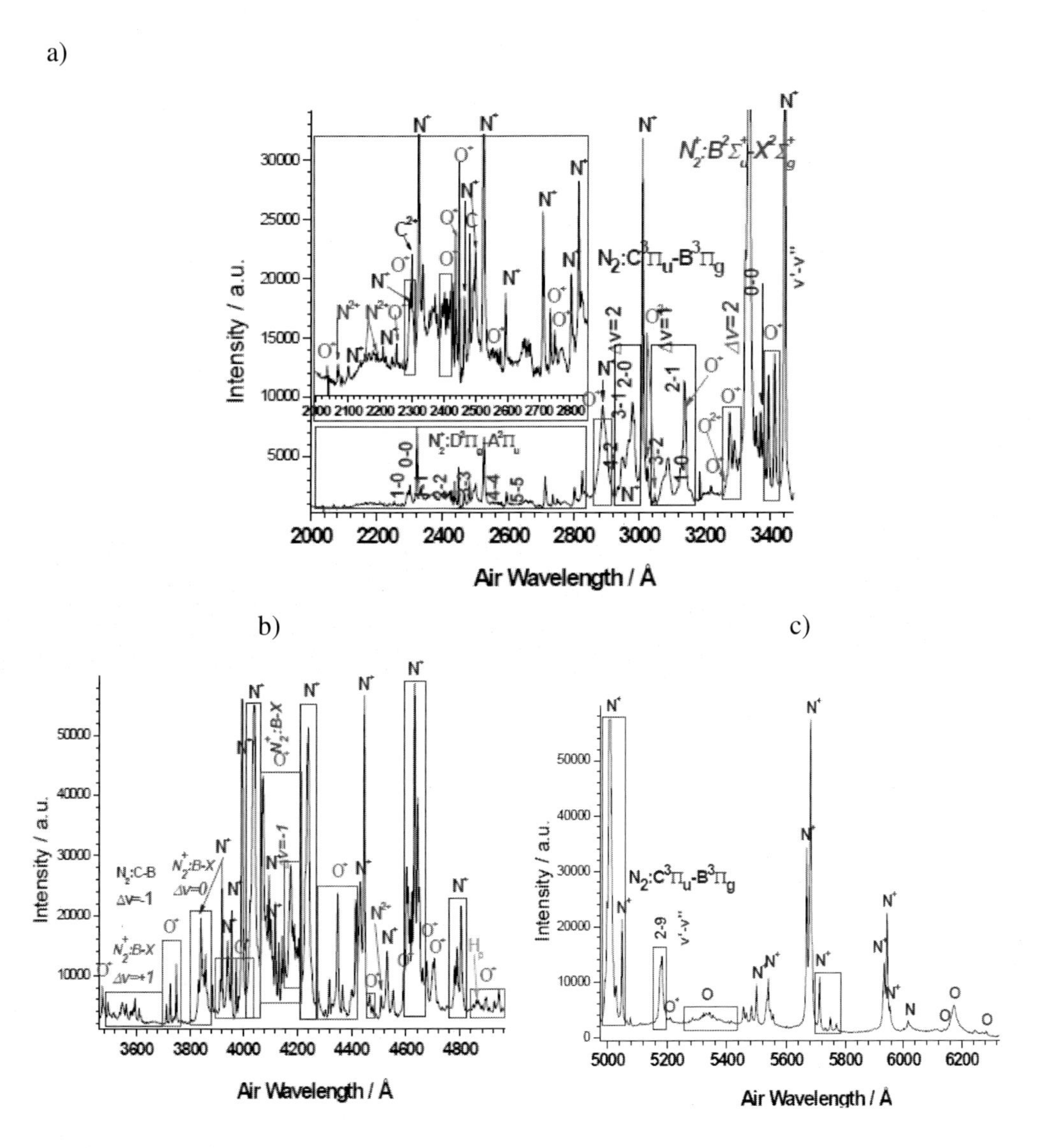

Figure 15. (Continued on next page).

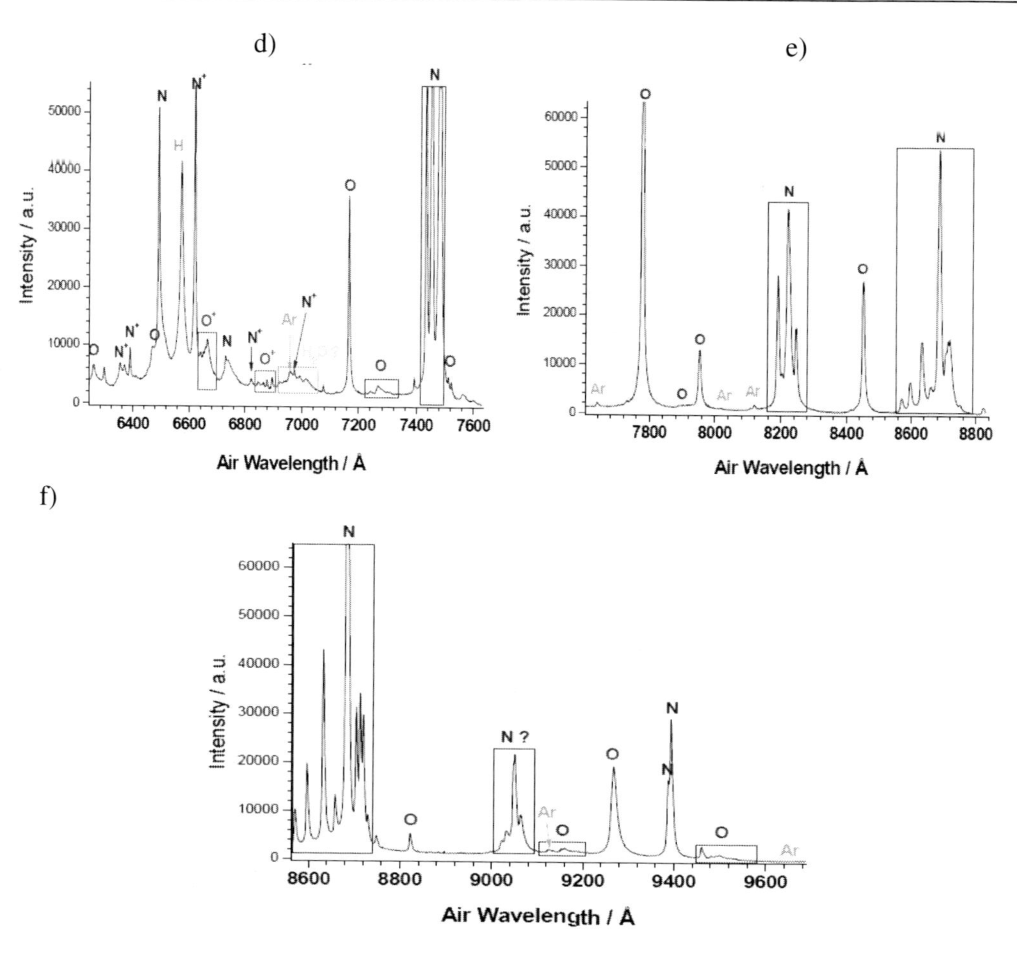

Figure 15. (a-f) Low-resolution LIB emission spectrum observed in the 2000-9690 Å region in ambient air at atmospheric pressure, excited by the TEA-CO2 laser at 10.591 μm and a power density of 2.2 GW×cm-2, and assignment of the atomic lines of N, O, C, C+, H, Ar, N+, O+, N2+, O2+, C2+ and molecular bands of N2+(B2Σu+-X2Σg+) and N2(C3Πu-B3Πg).

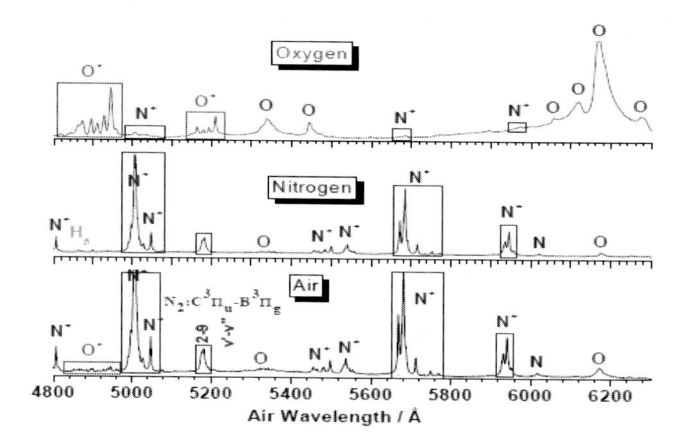

Figure 16. Low-resolution LIB emission spectra in air, nitrogen and oxygen, excited by the TEA-CO_2 laser at 10.591 μm and a power density of ~1 GW×cm^{-2}, and assignment of the atomic lines of N, O, H, N^+, O^+ and 2-9 molecular band of $N_2(C^3\Pi_u\text{-}B^3\Pi_g)$.

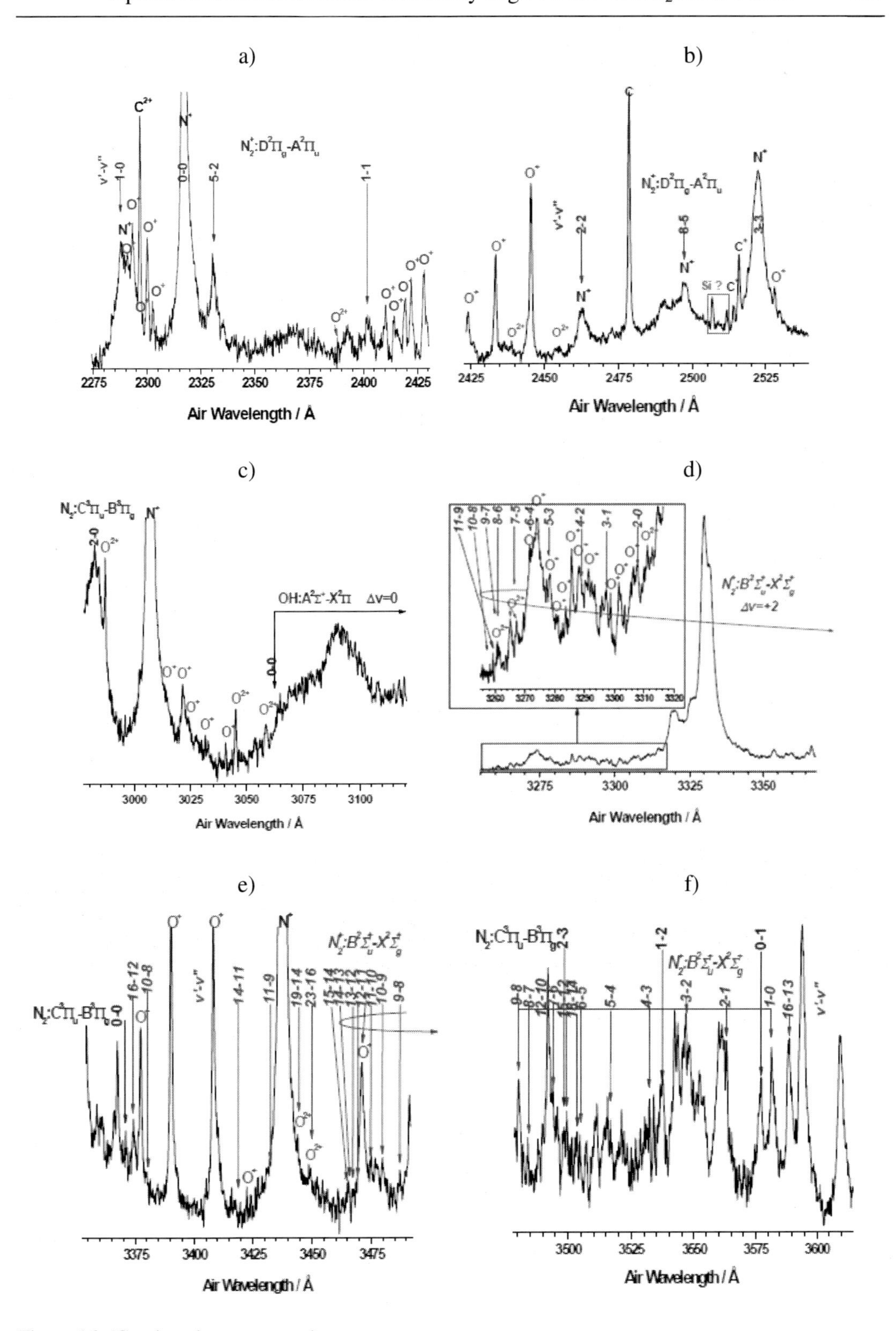

Figure 16. (Continued on next page).

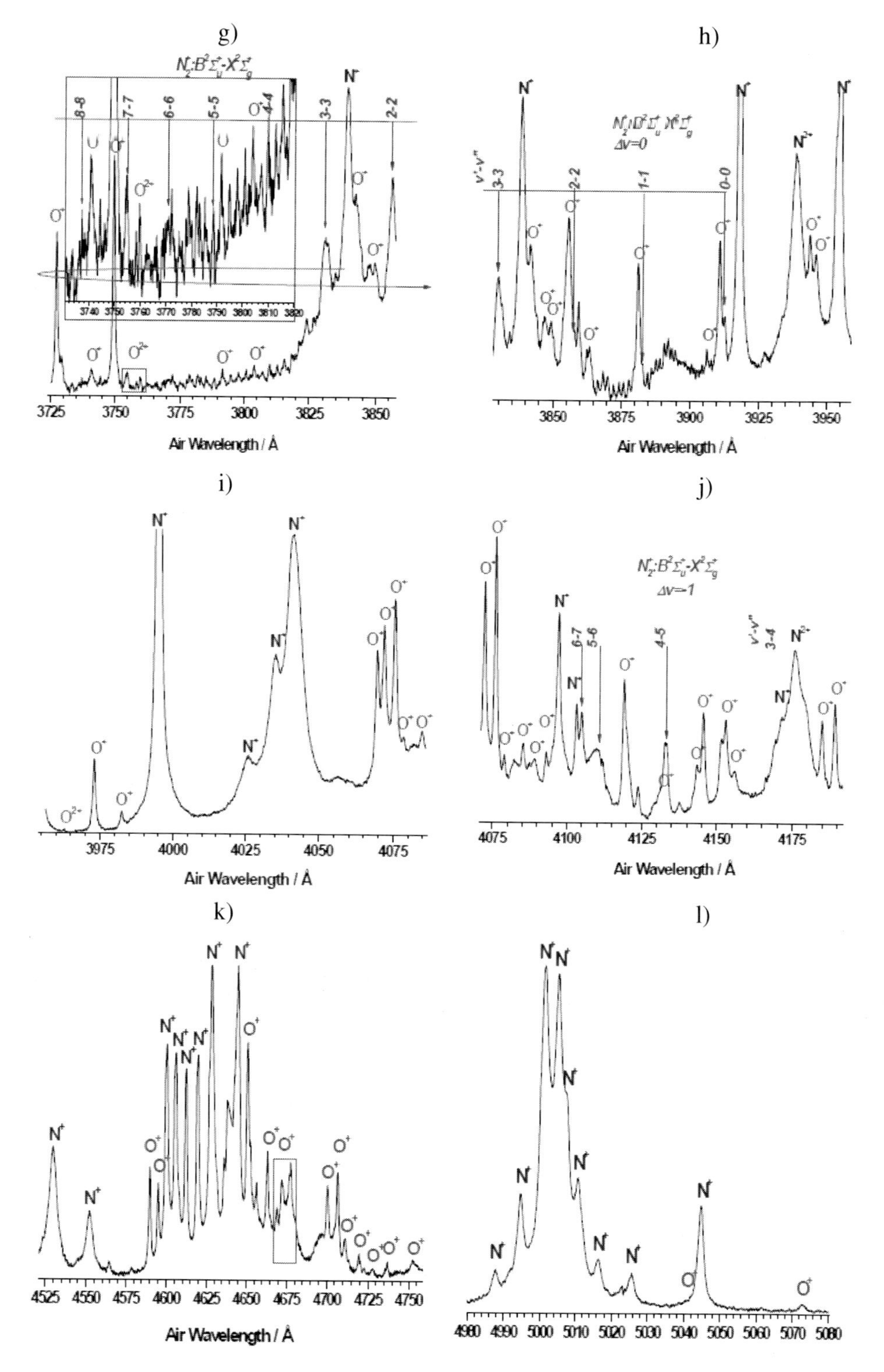

Figure 16. (a-l) High-resolution LIB emission spectrum observed in several spectral regions in ambient air at atmospheric pressure, excited by the CO_2 laser at 9.621 μm with a power density of 5.36 GW× cm^{-2}, and assignment of some atomic lines of N^+, O^+, N^{2+}, O^{2+}, and band heads of the molecular bands of $N_2^+(B^2\Sigma_u^+-X^2\Sigma_g^+)$, $N_2(C^3\Pi_u-B^3\Pi_g)$, $N_2^+(D^2\Pi_g-A^2\Pi_u)$ and $OH(A^2\Sigma^+-X^2\Pi)$.

The spectral features clearly show the complexity of the relaxation process. In Figures 16(a-l), a rather complex structure is observed, in consequence of the overlapping between rovibrational lines of different molecular band systems and atomic/ionic lines.

Figure 16(a) displays the overlapping between some bands of the $N_2^+(D^2\Pi_g\text{-}A^2\Pi_u)$ system and some lines of N^+, O^+, C^{2+} and O^{2+}. In this spectrum the predominant emitting species is the triplet structure of N^+ $2s^22p(^2P^0)4d$ $^3F^0_{3,2,4} \rightarrow 2s^22p(^2P^0)4d$ $^3D_{2,1,3}$ around 2317 Å overlapped with the 0-0 band of the N_2^+(D-A) system. In the spectrum of Figure 16(b), the most intense emitting species is the line of C $2s^22p(^2P^0)3s$ $^1P^0{}_1 \rightarrow 2s^22p^2$ $^1S_0$ at 2478.56 Å, and several ionic lines of O^+ $2s^22p^2(^3P)4p$ $^2D^0{}_{5/2} \rightarrow 2s^22p^2(^1D)3s$ $^2D_{5/2}$ at 2425.56 Å, O^+ $2s^22p^2(^1D)3p$ $^2D^0{}_{3/2} \rightarrow 2s^22p^2(^3P)3s$ $^2P_{1/2}$ at 2433.53 Å, and the doublet O^+ $2s^22p^2(^1D)3p$ $^2D^0{}_{3/2,5/2} \rightarrow 2s^22p^2(^3P)3s$ $^2P_{3/2}$ at 2444.25 Å and 2445.54 Å, respectively.

Several medium intensity ionic lines of N^+ also overlapped with different bands of N_2^+(D-A) and many weak lines of O^{2+} and O^+ are also present. In the spectrum of Figure 16(c), the most intense emitting species is the line of N^+ $2s^22p(^2P^0)4s$ $^1P^0{}_1 \rightarrow 2s^22p(^2P^0)3p$ $^1P_1$ at 3006.83 Å. This Figure displays the overlapping between the N_2(C-B) v'=2-v"=0 band, OH(A-X) Δv=0 sequence and many weak lines of O^+ and O^{2+}. In the spectrum of Figure 16(d), $N_2^+(B^2\Sigma_u^+\text{-}X^2\Sigma_g^+)$ Δv=2 sequence and many weak lines of O^+ and O^{2+} are recorded. This spectrum shows the reversal of the bands from v'=11, which is due to the overlap between high vibrational quantum number bands with low quantum number bands.

So, the first vibrational bands of N_2^+(B-X) (2-0, 3-1, 4-2 ... and 10-8) are shaded to the violet and after reversal (11-9, 12-10, 13-12 ...) are shaded to the red. In the spectrum of Figure 16(e), the most intense emitting species is the line of N^+ $2s^22p(^2P^0)3p$ $^1S_0 \rightarrow$ $2s^22p(^2P^0)3s$ $^1P^0{}_1$ at 3437.15 Å. Many medium intensity ionic lines of O^+ and O^{2+}, weak molecular bands of N_2(C-B) and N_2^+(B-X) are also present. Surprisingly, the relative intensity of the 0-0 band head in the N_2(C-B) system is very weak. In the spectrum of Figure 16(f), many medium intensity rovibrational molecular bands of N_2(C-B; Δv=-1 sequence) and N_2^+(B-X; mainly Δv=1 sequence) are observed. In the spectrum of Figure 16(g), the most intense emitting species are the lines of O^+ $2s^22p^2(^3P)3p$ $^4S^0{}_{3/2} \rightarrow 2s^22p^2(^3P)3s$ $^4P_{3/2}$ at 3727.32 Å, O^+ $2s^22p^2(^3P)3p$ $^4S^0{}_{3/2} \rightarrow 2s^22p^2(^3P)3s$ $^4P_{5/2}$ at 3749.49 Å, N^+ $2s^22p(^2P^0)4s$ $^3P^0{}_2$ $\rightarrow 2s^22p(^2P^0)3p$ $^3P_2$ at 3838.37 Å and N_2^+(B-X) Δv=0 sequence. Several weak intensity ionic lines of O^+ and O^{2+} also overlapped with different bands of N_2^+(B-X) are also present. As in the spectrum of Figure 16(d) a reversal of the bands for high vibrational levels is produced. So, the first vibrational bands of N_2^+(B-X) (0-0, 1-1, 2-2, ...) are shaded to the violet and after reversal are shaded to the red.

In the spectrum of Figure 16(h), several strong intensity ionic lines of N^+, O^+ and N^{2+} also overlapped with different bands of N_2^+(B-X) Δv=0 sequence. The relative intensity of the 0-0 band head at 3914.9 Å in the N_2^+(B-X) system is quite weak and partially overlapped with one O^+ line. Nevertheless, the 0-0 band is the most intense of their band sequence. In this spectrum, many weak intensity rovibrational molecular bands of N_2^+(B-X) can be appreciated. The strong emission observed in Figures 16(i-l) is mainly due to the relaxation of excited ionic fragments N^+ and O^+.

In Figure 16(j) various ionic lines overlap with molecular bands of N_2^+(B-X) transitions and produce rather complex structure, but this high-resolution spectra allow for a precise attribution of almost all observed transitions. The six lines of single ionized nitrogen between 4600-4650 Å correspond to the multiplet structure of N^+ $2s^22p(^2P^0)3p$ $^3P^0{}_{J'} \rightarrow 2s^22p(^2P^0)3s$

$^3P^0_{J''}$. As it can see, the LIB of air includes mainly contributions of both nitrogen and oxygen. It should be noted that the nitrogen and oxygen line intensities maintain the proportions of the air composition.

On the other hand, excitation temperatures of 23400 ± 700 K and 26600 ± 1400 K were estimated by means of N^+ and O^+ ionic lines, respectively. Electron number densities of the order of (0.5-2.4)×10^{17} cm^{-3} and (0.6-7.5)×10^{17} cm^{-3} were deduced from the Stark broadening of several ionic N^+ and O^+ lines, respectively.

LIBS spectra obtained by laser irradiation at the different wavelengths are compared in Figure 17. These high-resolution LIB emission spectra in air were obtained at atmospheric pressure, excited by two TEA-CO_2 laser wavelengths at 10.591 μm (I_W=6.31 GW×cm^{-2}) and 9.621 μm (I_W=5.36 GW×cm^{-2}).

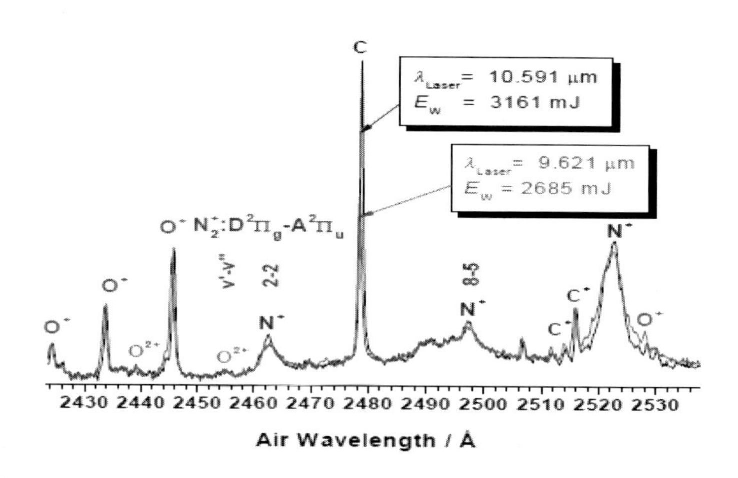

Figure 17. High-resolution LIB emission spectra in air at atmospheric pressure, excited by two TEA-CO_2 laser wavelengths at 10.591 μm (I_W=6.31 GW/cm^2) and 9.621 μm (I_W=5.36 GW/cm^2).

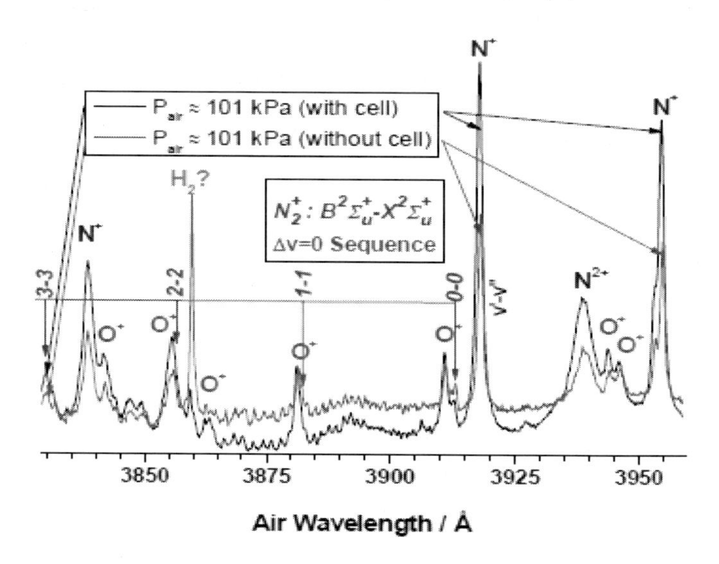

Figure 18. High-resolution LIB emission spectra in air at atmospheric pressure, excited by the TEA-CO_2 laser line at 9.621 μm (I_W=5.36 GW/cm^2) recorded with cell and without cell.

The spectral range was chosen in order to detect the emission lines of different atomic, single and double ionized species (C, C^+, N^+, O^+, O^{2+}). Both spectra have been obtained after the incidence of only one laser pulse. The relative intensities for different species practically do not change with the laser wavelength. The atomic C line at 2478.56 Å was enhanced when the LIBS is induced by the CO_2 laser at 10.591 μm. This fact is probably due to a self-absorption process in such resonance line.

Figure 18 shows high-resolution LIB emission spectra in air at atmospheric pressure, excited by the CO_2 laser line at 9.621 μm (I_W=5.36 GW/cm^2), recorded with cell and without cell. We see similar intensities for O^+ and different bands of N_2^+(B-X) Δv=0 sequence in both spectra. However, the intensity of N^+ and N^{2+} lines was increased when the LIB is with cell.

Also, there is a line in the spectra at ~3860 Å, possible due to H_2, which intensity increases without cell. This fact is probably caused by differences in relative humidity of air in both situations.

To see the effect that the laser irradiance has on the air breakdown measurements were also carried out at different laser intensities. Low-resolution LIB emission spectra of the air plasma at atmospheric pressure, excited by the TEA-CO_2 laser wavelength at 9.621 μm, as a function of the laser intensity is shown in Figure 19. These spectra were recorded after the incidence of only one pulse of the TEA-CO_2 laser. An increase of atomic/ionic emission intensity with increasing the laser irradiance was observed. Also the background increases with the laser power. At higher laser power densities (6.31-4.00 GWxcm^{-2}), the spectral lines are more broadened than at lower power densities as a result of the high pressure associated with the plasma. It is assumed that at higher laser fluence the LIB plasma is more energetic and more ionized. In order to see the effect of the laser intensity on molecular band emission, high-resolution LIB emission spectra of the air plasma at atmospheric pressure, (excited by the TEA-CO_2 laser wavelength at 9.621 μm) as a function of the laser intensity is shown in Figure 20. The assignment of this spectrum can be found in Figure 4(f). An increase of molecular band intensity of $N_2^+(B^2\Sigma_u^+-X^2\Sigma_g^+)$ and $N_2(C^3\Pi_u-B^3\Pi_g)$ with increasing the laser intensity was observed.

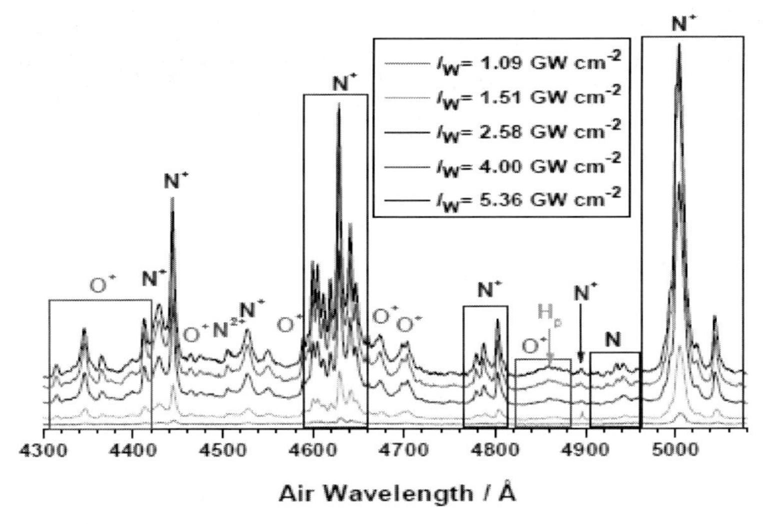

Figure 19. Low-resolution LIB emission spectra in air at atmospheric pressure observed in the 4300-5080 Å region, excited by the TEA-CO_2 laser wavelength at 9.621 μm, as a function of the laser power density.

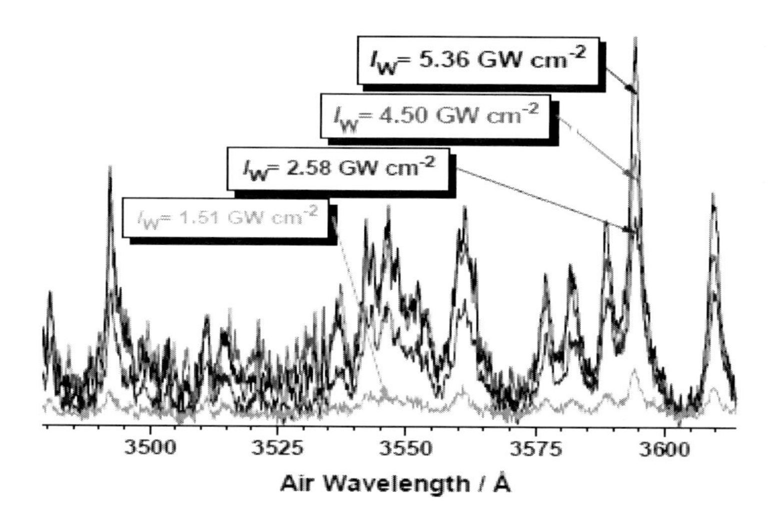

Figure 20. High-resolution LIB emission spectra in air at atmospheric pressure, excited by the TEA-CO_2 laser wavelength at 9.621 μm, as a function of the laser intensity.

Nanosecond TEA CO_2-laser produced plasma emission has been characterized as a function of air pressure. Figure 21 shows LIB emission spectra in the pressure interval from 32 to 101 kPa, excited by the TEA-CO_2 laser (10.591 μm) at a power density of 6.31 GWxcm^{-2}. As can be seen in Figure 21, the intensities of different spectral lines of C, N^+, O^+, O^{2+} and N_2^+(D-A) molecular bands increase with increasing pressure, reach a maximum at about 79.6 kPa, and then decrease with higher pressures. The measurements indicate enhancement due to a longer lifetime of the plasma expanding to a larger size at lower air pressures (p<79.6 kPa). Further increase in air pressure above to ~ 79.6 kPa resulted in a decrease in signal intensity, as a result of a reduction of collisional excitation of the emission lines which occurs when the plasma plume expands into the air atmosphere.

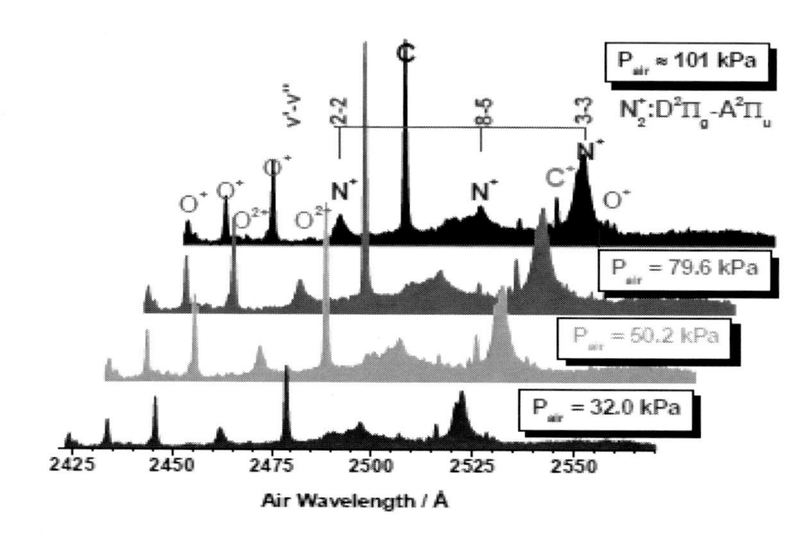

Figure 21. High-resolution LIB emission spectra at various air pressures, excited by the TEA-CO_2 laser (10.591 μm) at a power density of 6.31 GWxcm^{-2}.

The detection of the N_2^+(B-X) bands is of particular interest since it provides an estimation of the effective vibrational and rotational temperatures. The emission intensities of the N_2^+ Δv=-1 and Δv=0 band sequences were analyzed in order to estimate the molecular vibrational temperature T_{vib} (see section 3.5). Two Boltzmann plots (Equation 3.17) of the band intensities against the vibrational energy at the laser irradiance 5.36 GWxcm^{-2} are given in Figure 22(a-b) along with the corresponding Franck-Condon factors. The estimated effective vibrational temperatures were T_{vib}=12100 ± 700 K and 12000 ± 900 K, respectively. On the other hand, the emission intensities of the 0-0 ($B^2\Sigma_u^+$-$X^2\Sigma_g^+$) band of N_2^+ were analyzed in order to estimate the effective rotational temperature T_{rot}. In $^2\Sigma$-$^2\Sigma$ transitions both electronic states belong to Hund´s (b) coupling case (ΔN=±1) [46-48]. Each line of the *R* and *P* branches are doublet since they separate in two sublevels with $J=N\pm½$ each one subscripted with i=1, 2 (1 for $J=N+½$ and 2 for $J=N-½$)

$$R_i(N) = \widetilde{\nu}_0 + F_i^{'}(N+1) - F_i^{''}(N), \tag{5.1}$$

$$P_i(N) = \widetilde{\nu}_0 + F_i^{'}(N-1) - F_i^{''}(N). \tag{5.2}$$

In Equations (5.1) and (5.2) the *Q*-branch forms satellite branches, whose intensities are much lower than the main ones, very close to the corresponding *R* and *P* branches lines (called *R*-form or *P*-form *Q* branches) with wavenumbers:

$$^{R}Q_{21}(N) = \widetilde{\nu}_0 + F_{i2}^{'}(N+1) - F_1^{''}(N), \tag{5.3}$$

$$^{P}Q_{12}(N) = \widetilde{\nu}_0 + F_{i2}^{'}(N-1) - F_2^{''}(N). \tag{5.4}$$

The subscripts 21 and 12 indicate the transition which takes place from a term of the F_2 series to one of the F_1 series or vice versa:

$$F_1(N = J - \frac{1}{2}) = B_v N(N+1) + ... + \gamma\frac{N}{2}, \tag{5.5}$$

$$F_2(N = J + \frac{1}{2}) = B_v N(N+1) + ... - \gamma\frac{N+1}{2} \tag{5.6}$$

In Equations (5.1)-(5.6) *N* is the total angular momentum excluding spin of nuclear rotation, $\widetilde{\nu}_0$ is the wavenumber of the pure vibrational transition, B_v is rotational spectroscopic constant and γ is the spin-rotation coupling constant. Moreover, in the case of a molecule with two identical nuclei such as N_2^+, relative line intensities are affected by the nuclear spin (I=1). The nuclear spin governs the intensities through the Pauli exclusion principle; all wavefunctions are antisymmetric with respect to the interchange of fermions (half-integer spin particle) and symmetric with respect to the interchange of bosons (integer spin particle). Then the ratio of the statistical weights of the symmetric and antisymmetric rotational levels is I+1/I for bosons or I/I+1 for fermions. In the case of N_2^+ the $B^2\Sigma_u^+$ electronic state has only odd J values and then the ratio of the statistical weights of the symmetric and antisymmetric rotational levels is 2/1.

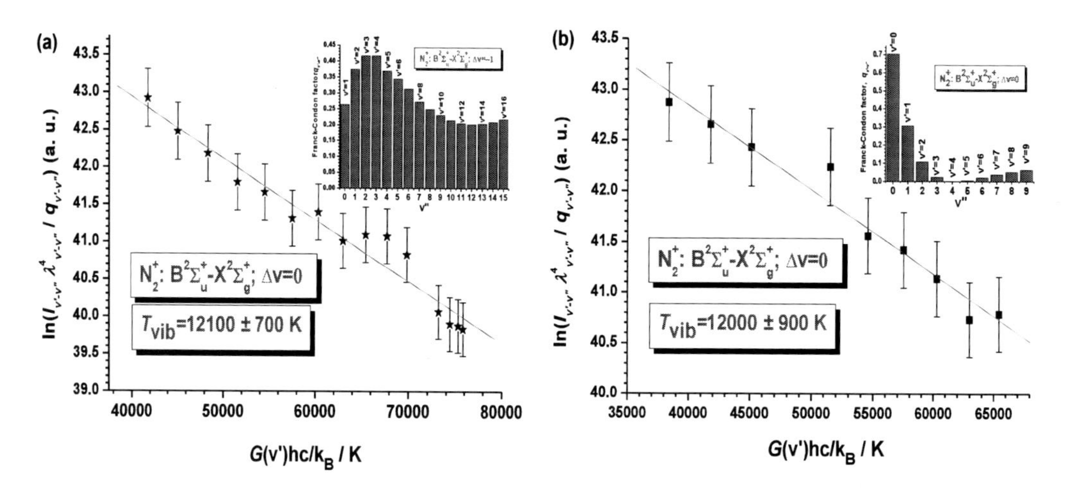

Figure 22. Left (a) panel: Linear Boltzmann plot of the N_2^+ (B-X) Δv=-1 band sequence intensity versus the normalized energy of the upper vibrational level; Right (b) panel: Linear Boltzmann plot of the N_2^+ (B-X) Δv=0 band sequence intensity versus the normalized energy of the upper vibrational level; Experimental conditions: laser power density of 5.36 GW cm^{-2} and atmospheric pressure. Plots also show linear fit to the data and the corresponding Franck-Condon factors.

The assignment of rotational spectrum of 0-0 $B^2\Sigma_u^+$-$X^2\Sigma_g^+$ band of N_2^+, recorded in experimental conditions indicated above, is shown in Figure 23(a). In Figure 23(b) we display the calculated fortrat diagram for P_1, P_2, R_1, R_2, $^RQ_{12}$ and $^PQ_{21}$ branches corresponding to this band. The alternation of the intensities of the rotational lines is observed.

To estimate the effective rotational temperature, we consider the J value for the maximum of the 0-0 band (B-X) of N_2^+ (J_{max}) (Section 3.5)

$$T_{rot} = \frac{2B_0 hc}{k_B}(J_{max} + \frac{1}{2})^2 , \qquad (5.7)$$

being B_0 the rotational constant for v'=0 and J_{max} the total angular momentum at the maximum. This effective rotational temperature is found to be T_{rot}=1900 ± 100 K. As in any gas, temperature in LIB plasma is determined by the average energies of the plasma species (e^-, N_2, O_2, Ar, H, N, O, C, OH, N_2^+, N^+, O^+, N^{2+}, O^{2+}, C^+, C^{2+} etc) and their relevant degrees of freedom (translational, rotational, vibrational, and those related to electronic excitation). Thus, LIB plasmas, as multi-component systems, are able to exhibit multiple temperatures. In LIB for plasma generation in the laboratory, energy from the laser electric field is first accumulated by the electrons and, subsequently, is transferred from the electrons to the heavy particles. Electrons receive energy from the electric field and, by collision with a heavy particle, lose only a small portion of that energy. That is why the electron temperature in plasma is initially higher than that of heavy particles. Subsequently, collisions of electrons with heavy particles (Joule heating) can equilibrate their temperatures, unless time or energy are not sufficient for the equilibration (such as in LIB and pulsed discharges). The temperature difference between electrons and heavy neutral particles due to Joule heating in the collisional weakly ionized plasma is conventionally proportional to the square of the ratio of the electric field E to the pressure p. Only in the case of small values of E/p do the temperatures of electrons and heavy particles approach each other. Numerous plasmas are characterized by multiple different temperatures related to different plasma particles and different degrees of freedom. Plasmas of this kind are usually called non-thermal plasmas. Although the relationship between different plasma temperatures in non-thermal plasmas can be quite sophisticated, it can be conventionally presented in LIB strongly ionized plasmas as T_e (electron temperature) > T_i (ions or excitation temperature) > T_{vib} > T_{rot}. In LIB non-thermal plasma studied here, electron temperature is about 10 eV, ions temperature ~2 eV, vibrational temperature ~1 eV, rotational temperature ~0.2 eV, whereas the gas temperature is close to rotational temperature.

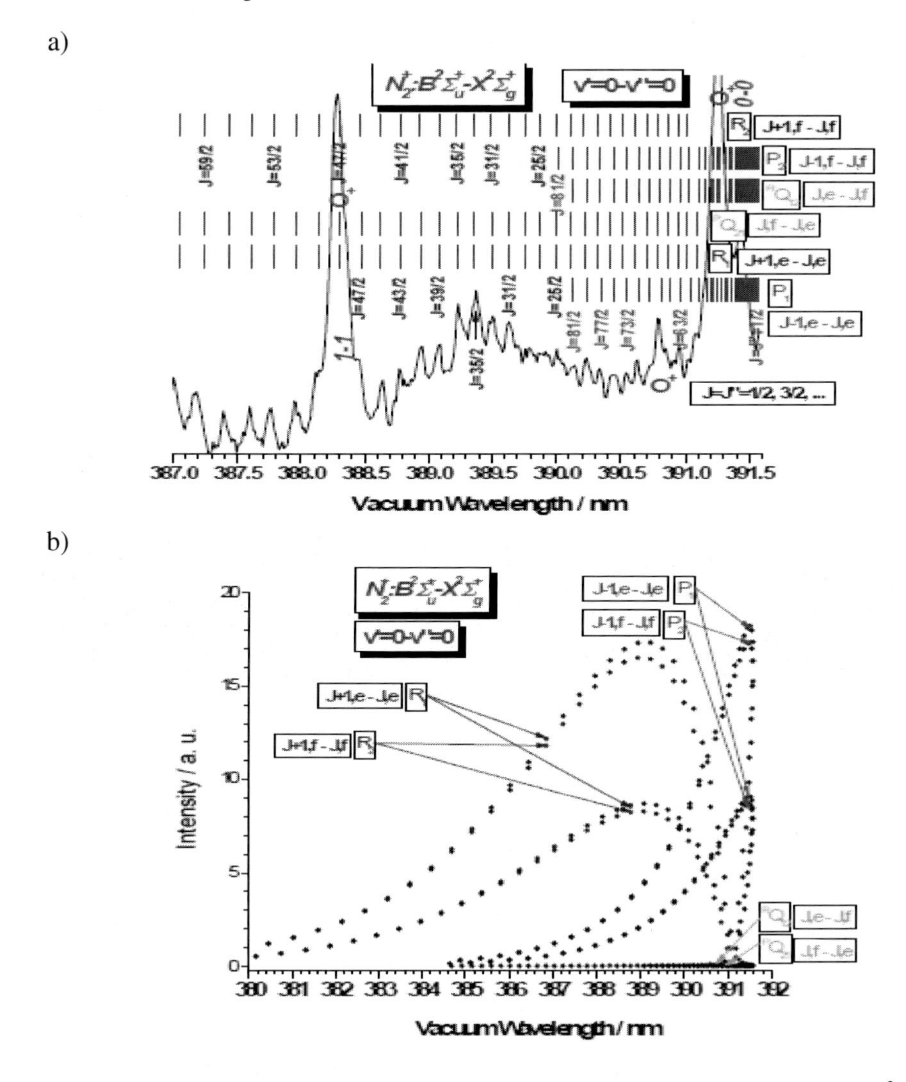

Figure 23. (a) Detailed LIB spectrum of air and partial rotational assignments of the 0-0 $B^2\Sigma_u^+$-$X^2\Sigma_g^+$ band of N_2^+. (b) Calculated fortrat diagram for P_1, P_2, R_1, R_2, $^RQ_{12}$ and $^PQ_{21}$ branches in the 0-0 band (B-X) of N_2^+ at T_{rot}=1950 K.

The strong atomic and ionic lines observed in LIB spectra of air indicate the high degree of excitation/ionization and the high rate of the N_2, O_2 dissociation processes achieved in the plasma. As we mentioned in section 2.2, when a high-power laser beam of intensity I_W interacts with a gas, electrons involving the formation of laser-induced plasma can be generated through two main processes: MPI and EII both followed by electron cascade. The ionization rate in MPI varies as I_W^n where n is the number of photons needed to strip off an electron. MPI is relatively improbable for nitrogen and oxygen atoms or molecules in the ground state [N($2s^22p^3$ $^4S^0{}_{3/2}$), O($2s^22p^4$ $^3P_2$), $N_2(X^1\Sigma_g^+)$ and $O_2(X^3\Sigma_g^-)$], since their high ionization potentials (14.534, 13.618, 9.567 and 12.070 eV, respectively), means that more than 100 photons are required for these processes. Besides, the probability of simultaneous absorption of photons decreases with the number of photons n necessary to cause ionization. In general, this probability is $W_{MPI} \propto \Phi_W^n \propto F_E^{2n}$. Calculations of MPI probability (Equation 2.4) for N, O, N_2 and O_2 give $W_{MPI}\cong 0$ s^{-1} for the CO_2 laser at λ=9.621 μm and I_W=4.5 GW×cm^{-2}. For example, for a 193 nm (ArF) at the laser intensity I_W=1 GW×cm^{-2} (n=2), the probability of MPI for O_2 gives W_{MPI}=1058 s^{-1}. EII consists on the absorption of light photon by free or quasifree electrons. These electrons in the focal volume gain sufficient energy, from the laser field through IB collisions with neutrals, to ionize mainly nitrogen and oxygen atoms, molecules or ions by inelastic electron-particle collision resulting in two electrons of lower energy being available to start the process again. EII is the most important process for the longer wavelengths used in this work. On the other hand, we have made experimental measurements of breakdown threshold laser intensities of N_2, O_2 and ambient air (Section 2.4). The threshold power density is dependent on the kind of laser, laser wavelength, pulse length, beam size of the focal volume, and gas pressure. Breakdown thresholds of solids and liquids are usually lower than for gases. Experimental threshold power densities for air is measured for the TEA-CO_2 laser at λ=10.591 μm. The threshold power densities for air at a given pressure are measured in the two following manners. First the cell was evacuated with the aid of a rotary pump, to a base pressure of 4 Pa that was measured by a mechanical gauge and then it was filled with air up to the desired pressure. The TEA-CO_2 laser was fired and its energy transmitted through the cell was increased until the breakdown was observed in 50% of laser pulses. The threshold was easily determined because it was always associated with the appearance of a blued bright flash of light in the focal region, with a cracking noise, and the abrupt absorption of the laser pulse transmitted through the focal region. Another way to measure the threshold was to induce a previous breakdown at a pressure over the desired value, later the pressure is lowered and the energy adjusted until the breakdown begins with some probability, usually around 50%. This method is similar to induce the breakdown with energy in excess and to attenuate the laser until the spark disappears.

In these cases it could be that initial free electrons have been produced by previous breakdowns and they are the seed of the avalanche process. In this last method the obtained threshold value is normally lower. The present experiments have shown that when high laser energy was used, air breakdown occurred easily and it was reproducible. When the laser energy was reduced to its threshold power density value, air breakdown became a sporadic event. Such sporadic behavior might be due to the difficulty of generating the seed electrons at the breakdown threshold values.

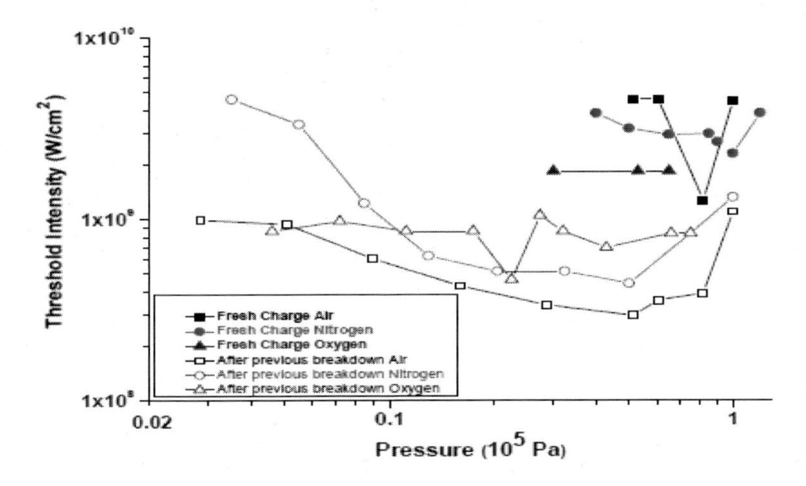

Figure 24. Experimental laser-induced breakdown thresholds excited by the TEA-CO_2 laser (10.591 μm) in air (square), nitrogen [18] (circle) and oxygen [20] (triangle) for different pressures. Solid: fresh charge (without previous breakdown); Open: after previous breakdown.

Figure 24 shows the measured breakdown threshold intensity for air, N_2 [18] and O_2 [20] as a function of pressure. We have measured the breakdown threshold intensity in air at atmospheric pressure finding 4.5×10^9 W×cm^{-2} (1.3×10^6 V×cm^{-1}) for an air fresh charge and 1.1×10^9 W×cm^{-2} (6.4×10^5 V×cm^{-1}) for an air non-fresh charge. As can be seen from Figure 24, if a breakdown has been previously induced in the gas (open symbol), the pressure range to induce the breakdown is bigger and the power density is lower than when no previous breakdown is induced (solid symbol). The number of free electrons is higher in conditions of previous breakdown than in the case of no previous breakdown, lowering the threshold for the plasma initiation. This has been observed by different authors. LIB threshold intensity of air at atmospheric pressure has been measured by Alcock *et al* [$\sim4\times10^{10}$ W×cm^{-2} (Ruby, 0.3472 μm) and $\sim10^{11}$ W×cm^{-2} (0.6943 μm)] [64], Haught *et al* ($\sim4\times10^{14}$ W×cm^{-2}) [65], De Michelis ($\sim4\times10^{10}$ Wx cm^{-2}) [66], Ireland *et al* ($\sim4\times10^{10}$ W×cm^{-2}) [67], Aaron *et al* ($\sim4\times10^{14}$ W×cm^{-2}) [68], Phuoc and White ($\sim2.5\times10^{12}$ W×cm^{-2}) [69], Tomlinson *et al* [$\sim2\times10^{11}$ W×cm^{-2} (Ruby, 0.69 μm) and $\sim7\times10^{10}$ W×cm^{-2} (Nd:YAG, 1.06 μm)] [70], Chan *et al* [$\sim1.5\times10^9$ W×cm^{-2} (CO_2, 10.6 μm, Λ(diffusion length)=4.8×10^{-3} cm), $\sim3\times10^9$ W×cm^{-2} (Λ=3.2×10^{-3} cm), and $\sim7\times10^9$ W×cm^{-2} (Λ=1.6×10^{-3} cm)] [30], Kawahara ($\sim7\times10^{10}$ W×cm^{-2}, Nd:YAG, 1.06 μm)] [71] and Zhuzhukalo *et al* [$\sim7\times10^{10}$ W×cm^{-2} (focal length = 1.4 m) and $\sim2\times10^{10}$ W×cm^{-2} (focal length =14 m)] [72]. Alcock *et al* [64] reported a decrease of the breakdown threshold intensity from $\sim9\times10^{10}$ to $\sim4\times10^{10}$ W×cm^{-2} at 0.3472 μm and $\sim2\times10^{11}$ to $\sim1\times10^{11}$ W×cm^{-2} at 0.6943 μm for air pressures between ~200 Torr and ~800 Torr. Chang *et al* [30] reported for several gases (O_2, N_2, air, He, and Ne) that the threshold power density decreases as the pressure increases and that it decreases as the focal volume increases. For air they reported a slow decrease with a minimum of the breakdown threshold intensity from $\sim1.5\times10^9$ to $\sim1\times10^9$ W×cm^{-2} at Λ=4.8×10^{-3} cm, $\sim3\times10^9$ to $\sim1.5\times10^9$ W×cm^{-2} at Λ=3.2×10^{-3} cm, and $\sim10\times10^9$ to $\sim5\times10^9$ W×cm^{-2} at Λ=1.6×10^{-3} cm for air pressures between ~200 Torr and ~10000 Torr. In this work, they used a CO_2 laser and the focal diameter range from 0.75×10^{-2} to 3×10^{-2} cm. Phuoc and White [69] reported a decrease of the breakdown threshold intensity from $\sim2\times10^{13}$ to $\sim2\times10^{12}$ W×cm^{-2} at 0.532 μm and $\sim8\times10^{12}$ to $\sim1\times10^{12}$ W×cm^{-2} at 1.064 μm for air pressures between 50 Torr and 3000 Torr. It has to be noted that we have obtained similar threshold power densities for air than those given by Chan [30], but lower values than reported in Refs. [64-72]. This fact can be related in part to the used focal length (24 cm) and beam size in the focal region (7.85×10^{-3} cm^2) that is one order of magnitude, at least, higher that the values commonly used in the literature, favoring the probability of existence of free electrons to seed the process and decreasing the threshold laser intensity due to the lack of the diffusion losses.

It has been established [3-10] that the threshold photon flux density or equivalently the threshold power density for MPI varies with $p^{-1/n}$, where p is the gas pressure and n is the number of simultaneously absorbed photons (see section 2.4). Therefore, MPI predicts a very weak dependence of the threshold power density on pressure. However, as we can see from Figure 24, the breakdown threshold power density in air versus pressure shows a minimum around 5×10^4 Pa if previous breakdown have existed and 8×10^4 Pa without previous breakdown.

Therefore, it can be seen from Figure 24 that the pressure dependence is not in harmony with MPI which predicts a very weak $p^{-1/n}$ dependence for the threshold power density, while it is in qualitative agreement with electron cascade. A minimum in the variation of the threshold power density versus pressure is predicted by the classical theory [2-3, 23]. In our experiments, a minimum in the threshold power density versus pressure curve (Figure 24) is observed. Therefore, starting from our experimental observations and calculations, we can conclude that although, the first electrons must appear via MPI or natural ionization, electron impact is the main mechanism responsible for the breakdown in air.

5.3.1. Temporal Evolution of the LIB Plasma

The absorption of light and heating of a gas is of primary significance for important practical problems (the fire ball of an explosion, the heating of artificial satellites during re-entry into the atmosphere, detection of environmental pollutants, ignition systems, laser machining, inertially confined fusion, etc.).

In this section, time-resolved OES analysis for the plasma produced by high-power tunable IR CO_2 pulsed laser breakdown of air is presented [21]. In these series of experiments, the CO_2 pulsed laser (λ=10.591 μm, 64 ns (FWHM), 47-347 J/cm^2) was focused onto a metal mesh target under air as host gas at atmospheric pressure [21]. It was found that the CO_2 laser is favorable for generating strong, large volume air breakdown plasma, in which the air plasma was then produced. While the metal mesh target itself was practically never ablated, the air breakdown is mainly due to electronic relaxation of excited N, O, C, H, Ar and ionic fragments N^+, O^+, N^{2+}, O^{2+}, C^+ and molecular band systems of $N_2^+(B^2\Sigma_u^+-X^2\Sigma_g^+$; $D^2\Pi_g-A^2\Pi_u)$, $N_2(C^3\Pi_u-B^3\Pi_g)$ and $OH(A^2\Sigma^+-X^2\Pi)$.

The difficulties arise on account of the imprecise definition of the extent of the focal region which, in turn lead to inadequate or inaccurate knowledge of the spatial- temporal characteristic of the beam intensity within the focal region for gases. On the other hand, it is well known that a pure gas plasma, due to the gas breakdown process, is produced when a TEA CO_2 laser is focused onto a metal sample at a gas pressure of around 1 atm, in which case practically all the laser irradiance is absorbed in the gas plasma [33, 73]. It is expected that, in some medium range pressures between 0.01 torr and 1 atm, both gas plasma and target plasmas are produced. In such a case, some interaction inevitable takes place between the gas and the target laser induced plasmas. The plasma is a mixture of electrons, atoms, molecules and ions, and mass from both the ablated target and the ambient gas. The interaction between the ejected mass (plume) and the surrounding air slows the expansion of the plasma. If a metal mesh is used as a target, low amount (if any) of ablated material will be formed and the spatial origin of gas breakdown process will be accurately defined independently of the laser fluence. After its formation in the vicinity of the metal mesh surface, the air breakdown plasma propagates towards the laser source at a supersonic speed. The shock wave heats up the surrounding air which is instantaneously transformed in strongly ionized plasma. Two different types of spectra were recorded: time-integrated and time-resolved. In the acquisition of time-integrated spectra, a good signal to noise ratio has been obtained averaging each spectrum over several successive laser pulses. Typically the signals from 20 laser pulses are averaged and integrated over the entire emission time. In time-resolved measurements, the delay t_d and width t_w times were varied. It was verified that the plasma emission was reproducible over more than 7 ablation events by recording the same spectrum several times. The temporal history of LIB air plasma is illustrated schematically in Figure 25. The time for the beginning of the CO_2 laser pulse is considered as the origin of the time scale (t=0). Inserts illustrate some emission spectra recorded at different delay and width times at two observed distances of z=2.5 and 5 mm. The temporal shape of the CO_2 laser pulse is also shown.

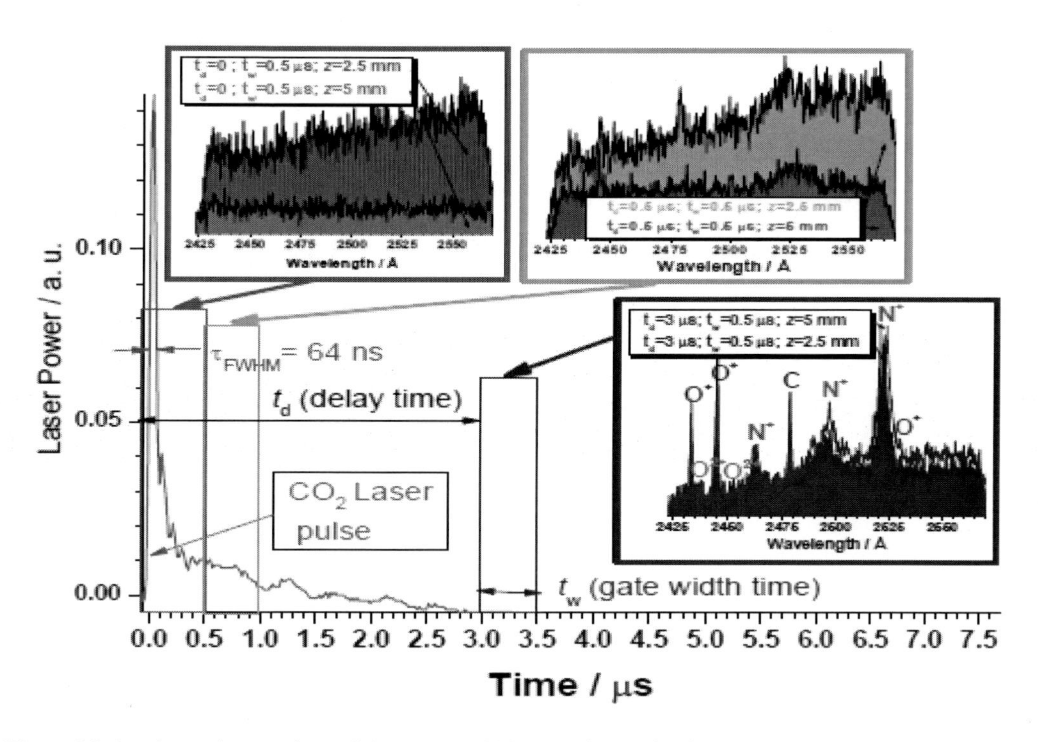

Figure 25. A schematic overview of the temporal history of LIB air plasma. Here t_d is the delay time and t_w is the gate width time during which the plasma emission is monitored. Insets plots illustrate some spectra observed at different delay times (0, 0.5 and 3 μs) for a fixed gate width time of 0.5 μs and z=2.5 and 5 mm. The temporal shape of the CO_2 laser pulse is also shown.

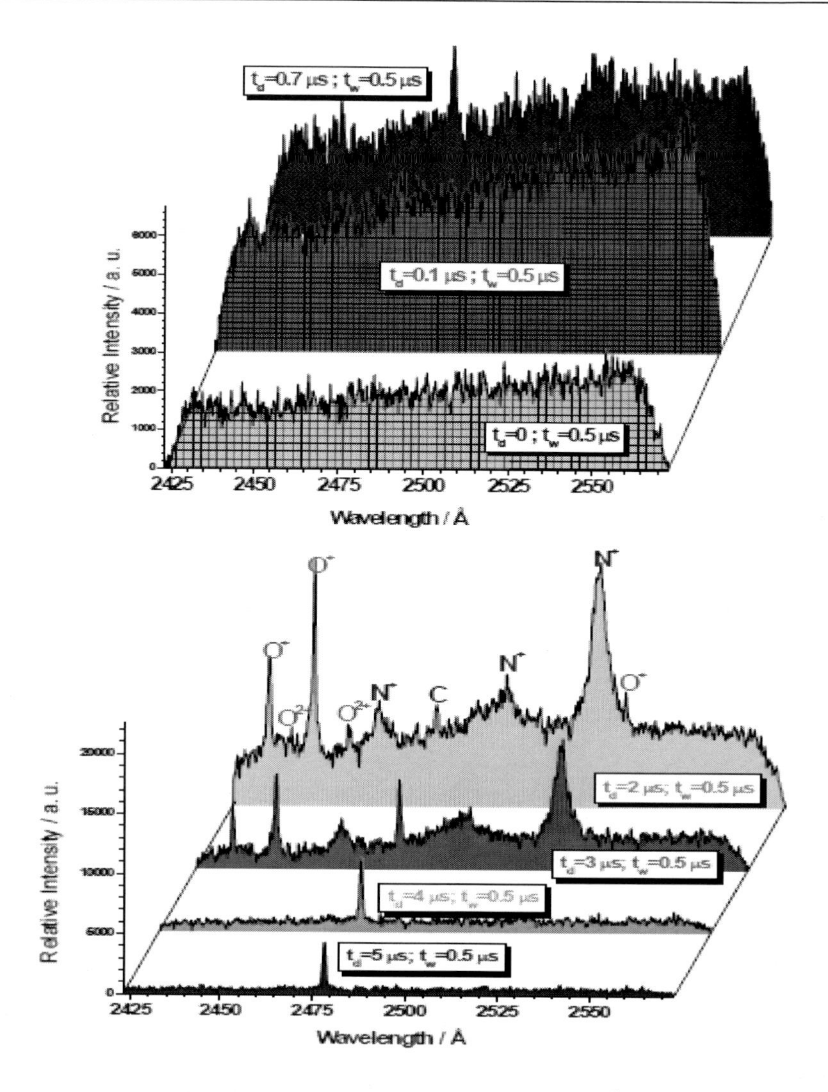

Figure 26. Time-resolved emission spectra from laser-induced (106 J/cm^2) air plasma observed in the region 2423-2573 Å monitored at 0, 0.1, 0.7, 2, 3, 4, and 5 μs gate delays for a fixed gate width time of 0.5 μs and z=2.5 mm.

The LIB spectra of air were measured at different delay and width times. In a first set of experimental measurements, the spectra have been obtained in the spectral region 2423-2573 Å, at a delay time ranging from 0 to 30 μs at 0.5 μs intervals and at an observed distance z from 1 to 10 mm. The spectral range was chosen in order to detect both single and double ionized oxygen species, several single ionized nitrogen lines and atomic carbon. Figure 26 illustrates the time-resolved evolution from laser-induced (106 J/cm^2) air plasma monitored at 0, 0.1, 0.7, 2, 3, 4, and 5 μs gate delays for a fixed gate width time of 0.5 μs and z=2.5 mm. One can see that after the laser pulse, the plasma emission consists in an intense continuum. This continuum radiation is emitted by the laser-induced plasma as a result of free-free and free-bound transitions. As seen in Figure 26 during the initial stages after laser pulse (t_d≤0.5 μs), continuum emission dominates the spectrum. As time evolves (0.5 μs≤t_d≤3.5 μs), N^+, O^+ and O^{2+} emissions dominate the spectrum. These ionic lines decrease quickly for higher delay times, being detected up to ~ 4.5 μs. The emission lines become progressively narrower as a consequence of the electron number density distribution. It points out that the electron density and excitation temperature must decrease during the plasma expansion. As the delay is increased (t_d>3.5 μs) C atomic emission line dominates the spectrum. This atomic line decreases for higher delay times, being detected up to ~ 30 μs. The maximum intensity of continuum and spectral lines is reached after a characteristic time, depending on the observation distance *z*.

Figure 27 shows time-resolved emission spectra from LIB (106 J/cm^2) in air collected at different distances (*z*=1 and 7.5 mm) and recorded at 2, 3, 4, and 5 μs gate delays for a fixed gate width time of 0.5 μs. By analyzing Figures 26 and 27 it is possible to see that C atoms are produced both in the ablated target and in the air breakdown. When the spectra were recorded near the target surface (*z*<2.5 mm), the main contribution of C atoms is due to the target surface. For *z*>2.5 mm, practically the spectrum is due to air breakdown. At far away distances from the metal mesh target surface, the plasma front arrives later than for close distances.

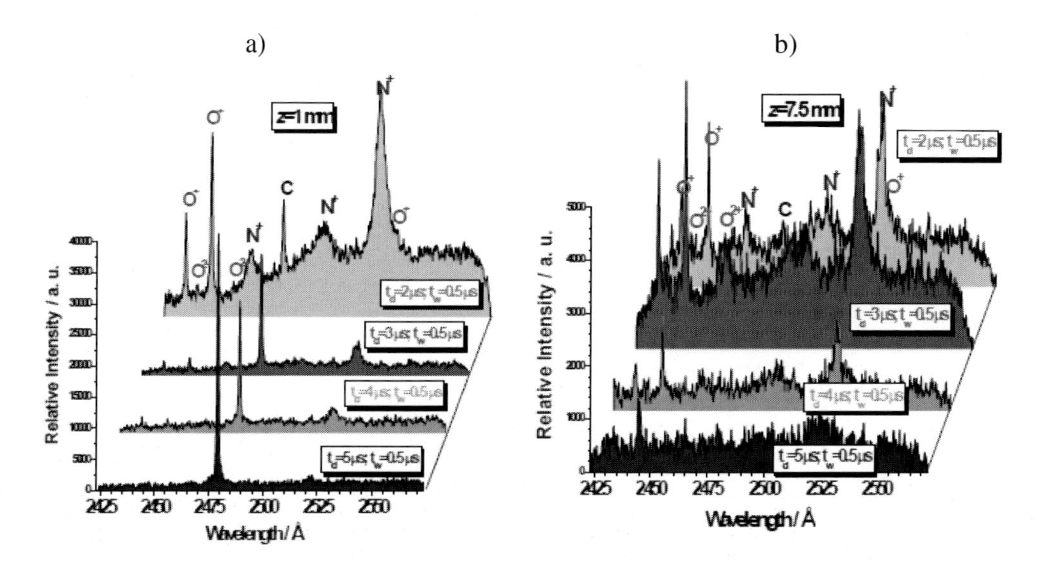

Figure 27. Time-resolved emission spectra from laser-induced (106 J/cm^2) air plasma observed in the region 2423-2573 Å monitored at 2, 3, 4, and 5 μs gate delays for a fixed gate width time of 0.5 μs and (a) *z*=1 mm; (b) *z*=7.5 mm.

Figure 28 displays several time-resolved emission spectra from LIB (106 J/cm^2) in air monitored at 1, 2.5, 5, and 7.5 mm observed distances and after the laser pulse (t_d<0.5 μs). In this Figure, one can see that at short time after the TEA CO_2 laser pulse, the plasma emission consists in an intense continuum that decreases with distance. Figure 29 shows time-resolved emission spectra from laser-induced air plasma (106 J/cm^2) recorded at t_d=3 μs with t_w=0.5 μs monitored at 1, 2.5, 5, 7.5 and 10 mm. In these recording conditions, the most intense plasma is monitored at 5 mm from the mesh target surface.

Figure 30 displays the temporal evolution of the LIBS air plasma (160 J/cm^2) in the spectral region 3725-3860 Å region monitored at 2.5, 3 and 4 μs gate delays for a fixed gate width time of 0.5 μs (*z*=1 cm) and time-integrated spectrum (t_d=0 and t_w>>30 μs). The inset plot shows the assignment of some ionic lines of N^+, O^+ and O^{2+} and band heads of different molecular bands of $N_2^+(B^2\Sigma_u^+-X^2\Sigma_g^+$; $\Delta v=0$ sequence) [see Figure 17(g)]. At early times (t_d<1 μs) (not shown), the plasma emission consists in a weak continuum. When the delay increases, some ionic lines of N^+, O^+ and O^{2+} and band heads of the molecular bands of $N_2^+(B^2\Sigma_u^+-X^2\Sigma_g^+)$ enhanced steeply as a consequence of the expansion and heating of the air plasma. At longer times (t_d>3 μs), the ion lines significantly decrease steeply in intensity as a consequence of the expansion and cooling of the plasma plume and its recombination into ground state ions. At t_d>4 μs, N^+ and O^+ ionic lines and N_2^+ rovibrational lines disappear. Figure 31 shows the temporal evolution of the LIBS air plasma (71 J/cm^2) in the spectral region 3830-3960 Å region monitored at 2, 3 and 4 μs gate delays for a fixed gate width time of 0.5 μs (z=1 cm) and time-integrated spectrum (t_d=0 and t_w>>30 μs).

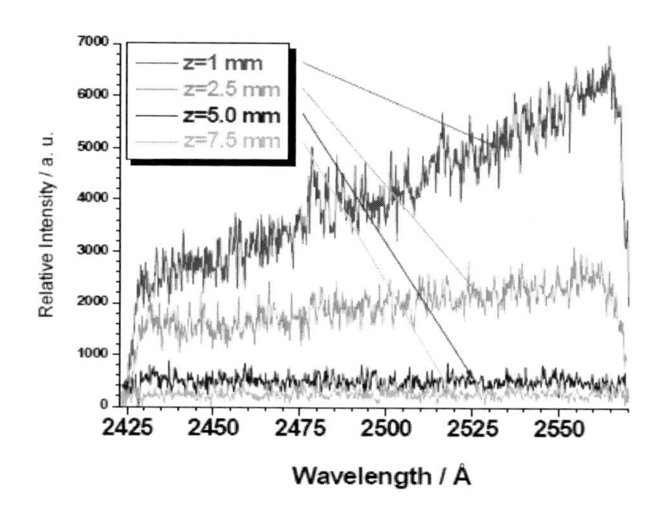

Figure 28. Time-resolved emission spectra from LIB (106 J/cm^2) in air at 1, 2.5, 5, and 7.5 mm along the plasma expansion direction (Z-axis) and recorded at t_d=0 and t_w=0.5 μs.

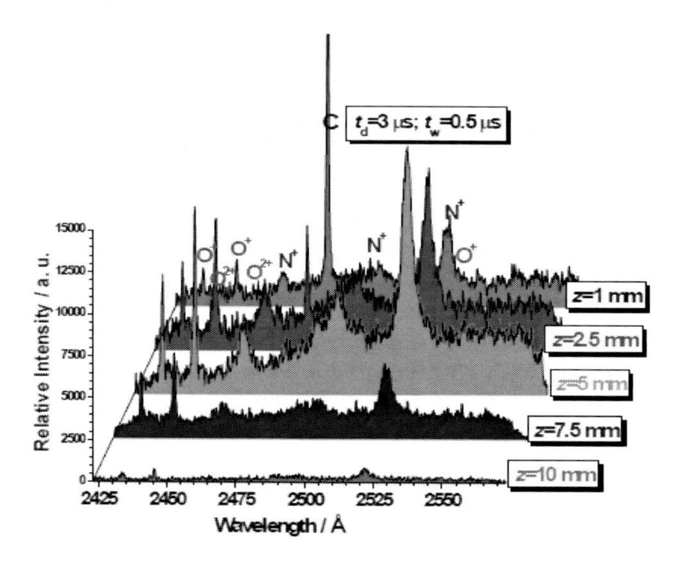

Figure 29. Time-resolved emission spectra from laser-induced air plasma (106 J/cm^2) observed in the region 2423-2573 Å at t_d=3 μs for t_w=0.5 μs monitored at 1, 2.5, 5, 7.5 and 10 mm along the plasma expansion direction.

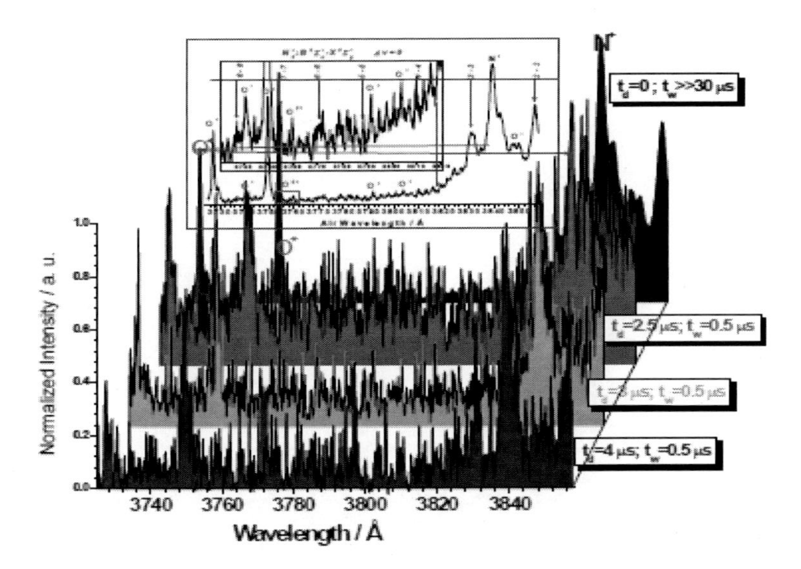

Figure 30. Time-resolved emission spectra from laser-induced (160 J/cm^2) air plasma observed in the region 3725-3860 Å region monitored at 2.5, 3 and 4 μs gate delays for a fixed gate width time of 0.5 μs (z=1 cm) and time-integrated spectrum (t_d=0 and t_w>>30 μs). The inset plot shows the assignment of some ionic lines of N^+, O^+ and O^{2+} and band heads of the molecular bands of $N_2^+(B^2\Sigma_u^+\text{-}X^2\Sigma_g^+)$.

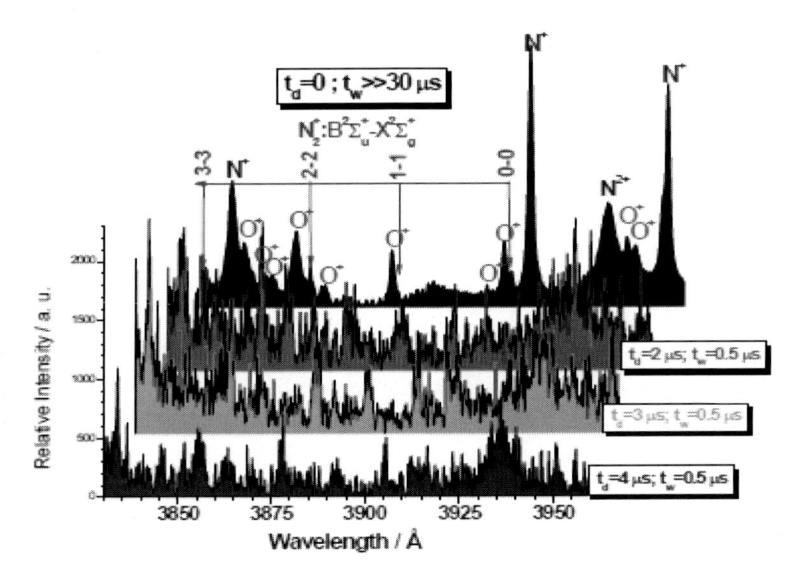

Figure 31. Time-resolved emission spectra from laser-induced air plasma (71 J/cm^2) observed in the region 3830-3960 Å monitored at 2, 3 and 4 μs gate delays for a fixed gate width time of 0.5 μs (z=1 cm) and time-integrated spectrum (t_d=0 and t_w>>30 μs). The assignments of some ionic lines of N^+, O^+ and N^{2+} and band heads of the molecular bands of $N_2^+(B^2\Sigma_u^+\text{-}X^2\Sigma_g^+)$ are indicated.

This plot also shows the assignment of some ionic lines of N^+, O^+ and N^{2+} and band heads of different molecular bands of $N_2^+(B^2\Sigma_u^+\text{-}X^2\Sigma_g^+$; Δv=0 sequence). At early times (t_d<1 μs) (not shown), the plasma emission consists in a continuum. When the delay increases, some ionic lines of N^+, O^+ and N^{2+} and band heads of the molecular bands of $N_2^+(B^2\Sigma_u^+\text{-}X^2\Sigma_g^+)$ enhanced as a consequence of the air plasma expansion. At longer times (t_d>3 μs), the ion lines significantly decrease steeply in intensity. At t_d>5 μs, N^+, N^{2+} and O^+ ionic lines and

N_2^+ rovibrational lines disappear. Figure 32 displays the temporal progress of the LIBS air plasma in the spectral window 4070-4195 Å observed at 2.5, 3 and 4 μs gate delays for a fixed gate width time of 0.5 μs (z=1 cm) and time-integrated spectrum (t_d=0 and t_w>>30 μs). The assignments of some ionic lines of N^+, O^+ and N^{2+} and band heads of different molecular bands of $N_2^+(B^2\Sigma_u^+-X^2\Sigma_g^+$; Δv=-1 sequence) are indicated.

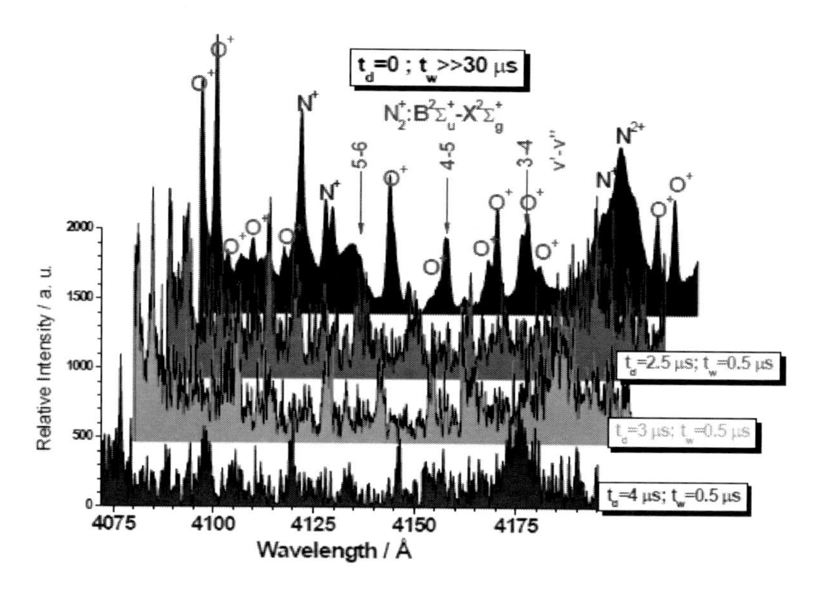

Figure 32. Time-resolved emission spectra from laser-induced air plasma (71 J/cm^2) observed in the region 4070-4195 Å monitored at 2.5, 3 and 4 μs gate delays for a fixed gate width time of 0.5 μs (z=1 cm) and time-integrated spectrum (t_d=0 and t_w>>30 μs). The assignments of some ionic lines of N^+, O^+ and N^{2+} and band heads of the molecular bands of $N_2^+(B^2\Sigma_u^+-X^2\Sigma_g^+)$ are indicated.

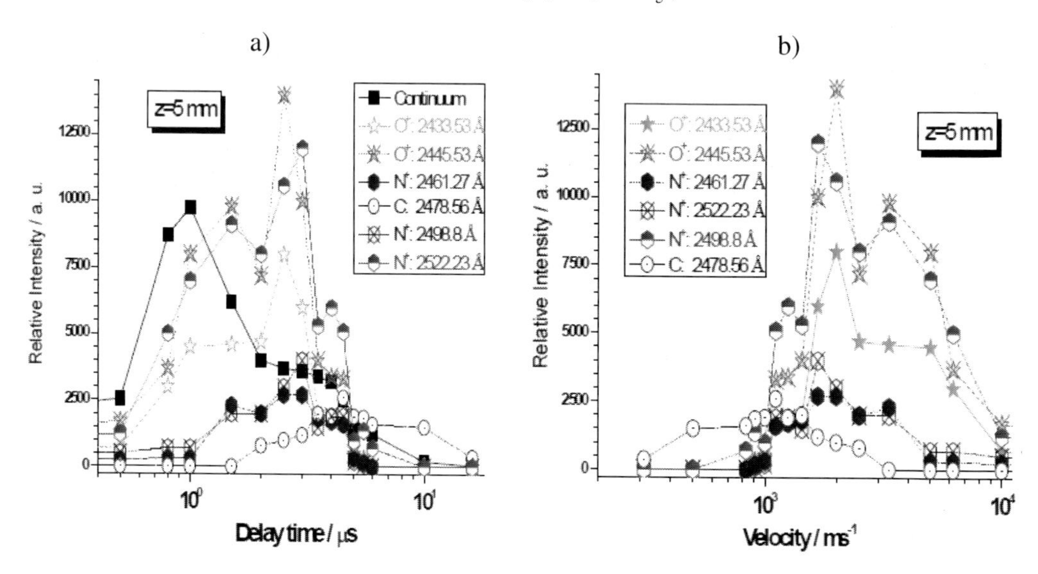

Figure 33. (a) Number density TOF distributions of continuum radiation, O^+(2433.53; 2445.54 Å), N^+(2461.27; 2478.56; 2522.23 Å) and C(2478.56 Å) lines as a function of delay time (fixed gate width time of 0.5 μs) for a laser fluence of 106 J/cm^2 and z=5 mm. (b) Velocity distributions derived from the experimental TOF profiles for the indicated species.

As before, at early times (t_d<1 μs), the plasma emission consists in a weak continuum. When the delay increases, some ionic lines of N^+, O^+ and N^{2+} and band heads of the molecular bands of $N_2^+(B^2\Sigma_u^+-X^2\Sigma_g^+)$ enhanced. At longer times (t_d>3 μs), the ion lines significantly decrease steeply in intensity. At t_d>4 μs, N^+, N^{2+} and O^+ ionic lines and N_2^+ rovibrational lines disappear.

5.3.2. Time of Flight, Velocity, Kinetic Energy and Electron Density

Space-and-time resolved OES measurements could be used to estimate plasma expansion rate and kinetic energy. The temporal evolution of spectral atomic, ionic and molecular line intensities at a constant distance from the target can be used to construct the time-of-flight (TOF) profile. TOF studies of the emission provide fundamental information regarding the time taken for a particular species to evolve after the laser-induced plasma has been formed. Specifically, this technique gives an indication of the velocity of the emitted species. A coarse estimation of the velocity for the different species in the plume can be inferred from the time resolved spectra by plotting the intensities of selected emission lines versus the delay time, and then calculating the velocity by dividing the distance from the target by the time where the emission peaks. This method for determination of plasma velocity should be used with care due to the superposition of both expansion and forward movements of the plasma plume. We assumed a plasma model consisting in two plasmas [74]: primary plasma that acts as initial explosion energy source and emits an intense continuum emission background for a short time just above the surface of the auxiliary target; secondary plasma expands with time around the primary plasma. The secondary plasma is formed by the excitation from the shock wave and by the emitting of atomic, ionic and molecular species characterized by a low background signal. Figure 33(a) displays the TOF profiles for the air breakdown experiments induced by CO_2 laser pulses (106 J/cm^2), of continuum radiation, O^+(2433.53 Å), O^+(2455.53 Å), N^+(2461.27 Å), N^+(2478.56 Å), N^+(2522.23 Å) and C(2478.56 Å) lines as a function of the delay time. We notice the appearance of a strong maximum for continuum background for a delay of 1 μs at z=5 mm. When t_d increases, the continuum drops steeply as a consequence of the reduction of electron density and temperature as the plume expands. For O^+ and N^+ ionic species and C atomic, the maxima appear for a delay of 2.5, 3 and 4.5 μs, respectively. The emission intensity of O^+ and N^+ ionic lines decreases more rapidly than the emission intensity of the C lines.

The time duration of ionic species was nearly 5 μs, while the time duration of C atomic emission was nearly 20 μs. The experimental TOF distributions $N(t)$ are essentially number density distributions. They are converted to flux distributions dN/dt by employing a correction factor z/t, where z means the flight distance along the plasma expansion and t is the delay time after the laser pulse incidence.

It should be mentioned that the estimation of velocity distributions assumes that the emitting species are generated on the assisting metal mesh target. The velocity distributions that are derived from these TOF distributions are display in Figure 33(b). At the laser fluence used in this series of experiments (106 J/cm^2) and z=5 mm, TOF distributions present different characteristics.

Thus, the velocity distributions of ionic species O^+(2445.54 Å) and N^+(2522.23 Å) are comparatively wider (~3.7 and ~4 km/s (FWHM), respectively) than the velocity distribution of carbon neutral species [~1.2 km/s (FWHM)]. The velocity distributions of O^+, N^+ and C lines species are centred at about 2, 1.7 and 1.1 km/s, respectively. From TOF spectra, the translational kinetic energy can be deduced [$KE=(1/2)m(z/t)^2$] by measuring the time t required to transverse the distance from the target to the detector z. The kinetic energy obtained for some species are plotted in Figure 34.

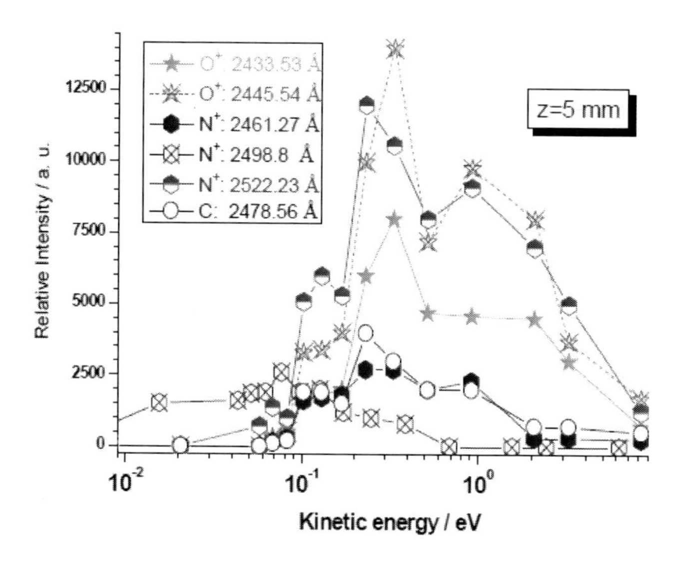

Figure 34. Kinetic energy distributions of O^+(2433.53 Å), O^+(2445.54 Å), N^+(2461.27 Å), N^+(2478.56 Å), N^+(2522.23 Å) and C(2478.56 Å) lines derived from the TOF spectra at z=5 mm.

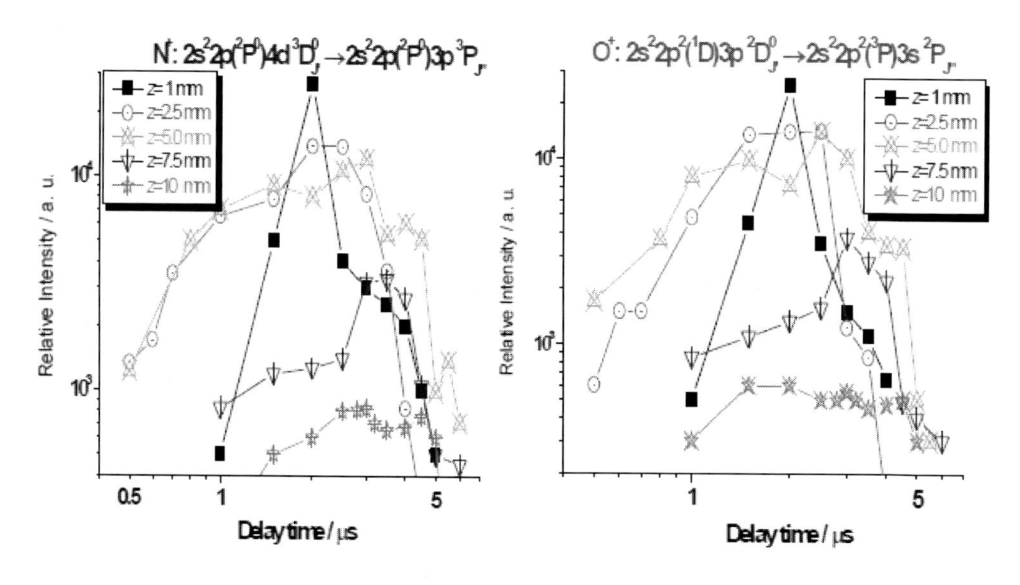

Figure 35. Number density TOF profiles of multiplet structures of N^+(~2522 Å) and O^+(~2445 Å) at 1, 2.5, 5, 7.5 and 10 mm as a function of delay time for a laser fluence of 106 J/cm^2.

We have observed small atomic and ionic average kinetic energies. As we have stated above, for a better understanding of the physical mechanisms underlying the plasma emission breakdown in air, LIB spectra were obtained by varying both the distance z (up to 10 mm) with respect to the auxiliary metal mesh and the laser energy. It is expected, in fact, that these two parameters would affect strongly the dynamic evolution of the plasma and the shock wave induced by the CO_2 laser. Different lines originating from atomic and ionic species of both nitrogen and oxygen were analyzed. O^+ doublet ($2s^22p^2(^1D)3p\ ^2D^0_{3/2,5/2} \rightarrow 2s^22p^2(^3P)3s\ ^2P_{3/2}$) at ~ 2445 Å and N^+ triplet ($2s^22p(^2P^0)4d\ ^3D^0_{1,2,3} \rightarrow 2s^22p(^2P^0)3p\ ^3P_{0,1,2}$) at ~ 2522 Å were chosen as representative of ionized lines. The energies of the lower levels of both multiplets are high (189068.514 for O^+ and 170572.61, 170607.89 and 170666.23 cm^{-1} for N^+) so that the self-absorption effect can be neglected. Figures 35 and 36 show TOF and velocity distributions of multiplet structures of N^+(~2522 Å) and O^+(~2445 Å) at different distances (1, 2.5, 5, 7.5 and 10 mm) as for a laser fluence of 106 J/cm^2 The temporal emission features are affected by the presence of strong continuum at short distances ($z \leq 3$ mm) and at early delay time. But at distances greater than 3 mm, the continuum radiation is considerably reduced and the interference of continuum on the TOF distributions is negligible. The spike observed in TOF and velocity profiles is the prompt signal that is used as a time maker. By the shift of the TOF peaks for each distance it is possible to calculate approximately the mean velocities of LIB along the propagation axis Z. The measured peak velocities of multiplet structures of N^+(~2522 Å) and O^+(~2445 Å) monitored at 1, 2.5, 5, 7.5 and 10 mm (for a laser fluence of 106 J/cm^2) are 0.5, 1, 1.7, 2.1 and 3.6 km/s and 0.5, 1, 2, 2.5 and 5 km/s, respectively. The peak velocities of N^+ and O^+ increase with the distance from the target surface. This is due to the initial acceleration of the ablated partials from zero velocity to a maximum velocity. Also we have studied OES of the air plasma by varying the laser energy. We observed that when the laser fluence is increased, the N^+ and O^+ TOF distributions broaden and move towards lower delay times. On the other hand, plasma temperature was determined from the emission line intensities of several N^+ and O^+ ionized lines observed in the laser-induced plasma of air for a delay time of 3 μs and a distance of z=5 mm. The obtained excitation temperatures, in the case of N^+ and O^+, were 23400 ± 900 K and 26600 ± 1300 K, respectively. For N^+ and O^+ lines, the Doppler line widths vary between 0.08-0.17 Å at 23400 K and 0.11-0.13 Å at 26600 K, respectively.

The N^+ triplet ($2s^22p(^2P^0)4d\ ^3D^0_{1,2,3} \rightarrow 2s^22p(^2P^0)3p\ ^3P_{0,1,2}$) at ~2522 Å was identified as candidate for electron-density measurements. By substituting the Stark line widths at different time delays in Eqn. (3.12) and the corresponding value of Stark broadening W (0.372 Å from Griem [41] at plasma temperature of 23400 K), we obtain the electron density. Figure 37 gives the time evolution of electron density and its first derivative with respect to time by setting the gate width of the intensifier at 0.5 μs. These values have been obtained by Stark broadening of the N^+ TOF curves at z = 5 mm and for a laser fluence of 106 J/cm^2. The initial electron density at 0.1 μs is approximately 1.3×10^{16} cm^{-3}. Afterwards, the density increases and reaches a maximum (1.7×10^{17} cm^{-3}) at ~0.8 μs, and then decrease as the time is further increased.

At shorter delay times (<0.1 μs), the line to continuum ratio is small and the density measurement is sensitive to errors in setting the true continuum level. For times >0.1 μs, the line to continuum ratio is within the reasonable limits and the values of electron density shown in Figure 37 should be reliable. After 6 μs, the electron density is about 2.7×10^{16} cm^{-3}. For a long time >6 μs, subsequent decreased N^+ emission intensities result in poor signal-to-noise ratios, and there exits a limitation in the spectral resolution.

The decrease of n_e is mainly due to recombination between electrons and ions in the plasma. These processes correspond to the so-called radiative recombination and three-body recombination processes in which a third body may be either a heavy particle or an electron.

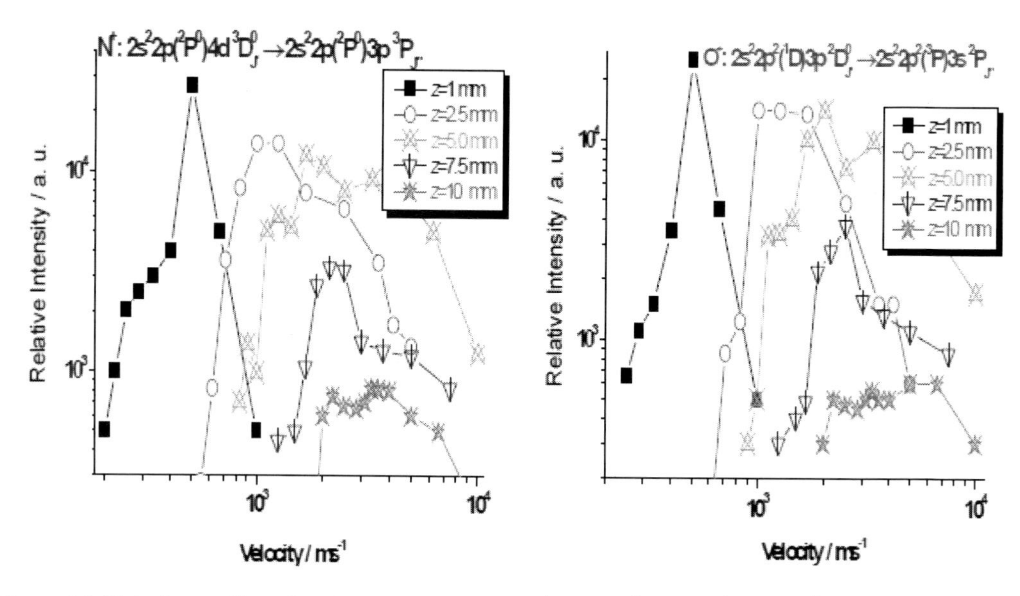

Figure 36. Velocity profiles of multiplet structures of N^+(~2522 Å) and O^+(~2445 Å) at 1, 2.5, 5, 7.5 and 10 mm as a function of delay time for a laser fluence of 106 J/cm^2.

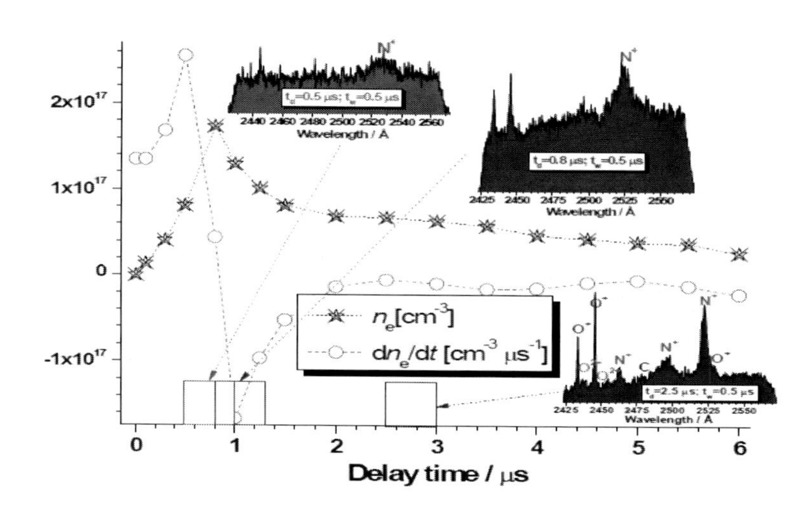

Figure 37. The temporal evolution of electron density n_e and dn_e/dt for different delay times from plasma ignition and z=5 mm. Inset plots illustrate some spectra observed at different delay times.

The electron number density n_e (cm^{-3}) in the laser induced plasma is governed by the kinetic balance equation 2.10. By considering the discussion reported in section 2.3, the equilibrium condition can be established at 0.8 μs (dynamical equilibrium) and t>2 μs (stationary equilibrium). For $t \leq 0.8$ μs the ionization prevails while for 0.8 μs<t<2 μs the three-body recombination dominates. In the case of TEA-CO_2 laser at a laser fluence of 106 J/cm^2 and z=5 mm, the recombination rate constant estimated is approximately 5x10^{-28} cm^6 s^{-1}

(Equation 2.11). The recombination time can be determined by the value of the rate constant of the recombination process as $t_{rec}=1/(n_e^2.k_{rec})$ obtaining $t_{rec}\approx0.4$-3 μs.

Optical emission accompanying TEA-CO_2 nanosecond LIB in air is very long lived (~20 μs) relative to the average radiative lifetimes of the excited levels that give rise to the observed emission lines. All of the emission lines of N, N^+, N^{2+}, O, O^+ and O^{2+} expected in the 2000-10000 Å wavelength range are observed, illustrating that the excited species giving rise to the optical emission are produced by non-specific mechanism during the TEA CO_2 LIB process. However, a direct excitation-de-excitation mechanism cannot explain the observed emission spectra. Electron impact ionization would explain the emission intensity variation with the time for N, N^+, N^{2+}, O, O^+ and O^{2+} species. On the other hand, the formation of the excited molecular species would happen in the gas phase by collisions between atomic or ionic species present in the plume at times far away from the plasma ignition.

The emission process at this plasma stage is divided into two different process associated, respectively with the shock formation and the plasma cooling. During the former, the atoms, molecules and ions gushing out from the laser focal region are adiabatically compressed against the surrounding gas. During the latter stage the temperature of the plasma and consequently the emission intensities of atomic lines and molecular bands decrease gradually.

CONCLUSION

This article reviews some fundamentals of LIBS and some experimental studies developed in our laboratory on N_2, O_2 and air gases using a high-power IR CO_2 pulsed laser. In this experimental study we used several laser wavelengths (λ=9.621 and 10.591 μm) and laser intensity ranging from 0.87 to 6.31 GW×cm^{-2}. The spectra of the generated plasmas are dominated by emission of strong atomic, ionic species and molecular bands. For the assignment of molecular bands a comparison with conventional emission sources was made. Excitation, vibrational and rotational temperatures, ionization degree and electron number density for some species were estimated by using different spectroscopic methods. The characteristics of the spectral emission intensities from different species have been investigated as functions of the gas pressure and laser irradiance. Optical breakdown threshold intensities in different gases have been measured experimentally. The physical processes leading to laser-induced breakdown of the gases have been analyzed. Plasma characteristics in LIBS of air were examined in detail on the emission lines of N^+, O^+ and C by means of time-resolved OES technique. The results show a faster decay of continuum and ionic spectral species than of neutral atomic and molecular ones. The velocity and kinetic energy distributions for different species were obtained from TOF measurements. Excitation temperature and electron density in the laser-induced plasma were estimated from the analysis of spectral data at various times from the laser pulse incidence. Temporal evolution of electron density has been used for the estimation of the three-body recombination rate constant.

ACKNOWLEDGMENTS

We gratefully acknowledge the support received in part by the DGICYT (Spain) Projects: MEC: CTQ2007-60177/BQU and MEC: CTQ2008-05393/BQU for this research.

REFERENCES

[1] Maker, P. D., Terhune, R. W., Savage, C. M. *Proc. 3rd Int. Conf. Quantum Electronics;* (Paris: Dunod) 1963, Vol. 2, 1559.

[2] Raizer, Y. P. Breakdown and heating of gases under the influence of a laser beam; *Soviet Physics: Uspekhi*. Vol. 8, Nº 5, 1966.

[3] Raizer, Y. P. *Laser-Induced discharge phenomena*; Consultants Bureau: New York. 1977.

[4] Bebb, H. B., Gold, A. Multiphoton Ionization of Hydrogen and Rare Gas Atoms; *Physics of Quantum Electronics*; ed. P. L. Kelly et al. McGraw-Hill: New York. 1966.

[5] De Michelis, C. *IEEE J. Quantum Electron*. 1969, 5, 188-202.

[6] Morgan, C. G. *Rep. Prog. Phys*. 1975, 38, 621-665.

[7] Bekefi, G. *Principles of Laser Plasma*; Wiley: New York. 1976.

[8] Lyman, J. L.,Quigley, G. P., Judd, O. P. *Multiple-Photon Excitation and Dissociation of Polyatomic Molecules*; Ed. Cantrell, C. D. Springer: Berlin., 1980.

[9] Rosen, D. I., Weyl, G. *J. Phys. D. Appl. Phys*., 1978, 20, 1264-1276.

[10] Gamal, D., Ye, E. *J. Phys, D. Appl. Phys*., 1988, 21, 1117-1120.

[11] Smith, D. C. *Laser induced gas breakdown and plasma interaction 38th Aerospace Sciences Meeting and Exhibit*. Reno: Nevada., 2000.

[12] Yueh, F. Y., Singh, J. P., Zhang, H. Encyclopedia of Analytical Chemistry (Laser-Induced Breakdown Spectroscopy, Elemental Analysis); ed. R. A. Meyers. Wiley: Chichester., 2000.

[13] Cremers, D. A., Radziemski, L. J. *Handbook of Laser-Induced Breakdown Spectroscopy*; Wiley: Chichester., 2006.

[14] Miziolek, A. W., Palleschi, V., Schechter, I. *Laser-Induced Breakdown Spectroscopy*; Cambridge University Press: Cambridge., 2006.

[15] Singh, J. P., Thakur, S. N. Laser-Induced Breakdown Spectroscopy; Elsevier:, New York. 2007.

[16] *Yong*. Ill, *L. Laser Induced Breakdown Spectrometry*; Nova Science Publishers: New York 2000.

[17] Marpaung, A. M., Kurniawan, H., Tjia, M. O., Kagawa, K. *J. Phys. D. Appl. Phys*., 2001, 34, 758-771.

[18] Camacho, J. J., Poyato, J. M. L., Diaz, L., Santos, M. *J. Phys. B: At. Mol. Opt. Phys*., 2007, 40, 4573-4590.

[19] Camacho, J. J., Poyato, J. M. L., Diaz, L., Santos, M. *J. Appl. Phys.,* 2007, 102, 103302-10.

[20] Camacho, J. J., Santos, M., Diaz, L., Poyato, J. M. L. *J. Phys. D. Appl. Phys*., 2008, 41, 215206-13.

[21] Camacho, J. J., Santos, M., Diaz, L., Juan, L. J., Poyato, J. M. L. *J. Appl. Phys.* 2010, 107, 083306-9.
[22] MacDonald, A. D. *Microwave Breakdown in Gases*; Wiley: New York. 1966.
[23] Raizer, Y. P. *Gas Discharge Physics*; Springer: Berlin, Heidelberg., 1991.
[24] Kopiczynski, T. L., Bogdan, M., Kalin, A. W., Schotwau, H. J., Kneubuhl, F. H. *Appl. Phys. B. Photophys. Laser Chem.*, 1992, 54, 526-530.
[25] Tozer, B. A. *Phys. Rev.*, 1965, 137, 1665-1667.
[26] Radziemski, L. J., Cremers, D. A., *Laser-induced plasma and applications;* New York: Dekker. 1989.
[27] Gurevich, A., Pitaevskii, L. *Sov. Phys. JETP.*, 1962, 19, 870-871.
[28] Capitelli, M., Capitelli, F., Eletskii, A. *Spectrochim. Acta B.*, 2000, 55, 559-574.
[29] Capitelli, M., Casavola, A., Colonna, G, Giacomo, A. D. *Spectrochim. Acta B.*, 2004, 59, 271-289.
[30] Chan, C. H., Moody, C. D., McKnight, W. B. *J. Appl. Phys.*, 1973, 44, 1179-1188.
[31] Smith, D. S., Haught, A. F. *Phys. Rev. Lett.*, 1966, 16, 1085-1088.
[32] Young, M., Hercher, M. *J. Appl. Phys.*, 1967, 38, 4393-4400.
[33] Barchukov, A. I., Bunkin, F. V., Konov, V. I., Lubin, A. A. *Sov. Phys. JETP*, 1974, 39, 469-477.
[34] Hermann, J., Boulmer-Leborgne, C., Mihailescu, I. N., Dubreuil, B. *J. Appl. Phys.*, 1993, 73, 1091-1099.
[35] Offenberger, A. A., Burnett, N. H. *J. Appl. Phys.*, 1972, 43, 4977-4980.
[36] Donaldson, T. P., Balmer, J. E., Zimmermann, J. A. *J. Phys. D: Appl. Phys.*, 1980, 13, 1221-1233.
[37] Meyerand, R. G., Haught, A. F. *Phys. Rev. Lett.*, 1964, 13, 7-9.
[38] Spitzer, L. *Physics of Fully Ionised* Gases; John Wiley: New York. 1962.
[39] Hermann, J., Boulmer-Leborgne, C., Dubreuil, B., Mihailescu, I. N. *J. Appl. Phys.,* 1993, 74, 3071-3079.
[40] Ready, J. F. *Effect of high power laser radiation*; Academic Press., New York. 1971.
[41] Griem, H. R. *Principles of plasma spectroscopy*; University Press: Cambridge., 1997.
[42] Griem, H. R. *Phys. Rev.*, 1962, 128, 515-523.
[43] Huddlestone, R. H., Leonard, S. L. *Plasma Diagnostic Techniques*; Academic Press: New York. 1965.
[44] Hutchinson, I. H. *Principles of plasma diagnostic*; University Press: Cambridge. 2002.
[45] NIST Atomic Spectra Database online at http://physics
[46] Herzberg, G. *Spectra of diatomic molecules*; Van Nostrand: New York. 1950.
[47] Steinfeld, J. I. *An introduction to modern molecular spectroscopy*; MIT Press: London. 1986.
[48] Bernath, P. F. *Spectra of atoms and molecules*; Oxford University Press: New York. 1995.
[49] Camacho, J. J., Pardo, A., Martin, E., Poyato, J. M. L. *J. Phys. B, At. Mol. Opt. Phys.*, 2006, 39, 2665-2679.
[50] Kovacs, I. *Rotational Structure in the Spectra of Diatomic Molecules*; Hilger: London. 1969.
[51] Drogoff, L. B., Margotb, J., Chakera, M., Sabsabi, M., Barthelemy, O., Johnstona, T. W., Lavillea, S., Vidala, F., Kaenela, V. Y. *Spectrochim. Acta Part B.*, 2001, 56, 987-1002.

[52] Lofthus, A., Krupenie, P. H. *J. Phys. Chem. Ref. Data.*, 1977, 6, 113-307.

[53] Herzberg, G. *Ergeb. Exakten Naturwiss.* 1931, 10, 207-284.

[54] Pannetier, G., Marsigny, G. L., Guenebaut, H. C. R. *Acad. Sci.* Paris., 1961, 252, 1753 1755

[55] Tanaka, Y., Jursa, A. S. *J. Opt. Soc. Amer.*, 1961, 51, 1239-1245.

[56] Meinel, A. B. *Astrophys. J.,* 1950, 112, 562-563.

[57] Ledbetter, J. W. Jr. *J. Mol. Spectrosc.*, 1972, 42 100-111.

[58] Huber, K. P., Herzberg, G. Molecular spectra an Molecular structure. IV. Constants of diatomic molecules; Van Nostrand Reinhold: New York. 1979.

[59] Laher, R. R., Gilmore, F. R. *J. Phys. Chem. Ref. Data.*, 1991, 20, 685-712.

[60] Gilmore, F. R. J. Quant. *Spectrosc. Radiat. Transfer.*, 1965, 5, 369-389.

[61] Aguilera, J. A., Aragon, C. *Spectrochim. Acta B.*, 2004, 59, 1861-1876.

[62] Martin, W. C., Zalubas, R. *J. Phys. Chem. Ref. Data.*, 1983, 12, 323-379.

[63] Sreckovic, A, Dimitrijevic, M. S., Djenize, S. *Astron. and Astroph.,* 2001, 371, 354-359.

[64] Alcock, A. J., Kato, K., Richardson, M. C. *Opt. Comm.*, 1968, 6, 342-343.

[65] Haught, A. F., Meyerand, R. G., Smith, D. C. *Physics of Quantum Electronics*; In: P. L. Kelley, B. Lax and P. E. Tannenwald (Eds.), MacGraw-Hill: New York. 1966. 509.

[66] De Michelis, C. *Opt. Comm.*, 1970, 2, 255-256.

[67] Ireland, C. L. M., Morgan, C. G. *J. Phys. D: Appl. Phys.*, 1973, 6, 720-729.

[68] Aaron, J. M., Ireland, C. L.,M. Morgan, C. G. *J. Phys. D: Appl. Phys.* 1974, 7, 1907-1917.

[69] Phuoc, T. X., White, C. M. *Opt. Comm.*, 2000, 181, 353-359.

[70] Tomlinson, R. G., Damon, E. K., Buscher, H. T. Physics of Quantum Electronics; In: P. L. Kelley, B. Lax and P. E. Tannenwald (Eds.), MacGraw-Hill., New York. 1966. 520.

[71] Kawahara, N., Beduneau, J. L., Nakayama, T., Tomita, E., Ikeda, Y. *Appl. Phys., B* 2007, 86, 605-614.

[72] Zhuzhukalo, E. V., Kolomiiskii, A. N., Nastoyashchii, A. F., Plyashkevich, L. N. *J. Quantum Electron.*, 1981, 11, 670-671.

[73] Marcus, S., Lowder, J. E., Mooney, D. L. *J. Appl. Phys.*, 1976, 47, 2966-2968.

[74] Kagawa, K., Yokoi, S. *Spectrochim. Acta, B* 1982, 37, 789-795.

In: Advances in Laser and Optics Research. Volume 10
Editor: William T. Arkin
ISBN: 978-1-62257-795-8

Chapter 5

Towards Shaping of Pulsed Plane Waves in the Time Domain via Chiral Sculptured Thin Films

Joseph B. Geddes III
CATMAS—Computational and Theoretical Materials Sciences Group
Department of Engineering Science and Mechanics
The Pennsylvania State University
University Park, PA, US

Abstract

We review our investigations concerning the propagation of ultrashort optical pulsed plane waves across chiral sculptured thin films (STFs) in the time domain. The phenomenon of pulse bleeding—i.e., the time domain manifestation of the circular Bragg phenomenon (CBP)—is explained, and its importance for the shaping of ultrashort pulses is investigated.

1. Introduction

As the generation of ultrashort optical pulses becomes more routine, efforts to more efficiently and inexpensively shape and manipulate such pulses rise in importance. In this chapter, we describe time–domain calculations that show how chiral sculptured thin films (STFs) could be used to sort and alter such ultrashort pulses in ways that depend on their amplitude, shape, duration, carrier polarization, and carrier phase. Chiral STF–based devices could be especially useful in manipulating pulses whose carrier plane wave is circularly polarized.

STFs are assemblies of parallel nanowires affixed to a substrate. They are fabricated via physical vapor deposition—either evaporation or sputtering—in vacuum, and their constituents can be semiconducting oxides, metals, or even polymers. The nanowire shapes, which can be bent and twisted during deposition into both continuous and discontinuous curvilinear shapes, endow these nanoengineered materials with their most useful optical properties. Patterning of substrates before deposition can also affect the shapes of the growing nanowires; such patterning raises the possibility of creating three dimensional STF architectures for photonic crystals and waveguides [1, 2]. The intensive research of the past ten years on STF morphology and optics has been recently summarized in book form [3].

In chiral STFs, the nanowires are helicoidal in shape. They are fabricated by evaporating material at an oblique angle onto a relatively cool substrate that rotates about an axis normal to its surface. The nanowire helices grow normal to the substrate, and their pitch depends on the deposition rate, rotation rate, and deposition angle. An electron micrograph of a chiral STF is presented in Figure 1.

When the pitch of the helices is constant or nearly so, chiral STFs exhibit the circular Bragg phenomenon (CBP). Suppose light is incident normally on a structurally right–handed chiral STF. The wavelengths of light that lie within a certain range, known as the Bragg regime, are largely reflected from the chiral STF if they are right circularly polarized (RCP). Left circularly polarized (LCP) light, in contrast, is largely refracted into the film, where it is either subsequently transmitted or absorbed, depending on the film properties. The CBP is the phenomenon that makes possible many of the chiral STF–based devices that have been designed and tested, including circular polarization handedness inverters, circular polarization filters, and spectral hole filters [4, 5, 6]. Advances in deposition technology have resulted in improved chiral STF properties [7], including increased optical activity [8].

Almost since the invention of the laser, researchers have sought to discover ways to use that device to create optical pulses of ever shorter duration [9]. Such pulses are used for many applications, e.g. in telecommunications and scientific research [10, 11]. Ultrashort pulses are modeled as a carrier plane wave, which we take to be of unit amplitude, multiplied by a pulse envelope. The envelope determines, in general, the pulse's total energy, shape, and duration, while the carrier plane wave sets the pulse's dominant wavelength and its polarization.

This chapter summarizes the key results—including the discovery of the spatiotemporal manifestation of the CBP and its effect in shaping of pulsed optical plane waves—obtained so far in using chiral STFs to manipulate pulsed plane waves in the time domain. Section 2. contains a description both of the chiral STF constitutive equations, the initial boundary value problem to be solved, and a finite–difference (FD) algorithm to solve it numerically. The following two sections describe the spatiotemporal evolution of optical pulsed plane waves. Section 3. describes this evolution when pulses are incident on chiral STF half–spaces, and Section 4. when pulses are incident on chiral STF slabs of finite thickness. Finally, we tie these results—many of which have been previously collected in a thesis [12]—together and suggest several applications for the work in Section 5.. The work concentrates on the phenomenology of reflection, refraction, and transmission of pulses by chiral STFs; quantification of pulse shaping effects will be the topic of future investigations.

2. Theory

In order to study the spatiotemporal evolution of an ultrashort optical pulse as it propagates through a chiral STF, mathematical models of both the pulse and the film are needed. In this section, we present the constitutive relations for a dielectric chiral STF, formulate the Maxwell equations in matrix form, and then discretize those equations for numerical solution.

At time $t = 0$, a pulsed plane wave, consisting of a monochromatic carrier plane wave amplitude–modulated by a pulse envelope, is launched from the plane $z = 0$ toward a chiral STF slab. The film occupies the region $z_0 \leq z \leq z_1$ (note that $z_0 > 0$). The pulse traverses

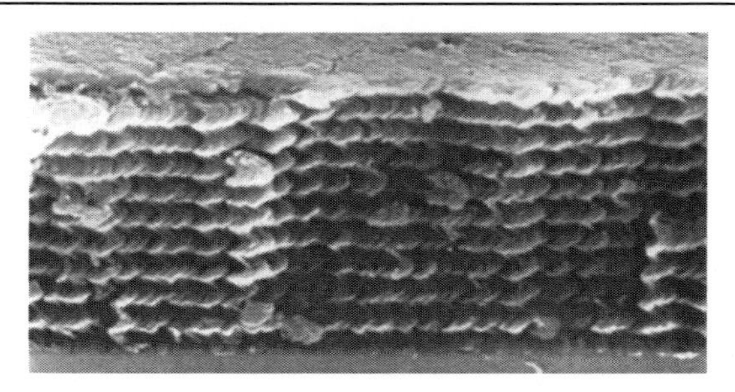

Figure 1. Scanning electron micrograph of a SiO chiral STF showing a cross section of its 10–turn helicoidal microstructure. The film is approximately 4 μm thick. *Courtesy: Russell Messier, The Pennsylvania State University*.

the vacuum half–space $z \leq z_0$, excites the chiral STF, whereupon it splits into reflected pulses in the half–space $z \leq z_0$ and transmitted pulses in the half–space $z \geq z_1$, also vacuous.

2.1. Chiral STF Constitutive Relations

As mentioned in Section 1., chiral STFs exhibit a continuously twisted, unidirectionally periodic microstructure, in this case along the z axis. Their nanowires, whose constituents are dispersive in bulk form in accord with causality requirements, possess cross sections that are nominally elliptical, and thus exhibit locally biaxial dielectric properties. The time–domain chiral STF constitutive relations encompass all these properties in the equations

$$\underline{D}\,(\underline{r},t) = \epsilon_0 \left(\underline{\underline{\epsilon}}_r * \underline{E}\right)(\underline{r},t)\,, \quad \underline{B}\,(\underline{r},t) = \mu_0\, \underline{H}\,(\underline{r},t)\,, \tag{1}$$

where ϵ_0 and μ_0 are the permittivity and the permeability of free space, respectively; $\underline{r}$ is the position vector; the electromagnetic fields $\underline{E}\,(\underline{r},t)$, $\underline{D}\,(\underline{r},t)$, $\underline{H}\,(\underline{r},t)$, and $\underline{B}\,(\underline{r},t)$ have their usual meanings [13]; while the operation $*$ denotes convolution with respect to time as follows:

$$\left(\underline{\underline{\epsilon}}_r * \underline{E}\right)(\underline{r},t) = \int_0^t \underline{\underline{\epsilon}}_r\left(\underline{r}\cdot\underline{u}_z, t'\right)\cdot \underline{E}\left(\underline{r}, t-t'\right)\, dt'\,. \tag{2}$$

Equations (1) contain the tacit assumption that chiral STFs exhibit only dielectric properties at optical wavelengths [14], an assumption that can be relaxed in future work. The relative permittivity dyadic is defined as

$$\underline{\underline{\epsilon}}_r\,(z,t) = \begin{cases} \underline{\underline{I}}\,\delta(t)\,, & z \notin [z_0, z_1] \\ \underline{\underline{S}}_z\,(z-z_0)\cdot \underline{\underline{S}}_y\,(\chi)\cdot\left[\epsilon_a(t)\,\underline{u}_z\underline{u}_z + \epsilon_b(t)\,\underline{u}_x\underline{u}_x + \epsilon_c(t)\,\underline{u}_y\underline{u}_y\right] & \\ \quad \cdot \underline{\underline{S}}_y^{-1}(\chi)\cdot \underline{\underline{S}}_z^{-1}(z-z_0)\,, & z \in [z_0, z_1] \end{cases}\,, \tag{3}$$

where the structural handedness of the chiral STF is described by the rotation dyadic

$$\underline{\underline{S}}_z\,(z) = \underline{u}_z\underline{u}_z + \left(\underline{u}_x\underline{u}_x + \underline{u}_y\underline{u}_y\right)\cos(\pi z/\Omega) + h\left(\underline{u}_y\underline{u}_x - \underline{u}_x\underline{u}_y\right)\sin(\pi z/\Omega) \tag{4}$$

with 2Ω as the structural period, and $h = +1$ for a structurally right–handed chiral STF, and $h = -1$ for a structurally left–handed one. The identity dyadic is denoted as $\underline{\underline{I}}$, while

$\delta(t)$ is the Dirac delta function and $\underline{u}_x$, $\underline{u}_y$, and $\underline{u}_z$ are the cartesian unit vectors. The tilt dyadic

$$\underline{\underline{S}}_y(\chi) = \underline{u}_y\underline{u}_y + (\underline{u}_x\underline{u}_x + \underline{u}_z\underline{u}_z)\cos\chi + (\underline{u}_z\underline{u}_x - \underline{u}_x\underline{u}_z)\sin\chi \tag{5}$$

is a function of the angle of rise χ.

We assume that the dielectric properties of our linear chiral STF are captured by a single–resonance Lorentz model [15], and thus

$$\begin{aligned} \epsilon_{a,b,c}(t) &= \delta(t) + \chi_{a,b,c}(t) \\ &= \delta(t) + p_{a,b,c}\left(\frac{2\pi c_0}{\lambda_{a,b,c}}\right)\sin\left(\frac{2\pi c_0}{\lambda_{a,b,c}}t\right)\exp\left(-\frac{c_0 t}{N_{a,b,c}\lambda_{a,b,c}}\right)\mathcal{U}(t)\,, \end{aligned} \tag{6}$$

where $\mathcal{U}(t)$ is the Heaviside step function, $c_0 = (\epsilon_0\mu_0)^{-1/2}$, and the oscillator strengths are $p_{a,b,c}$. The absorption bands, which occur around the resonance wavelengths $\lambda_{a,b,c}\left(1 + N_{a,b,c}^{-2}\right)^{-1/2}$, become narrower as $N_{a,b,c}$ increase. Additional terms may be added to (6) to model nonlinear dielectric responses, as illustrated elsewhere [16].

2.2. Derivation of Matrix Partial Differential Equation

Upon abbreviating partial differentiation $\partial/\partial_x = \partial_x$, $\partial/\partial_t = \partial_t$, and so forth, we insert (1) and (3) into the Maxwell curl equations

$$\nabla \times \underline{E}(\underline{r}, t) = -\partial_t \underline{B}(\underline{r}, t)\,, \tag{7}$$
$$\nabla \times \underline{H}(\underline{r}, t) = \partial_t \underline{D}(\underline{r}, t)\,, \tag{8}$$

with no source current densities present. For propagation in the z direction, $\partial_x \equiv \partial_y \equiv 0$ and $\nabla \equiv \underline{u}_z\,\partial_z$, and we obtain the following six differential equations:

$$\partial_z E_x(z,t) = -\mu_0\partial_t H_y(z,t)\,, \tag{9}$$
$$\partial_z E_y(z,t) = \mu_0\partial_t H_x(z,t)\,, \tag{10}$$
$$\partial_z H_x(z,t) = \epsilon_0\partial_t\left[\underline{u}_y\cdot\left(\underline{\underline{\epsilon}}_r * \underline{E}\right)(z,t)\right], \tag{11}$$
$$\partial_z H_y(z,t) = -\epsilon_0\partial_t\left[\underline{u}_x\cdot\left(\underline{\underline{\epsilon}}_r * \underline{E}\right)(z,t)\right], \tag{12}$$
$$0 = \partial_t\left[\underline{u}_z\cdot\left(\underline{\underline{\epsilon}}_r * \underline{E}\right)(z,t)\right], \tag{13}$$
$$0 = \partial_t H_z(z,t)\,. \tag{14}$$

Equation (14) implies $H_z(z,t) \equiv H_z(z)$, and the initial conditions to be prescribed later ensure that $H_z(z,t) = 0$. Therefore, we define the column 5–vector

$$[\underline{F}](z,t) = [E_x(z,t),\, E_y(z,t),\, H_x(z,t),\, H_y(z,t),\, E_z(z,t)]^T \tag{15}$$

to contain the five nonzero components of the electromagnetic field, with superscript ^{T} denoting the transpose. Then we write (9)–(13) compactly in matrix form as

$$[\underline{J}]\,\partial_z[\underline{F}](z,t) = \left[\underline{\underline{Q}}\right]\partial_t[\underline{F}](z,t) + \epsilon_0\partial_t\left\{\left([\underline{A}] * [\underline{F}]\right)(z,t)\right\}. \tag{16}$$

In (16), the 5×5 matrix $\left[\underline{\underline{A}}\right](z,t)$ contains the constitutive properties. It is identically null–valued for $z \notin [z_0, z_1]$, while

$$\left[\underline{\underline{A}}\right](z,t) = \begin{bmatrix} 0 & 0 & 0 & 0 & 0 \\ 0 & 0 & 0 & 0 & 0 \\ A_{31}(z,t) & A_{32}(z,t) & 0 & 0 & A_{35}(z,t) \\ -A_{41}(z,t) & -A_{31}(z,t) & 0 & 0 & -A_{45}(z,t) \\ A_{45}(z,t) & A_{35}(z,t) & 0 & 0 & A_{55}(z,t) \end{bmatrix}, \quad z \in [z_0, z_1], \tag{17}$$

with

$$A_{31} = \left(\epsilon_a \sin^2\chi + \epsilon_b \cos^2\chi - \epsilon_c\right) \sin\frac{\pi(z-z_0)}{\Omega} \cos\frac{\pi(z-z_0)}{\Omega}, \tag{18}$$

$$A_{32} = \left(\epsilon_a \sin^2\chi + \epsilon_b \cos^2\chi\right) \sin^2\frac{\pi(z-z_0)}{\Omega} + \epsilon_c \cos^2\frac{\pi(z-z_0)}{\Omega} - \delta(t), \tag{19}$$

$$A_{35} = (\epsilon_b - \epsilon_a) \sin\chi \cos\chi \sin\frac{\pi(z-z_0)}{\Omega}, \tag{20}$$

$$A_{41} = \left(\epsilon_a \sin^2\chi + \epsilon_b \cos^2\chi\right) \cos^2\frac{\pi(z-z_0)}{\Omega} + \epsilon_c \sin^2\frac{\pi(z-z_0)}{\Omega} - \delta(t), \tag{21}$$

$$A_{45} = (\epsilon_b - \epsilon_a) \sin\chi \cos\chi \cos\frac{\pi(z-z_0)}{\Omega}, \tag{22}$$

$$A_{55} = \epsilon_a \cos^2\chi + \epsilon_b \sin^2\chi - \delta(t); \tag{23}$$

the other two 5×5 matrixes in (16) are

$$\left[\underline{\underline{J}}\right] = \mathrm{diag}\,[1,1,1,1,0], \tag{24}$$

$$\left[\underline{\underline{Q}}\right] = \begin{bmatrix} 0 & 0 & 0 & -\mu_0 & 0 \\ 0 & 0 & \mu_0 & 0 & 0 \\ 0 & \epsilon_0 & 0 & 0 & 0 \\ -\epsilon_0 & 0 & 0 & 0 & 0 \\ 0 & 0 & 0 & 0 & \epsilon_0 \end{bmatrix}. \tag{25}$$

Note that for calculations involving chiral STFs exhibiting magnetic properties, a 6–vector formalism is needed.

2.3. Finite–Difference Algorithm

An analytical solution for (16) is not known, and thus we employed a FD scheme. Upon discretization of both space and time as $z_i = i\Delta z$, $(i = 0,1,2,3,\ldots)$, and $t_n = n\Delta t$, $(n = 0,1,2,3,\ldots)$, replacing derivatives with central differences, and using the leapfrog method, (16) transforms into the matrix difference equation

$$\begin{aligned} \left[\underline{\underline{J}}\right] \left(\frac{\left[\underline{F}\right]_{i+1}^{n} - \left[\underline{F}\right]_{i-1}^{n}}{2\Delta z}\right) &= \left[\underline{\underline{Q}}\right] \cdot \left(\frac{\left[\underline{F}\right]_{i}^{n+1} - \left[\underline{F}\right]_{i}^{n-1}}{2\Delta t}\right) \\ &+ \epsilon_0 \sum_{m=1}^{n-1} \left[\underline{\underline{A}}\right]_{i}^{m} \cdot \left(\frac{\left[\underline{F}\right]_{i}^{n-m+1} - \left[\underline{F}\right]_{i}^{n-m-1}}{2}\right) \\ &+ \frac{\epsilon_0}{2} \left[\underline{\underline{A}}\right]_{i}^{n} \cdot \left(\left[\underline{F}\right]_{i}^{1} - \left[\underline{F}\right]_{i}^{0}\right). \end{aligned} \tag{26}$$

In (26),

$$\underline{[F]}_i^n = \underline{[F]}(z_i, t_n)\,, \quad \underline{\underline{[A]}}_i^n = \underline{\underline{[A]}}(z_i, t_n) \tag{27}$$

and $\beta = c_0 \Delta t / \Delta z < 1$ for stability [17]. On solving (26) for $\underline{[F]}_i^{n+1}$, we obtain the formula

$$\begin{aligned} \underline{[F]}_i^{n+1} &= \underline{[F]}_i^{n-1} + \beta \underline{\underline{[V]}} \cdot \left(\underline{[F]}_{i+1}^n - \underline{[F]}_{i-1}^n\right) \\ &- \underline{[C]}_i^n - \underline{\underline{[W]}}_i^n \cdot \left(\underline{[F]}_i^1 - \underline{[F]}_i^0\right) \Delta t\,, \end{aligned} \tag{28}$$

whose last two terms must be calculated only if $z_i \in [z_0, z_1]$. In (28), the matrixes $\underline{\underline{[W]}}(z,t) = \epsilon_0 \left[\underline{\underline{Q}}\right]^{-1} \underline{\underline{[A]}}(z,t)$ and $\underline{\underline{[V]}} = c_0^{-1} \left[\underline{\underline{Q}}\right]^{-1} \underline{\underline{[J]}}$. Due to the initial conditions we impose later, the final terms in (26) and (28) are identically null–valued and thus need not be computed.

The bulk of the convolution is contained in the term

$$\underline{[C]}_i^n = \sum_{m=1}^{n-1} \underline{\underline{[W]}}_i^m \cdot \left(\underline{[F]}_i^{n-m+1} - \underline{[F]}_i^{n-m-1}\right) \Delta t\,, \tag{29}$$

and direct calculation of it accounts for most of the computational time and memory requirements. A quicker algorithm is needed to handle the largest simulations. Such an algorithm, which exploits the chosen Lorentzian properties, and inspired by methods devised for simpler constitutive relations [18, 19], is described in detail elsewhere [20].

2.4. Initial and Boundary Conditions

For a numerical solution to be obtained, (28) requires initial and boundary conditions. The initial condition

$$\underline{[F]}_i^0 = \underline{[0]} \ \forall\, i \geq 1 \tag{30}$$

ensures that the electromagnetic field is null–valued everywhere at time $t = 0$ except possibly at $z = 0$. The boundary condition

$$\underline{[F]}_0^n = g(t_n)\, \underline{[\varphi]}(t_n) \tag{31}$$

contains a pulse envelope $g(t)$ that modulates the amplitude of a carrier plane wave $\underline{[\varphi]}(t)$. The column vector $\underline{[\varphi]}(t)$ is defined as

$$\left[\underline{\varphi}_{\pm}\right](t) = \left[\cos(\omega_{\text{car}} t + \phi), \pm \sin(\omega_{\text{car}} t + \phi), \mp \eta_0^{-1} \sin(\omega_{\text{car}} t + \phi), \eta_0^{-1} \cos(\omega_{\text{car}} t + \phi), 0\right]^T, \tag{32}$$

where $\left[\underline{\varphi}_{+}\right](t)$ represents a LCP plane wave, $\left[\underline{\varphi}_{-}\right](t)$ represents a RCP plane wave, both with carrier wavelength λ_{car} and carrier angular frequency $\omega_{\text{car}} = 2\pi c_0 / \lambda_{\text{car}}$. The pulse envelope

$$g(t) = A_{\text{p}} \frac{t}{\tau_{\text{p}}} \exp\left(\frac{-2t}{\tau_{\text{p}}}\right). \tag{33}$$

indicates a pulse with a rounded shape whose duration depends on the parameter τ_{p}; ϕ denotes the carrier phase.

2.5. Parameter Values

For the calculations reported in this chapter, we chose $h = +1$, i.e. the chosen chiral STFs are structurally right–handed. The oscillator strengths $p_a = 0.40, p_b = 0.52, p_c = 0.42$, angle of rise $\chi = 20°$, and structural half–period $\Omega = 200$ nm; we chose resonance wavelengths $\lambda_a = \lambda_c = 280$ nm, $\lambda_b = 290$ in the ultraviolet regime [15]. These choices fix the center wavelength of the Bragg regime λ_{Br} at approximately 516 nm [21]. Some variables, including $N_{a,b,c}$, A_{p}, τ_{p}, and ϕ, were picked to suit individual cases; their values are listed in the discussions of those cases. The variables Δz and β, and hence Δt, were also chosen based on the computational needs of each case.

3. Half–Space Problems

To examine the CBP in the time domain, we first consider problems in which an ultrashort optical pulsed plane wave falls normally upon a chiral STF half–space—i.e., we increase z_1 so that it is beyond the domain of computation.

3.1. Pulse Bleeding and the Light Pipe

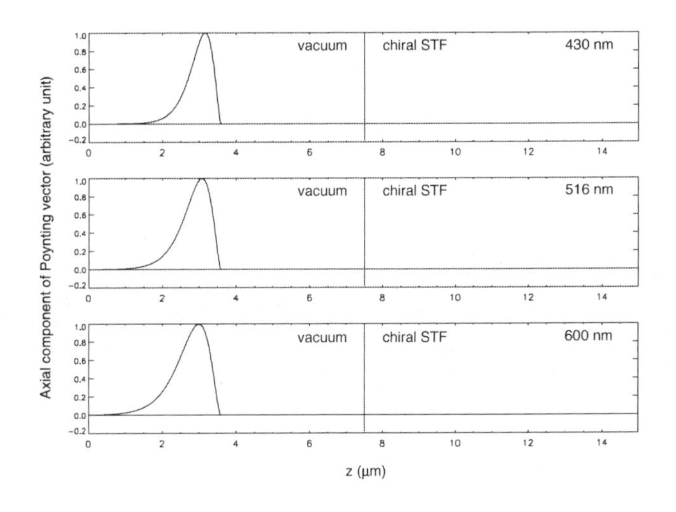

Figure 2. Plots of $P_z(z,t) = \underline{u}_z \cdot \underline{P}(z,t)$ at $t = 12.0$ fs, with $\lambda_{\mathrm{car}} = 430, 516, 600$ nm (top to bottom) showing incident pulses. Note that the pulse shape, as defined by P_z, is the same regardless of the circular polarization state of the carrier plane wave. *Adapted from Geddes and Lakhtakia [22].*

For these calculations [22], $N_{a,b,c} = 100$, $A_{\mathrm{p}} = 1$ V/m, $\tau_{\mathrm{p}} = 2\lambda_{\mathrm{car}}/c_0$, and $\phi = 0$. The vacuum / chiral STF interface was positioned at $z_0 = 7.5$ μm, and we selected $\Delta z = 5$ nm and $\beta = 0.9$ so that $\Delta t = 0.015$ fs. We examined pulsed plane waves with carrier wavelengths of $\lambda_{\mathrm{car}} = 430, 516$, and 600 nm, i.e., below, within, and above the Bragg regime. Note that even the spectrum of a pulse whose carrier wavelength lies outside the

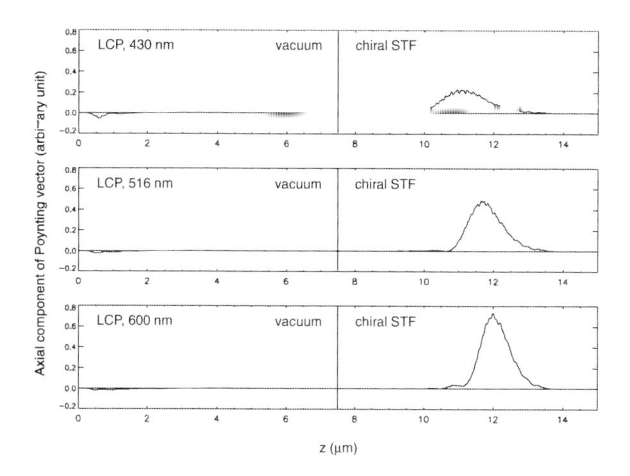

Figure 3. Plots of P_z at $t = 49.5$ fs for $\lambda_{\text{car}} = 430$, 516, and 600 nm resulting from incident pulses whose carrier plane waves were LCP. *Adapted from Geddes and Lakhtakia [22].*

Bragg regime may overlap with it. Snapshots of the axial component of the Poynting vector, a measure of power per unit area, or

$$P_z(z,t) = \underline{u}_z \cdot [\underline{E}(z,t) \times \underline{H}(z,t)] \,, \tag{34}$$

are plotted at two different instants of time in Figures 2–5. Note that when $P_z > 0$, energy flows in the $+z$ direction; and when $P_z < 0$, energy flows in the $-z$ direction. In Figure 2, at time $t = 12.0$ fs, only the incident pulses are captured in the snapshots, and the pulse shapes, which are independent of the polarization state of the carrier plane wave, are clearly visible. Figures 3–5 show P_z at time $t = 49.5$ fs, after the incident pulse has reached the vacuum / chiral STF interface. In those plots, part of each incident pulse has entered the chiral STF and become a refracted pulse, and part of each incident pulse has turned back from the film and become a reflected pulse. In Figure 3, the reflected pulses are of short duration and contain little energy because the incident pulses had LCP carrier plane waves. Thus, the CBP was not manifested by the structurally right–handed chiral STF. However, in Figure 4, and when $\lambda_{\text{car}} = 516$ nm, the reflected pulse is of much longer duration and hence contains more energy; the CBP is manifested in this case. Furthermore, inspection of the magnified views of P_z near the vacuum / chiral STF interface at $t = 49.5$ fs, as shown in Figure 5, reveals that there is a continuous flow of energy across the interface when the conditions for the CBP are present, but not if otherwise. This flow of energy through a so–called light pipe, and termed pulse bleeding, is the spatiotemporal fingerprint of the CBP in the time domain. Fourier transformation of the time–domain results recovers the more familiar frequency domain results [22], and the pulse bleeding phenomenon is apparent for post–resonant Bragg regimes too [23].

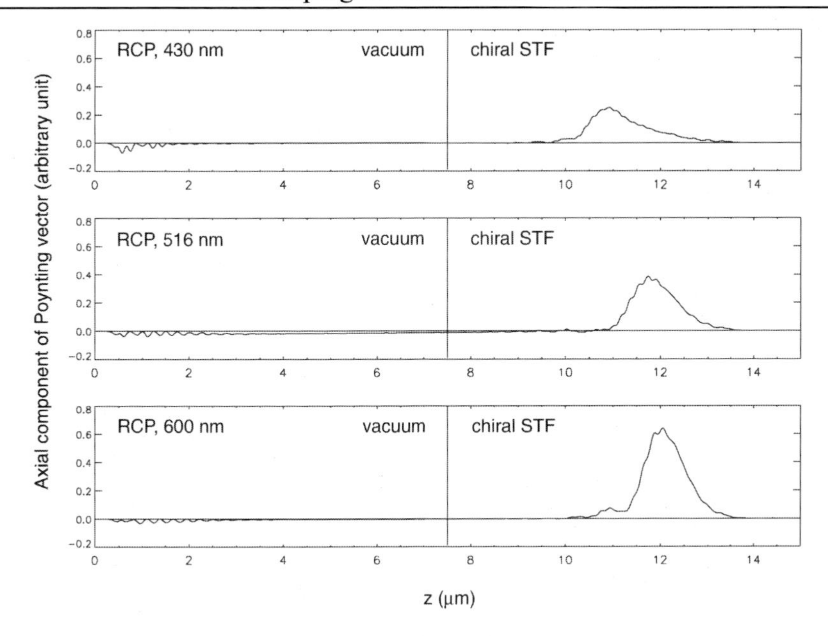

Figure 4. Same as Figure 3, except resulting from incident pulses whose carrier plane waves were RCP. *Adapted from Geddes and Lakhtakia [22].*

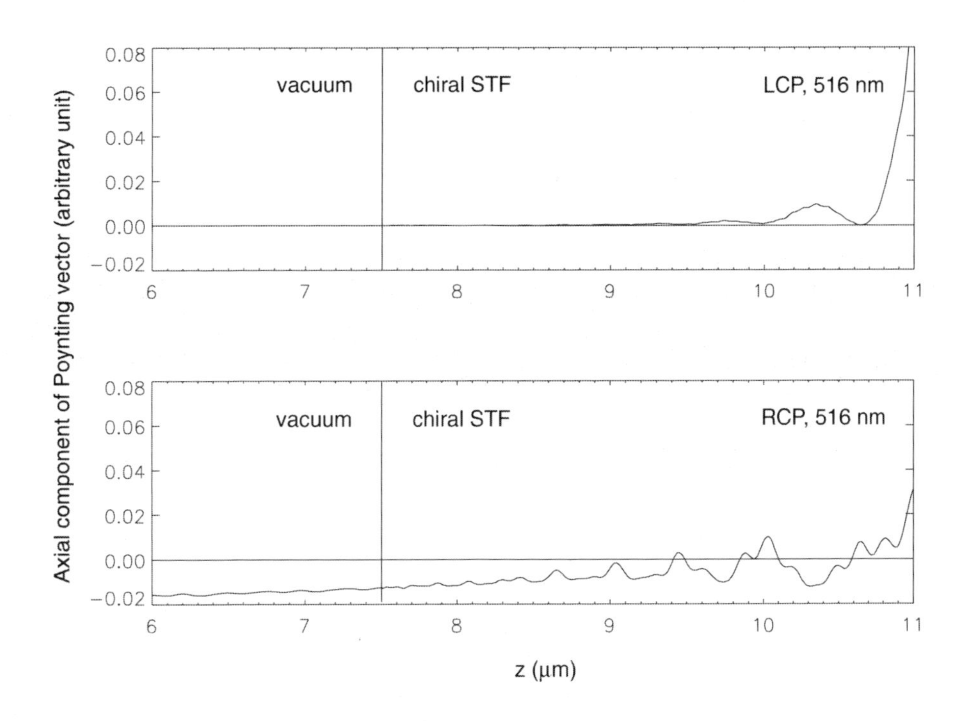

Figure 5. Magnified views of P_z in the region containing the vacuum / chiral STF interface from Figures 3 and 4, when $\lambda_{\rm car} = 516$ nm. Note that the light pipe is clearly absent when the incident pulses had LCP carrier plane waves (top), but present if otherwise (bottom). *Adapted from Geddes and Lakhtakia [22].*

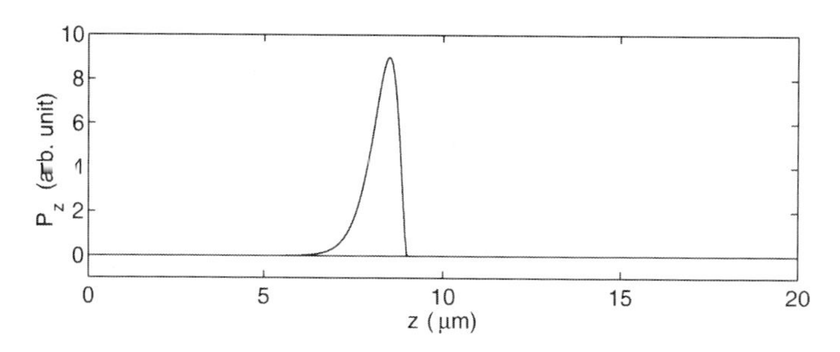

Figure 6. Plot of P_z showing the incident pulse shape at $t = 30.1$ fs. The shape is independent of both the polarization of the carrier plane wave and the carrier phase. The vacuum / chiral STF interface lies at $z = 20$ μm, and hence is not shown. *Adapted from Geddes and Lakhtakia [16].*

3.2. Effects of Carrier Phase

Having clarified the role of λ_{car} and carrier plane wave polarization in the manifestation of the time domain CBP, we now turn to the effects of carrier phase. For these calculations [16], $N_{a,b,c} = 100$, $A_{\mathrm{p}} = 10^6$ V/m, and $\tau_{\mathrm{p}} = 3.44$ fs; the boundary between vacuum and film lies at $z_0 = 20$ μm. A snapshot of the incident pulse[1], whose shape is independent of carrier phase, wavelength, and polarization, is shown in Figure 6. Plots of P_z for reflected pulses at $t = 132$ fs for three carrier wavelengths ($\lambda_{\mathrm{car}} = 430, 516$, and 600 nm), two carrier phases ($\phi = 0$ and $\pi/2$), and two carrier polarizations (LCP and RCP) are given in Figures 7 and 8. In the former, P_z at the leading edges of the reflected pulses increases with decreasing λ_{car}, presumably because of the greater impedance mismatch between vacuum and film at lower wavelengths. Pulse bleeding is apparent in all reflected pulses when the incident carrier plane wave is RCP (Figure 8), because the spectrums of the incident pulses are wide enough to overlap the Bragg regime even when λ_{car} lies outside it. The rapid oscillations—which depend on the carrier phase ϕ—in P_z in the leading portions of those pulses are due to interference between LCP and RCP components in each reflected pulse—i.e., even though the carrier plane wave of an incident pulse is pure RCP, the reflected pulse contains both LCP and RCP components. The LCP component is created—at least largely—by reflection due to impedance mismatch near the surface of the film. The reflected pulses' long tails contain mostly RCP light due to pulse bleeding but little LCP light, and hence large oscillations in P_z are not observed there. The refracted pulse shapes in Figures 9 and 10 differ little with carrier phase, but notice that the refracted pulses carry more energy with increasing λ_{car} due to the presence of absorption resonances at ultraviolet wavelengths. These phenomenons, and reflection and refraction of pulses from nonlinear chiral STFs, were examined in more detail elsewhere [16], along with their connections to the interrelationship between phase, length, and time in optical nanotechnology [24].

[1]Note that though the scaling of the arbitrary units may be different from subsection to subsection in this chapter, comparisons of power densities are appropriate within each subsection.

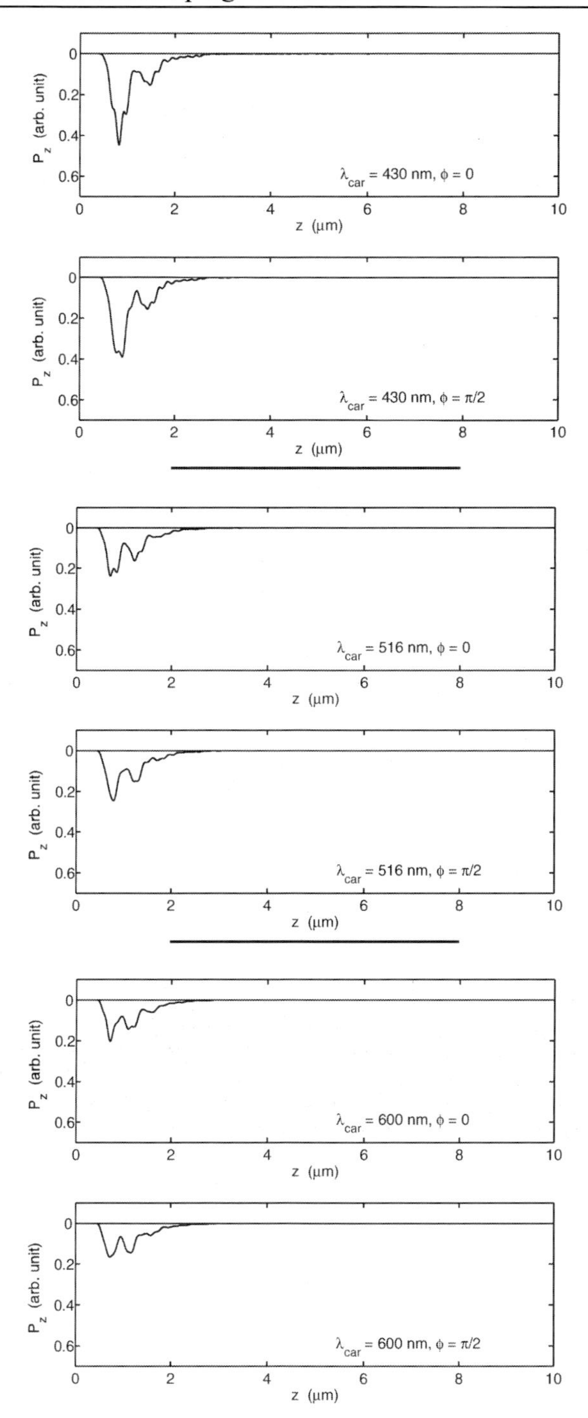

Figure 7. Plots of P_z at $t = 132$ fs, which show pulses reflected by a chiral STF. The carrier plane wave of the incident pulse was LCP; the carrier phase is either $\phi = 0$ or $\phi = \pi/2$; the carrier wavelength is $\lambda_{car} = 430$ nm (top), $\lambda_{car} = 516$ nm (middle), or $\lambda_{car} = 600$ nm (bottom). The vacuum / chiral STF interface at $z = 20\ \mu$m is not depicted. *Adapted from Geddes and Lakhtakia [16].*

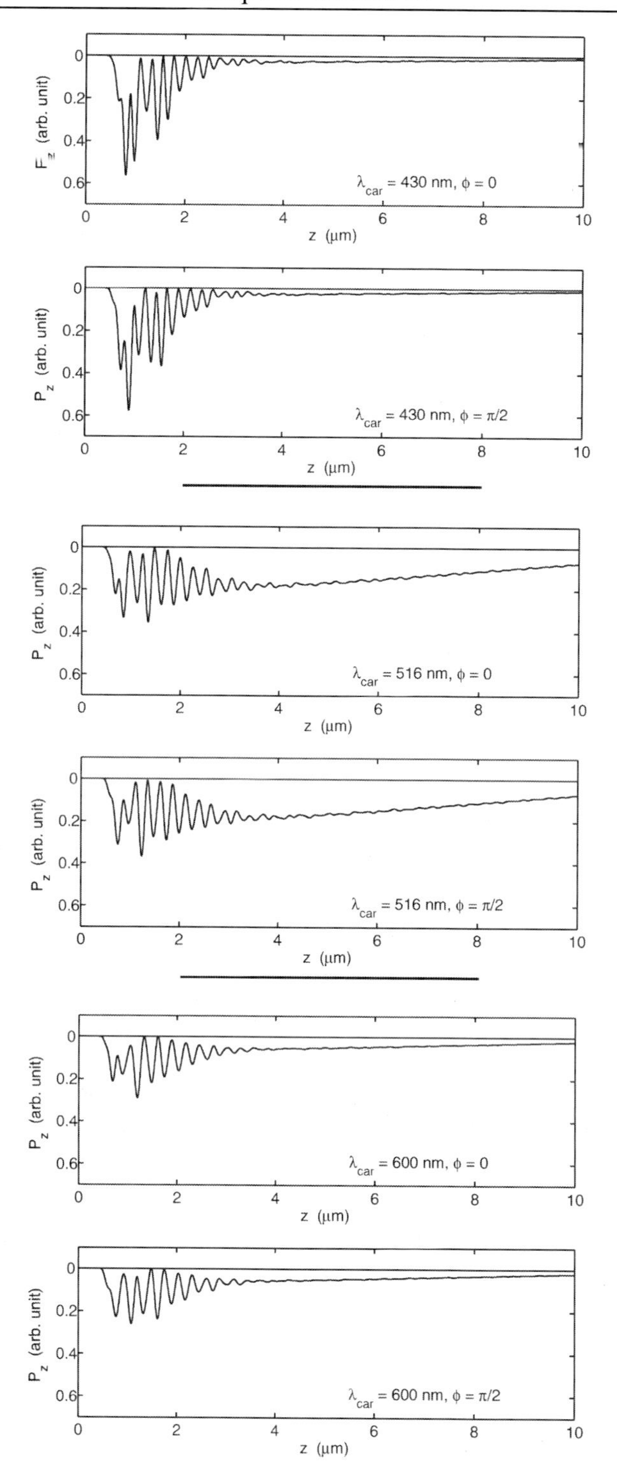

Figure 8. Same as Figure 7, except that the carrier plane wave of the incident pulse was RCP. *Adapted from Geddes and Lakhtakia [16].*

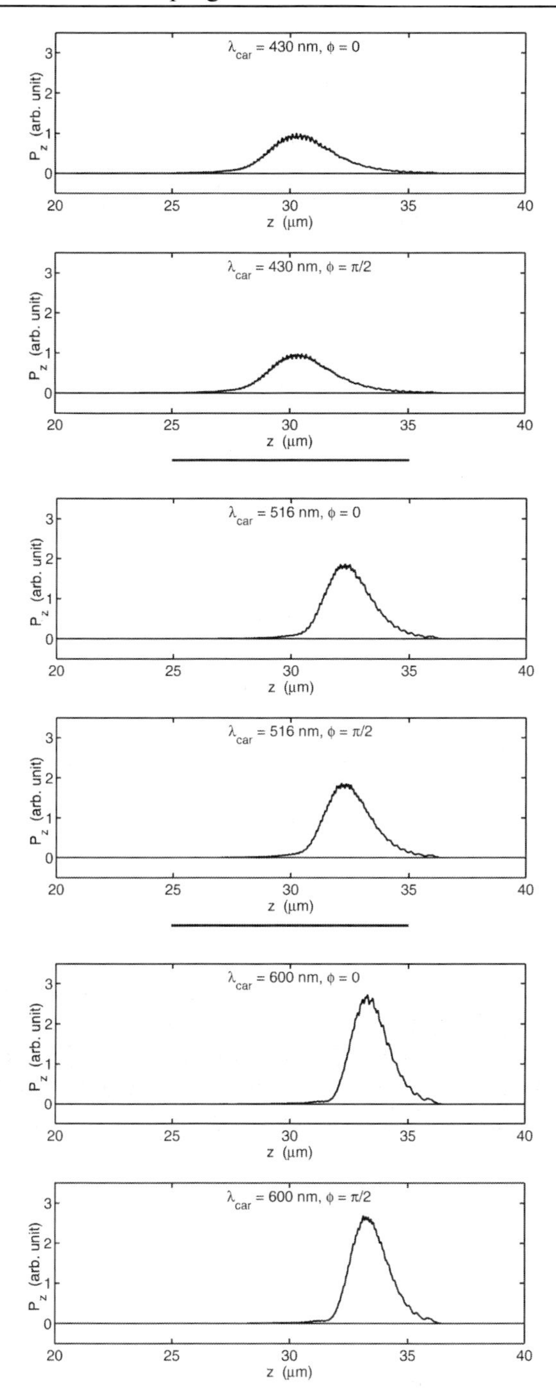

Figure 9. Plots of P_z at $t = 132$ fs, which show pulses refracted into a chiral STF. The carrier plane wave of the incident pulse was LCP; the carrier phase is either $\phi = 0$ or $\phi = \pi/2$; the carrier wavelength is $\lambda_{\text{car}} = 430$ nm (top), $\lambda_{\text{car}} = 516$ nm (middle), or $\lambda_{\text{car}} = 600$ nm (bottom). The vacuum / chiral STF interface at $z = 20$ μm is not depicted. *Adapted from Geddes and Lakhtakia [16].*

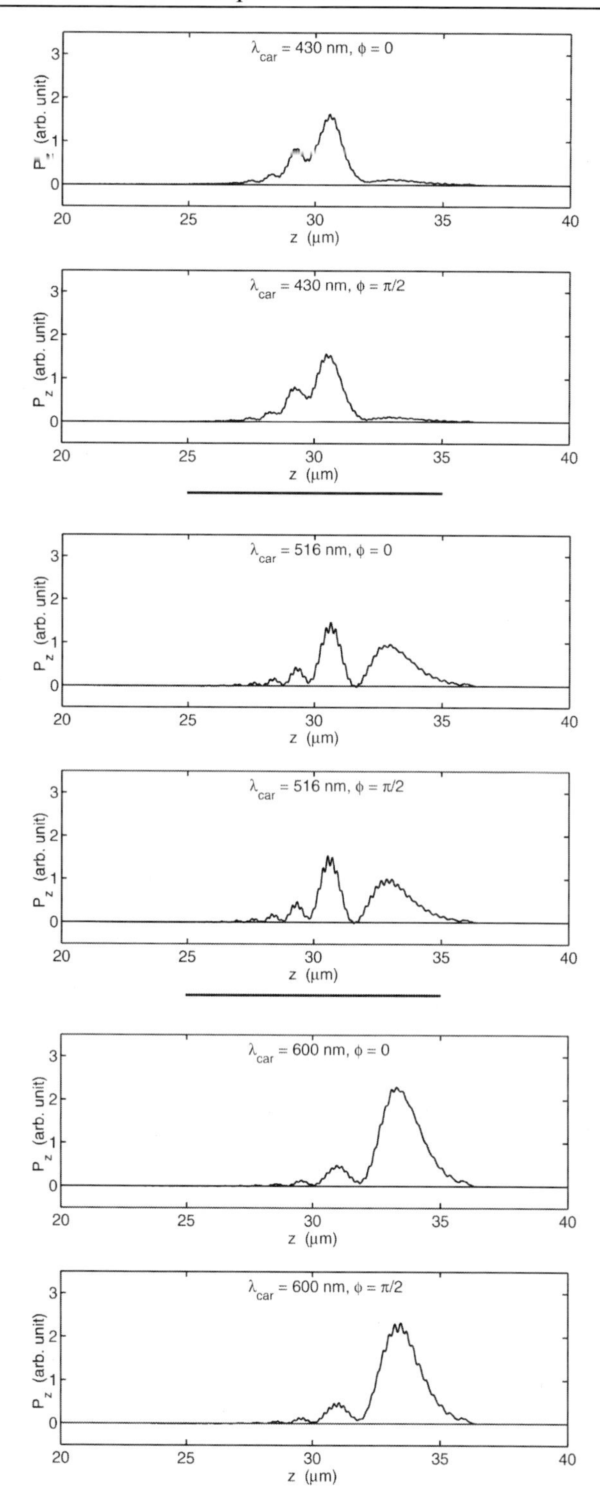

Figure 10. Same as Figure 9, except that the carrier plane wave of the incident pulse was RCP. *Adapted from Geddes and Lakhtakia [16].*

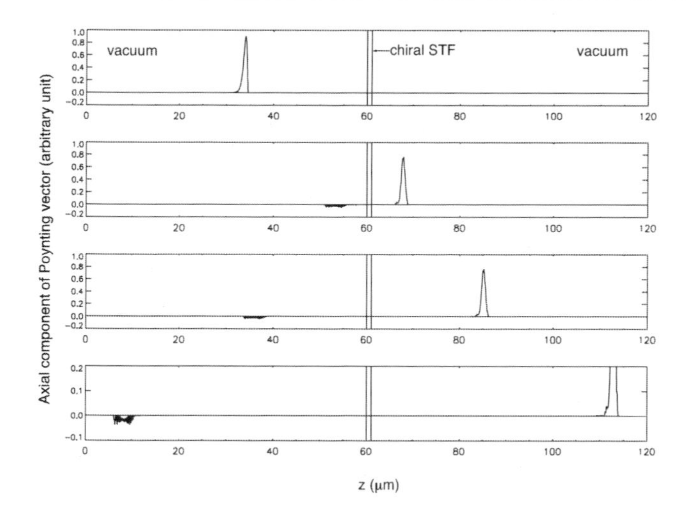

Figure 11. Snapshots of P_z at $t = 115.3, 230.6, 288.2$, and 380.4 fs—showing the spatiotemporal evolution of a narrow–extent pulse modulating a RCP carrier plane wave which impinges on a structurally right–handed chiral STF of thickness $z_1 - z_0 = 5\Omega$. The incident pulse strikes the film from the left, and gives rise to a reflected pulse (left of the film) and a transmitted pulse (right of the film). *Adapted from Geddes and Lakhtakia [20].*

4. Finite–Thickness Problems

In Section 3., FD computations elucidated the time–domain mechanism responsible for the CBP in chiral STFs—pulse bleeding through a light pipe. In this section, the phenomenon of pulse bleeding is used to explain the distortion of optical pulses as they traverse a chiral STF slab.

4.1. Ripening of the CBP

We now examine the ripening of the CBP, familiar from frequency domain research [25], as the thickness of a chiral STF slab increases. For these calculations [20], $N_{a,b,c} = 10^5$, $A_p = 1$ V/m, $\tau_p = 2\lambda_{car}/c_0$, and $\phi = 0$; the value $z_0 = 60$ μm is fixed, while the slab thickness $z_1 - z_0$ was varied. We chose the grid size $\Delta z = 4$ nm and stability parameter $\beta = 0.9$ to fix $\Delta t = 0.012$ fs. Two cases are shown: in one $z_1 - z_0 = 5\Omega$; in the other $z_1 - z_0 = 100\Omega$. Plots of P_z are given at $t = 115.3, 230.6, 288.2$, and 380.4 fs to show the spatiotemporal evolution of each pulse. Conditions are right for the CBP to occur in both Figures 11 and 12, but the transmitted pulse at $t = 380.4$ fs in the former is distorted little in comparison to the incident pulse at $t = 115.3$ fs, while much distortion occurs in the latter case. This is because in the latter case, there are many more structural periods for the refracted pulse to pass through as it traverses the film, and hence pulse bleeding and distortion of the refracted pulse can occur for a longer time. Thus the transmitted pulse in Figure 12 possesses little energy compared to the transmitted pulse in Figure 11 in addition to its shape being more distorted. Note by examining Figure 13 that an incident pulse whose carrier plane wave is LCP passes through a 100Ω thick chiral STF having undergone some absorption, but

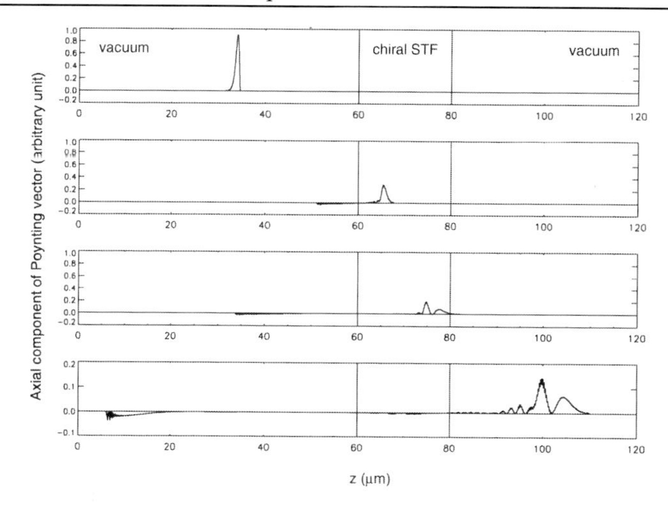

Figure 12. Same as Figure 11 except that the chiral STF is 100Ω thick. *Adapted from Geddes and Lakhtakia [20].*

not nearly the distortion experienced by an incident pulse whose carrier plane wave is RCP. Thus the CBP is at least partly responsible for the distortion of ultrashort pulses that traverse chiral STF slabs. More details about this process are available elsewhere [20].

4.2. Videopulse Propagation

Finally in this section, we examine the propagation of videopulses through chiral STF slabs. Videopulses are optical pulses of such short duration that they may encompass a few or even less than one optical cycle. For these calculations, $N_{a,b,c} = 10^5$, $A_\mathrm{p} = 0.20$ V/m, $\tau_\mathrm{p} = 0.688$ fs, and $\phi = 0$; the boundaries of the chiral STF were delineated by $z_0 = 8\ \mu$m and $z_1 = 12\ \mu$m. Figures 14–16 show the results. Even for incident videopulses of such short duration as these—about $1\frac{1}{2}$ optical cycles with $\lambda_\mathrm{car} = 516$ nm—the phenomenon of pulse bleeding is apparent. Notice however, that both incident pulses with LCP and RCP carrier plane waves are quite distorted when they exit the chiral STF into the half–space $z > z_1$, an observation that indicates that while the CBP is important for shaping optical pulses, it is not the only effect which does so. More details about videopulse propagation through chiral STF slabs is given elsewhere [26].

5. Conclusion

Based on the results presented in Sections 3.–4., we can draw the following conclusions regarding the propagation of ultrashort optical pulsed plane waves through chiral STFs:

- the phenomenon of pulse bleeding, a polarization–dependent transfer of energy from a refracted pulse to a reflected pulse across a light pipe, is the spatiotemporal fingerprint of the CBP in the time domain (even for videopulses), and

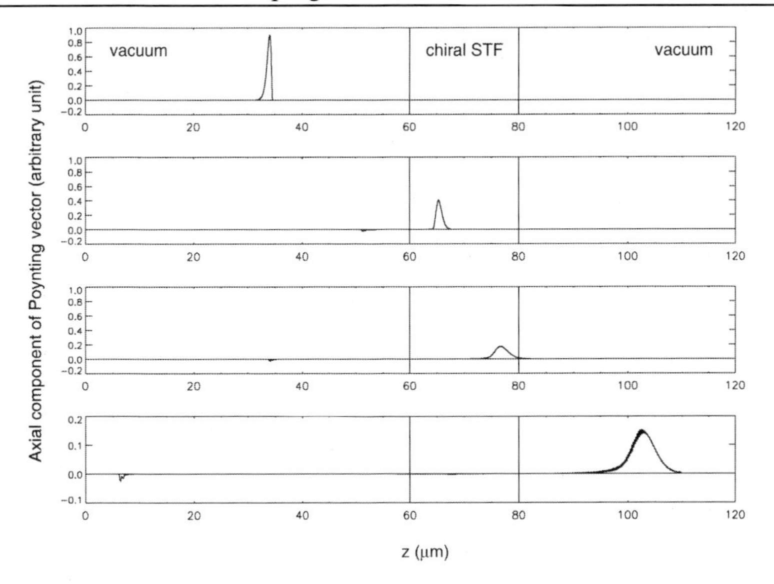

Figure 13. Same as Figure 12 except that the incident carrier plane wave is LCP. *Adapted from Geddes and Lakhtakia [20].*

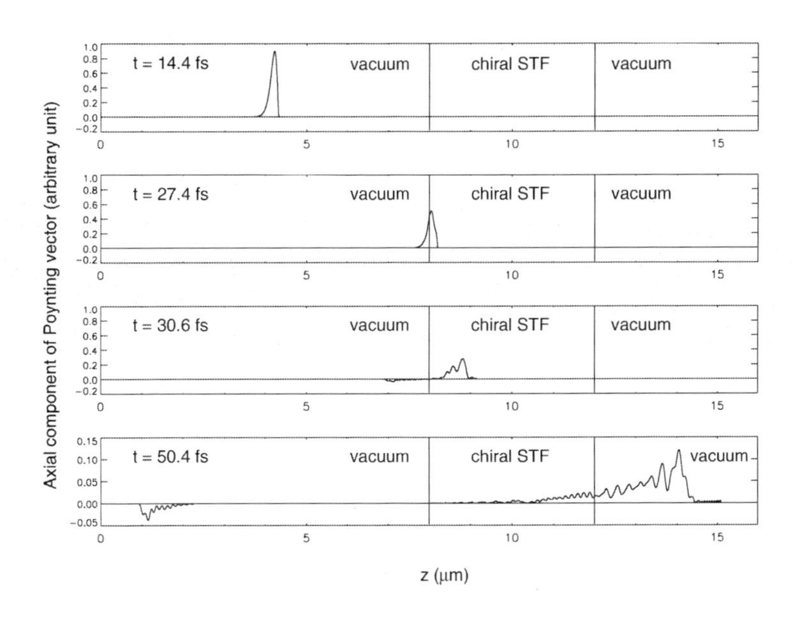

Figure 14. Plots of P_z at $t = 14.4, 27.4, 30.6$, and 50.4 fs showing the tranformation of an incident pulse into reflected and transmitted pulses. The incident pulse, shown at left in the first panel, has a LCP carrier plane wave. *Adapted from Geddes and Lakhtakia [26].*

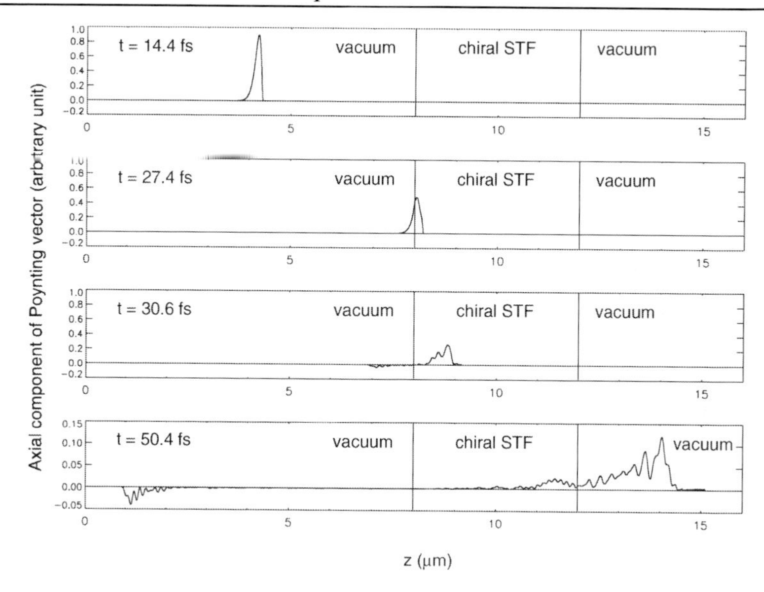

Figure 15. Same as Figure 15, except that the incident pulse possesses a RCP carrier plane wave. *Adapted from Geddes and Lakhtakia [26].*

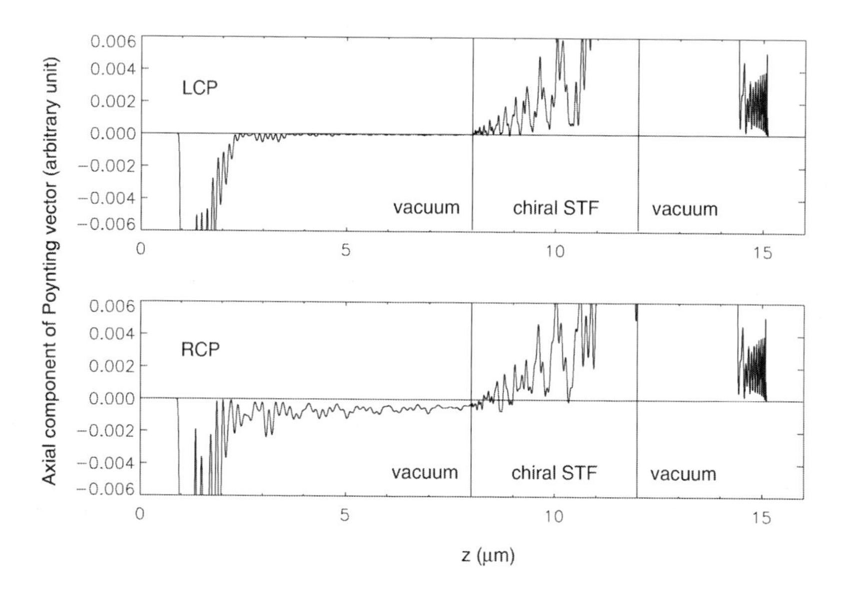

Figure 16. Magnified views of the bottom panels (t = 50.4 fs) of Figures 14 and 15 illustrating the pulse bleeding phenomenon for videopulses. *Adapted from Geddes and Lakhtakia [26].*

- pulse bleeding is responsible for at least some of the pulse shaping effects observed when ultrashort optical pulses traverse chiral STFs, but not all.

The mathematical analogy between the constitutive relations of chiral STFs and cholesteric liquid crystals (CLCs) implies that many of the conclusions of this chapter apply also to CLCs [27, 28, 29].

Many lines of research in time–domain pulse propagation through STFs and other complex materials, e.g. isotropic chiral materials [30], remain unexplored. To advance our understanding of such pulse propagation and use our knowledge to design pulse shaping devices, the phenomenology described in this chapter must be extended to include quantitative techniques for measuring pulse distortion.

Acknowledgments

I thank Akhlesh Lakhtakia for his patient guidance in the conduct of the research reviewed in this chapter. Computer facilities were supplied, in part, by the Pittsburgh Supercomputing Center. I am grateful for the support of an National Science Foundation Graduate Research Fellowship, a SPIE Educational Scholarship, and a Proctor & Gamble Summer Undergraduate Research Fellowship.

About the Author

Joseph B. Geddes III graduated from the Pennsylvania State University with bachelors and masters degrees in 2001, and with a doctoral degree in 2006—all in Engineering Science and Mechanics. He won the Schreyer Honors College Dean's Research Award and the Xerox Research Award in 2001 for his BS and MS theses, respectively, and he was awarded a NSF Graduate Fellowship and a SPIE Educational Scholarship. His work centers on the electromagnetics of complex materials, and he has published twelve journal papers in that field. To date, his research has focussed on the time–domain responses of sculptured thin films to ultrashort optical pulses.

References

[1] M. W. Horn, M. D. Pickett, R. Messier, and A. Lakhtakia. Blending of nanoscale and microscale in uniform large–area sculptured thin film architectures. *Nanotechnology*, 15:303–310, 2004.

[2] M. W. Horn, M. D. Pickett, R. Messier, and A. Lakhtakia. Selective growth of sculptured nanowires on microlithographic substrates. *J. Vac. Sci. Technol. B*, 22:3426–3430, 2004.

[3] A. Lakhtakia and R. Messier. *Sculptured Thin Films: Nanoengineered Morphology and Optics*. SPIE Optical Engineering Press, Bellingham, WA, USA, 2005.

[4] I. J. Hodgkinson, A. Lakhtakia, and Q. h. Wu. Experimental realization of sculptured–thin–film polarization–discriminatory light–handedness inverters. *Opt. Eng.*, 39:2831–2834, 2000.

[5] Q. Wu, I. J. Hodgkinson, and A. Lakhtakia. Circular polarization filters made of chiral sculptured thin films: Experimental and simulation results. *Opt. Eng.*, 39:1863–1868, 2000.

[6] I. J. Hodgkinson, Q. h. Wu, A. Lakhtakia, and M. W. McCall. Spectral–hole filter fabricated using sculptured thin–film technology. *Opt. Commun.*, 177:79–84, 2000.

[7] I. Hodgkinson and Q. h. Wu. Serial bideposition of anisotropic thin films with enhanced linear birefringence. *Appl. Opt.*, 38:3621–3625, 1999.

[8] I. Hodgkinson, Q. h. Wu, B. Knight, A. Lakhtakia, and K. Robbie. Vacuum deposition of chiral sculptured thin films with high optical activity. *Appl. Opt.*, 39:642–649, 2000.

[9] T. R. Gosnell and A. J. Taylor, editors. *Selected Papers on Ultrafast Laser Technology*. SPIE Optical Engineering Press, Bellingham, WA, USA, 1991.

[10] W. H. Knox. Ultrafast technology in telecommunications. *IEEE J. Sel. Topics Quantum Electron.*, 6:1273–1278, 2000.

[11] N. Bloembergen. From nanosecond to femtosecond science. *Rev. Modern Phys.*, 71:S283–S287, 1999.

[12] J. B. Geddes III. Traversal of optical pulses through dielectric thin–film helicoidal bianisotropic mediums. Master's thesis, The Pennsylvania State University, 2001.

[13] J. D. Jackson. *Classical Electrodynamics*. John Wiley & Sons, Inc., New York, NY, USA, 3rd edition, 1999.

[14] V. C. Venugopal and A. Lakhtakia. Sculptured thin films: Conception, optical properties, and applications. In O. N. Singh and A. Lakhtakia, editors, *Electromagnetic Fields in Unconventional Materials and Structures*, chapter 5, pages 151–216. Wiley Interscience, New York, NY, USA, 2000.

[15] C. Kittel. *Introduction to Solid State Physics*. Wiley Eastern, New Delhi, India, 1974.

[16] J. B. Geddes III and A. Lakhtakia. Effects of carrier phase on reflection of optical narrow–extent pulses from axially excited chiral sculptured thin films. *Opt. Commun.*, 225:141–150, 2003.

[17] N. Gershenfeld. *The Nature of Mathematical Modeling*. Cambridge University Press, Cambridge, UK, 1999.

[18] R. Luebbers, F. P. Hunsberger, K. S. Kunz, R. B. Standler, and M. Schneider. A frequency–dependent finite–difference time–domain formulation for dispersive materials. *IEEE Trans. Electromagnetic Compat.*, 32:222–227, 1990.

[19] R. J. Luebbers, F. Hunsberger, and K. S. Kunz. A frequency–dependent finite–difference time–domain formulation for transient propagation in plasma. *IEEE Trans. Antennas Propagat.*, 39:29–34, 1991.

[20] J. B. Geddes III and A. Lakhtakia. Time–domain simulation of the circular Bragg phenomenon exhibited by chiral sculptured thin films. *Eur. Phys. J. Appl. Phys.*, 14:97–105, 2001. *Erratum*: 16:247, 2001.

[21] A. Lakhtakia. Spectral signatures of axially excited slabs of dielectric thin–film helicoidal bianisotropic mediums. *Eur. Phys. J. Appl. Phys.*, 8:129–137, 1999.

[22] J. B. Geddes III and A. Lakhtakia. Reflection and transmission of narrow–extent pulses by axially excited chiral sculptured thin films. *Eur. Phys. J. Appl. Phys.*, 13:3–14, 2001. Erratum: 16:247, 2001.

[23] J. Wang, A. Lakhtakia, and J. B. Geddes III. Multiple Bragg regimes exhibited by a chiral sculptured thin film half–space on axial excitation. *Optik*, 113:213–222, 2002.

[24] A. Lakhtakia and J. B. Geddes III. Nanotechnology for optics is a phase–length–time sandwich. *Opt. Eng.*, 43:2410–2417, 2004.

[25] V. C. Venugopal and A. Lakhtakia. On selective absorption in an axially excited slab of a dielectric thin–film helicoidal bianisotropic medium. *Opt. Commun.*, 145:171–187, 1998. Erratum: 161:370, 1999.

[26] J. B. Geddes III and A. Lakhtakia. Videopulse bleeding in axially excited chiral sculptured thin films in the Bragg regime. *Eur. Phys. J. Appl. Phys.*, 17:21–24, 2002.

[27] J. B. Geddes III, M. W. Meredith, and A. Lakhtakia. Circular Bragg phenomenon and pulse bleeding in cholesteric liquid crystals. *Opt. Commun.*, 182:45–57, 2000.

[28] M. W. Meredith and A. Lakhtakia. Time–domain signature of an axially excited cholesteric liquid crystal. Part I: Narrow–extent pulses. *Optik*, 111:443–453, 2000.

[29] J. B. Geddes III and A. Lakhtakia. Time–domain signature of an axially excited cholesteric liquid crystal. Part II: Rectangular wide–extent pulses. *Optik*, 112:62–66, 2001.

[30] A. Lakhtakia, editor. *Selected Papers on Natural Optical Activity*. SPIE Optical Engineering Press, Bellingham, WA, USA, 1990.

In: Advances in Laser and Optics Research. Volume 10 ISBN: 978-1-62257-795-8
Editor: William T. Arkin

Chapter 6

INTEGRATED OPTICS ON SILICON

Ulrich Hilleringmann*
Center of Optoelectronics and Photonics, Sensor Technology Department,
University of Paderborn, Paderborn, Germany

ABSTRACT

This review article describes various kinds of processes to integrate optical waveguides on silicon substrates, suitable for the visible to the near infrared spectral range. Different types of materials for the light guiding film and set ups for the waveguide structures are compared according to the propagation loss and the integration density. The link of the strip loaded rib waveguides to photo detectors, integrated in the same substrate, is performed by butt- or leaky wave coupling. Two kinds of processing for the monolithic integration of waveguides, photo detectors and CMOS circuits are presented, and an application as a pressure sensor is given. The last part of the article discusses optical waveguides in silicon suitable for data transmission in the 1.55 μm spectral range. Doped silicon waveguides and silicon-on insulator waveguides are presented as light guiding films. The capability of the silicon technology is demonstrated by a capacity controlled all-silicon modulator for 10 GHz.

1. INTRODUCTION

Although silicon is a semiconducting material with an indirect band gap, there is a huge interest in silicon technology for integrated optics [1]. In contrast to the compound semiconductors, it is a very cheap high quality material with excellent mechanical properties. Today the simulation of silicon devices and the processing of silicon wafers is well understood. Lithography, layer deposition and dry etching can be controlled with nanometer scale perfection. This has lead to strong research in integrated optics on silicon. First interests

* Corresponding author: Ulrich Hilleringmann. Center of Optoelectronics and Photonics, Sensor Technology Department, University of Paderborn, Warburger Str. 100, D-33098 Paderborn, Germany.

were located on the visible spectral range only, but today the infrared radiation at 1.55 μm used for data transmission in glass fibers is of strong interest, too.

The only property missing for applications in integrated optics is the light emission from silicon. The attempts to use porous silicon [2] or radiation caused by energetic electrons in *p-n* junctions (Bremsstrahlung) failed because of the very low efficiencies (far below 0.1%) [3]. Attempts to integrate silicon light emitting diodes (LEDs) with sufficient efficiencies failed up to now [4]. Nevertheless, only silicon is suitable for the integration of optical waveguides, photo detectors and high density CMOS circuits on one chip [5]. As already demonstrated, integrated optics on silicon is capable for high performance Mach-Zehnder interferometers and Michelson interferometers [6], which result in many other sensor applications like gas or chemical sensors [7]. Most common are silicon oxinitride layers as light guiding films, but during the last years, materials with higher refractive index (RI) like silicon nitride or aluminium oxide are of growing interest. On the other hand, integrated optics using silicon itself as a light guiding film is applicable for wavelengths above 1.2 μm [8]. Silicon waveguides can give high quality passive devices in optical networks. Either early *n*-doped silicon waveguides or today's SOI-waveguides with high RI contrast result in excellent device characteristics.

2. Integrated Optics on Silicon in the Visible Range

Waveguides typically consist of a core with a higher RI n_1 than the surrounding cladding layer n_2. This structure is transferred to the planar technology for silicon wafer processing using different types of inorganic films, which are standard materials in MOS technology.

Silicon technology offers different transparent materials with a wide range of RIs which can be used for light guiding films and waveguide structures. The most common material for the cladding layers is silicon dioxide (SiO_2). It is normally used as a dielectric layer in MOS transistors. The RI is about 1.46, and the absorbance rate is extremely low in the visible spectral range. SiO_2 can be deposited by thermal oxidation of silicon or by chemical vapour deposition (CVD) using low pressure (LPCVD) [9,10] or plasma enhanced CVD (PECVD) [11,12].

Another high quality material is silicon nitride (Si_3N_4), which is often used as a diffusion mask for local oxidation of silicon in MOS processing. Its RI is about 2.0. The deposition is usually performed by LPCVD or PECVD; both methods give high quality films.

Beside silicon oxide and nitride, transparent films of aluminium oxide [13] or nitride can be deposited by sputtering. These films may be suitable for waveguides, however, there are only very few publications about aluminium based waveguides. The disadvantage of these aluminium based layers is the missing availability of a dry etching process for materials.

Most waveguides presented in literature use silicon oxinitride (SiON), a mixture of SiO_2 and Si_3N_4. The RI of this material can easily be adjusted in the range from 1.46 to 2.0 during the layer deposition in CVD processes by the control of the deposition gas mixture. Figure 1 illustrates the dependence of the RI of the film on the gas composition during the deposition process.

Optical waveguides on silicon can use different types of setup. Figure 2 depicts the most common structures of waveguides suitable for monolithic integration. The rib waveguide

[Figure 2(a)] consists of a low refracting layer which is covered by a layer with a RI as the light guiding film. The top layer is partly etched to form a rib. The electromagnetic wave can propagate in the area of the rib only; outside the rib no mode can exist due to the reduced film thickness. Because of the roughness of the etched surface, the propagation loss of this kind of waveguide is large [15].

Figure 2(b) type of waveguides yield less propagation loss because the light guiding film is not etched itself. Only the cover layer is structured to ribs. This kind of waveguide can easily be integrated on silicon, but propagation loss is not optimized due to the roughness of the vertical rib side walls caused by dry etching. The strip loaded film waveguide [Figure 2(c)] exhibits very low propagation loss. The surface cladding layer is deposited on top of the light guiding film, and the etching process to form the strip does not affect the quality of the light guiding layer. So, the propagation loss is very low. The integration process is as simple as in case of the rib loaded waveguide.

Figure 2(d) shows a core waveguide. Its shape can be symmetric or rectangular. The propagation loss will be determined by the quality of etching process applied to form the core layer. Neither wet etching nor dry etching will result in an absolutely smooth side wall of the core, and even nanometer scale roughness will cause large propagation loss.

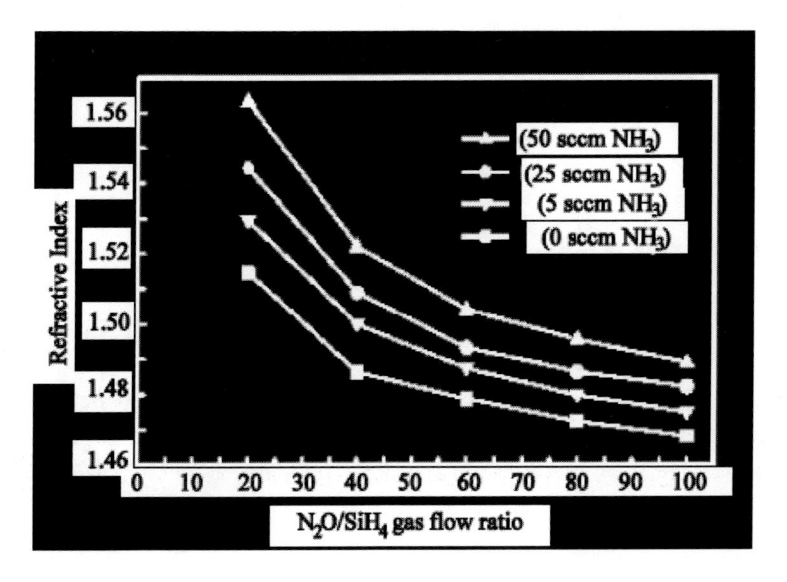

Figure 1. Refractive index of the deposited silicon oxinitride film depending on the ammonia flow for different ratio of N_2O/SiH_4 [14].

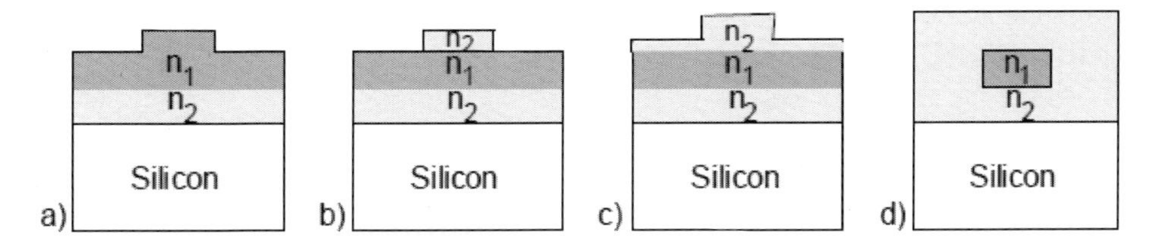

Figure 2. (a) Rib waveguide, (b) rib loaded film waveguide, (c) strip loaded film waveguide, and (d) core waveguide.

As a result, the propagation loss of an integrated waveguide will be high in all cases when the core layer is treated by an etching process. So, only the type of waveguide according to Figure 2(c) is suitable for high quality applications.

Figure 3 represents the structure of a typical monomode SiON waveguide for integrated optics on silicon in the visible spectral range. It consists of 2 μm of thermal oxide ($n_2 = 1.46$), 0.5 μm of SiON ($n_1 = 1.52$), and 0.5 μm of deposited silicon oxide. The top layer is etched by reactive ion etching to form a rib of 0.4 μm for light confinement in the SiON film below. The width of the rib is about 3 μm for monomode applications. The buffer oxide layer of 2 μm is usually grown by thermal oxidation in wet atmosphere at a temperature of about 1025 °C to 1100 °C. The SiON film deposition uses LPCVD or PECVD techniques. In the case of LPCVD, a gas mixture of SiH_2Cl_2, NH_3 and N_2O is thermally decomposed at about 920 °C. In the case of PECVD, the deposition starts at 350 °C applying a gas mixture of SiH_4, NH_3 and N_2O [16]. In both cases, the rate of NH_3 to N_2O controls the RI in the deposited layer. Typical deposition rates vary between 8 nm/min and 50 nm/min depending on the equipment. Directly after the SiON deposition, the capping layer of SiO_2 grows on top by simply changing the deposition atmosphere of the CVD process. The NH_3 flow stops and the N_2O flow is increased. As the last step, optical lithography defines the rib size which is transferred into the top oxide cladding layer. In the case of infrared light, the oxide layer thickness below the SiON film must be increased to about 8 to 10 μm to prevent leaky wave coupling to the silicon substrate. The SiON core layer is typically about 2 μm thick, and the surface cladding layer thickness is about 7 μm. The kind of waveguide used is of the ridge type, with an SiON rib height of 1.3 μm, as illustrated in Figure 3(b).

3. Characteristics of Waveguides in the Visible Range

The waveguides are typically tested at 633 nm using a He-Ne laser. The light is coupled into the waveguide by a lens using a (x, y, z, θ) stage for proper alignment. The propagation loss of integrated waveguides on silicon depends on the thermal treatment of the oxynitride layer during and after deposition. It is about 1.1 dB/cm [17] to 1.8 dB/cm [18] at 633 nm for PECVD oxynitride layers, but even less than 0.25 dB/cm are claimed [19]. This is negligible with regard to the dimensions of a chip.

The integration of optical devices needs curvatures to form beam splitters, interferometers and couplers. Small radii result in radiation losses at the bending of the waveguide. But, due to the small size of a chip, it is necessary to prevent large radii for curvatures. So, test structures with different sizes of curvatures have been integrated and tested. Figure 4 shows the scattered light photograph of different waveguide curvatures using SiON (left) and Si_3N_4 (right) light guiding layers with a RI of 1.52 respective 2.0.

Concerning the radiation loss in the curvatures, a minimum radius of about 200 μm is necessary in the SiON/SiO_2 system. In the case of nitride waveguides on silicon, even a radius of 20 μm is applicable, making thereby possible to increase the integration density by increase of the RI. The radiation loss in curvatures can be calculated as described in [20]. Figure 4 indicates that it is impossible to make small optical chips even with a relative high difference in the RI like SiON/SiO_2 waveguides with a $\Delta n/n$ of about 4%. Only a large index step enables highly integrated optical chips, e.g. the system Si_3N_4/SiO_2 with $\Delta n/n = 36\%$.

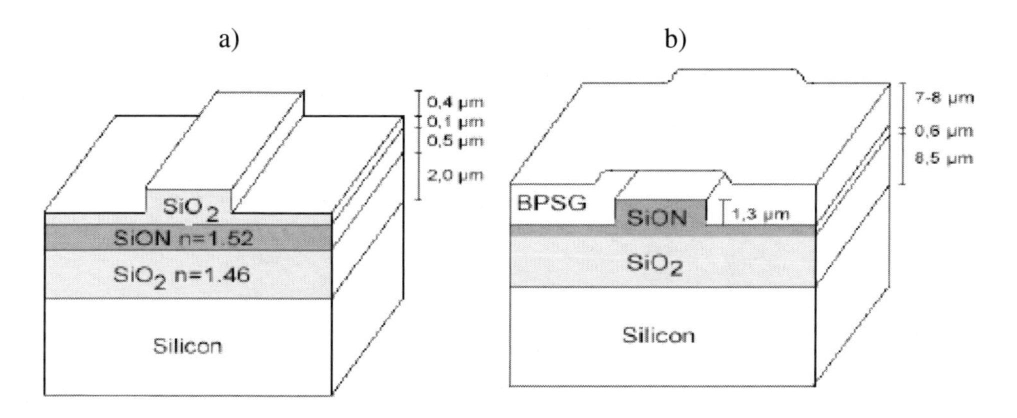

Figure 3. Structure of a typical SiON waveguide for (a) visible light, and (b) infrared light at 1.55 µm, as used in integrated optics on silicon.

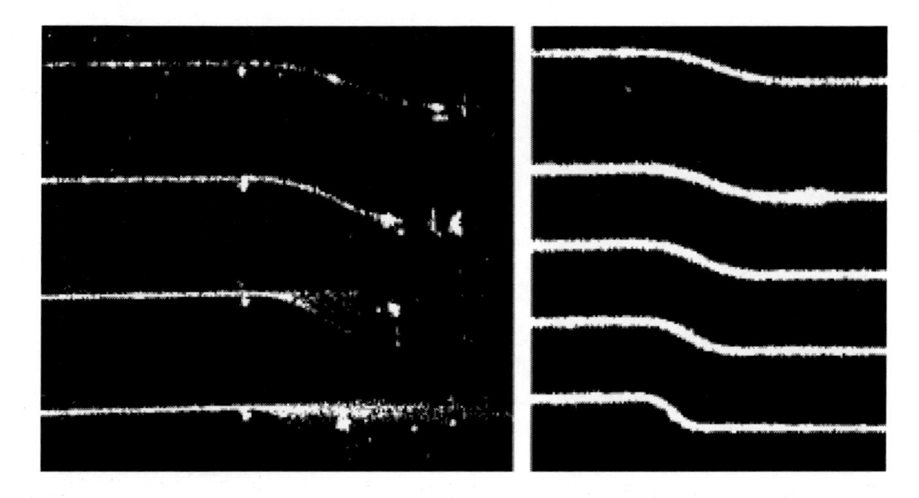

Figure 4. Different waveguide curvatures with radii of 500 µm, 200 µm, 100 µm, 50 µm, and 20 µm (left only), left for SiON (n = 1.52), right for Si_3N_4 (n = 2.0) (λ = 633 nm).

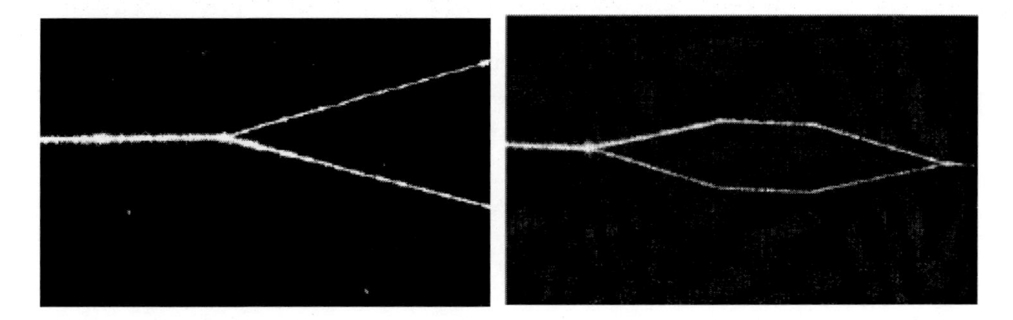

Figure 5. Mirror as a beam splitter, and Mach-Zehnder interferometer, consisting of two triangle shaped mirrors as beam splitters; four rectangular mirrors for light deflection.

As an alternative, the curvatures may be replaced by mirrors consisting of a transition from the SiON waveguide to ambient air. In this case, a hole with rectangular or triangular geometry is etched into the light guiding film structure perpendicular to the wafer surface.

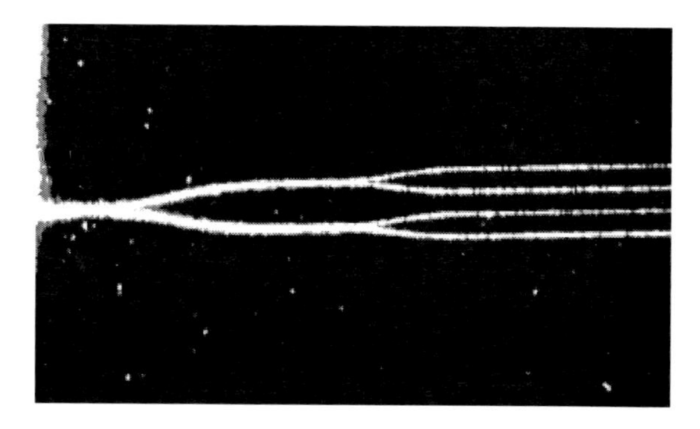

Figure 6. One to four beam splitting by three identical Y splitters with curvatures of about 1000 μm.

As depicted by the scattered light photograph of the Mach-Zehnder interferometer in Figure 5, the light deflection at the mirrors and the beam splitting by a triangular shaped mirror operate well. Nevertheless, each mirror results in a loss of about 1 dB.

Figure 6 depicts a beam splitter consisting of Y-splitters. The intensity is divided nearly uniformly into four waveguides. The propagation loss of the splitter strongly depends on the sharpness of the tip between the ribs etched in the center of the Y. Only high resolution lithography like e-beam guarantees low-loss beam splitting.

4. Integrated Optical Devices

Different kinds of switches have been integrated in SiON waveguides by applying micromechanical elements [21] or thermo-optical actuators [22]. A 1-to-8 switch using an arrayed waveguide grating wavelength multiplexer with thermal control is presented in [23]. It achieves high optical isolation of about 29 dB at $\lambda = 1.55$ μm between the active channel and the off state channels. Switching can be performed in 1 ms, whereas the power to control all switches simultaneously is about 6 W. In [18], an arrayed waveguide grating demultiplexer has been integrated in SiON technique in a very compact size of 5 mm × 2 mm. SiON add-drop filters for wavelength-division multiplexing have been demonstrated in [24]. The devices show high performance over a large spectral range. Today silicon devices for wavelength division multiplexing are based on resonant couplers with SiON waveguides [25]. The devices have a total loss of 6 dB from fiber-to-fiber, and the crosstalk for a single wavelength is less than –55 dB.

5. Optoelectronic Systems on Silicon

Integrated optics on silicon offers the advantage of monolithic optoelectronic chips with waveguide structures for sensor applications, and CMOS circuits for signal processing on the same substrate [26]. One condition for these applications is an interface between the optical and the electronic part of the system, e.g. a photodiode for the conversion of the optical information into an electrical signal. The integration of photodiodes is a standard in silicon MOS processing, as each *p-n* junction generates charge carriers in case of incident light. The main problem rises out of the thick oxide layer below the light guiding SiON film, which separates the guided photons in the SiON film from the semiconducting silicon substrate. Due to this oxide, the propagating electromagnetic wave cannot generate charge carriers in an integrated diode. There are different ways to get the light guiding film in contact to within silicon for signal transfer:

- applying photodiodes integrated in a mesa structure
- forming a planar surface between the diode surface and the thermal oxide surface
- tapering the oxide layer down to the silicon surface
- deflecting the light out of the waveguide to the silicon surface

5.1. Butt Coupling

Butt coupling uses a silicon mesa structure for the integration of the photodetector. The mesa rises out of the oxide layer, so the SiON waveguide film deposition takes place beside and on top of the mesa. So, the propagating light hits the side wall of the diode mesa, and can easily penetrate into the active silicon area, as depicted in Figure 7. Efficiencies of 84% have been demonstrated [27]. The main difficulty in this kind of coupling is the processing of MOS transistors on the same chip. Due to the necessity of mesa structure for photo diode, all MOS transistors are integrated in mesas. This causes a complete change in MOS processing at least, which is not possible due to its complexity.

5.2. Leaky Wave Coupling

To avoid the mesa structures rising out of the oxide layer, another structure has been developed applying leaky wave coupling of waveguides and integrated photo diodes (Figure 8). In this case, the silicon oxide layer below the light guiding SiON film and the silicon surface are on the same level. As such, no silicon mesa rises out of the surface. MOS processing is affected only weak. The main difficulty of this process is to get a smooth transition from the oxide layer to the active silicon area. This is impossible without a chemical mechanical polishing (CMP) step. CMP is a complex and expensive process step, which easily causes defects reducing the yield. Without CMP, small steps occur at the transition. They result in reflection and scattering of the electromagnetic wave, reducing the effective intensity in the waveguide.

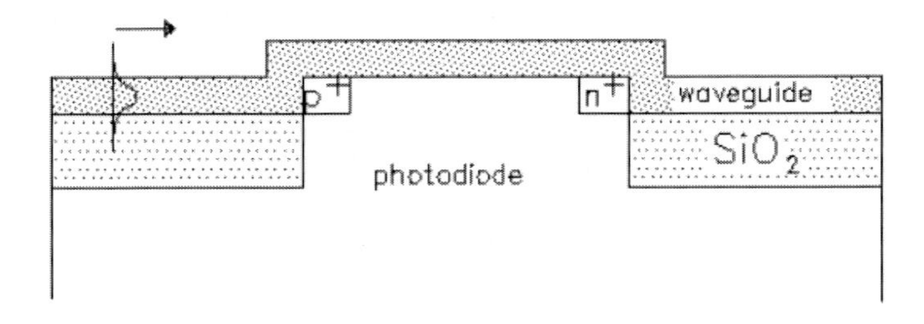

Figure 7. Butt coupling of an integrated waveguide to a photo diode on silicon.

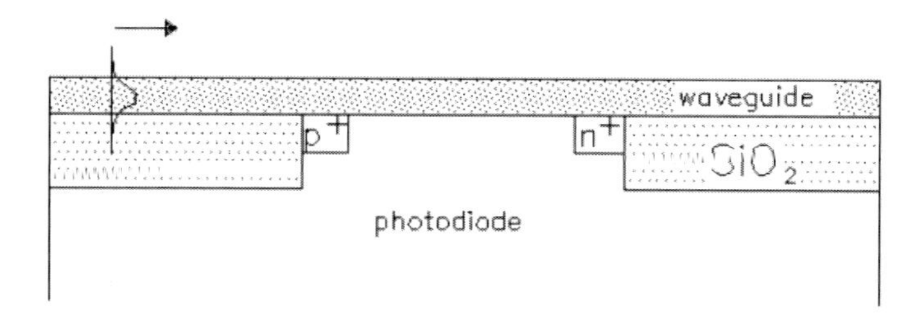

Figure 8. Leaky wave coupling of waveguide and photo diode applying a smooth transition from the oxide layer to the active silicon surface.

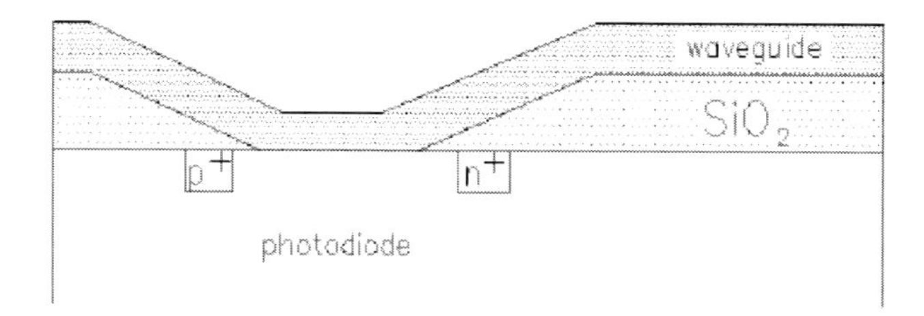

Figure 9. Coupling of waveguide and photo diode by an extended taper.

5.3. Taper Coupling

Another coupling technique uses an extended taper in the oxide layer below the light guiding film (Figure 9). Due to the small RI difference between the cladding layers and the SiON, the taper must be very smooth to prevent optical leakage. The processing of an extended taper is possible, but it is not easy to do. Even in the well developed silicon technology, it is expensive to etch a suitable structure for this kind of coupling. Additionally, the leaky wave coupling starts at the point of reduced oxide thickness. So, charge carriers are generated outside the photo detector, which results in low frequency operation due to charge carrier diffusion.

5.4. Coupling by Integrated Mirrors

The most promising technique for coupling of waveguides and photodiodes is a special kind of butt coupling using an integrated mirror to deflect the light out of the waveguide into the photo diode (Figure 10). Again a taper is necessary to get a sloped wall with a slope of about 45°. The propagating electromagnetic wave hits the mirror surface, consisting of the transition SiON to ambient air, and is reflected by total reflection. This technique is applicable on nearly any MOS circuit, because there is no special need for a planar surface at the transition from the silicon oxide to the active silicon area. The advantage of the mirror coupling technique results in the modular processing of the MOS structures and the waveguide layers. It can easily be implemented in any MOS technology. Figure 11 shows the scattered light photograph of a waveguide coupled to a photodiode by applying the process of mirror coupling.

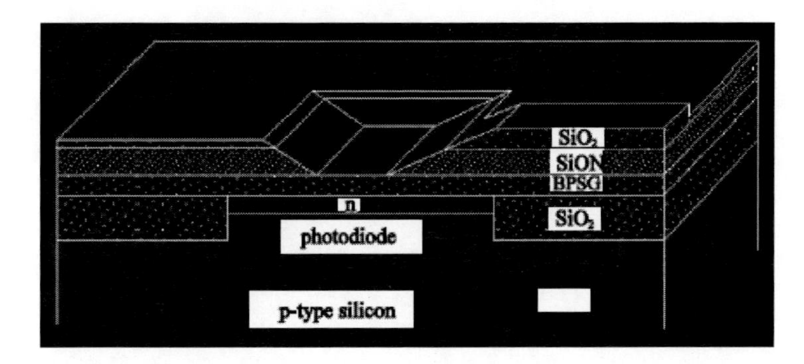

Figure 10. Coupling of a waveguide to an integrated mirror to deflect the light into the photo diode.

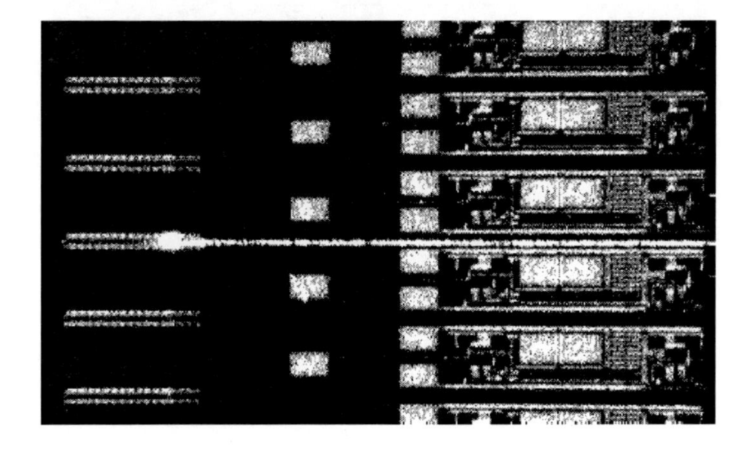

Figure 11. Scattered light photograph of a waveguide coupled to a photodiode applying mirror coupling.

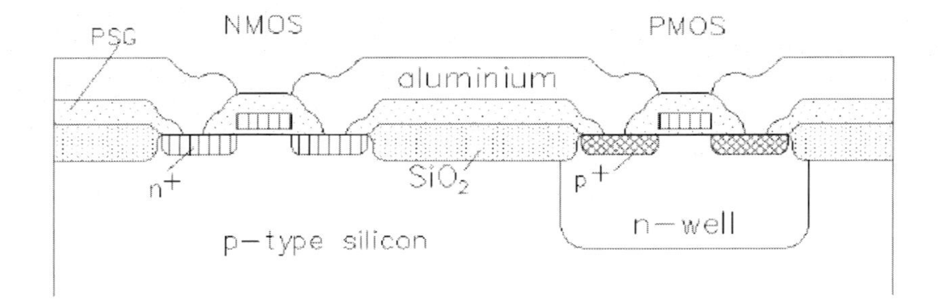

Figure 12. Cross-section of a normal CMOS structure in LOCOS technique.

6. MONOLITHIC INTEGRATION OF WAVEGUIDES AND MOS CIRCUITS

Today's MOS processes apply local oxidation techniques (Figure 12) or shallow trench isolation to separate the active transistor areas by silicon oxide [28]. These processes typically have a soft or step less transition from the oxide area to the active silicon surface, but there is no possibility to extend the active silicon area to form mesa structures for butt coupling.

The Side Wall Mask Isolated LOCal Oxidation of Silicon (SWAMI-LOCOS) technique [29] offers a simple way to get mesa structures in an oxidized silicon wafer. As depicted in Figure 13, after deposition and etching of the standard LOCOS mask, a deep silicon etch removes the crystalline material outside the active device areas. The depths of this etch defines the kind of coupling – in the case of 55% of the oxide thickness grown afterwards, it results in leaky wave coupling. In the case of about 75% of the thermal oxide thickness needed for the separation of the light guiding SiON film from the silicon substrate, butt coupling of the waveguides to the diodes results.

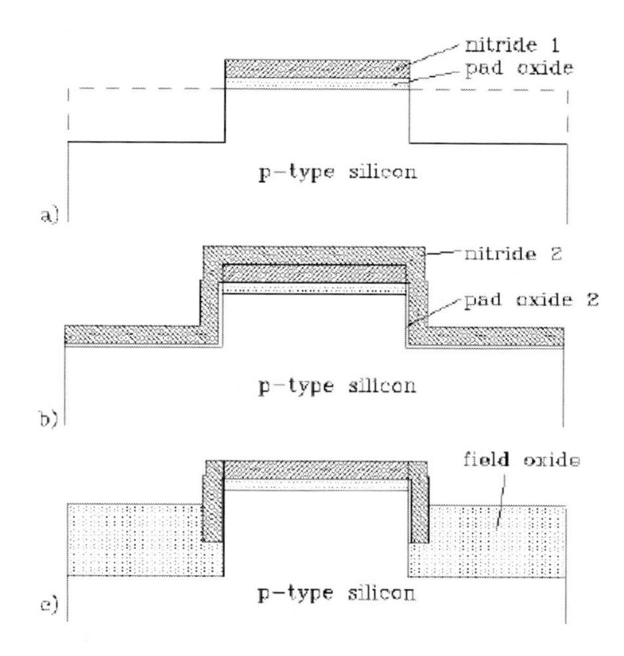

Figure 13. Formation of mesa structures using SWAMI-LOCOS technique.

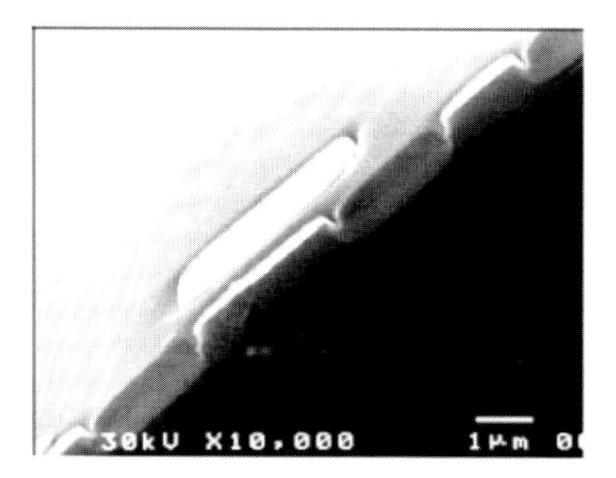

Figure 14. Surface planarity after local oxidation applying SWAMI technique.

Next, another nitride layer deposition masks the vertical walls of the etched mesas. This layer is etched back anisotropically by reactive ion etching. So, only the side walls and the top of the mesas are protected against the thermal oxidation step. During the thermal oxidation of the silicon, the oxide grows outside the mesa structure. After removal of the nitride mask, the surface is nearly plane, or a mesa rises out of the oxide layer. The planarity of the surface after 55% of the oxide thickness for silicon etching is depicted in Figure 14.

The only disturbance of the flatness is caused by the side wall passivation nitride around the active silicon areas, which is removed during surface nitride etching.

In the following section, two integration principles are explained for monolithic integration of optical waveguides and CMOS circuits on one silicon chip applying SWAMI-LOCOS.

7. LAYER-ADJUSTED MONOLITHIC INTEGRATION TECHNIQUE

The layer-adjusted monolithic integration technique uses the field oxide and the intermetal dielectric layer of the MOS process for the waveguide structure by the adjustment of the layer thicknesses and their RIs. If the field oxide in the MOS process is extended to 2 μm, and the SiON layer replaces the standard PSG or BPSG to cover the polysilicon, the typical strip loaded film waveguide structure is available in any MOS process. Instead of the simple LOCOS technique, the process must use SWAMI-LOCOS to form silicon mesas or a planar surface. So, as a starting point of MOS transistor integration, a field oxide of 2 μm covers the wafer surface outside the active transistor areas. Then MOS transistors can be integrated by gate oxidation, polysilicon gate deposition and ion implantations for the drains and sources of the devices. Afterwards, the SiON deposition for the film waveguide can start. It replaces standard PSG or BPSG. As the reflow characteristics of SiON are nearly zero, the aluminium metallization is difficult in the area of the polysilicon gates.

The light guiding film covers the mesa structure of the integrated photo diode, and the waveguide hits the diode perpendicular from its side.

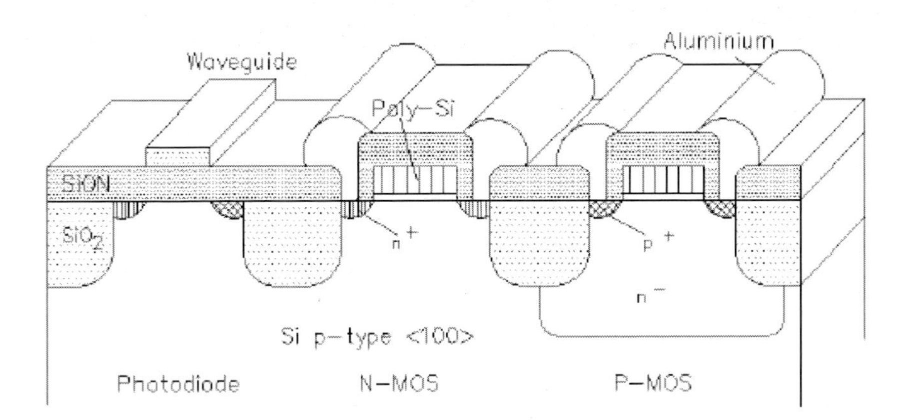

Figure 15. Cross-section of the combined monolithic integration process for waveguides and CMOS components on the same chip using either leaky wave coupling or butt coupling, depending on the step at the transition from the SiO_2 to the active silicon area.

The light guiding film covers the mesa structure of the integrated photo diode, and the waveguide hits the diode perpendicular from its side. The SiON film serves as the polysilicon/intermetal dielectric layer, and after contact opening the metallization process starts. Figure 15 illustrates a cross-section of this monolithic integration process.

The integration of the CMOS structures and waveguides using this kind of processing requires no additional layers, but the quality of the waveguides and of the transistors is rather low. Due to the reactive ion etching of the aluminium wires on top of the SiON film, the surface roughness increases, resulting in high propagation loss. This may be prevented by a high selective etching process for the metallization, e.g. by wet etching. Additionally, it is nearly impossible to get a smooth transition from the field oxide to the active diode area.

On the other hand, the parameters of the electronic devices depict large leakage currents in the *p-n* junctions. Even in the off-state, MOS transistors leakage is high, because of the boron segregation below the field oxide. This causes channels for a current flow below the oxide. In conclusion, this kind of monolithic integration results in low quality optical and electrical devices, although the process flow does not need any additional layers. So, the process flow must be developed completely new to get sufficient yield in electrical and optical devices.

Figure 16 shows a scattered light photograph of waveguides coupled to three identical photo diodes aligned in a chain. The diodes were integrated using the layer adjusted monolithic integration technique with butt coupling, as explained before. Top diodes consist of 25 μm active silicon below the waveguide, the bottom diodes have only 3 μm active diode area. Optical power input is 5 mW at 633 nm. The intensive spot at each transition from the field oxide to the active diode area characterizes a nonideal transition of the waveguide to the diode. Due to a non-perpendicular diode side wall caused by the SWAMI-LOCOS processing, a large amount of the light is scattered out of the waveguide, and only a part of the intensity is guided. Figure 17 depicts the photo currents at the three diodes coupled to one waveguide for different diode lengths. The data are normalized to the current of the first diode due to the unknown intensity coupled into the waveguide. In short diodes, only a part of the intensity couples into the diodes, and results in charge carrier generation.

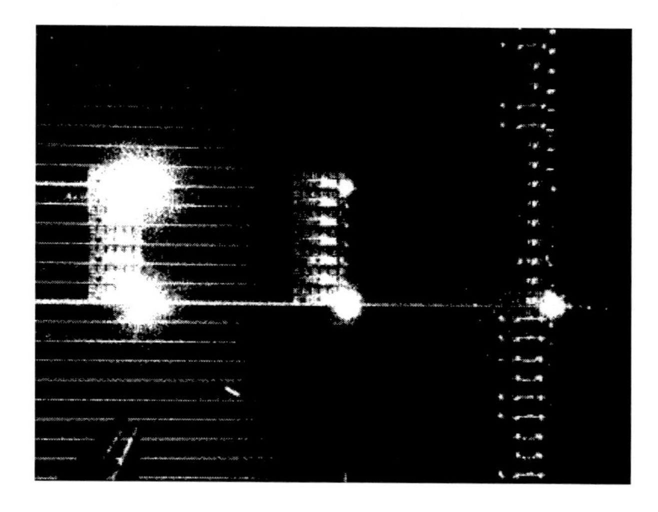

Figure 16. Waveguides coupled by butt coupling to three identical diodes with a length of: top 25 μm, bottom 3 μm.

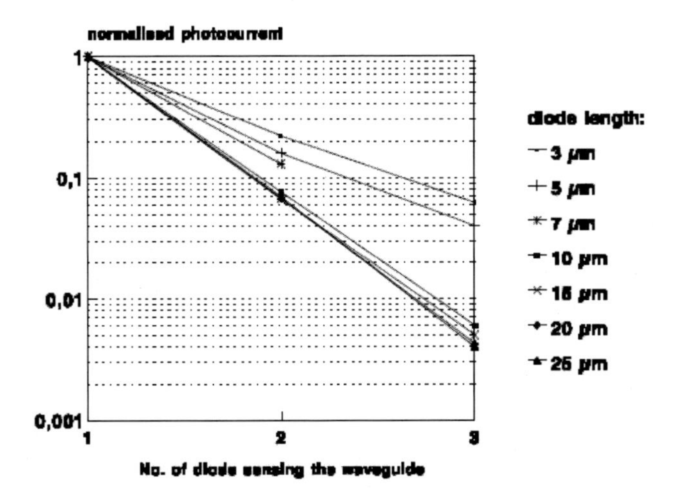

Figure 17. Photocurrent of three diodes in a chain in dependence on the active diode length.

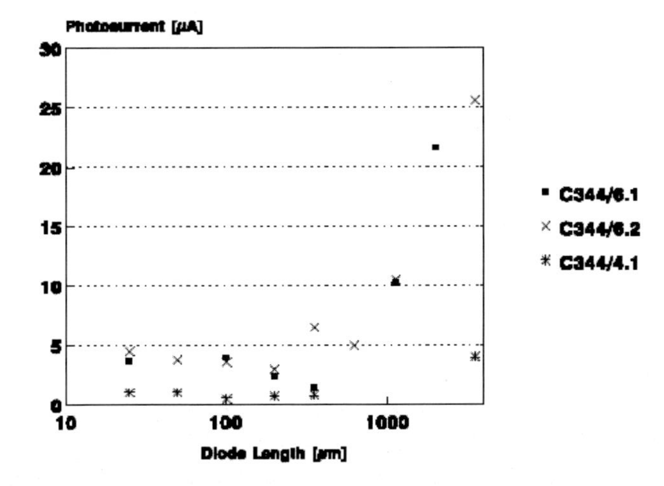

Figure 18. Photocurrent of diodes of different lengths coupled to a waveguide by leaky wave coupling.

Thus, the intensity at the second diode is much higher, and results in a relatively high photocurrent. With increasing diode length, the absorption raises, that means decreasing photocurrents at the second and third diodes. Figure 18 indicates the coupling efficiency for leaky wave coupling. The photocurrents measured at the diodes increase at about 1000 μm diode length; below this value there are no clear tendencies detectable. The reason is the quality of the transition of the waveguide to the diode surface. Even a step of 40 nm causes a partial butt coupling. So, the photocurrents of the short diodes do not show the leaky wave photocurrent, but mainly the butt coupling caused photocurrents.

8. Modular Processing

Modular processing of waveguides and MOS transistors starts with the integration of the electronic components. The only modification of the standard MOS process is the replacement of the standard LOCOS by SWAMI-LOCOS to get a plane wafer surface after standard field oxide growth. After transistor integration, the wafers may be tested electrically, and only if all parameters are right, the waveguide deposition can be added. Waveguides need 2 μm of isolation oxide below the light guiding film.

So, an additional oxide deposition at low temperature must be performed, e.g. by PECVD, to increase the isolation layer to 2 μm. Then the SiON film and the rib layer are deposited by PECVD. As the last step, the rib is defined by lithography and etched into the top layer by reactive ion etching.

The modular process uses PECVD layers for waveguides due to the very low deposition temperature. LPCVD is not applicable, because a temperature above about 450 C will destroy the electronic structures by the formation of an aluminium silicon alloy in the junction areas.

Nevertheless, after layer deposition, the MOS structures are operating like standard transistors. There is no leakage in the *p-n* junctions, the field areas of the circuits or in the off-state of the transistors. The waveguide structures are of high quality too, because there is no reactive ion etching process on top of the SiON film.

Only the coupling efficiency is low because of a rather thick oxide film of more than 1.3 μm between the light guiding SiON film and the diode surface. Leaky wave coupling is very low, so the electrical diode output signal is small. To increase the coupling efficiency, it is necessary to etch a mirror into the light guiding film to get butt coupling by a mirror (Figure 19). The mirror etch uses a mixture of CHF_3 and oxygen during reactive ion etching; the amount of oxygen determines the slope of the mirror in relation to the wafer surface. At about 18% oxygen content in the plasma, during the dry etching, an angle of 45° will result (Figure 20). Figure 21 shows a photograph of mirror etched into a waveguide. The coupling efficiency depends on the etching depth strongly (Figure 22). If only the rib in the surface cladding is etched, nearly no photocurrent is generated in the photo diode. If the light guiding SiON film is etched completely, almost all light is coupled into the diode.

The modular integration technique has been used for microsystem integration consisting of waveguides, photo detectors and amplifiers on one silicon chip [30]. It is applicable to frequencies up to 20 MHz, limited by the features of the CMOS compatible photo detector.

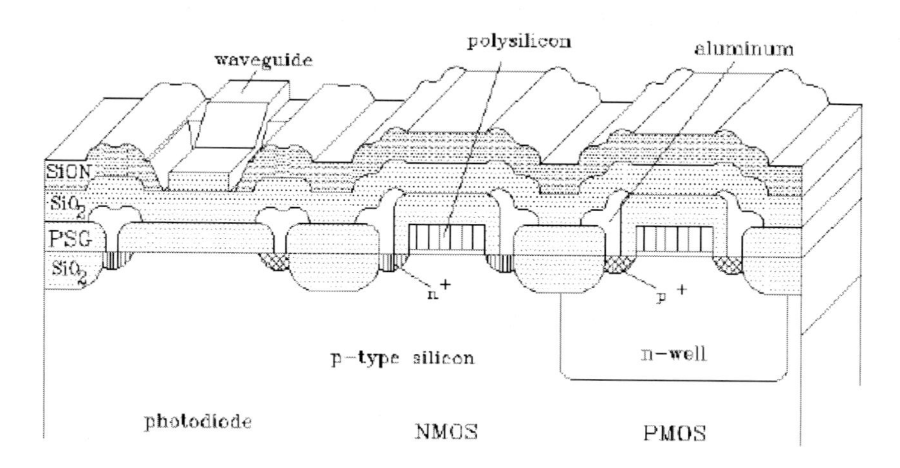

Figure 19. Modular integration of CMOS components and waveguides using PECVD waveguides after finishing the CMOS integration process.

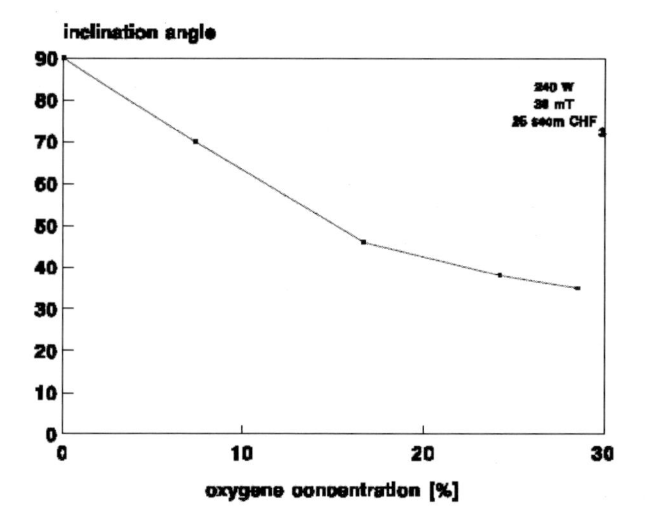

Figure 20. Slope of the mirror opening in dependence on the oxygen content of the plasma during reactive ion etching.

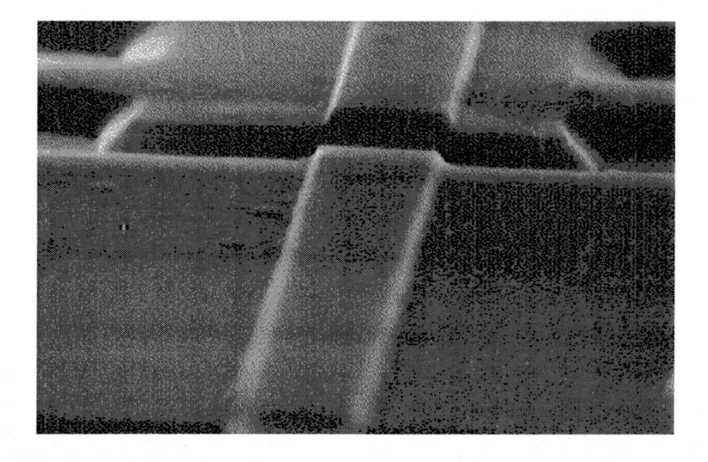

Figure 21. Mirror etched into a waveguide.

9. MICROMECHANIC ADD –ONS FOR SENSOR APPLICATIONS

Integrated optics give the opportunity to detect small elongations or bendings of membranes by interferometry [31]. It enables the fabrication of a microsystem consisting of a pressure sensor with optical readout and microelectronic circuitries, as presented in Figure 23. The main component is a Mach-Zehnder interferometer in SiON technique. Its measurement branch is located on top of a membrane, whereas the reference branch is placed beside the sensitive area. The membrane bending causes a change in the effective RI of the light guiding film, due to the mechanical stress. So, the phase of the propagating light in the sensitive branch is correlated to the external pressure applied to the membrane. The result is a small difference in the optical paths of the measurement and the reference branch of the interferometer, which causes an intensity change at the interferometer output.

Applying the modular technology for integrated optics on silicon (as explained above), it is possible to transform the optical information into an electrical signal by a photodetector and a MOS-amplifier, all integrated on the same chip.

So, the magnitude of the external pressure is directly related to the interferometer output signal, respective to the photocurrent at the photodetector. The integration of such a microsystem can use the modular processing of waveguides and microelectronics with slight modifications for the mechanical components only.

In this case, the micromechanic structures like cantilevers and membranes may be integrated after the test of the microelectronic circuits, prior to waveguide deposition. The add-ons require one resist mask and two reactive ion etching steps [32].

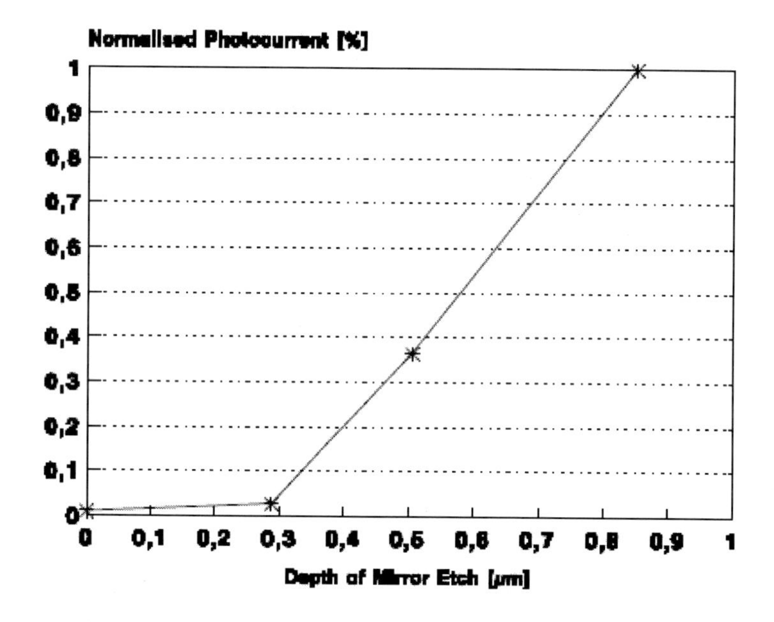

Figure 22. Photocurrent measured at a diode in dependence on the etch depth of the coupling mirror.

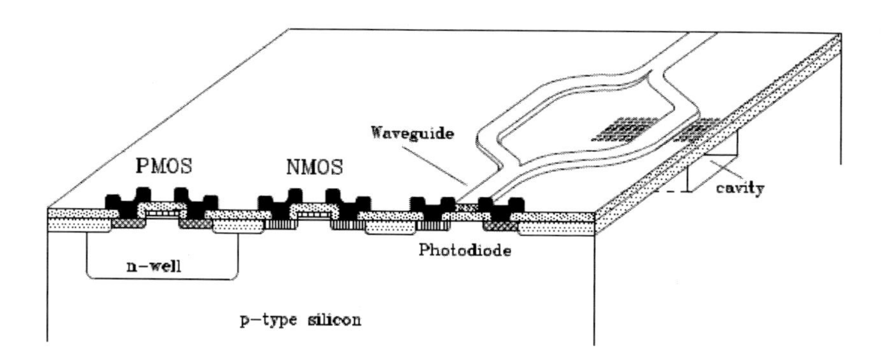

Figure 23. Cross-section of an integrated microsystem for pressure sensing, consisting of a membrane, a Mach-Zehnder interferometer, digital and analogue CMOS circuits, built on one silicon chip.

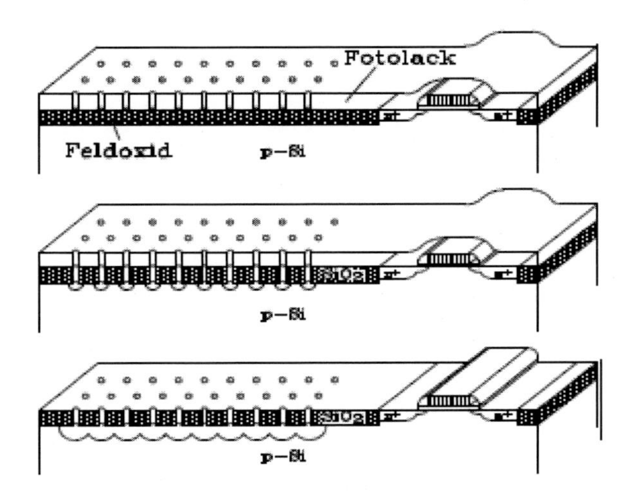

Figure 24. Flow chart for the integration of a membrane.

The photoresist defines an array of holes with diameters of about 1 μm on top of the field oxide in the membrane area (Figure 24). These openings are etched down to the silicon surface anisotropically, across the whole field oxide of the MOS structures.

In the next step, the silicon substrate, which is buried below the dielectric films, is etched isotropically through the oxide holes using an SF_6 plasma. This is highly selective to oxide and photoresist. Due to the undercut of the oxide layer, small cavities arise below each opening. Finally, they grow up to one large cavity closed by a porous oxide membrane [33].

To get a closed cavity for pressure reference, a nearly conformal PECVD oxide film deposition of about 1.0 μm thickness is performed. In addition, this oxide serves as an additional layer to increase the optical insulating layer for the SiON waveguides. On top of this oxide, the light guiding film and the capping oxide is deposited by PECVD. Figure 25 shows the result of these depositions. Due to the low conformity of the deposition processes, the openings were not closed completely. Their cross-sections were reduced, but the membrane is still not sealed. A more conformal deposition process with ozone would help to solve this problem.

The process concludes by rib etching to form the waveguides. It results in an optoelectronic/micromechanic chip, which is capable of pressure sensing. To increase the sensitivity, it is possible to integrate some membranes in one interferometer. Figure 26 depicts a scattered light photograph of a microsystem consisting of four membranes in a Mach-Zehnder interferometer. The interferometer output hits a photo detector; its signal is amplified by an operational amplifier integrated on the same chip. The whole chip size is about 14 mm × 0.5 mm. The only external element is the laser diode.

In this monolithic integration process, all steps were done from the front side wafer surface. Only materials typically used in MOS-technology, were applied for preparing the structures. The tested microsystems for pressure sensing consist of one or four membranes of 120 μm × 120 μm or 200 μm × 200 μm size, crossed by the sensitive branch of a Mach-Zehnder interferometer of 2500 μm length.

A photobipolar transistor detects the output intensity of the interferometer; this signal is amplified by a transimpedance amplifier.

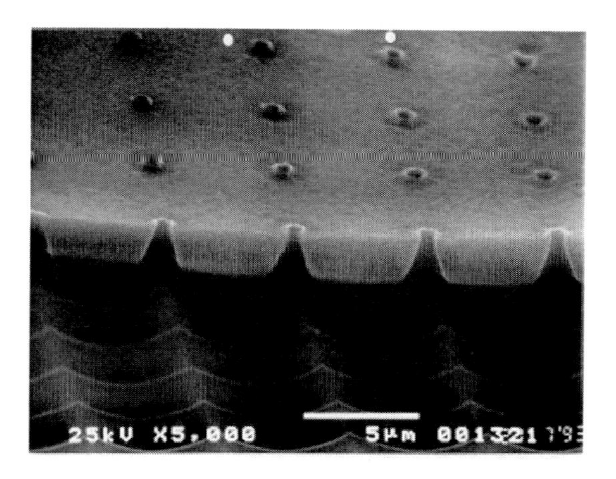

Figure 25. SEM cross-section of the membrane area after isotropic silicon etch and layers depositions.

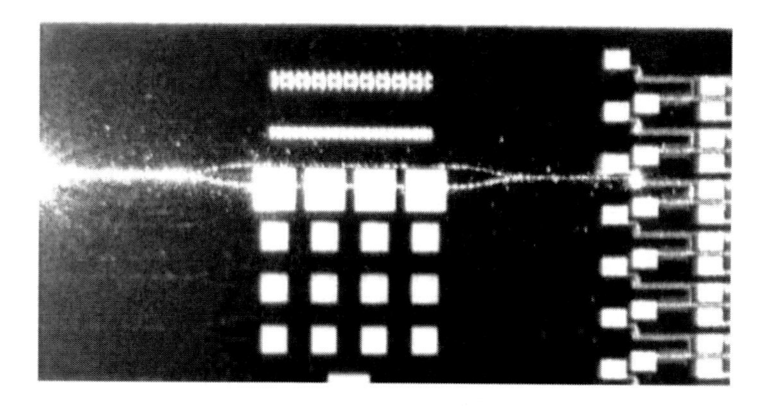

Figure 26. Top view of the monolithically integrated optoelectronic microsystem with micromechanic pressure sensor, and mirror coupling of waveguide and photodetector.

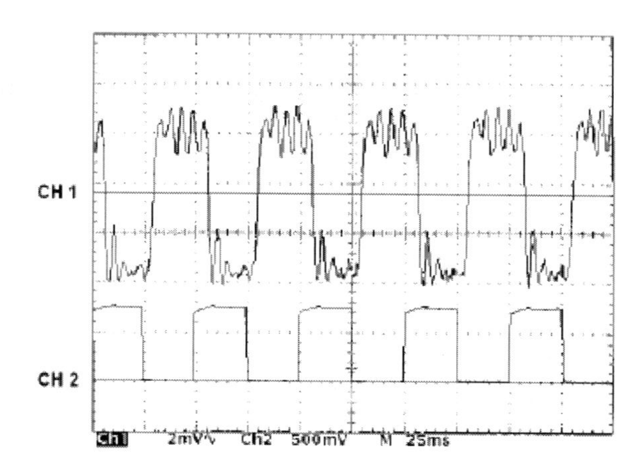

Figure 27. Top: photo detector output signal at dynamic air flow modulation, bottom: valve control voltage.

In Figure 26, He-Ne laser light, used for the parameter extraction, is coupled into the waveguide structure. The scattered light photograph shows a Mach-Zehnder interferometer crossing four membranes of 200 μm × 200 μm each. To rate the sensitivity of the optomechanical pressure sensor, the photocurrent was measured directly at the photodetector output, without any amplification. Membranes were stimulated by a pulsed stream of compressed air. Figure 27 shows the dynamic output signal of the photodetector at 20 Hz, applying a pressure of 400 mbar to a system consisting of four membranes of 200 μm × 200 μm each. The amplitude of about 5.5 mV results in a signal rate of 13 μV/mbar. This signal may be amplified by the transimpedance amplifier integrated on the same chip. There is a small delay time between the valve voltage (CH2) and the output voltage swing (CH1), caused by the response time of the valve and the compressed air spread. The output signal is modulated by a frequency of about 180 Hz. This may be a resonance frequency of the whole microsystem, because the etch openings in the membranes under test were not completely closed. According to the theory of plates and shells [34], the membrane resonance frequency has to be much higher. Nevertheless, the resonance depends on the membrane size.

10. Infrared Light Waveguides in Silicon

In the visible range, it is necessary to use the above mentioned waveguides consisting of dielectric layers because of the strong absorption of silicon. But, at $\lambda = 1.3$ to 1.55 μm, silicon itself is transparent, and therefore, no absorption of the electromagnetic field occurs. The RI of silicon at 1.55 μm is 3.5 for undoped silicon, and only a very little higher value for n-type silicon. So, silicon can be used as the waveguide or the cladding layer itself. The typical structures of silicon waveguides is formed by phosphorous diffusion or implantation into a silicon wafer, as depicted in Figure 28(a). During the last years, interest in silicon on insulator (SOI) waveguides has been growing. An SOI waveguide consists of a 1.5 μm thick crystalline silicon layer on top of a buried oxide film. Beam confinement in horizontal direction is performed by a polysilicon rib deposited on top.

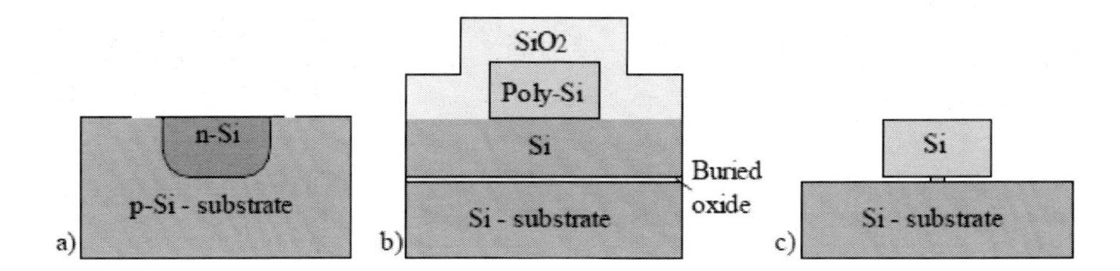

Figure 28. Cross-sections of silicon based waveguides: (a) doped surface core waveguide, (b) rib waveguide, and (c) column supported waveguides.

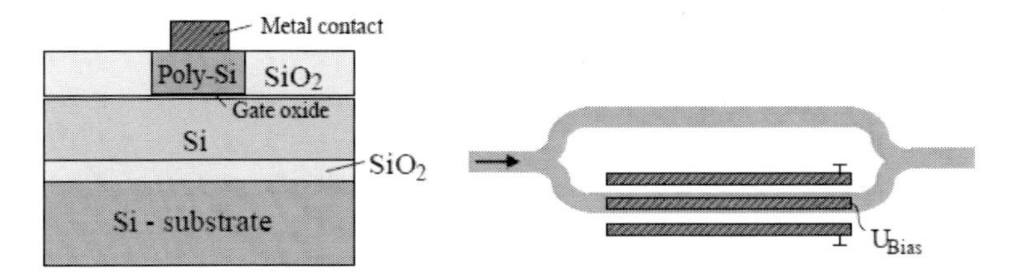

Figure 29. Silicon phase shifter and the complete Mach-Zehnder interferometer modulator.

In vertical direction, the buried oxide and a surface oxide serve as claddings. Another kind of waveguide consists of a rectangular silicon core, which is located on a few supporting oxide columns only. So, the ambient serves as the cladding. Figure 28 presents their cross-sections. These kinds of waveguides give the opportunity to integrate modulators even in silicon [35]. Although there is no linear electrooptical effect, and the Kerr effect is very small [36] too. A modulation up to about 1 MHz is possible by the thermo-optic effect [37]. Modulation frequencies up to the GHz range have been predicted by the use of the free carrier plasma dispersion effect in forward biased *p-i-n* diode devices [38]. In these devices, the operating speed is limited by the long carrier lifetime in silicon. Another approach is a modulation of the RI in the silicon waveguide by a MOS capacitor integrated in a Mach-Zehnder interferometer. A voltage applied to the electrodes of the capacitor causes a charge carrier accumulation below the oxide at the waveguide surface. It causes a RI modulation leading to a phase shift, which results in an intensity modulation at the interferometer output. A cross-section of the phase shifter is shown in Figure 29. It has been demonstrated in [39] that a length of the device of about 8 mm will cause a phase shift of about 180° at an applied voltage of 10 V and a gate oxide thickness of 6 nm. The maximum operation speed in silicon modulators, using the principle like the device shown above, is 10 GHz reported up to now [40]. Optimising the design, even higher speed has been predicted by the authors. Although this is much lower as switching of today's $LiNbO_3$ devices, it is a promising modulation technique for data transmission.

CONCLUSION

It has been demonstrated that the indirect band gap semiconductor silicon is a valuable material for integrated optics in the visible and near infrared spectral range. In the visible range, SiON and Si_3N_4 waveguides can be integrated on silicon in a CMOS compatible technique using a special kind of local oxidation process. Different kinds of sensors demonstrate the capabilities of monolithic system integration with optical waveguides, photo detectors and CMOS circuits on one chip.

At a wavelength of 1.55 μm, silicon can be used as a light guiding medium itself. Although there is no electro-optic effect in silicon, an intensity modulator, based on a Mach-Zehnder interferometer with capacitive charge carrier accumulation, demonstrates 10 GHz operating frequency. This opens up a wide area of further applications for integrated optics on silicon.

ABOUT THE AUTHOR

Ulrich Hilleringmann was born in Germany in 1958. He studied physics from 1978 to 1984 at University of Dortmund, Germany. From 1984 to 1985 he was with the Fraunhofer Institute of Microelectronic Circuits and Systems, Duisburg, Germany. Afterwards Hilleringmann changed to the Department of Electrical Engineering at University of Dortmund, where he wrote a thesis on "Laser Recrystallization of Silicon" in 1988. In 1994 he received the "venia legendi" at the same department with the corresponding thesis

"Integrated Optics on Silicon". Since 1999 he has been a Professor at the Institute of Electrical Engineering, Sensor Technology Department, University of Paderborn.

His main interests are silicon technology, nanometer scale devices, integrated optics on silicon, organic transistors and microsystem technologies.

REFERENCES

[1] Soref, R. A. *Proc. IEEE* 1993, 81, 1687–1706.

[2] Richter, A., Steiner, P., Kozlowski, F., Lang, W. *IEEE Trans. Electron. Dev. Lett.* 1991, 12, 691–692.

[3] Tsuchiya, T., Nakajima, S. *IEEE J. Solid-State Circuits* 1985, 20, 325–332.

[4] Ng, W. L., Lourenco, M. A., Gwilliam, R. M., Ledain, S., Shao, G., Homewood, K. P. *Nature* 2001, 410, 192–194.

[5] Hilleringmann, U., Goser, K. *IEEE Trans. Electron. Dev.* 1995, 42, 841–846.

[6] Yaffe, H. H., Henry, C. H., Kazarinov, R. F., Milbrodt, M. A. *J. Light. Tech.* 1994, 12, 64–67.

[7] Veldhuis, G. J., van der Veen, L. E. W., Lambeck, P. V. *J. Light. Tech.* 1999, 17, 857–864.

[8] Rickman, A. G., Ree, G. T., Namavar, F. J. *Light. Tech.* 1994, 12, 1771–1776.

[9] Modreanu, M., Cosmin, P. *Int. Semiconductor Conf. Proc. CAS'95*, 1995, 323–326.

[10] Peters, D., Fischer, K., Müller, J. *Sensors and Actuators A* 1991, 26, 425–431.

[11] Ay, F., Aydmli, A., Roeloffzen, C., Driessen, A. *IEEE Lasers and Electro-Optics Society 2000 Annual Meeting, LEOS* 2000, 2, 760–761.

[12] Kapser, K., Wagner, C., Deimel, P. P. *IEEE J. Phot. Tech. Lett.* 1991, 3, 1096–1098.

[13] Smit, M. K., Acket, G. A., Van der Laan, C. J. *Thin Solid Films* 1986, 138, 171–181.

[14] Bona, G.-L., Germann, R., Offrein, B. J. *IBM J. Res. and Dev.* 2003, 47, 239–249.

[15] Zhang, A., Chan, K. T. *Proc. of the 6th Chinese Symposium on Optoelectronics* 2003, 124–127.

[16] Hoffmann, M., Voges, E. *Tech. Dig. 10th Int. Conf. on Integrated Opt. and Opt. Fiber Commun. (IOOC '95)* 1995, 5, 33–36.

[17] *Agnihotri, O. P., Tyagi, R., Kato, I. Jpn. J. Appl. Phys. 1997,* 36, 6711–6713.

[18] Schauwecker, B., Przychrembel, G., Kuhlow, B., Radehaus, C. *IEEE J. Phot. Tech. Lett.* 2000, 12, 1645–1646.

[19] Bezzaoui, H., Baus, A., Voges, E. Micro System Technologies; Springer: Berlin, 1990; 283–288.

[20] Pregla, R., Ahlers, E. *Electron. Lett.* 1994, 30, 1478–1479.

[21] Hoffmann, M., Kopka, P., Groß, T., Voges, E. *Electron. Lett.* 1998, 34, 207–208.

[22] House, A., Whiteman, R., Kling, L., Day, S., Knights, A., Hogan, D., Hopper, F., Asghari, M. *Proc. Opt. Fiber Commun. Conf.* 2003, 2, 449–450.

[23] Flück, E., Horst, F., Offrein, B. J., Germann, R., Salemink, H. W. M., Bona, G.-L. *IEEE J. Phot. Tech. Lett.* 1999, 11, 1399–1401.

[24] Rotolo, S., Tanzi, A., Brunazzi, S., DiMola, D., Cibinetto, L., Lenzi, M., Bona, G.-L., Offrein, B. J., Horst, F., Germann, R., Salemink, H. W. M., Baechtold, P. H. *J. Light. Tech.* 2000, 18, 569–578.

[25] Offrein, B. J., Germann, R., Horst, F., Salemink, H. W. M., Beyeler, R., Bona, G. L. *IEEE J. Select. Topics in Quantum Electron.* 1999, 5, 1400–1406.

[26] Zimmermann, H. *Integrated Silicon Optoelectronics*; Springer: Berlin, 2000.

[27] Wunderlich, S., Schmidt, J. P., Müller, J. *Appl. Opt.* 1992, 31, 4186–4189.

[28] Sze, S. M. *VLSI-Technology;* McGraw-Hill: NY, 1988.

[29] Kahng, D., Shankoff, T. A., Sheng, T. T., Haszko, S. E. *J. Electrochem. Soc.* 1980, 127, 2468–2471.

[30] Braß, E., Hilleringmann, U., Goser, K. *IEEE J. Solid-State Circuits* 1994, 29, 1006–1010.

[31] Fischer, K., Hoffmann, R., Müller, J. *Micro System Technologies '91* (Eds. Krahn, R., Reichl, H.) VDE: Berlin; 1991, 472–481.

[32] Adams, S., Hilleringmann, U., Goser, K. *Micro System Technologies '92* (Eds. Krahn, R., Reichl, H.) VDE: Berlin, 1992, 217–226.

[33] Adams, S., Hilleringmann, U., Goser, K. ESSDERC '92, *Microelectronic Engineering* 1992, 19, 191–194.

[34] Timoshenko, S., Woinowsky-Krieger, S. *Theory of Plates and Shells*; McGraw-Hill: NY, 1959.

[35] Reed, G. T., Png, C. E. *J. Mat. Today,* January 2005, 40–50.

[36] Soref, R. A., Bennett, B. R. *IEEE J. Quantum Electron.* 1987, 23, 123–129.

[37] Cocorullo, C., Iodice, M., Rendina, I., Sarro, P. *IEEE J. Phot. Tech. Lett.* 1995, 7, 363–365.

[38] Tang, C. K., Reed, R. T. *Electron. Lett.* 1995, 31, 451–452.

[39] Liao, L., Liu, A., Jones, R., Rubin, D., Samara-Rubio, D., Cohen, O., Salib, M., Paniccia, M. *IEEE J. Quantum Electron.* 2005, 41, 250–257.

[40] Liao, L., Samara-Rubio, D., Morse, M., Liu, A., Hodge, D., Rubin, D., Keil, U. D., Franck, T. *Opt. Exp.* 2005, 13, 3129–3135.

In: Advances in Laser and Optics Research. Volume 10
Editor: William T. Arkin
ISBN: 978-1-62257-795-8

Chapter 7

COMBINING ENERGY EFFICIENCY WITH AESTHETIC APPEAL USING ADVANCED OPTICAL MATERIALS

Geoff Smith*
Department of Applied Physics and Institute of Nanotechnology
University of Technology, Sydney (UTS), Australia

ABSTRACT

Buildings, internal spaces and urban environments, which are not only very energy efficient, but also pleasant to occupy and look at, are needed to accelerate the rate of reduction worldwide in CO2 emissions, and to raise living standards. New materials, whose optical properties are tailored to this task, are making this goal possible. Lighting, visual, and thermal control properties of windows, skylights, paints and lighting systems are treated. These three functions must be considered together, not independently. The way a range of new polymeric and nano-structured materials enable the optimised spectral management and distribution of incoming solar or lamp radiation, and outgoing thermal radiation, is presented.

Most of the materials discussed are composites whose properties can be varied in a controlled way, if the physics is understood. The basic models linking structure, composition and optical response are presented, for example, from among four classes of optical materials: opaque, transparent, translucent, and the group – light piping, mixing and homogenisation. Issues for plasmon resonant nanoparticles in visibly transparent glazing include resonance tuning, and differences in short wavelength (Rayleigh) and plasmon resonance scattering. For energy efficient glare control and uniform daylighting, for integration of light piping and continuous illumination, and for short length light mixing and homogenisation, the use of dopants, which are close in refractive index (RI) to the host, are analysed. These are found to be ideal to use with many emerging LED applications and daylight. For LED's they enable very energy efficient white light lamps based on RGB arrays at low cost, and also, enable uniform illumination despite the small area of the source. Finally, some special spectrally selective paints are discussed; those that can be deep colours or even black and still either reflect or absorb much solar energy

* Corresponding author: Geoff Smith. School of Physics and Advanced Materials and Institute of Nanotechnology University of Technology, Sydney (UTS) Broadway NSW 2007, Australia.

(whichever is desired), and those that enable efficient radiative cooling or water collection from the atmosphere.

1. INTRODUCTION AND BACKGROUND

1.1. Energy Savings and Building Ambience

Energy efficiency in interior spaces concerns supply of lighting needs and maximising thermal comfort, with minimum use of electrical power from the grid, and of fossil fuels. It is relevant to all classes of buildings, and also, to transport. The world is faced with two apparently conflicting demands right now, a rapid growth in demand for better living standards and lifestyles, and an urgent need to cut greenhouse gas emissions. If this "conflict" can be eliminated or softened, then the process of scaling back our negative impact on the environment will accelerate. If it cannot, all living standards are at risk in the long term. Such a changeover, in common with past technology driven shifts in human activity, will also generate wide ranging opportunities for economic growth in all regions of the world. There is much new science needed to optimise these technologies, and optics is playing a central role. Examples of two science based systems for better use of natural lighting are in Figure 1.

The human element, not just energy, is thus central to this study, and impacts on the science needed. Our perception of how a space feels, looks and how it performs, are strongly influenced by the optical and radiative properties of the materials at its surfaces. These surfaces may be opaque, transparent or translucent.

Figure 1. A spectrally selective roof window and an angular selective mirror light pipe – two ways of combining daylighting with thermal control. They exemplify building technology based on new optical materials. (Courtesy Skydome Skylight Systems Pty Ltd, www.skydome.com.au).

Optical characteristics of the radiation sources contributing to light and energy fluxes in the space must also be taken into account when choosing and designing these surfaces. The sun, and various lamps are thus of interest in this study. These characteristics also depend on sky conditions and geographic location, the luminaires which house the lamps, and light pipes or other devices which are used to distribute the light or reduce solar heat gain.

Lighting is a core human need, but the complexities of lighting science and, in particular, the many difficulties in using our abundant daylight resource, especially in warm climates, has meant that lighting technology has evolved slowly, and there remains much scope for better lighting in terms of energy efficiency, greater use of daylight, visual comfort, decorative appeal and versatility. Quoting from Balcomb [1] "In building design strategies the use of natural light to replace artificial light - fills a unique niche. It stands alone as the most important design issue". Specially engineered optical materials provide us with the ability to control thermal energy and light flows between the environment and interior spaces, in combination with the ability to control the ambience, appeal and enjoyment of these spaces. Issues to be considered include colour and colour rendering, glare, the exterior view from inside, the exterior appearance from outside, light distribution, lighting dynamics, functionality, occupant productivity and health, and design flexibility. Dealing with so many issues at once appears formidable. We will show in this article that, starting with a systematic basic approach to what is needed in spectral, directional, and scattering properties of materials, then looking at how such properties can be realised in novel composite or structured materials, makes an overall approach to these issues feasible and has led to much recent progress and new products. While the science is novel and sometimes complex, we are only interested in technology which is easy to realise in large volumes or areas, and hence, is cost effective. Examples which have come up in my own and other solar materials research groups in recent years, may help to clarify the nature of these goals.

- A good performance with partially coloured, as opposed to the usual black, solar selective absorbers for better roof integration?
- A visibly black surface which is a reasonable solar reflector?
- Arbitrary coloured completely uniform sources of light, including white, from mixing output from multiple LEDs with almost no energy loss?
- A simple transmitting surface which blocks light from some directions of incidence while transmitting others, without distorting the view?
- Use of those components of solar radiation which are not wanted in a window or skylights transmission, to produce useful heat or electric power in the window itself?
- Reduced cooling loads or condense useful amounts of atmospheric water with special radiative coatings?
- Simple extraction of selected incident radiation components, then channelling and concentrating them for various remote uses.

1.2. Optical Properties and Materials

In this field, required optical properties must be evaluated according to a subjective human critique, as well as energy performance. Much related recent research by designers and lighting scientists appeared in the literature [2, 3], but detailed linking of measurable physical parameters to subjective response is still in its infancy. Standards and regulations must often be satisfied. They rely on measurable quantities, are based on experience, and are typically defined by task or activity requirements and safety, rather than overall appeal or energy issues. Energy fluxes, luminance (candela), illuminance (lux), and colour can each be measured using standard instruments. They can also be simulated in terms of material properties, radiation sources and the geometric properties of the spaces, using programs such as Radiance [4]. As an example, we used Radiance to simulate daylighting in the 2000 Sydney Olympic Stadium using the measured properties of four differently doped translucent panels distributed over the large roof area. The needs of spectators, the media and the athletes had all to be considered, in thermal and glare terms. Such subjective issues inside buildings include our response to the interplay between temperature, humidity and air flows; glare, the dynamics of light and temperature, lighting patterns, visual clarity, and modification of external views. Display lighting performance is primarily subjective and depends on brightness, homogeneity, colour, contrast when daylight or other strong sources are present, and field of view. Speciality lighting in hospitals, for safety on stairs and in aisles, for signalling, and for inside refrigerators and coolers, can also be made optimal in terms of energy use, appeal and functionality. Paints also have an important role to play, and interesting recent developments can help in their energy performance [5, 6] while broadening decorative options. While the focus in this article is on passive environmental control systems, some of the material and optical properties we analyse can also be adapted for use in active energy saving systems such as solar water heaters, air heaters, radiative cooling, water distillation and improved solar cells. Indeed, it will be seen that it is possible to even combine some of these active functions with the passive control of lighting and heat flows, for a very high degree of multi-functionality in a single building element.

1.3. Polymer Composites and Nano-Structured Materials

Two rapidly developing and overlapping areas of materials and optical science are helping to achieve these goals – new polymer optical systems, and nano-materials. They can combine attractive cost with new approaches to saving energy. Flexible polymer light guides and polymer rods, both with unusual optical characteristics are one feature in this work. Nanoparticle doped polymers are another. Coatings based on vacuum and pyrolytic deposited thin films have well established capabilities in the windows area [7], while important developments are also occurring with switchable films, some of which also rely on nanostructures. In general, however, the established technologies have less chance of sizeable cost reductions. In the area of energy efficient windows, single and multi-layer thin film systems with low thermal emittance are now commonplace, and often mandatory in colder climates, where minimising heat loss is paramount. A major need is for simpler and cheaper glazing and daylighting systems, optimised for cooling dominated climate zones. An example of warm climate needs is in the rapid growth in uptake of air conditioning. It is causing emerging summer electricity demand problems in many parts of Australia and Asia, and is an established problem in California. Comfort is possible with much lower loads on these cooling units, often without them.

1.4. Optical Phenomena of Interest for Energy Efficient Materials

The optical and material physics, needed for analysis and development of both energy efficiency and aesthetic appeal, include scattering of nanoparticles and microparticles, surface plasmon resonance on conducting nanoparticles, nanostructured conductor surface optics, optical homogenisation in nanostructured materials, and light redirection and transport mechanisms intrinsic to materials. Mirrors and refractive optics are often used for redirection, but except possibly for micro-optic systems on surfaces, we have problems for building systems, such as a need for tracking and cleaning, which we wish to avoid. Micro-optics is currently too expensive for the scales we are after, but that may change. We will also touch on other materials which are difficult to realise cheaply at present, but further research could change that. Phenomena including optical anisotropy [8], and one-dimensional photonic crystal multi-layers [9] apply in these. Details will not be given on those thin films for energy efficiency, which have been well reported elsewhere, except to draw some comparisons. Switchable films also have an extensive review literature [7,10] and recent developments, such as gasochromics [11], which have reached the product stage, are relevant, but will not be reviewed. We will conclude with a look ahead based on the question – What are some possible "ultimates" in building skin technology, especially with paint, polymers, windows and skylights? What degree of multi-functionality can we achieve through attention to both using and controlling solar and thermal radiation in each of four spectral zones UV, visible, near infra red and thermal infra red, for different times of day and year? This then leads to a final question – can we cost effectively supply all our energy needs from the walls and roofs of buildings, while keeping them attractive or even looking much the same as they always have?

1.5. Spectral and Directional Issues: The Sun and the Sky

The characteristics of incoming solar energy and daylight, and of outgoing thermal radiation, which we must account for in our materials design, are determined by the sun and the sky conditions. The solar spectrum at the surface of the earth is illustrated using spectral irradiance as a function of wavelength in Figure 2. For discussion of spectral management, this spectrum can be conveniently divided into three spectral regions UV (300–400 nm), visible (400–750 nm), and near infra red (NIR) (750–2500 nm), as shown in Figure 2. In Figure 2, to aid discussion, we have further subdivided the NIR, because little solar energy comes in beyond 1300 nm. Thus, for NIR blocking mechanisms which are narrow band, as with selected nanoparticles or some organic dyes, we should centre them around 1000 nm. If the skin is directly irradiated, as often occurs in a car, some argue [12] that human comfort depends also on wavelengths beyond 1300 nm because skin absorptance is much higher there. This is due to our body's absorption bands due to water, but atmospheric water has the same absorption bands which are what makes the solar intensity weak in these spectral zones.

Material optical properties modulate the incoming radiation in these different spectral zones. The amount of solar heat transmitted is T_{sol}, reflected is R_{sol}, or absorbed is $A_{sol} = 1 - T_{sol} - R_{sol}$, and the amount of visible radiation transmitted is T_{vis}. T_{sol} and T_{vis} are defined in equations (1) and (2), respectively

$$T_{sol} = \int_0^\infty d\lambda S(\lambda)T(\lambda) \tag{1}$$

$$T_{vis} = \int_0^\infty d\lambda S(\lambda)T(\lambda)Y(\lambda) \tag{2}$$

with $T(\lambda)$ as the total or global spectral transmittance, $S(\lambda)$ as the total solar energy spectral irradiance incident *on the surface of interest* at wavelength λ, and $Y(\lambda)$ as the photopic sensitivity of the eye. Reflectance counterparts are found by exchanging R for T in eqs. (1) and (2). $S(\lambda)$ is most commonly taken as that for air mass 1.5 (Figure 2) (reference spectra and tables downloadable at http://rredc.nrel.gov/solar/spectra/am1.5 or as the ASTM G-173-03 standard). If T is for normal incidence, as is common when comparing different glazing products, then the implicit assumption is that the radiation is specular not diffuse. To simulate performance in actual installations, we must know T and S for all angles, or directions of incidence.

For light fittings, we simply change the function $S(\lambda)$ in equation (2), to that produced by the lamp, or the lamp and its associated reflectors. The ways of representing the directional dependence of existing specular glazing products with algorithms for easy simulation has received a lot of interest in recent years, and is largely resolved [13,14]. Diffuse materials require much more work to reach this utility in simulation. They require a BRTF (Bi-Directional Reflectance-Transmittance Function) which connects incoming radiance for each direction with outgoing radiance in each direction.

There has been some recent progress [15,16] but all too often made a Lambertian or "ideal diffuser" approximation, which is far from the behaviour of most commonly used diffusers in daylighting and lighting. The ideal situation would be a first principle scattering model for each generic system, rather than empirical models. Then changes in *BRTF* with say dopant concentration can be readily explored. In limited cases only, we can do this, since multiple scattering is itself usually too complex. One such case is the special diffusers discussed in Section 4. In that case, the ability to generate output light profiles from first principles for each particle has enabled much progress with the related technologies.

In discussing total heat management, we must include the thermal IR (or black body range) as it applies to radiating surfaces at moderate temperatures up to ~120 °C, as needed in buildings and some solar devices. The thermal emittance ε is the most widely used parameter for studying radiative losses as defined in equation (3).

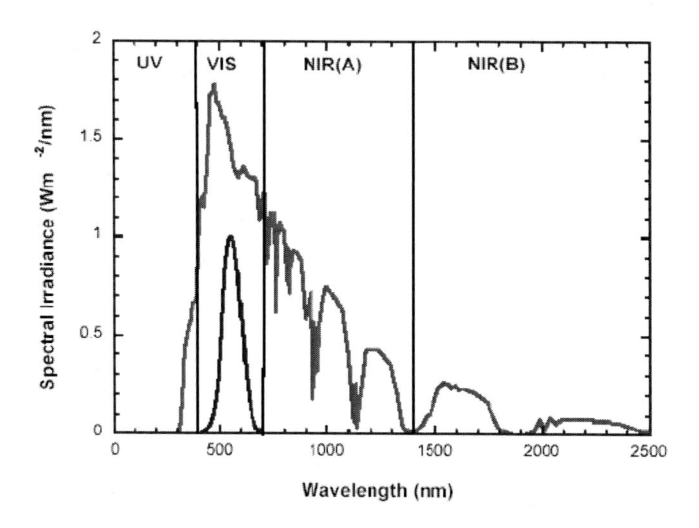

Figure 2. Air mass 1.5 solar spectral irradiance showing useful sub-divisions for spectral management purposes.

However, for many situations, when any surface "sees" the sky to any significant extent, one can subdivide the thermal IR band further to better engineer radiative losses. This is because we want to either channel (to cool) or block (to retain heat) radiation loss to outer space which occurs in the 8–13 μm band. The parameter of main interest is then ε_A (atmospheric emittance) as given (un-normalized) in equation (4).

$$\varepsilon = \int_0^\infty [1 - R(\lambda)]S_{bb}(\lambda)d\lambda \tag{3}$$

$$\varepsilon_A = \int_0^\infty [1 - R(\lambda)][1 - A_{atmosphere}(\lambda)]S_{bb}(\lambda)d\lambda \approx \int_{8\mu m}^{13\mu m} [1 - R(\lambda)]S_{bb}(\lambda)d\lambda \tag{4}$$

here $A_{atmosphere}$ gives the atmospheric absorption and S_{bb} the black body intensity. For best cooling, when the sky is involved, it is thus actually best to have a high value of R at all wavelengths except in the 8–13 μm band, R should be very low [17], since the atmospheric absorption is high elsewhere. This is only for clear skies. The next best for cooling is to have the more common occurrence where $R(\lambda)$ is low across the whole 2.5–30 μm band, and high across the solar spectrum. This radiation band is the complete black body region at typical temperatures in windows, walls and roofs.

The thermal radiation band starting at 2.5 μm, has almost negligible overlap with the solar spectrum. This is where the original solar spectral selective concept originated for use on thermal collectors. For day-night radiative cooling, as just noted, the exact opposite qualitative spectral response to selective absorbers is required. Today we broaden the concept of spectral selectivity to include windows, and require one or more large qualitative switches in optical properties near some critical wavelength such as 750 nm where the human eye ceases to sense, or around 2.5 μm, where solar radiation vanishes and thermal radiation starts, or in the range 8–13 μm for sky cooling. An example is a switch from high to low transmittance in a window at 750 nm. Spectral and angle of incidence dependence of key spectral properties, which control energy flows, are transmittance $T(\lambda,\theta)$, reflectance $R(\lambda,\theta)$ and absorptance $A(\lambda,\theta)$; these must be known to track both selectivities. Often wavelength and angle of incidence impacts are not independent since the physical mechanisms, such as anisotropy in surface plasmon resonance, can cause both special angular effects and spectral shifts. Normal incidence is assumed hereon if we just define integral properties without reference to direction. For materials which are labelled "angular selective" spectral shifts of properties with incident angle or plane of incidence may also be quite large and polarisation sensitive.

2. Materials and Applications

It is useful to first catalogue the various materials and systems of interest according to four generic optical and visual groupings – (i) transparent, (ii) opaque, (iii) translucent, and (iv) light redirection, capture and transport. In later sections, selected promising examples of each will be discussed in detail. For basic systems including clear or tinted glass and polymer, with and without simple coatings, the underlying optical science is either quite mature or relatively simple, and we will not go into detail on these. It is, however, still useful to see these established systems placed into the context of the newer energy efficiency approaches, which are our focus. They all appear in Table 1 and examples of spectra and light distribution characteristics pertinent to a selection follow in figures 3, 4, 5 and 6. Some of the opaque systems in Table 1a have a long scientific history, but interesting and useful new developments and possibilities are still emerging.

It can be seen in Table 1 that energy efficient materials and advanced lighting materials are now a large and growing field. In column 1, the wavelengths, around which qualitative shifts occur, are included. New opportunities are expanding rapidly with developments in nano-materials, polymer optics, and new light sources such as LEDs.

Table 1. Examples of multi-purpose optical systems which combine energy efficiency with functionality and subjective appeal

(1a). Transparent – view or image preserving systems with low or negligible haze

Optical function	Generic systems	Specific examples
Block UV, admit daylight and provide clear view, high solar heat gain (400 nm)	Modified clear polymer, glass sheets, and glass/polymer laminates	Organic additives in polymer sheets and films, nanoparticles of ZnO or TiO_2, thin films SnO_2, ITO [18,19]
Reduce solar heat gain, admit daylight and provide clear view (750 nm)	NIR absorbing additives which do not scatter, 1D photonic crystal using bi- and tri-polymer unit cell	Iron doped glass, polymers doped with ITO, ATO or LaB_6 nanoparticles [20,21,22], PMMA/polyester multilayers [9,23]
Optical function	Generic systems	Specific examples
Reduce solar heat gain, admit daylight, provide clear view, and improved thermal insulation (750 nm, 2500 nm)	Thin films and coatings which reflect both NIR and black body IR energy	Polymers doped with ITO, ATO or LaB_6 nanoparticles and exterior coat of a transparent conductor, TiO_2/Ag/ TiO_2/... thin film multi-layers [7], TiN thin films and multi-layers [24,25]
Block UV, admit daylight and provide clear view, high solar heat gain and improved thermal insulation (400 nm, 2500 nm)	Transparent coatings which reflect only thermal IR (low emittance coatings) on glass or polymer	Pyrolytic SnO_2 on glass [26], ITO on glass or polymers [18,19]
Block solar heat and direct sunlight from high incidence angles, admit diffuse daylight, provide clear view, redirect daylight	Angular selective thin films, angular selectivity by total internal reflection in a structured polymer, microstructured surfaces, holographic films	Oblique columnar cermet films (Ag/SiO_2, Cr/Cr_2O_3) [8,27,28], anisotropically microstructured polymer surfaces by lithography [29,30], laser cut polymer sheets [31], holographic sheets [32]

(1b). Opaque systems – exterior surfaces in buildings and vehicles

Optical function	Generic systems	Specific examples
High solar absorptance and low thermal emittance (2500 nm)	Thin films on metal substrates, nanostructured metal surfaces	Ni doped anodic alumina on aluminium, Mo/WO_3 [33], Cr/Cr_2O_3 cermet films on metal [34], surface nanopatterned Ni or Mo [35]
Coloured and black layers which reflect the NIR solar heat component and have high thermal emittance (750 nm, 2500 nm)	Thin insulating films on metal flakes and dyed paint layers on metal	Thin films of SiO2, or the bi-layer Fe2O3/SiO2 on Al, visibly absorbing dye layers which are NIR transparent in binder on metal [5, 6]
Coloured layers which have high solar absorptance and low thermal emittance (2500 nm)	Nano-thin insulating layers on some metals, mixed pigments in thin layers on metal	Anodised or oxide coated stainless steel (blue or violet only) [36], green and black pigments in thin binder layer on Al [37,38]
High solar reflectance with (a) high thermal emittance (i.e. low black body IR reflectance-2500 nm) or (b) high emittance confined to the "sky window" 8–13 □m (8 □m)	(a) Thermal IR absorbing white pigments in dense paint (b) thin films or paints which are transparent except for high absorptance in the "sky window" (on metal)	(a) Dense TiO2 pigmented layers [39], ZnO, ZnS pigmented polymer or binder layers [40,41], (b) silicon oxynitride films [42]

(1c). Translucent – high light transmittance, non- and partial view preserving

Optical functions	Generic systems	Specific examples
High forward transmittance, negligible backscattering with controlled light spreading	Large TRIMM (transparent RI matched microparticles) in polymer sheet, certain co-polymers	Cross linked PMMA spheres in PMMA sheet [43,44], co-polymers with close index [45]
Block solar heat and direct sunlight from high incidence angles, admit diffuse daylight	Microstructured polymers	Laser cut polymer sheets [46], microstructured polymer surfaces [29,30]

(1d). Light capture, redirection, transport and emission

Optical functions	Generic systems	Specific examples
Light transport by total internal reflection over short and long distance with controlled emission rates from the guide	TRIMM microparticles in clear flexible polymer light guides or side lit flat polymer layers	Cross linked PMMA spheres in flexible acrylic cylindrical pipes and rectangular sheets at doping levels controlled according to transport distance [47,48]
Multi-source light mixing with homogenisation and low loss to side or back	TRIMM microparticles in clear polymer rods	Low concentration of PMMA microparticles in clear PMMA rods or clear rods plus doped sheets [49,47]
Fluorescence conversion and its emission trapping in part of a light guide	Dye doped polymer sheet optically bonded to clear polymer light guide	BASF Lumogen dye doped PMMA sheet optically glued to a PMMA ribbon light guide [50,51]

Table 1d. (Continued)

Optical functions	Generic systems	Specific examples
Near field coupling of (a) nanoparticle scattered radiation (b) plasmon resonant emission (c) fluorescent dye molecule emission into a close coupled waveguide	Oxide, metal or compound conductor nanoparticles deposited onto a thin transparent film or semiconductor layer to achieve (a) greater absorption in thinner underlying layer or (b) lateral transport to layer edges	Silica, Ag, LaB6 nanoparticles onto an SiO2 or Si layer. Film thickness controls waveguide modes and for plasmons, particle resonance wavelengths [52].

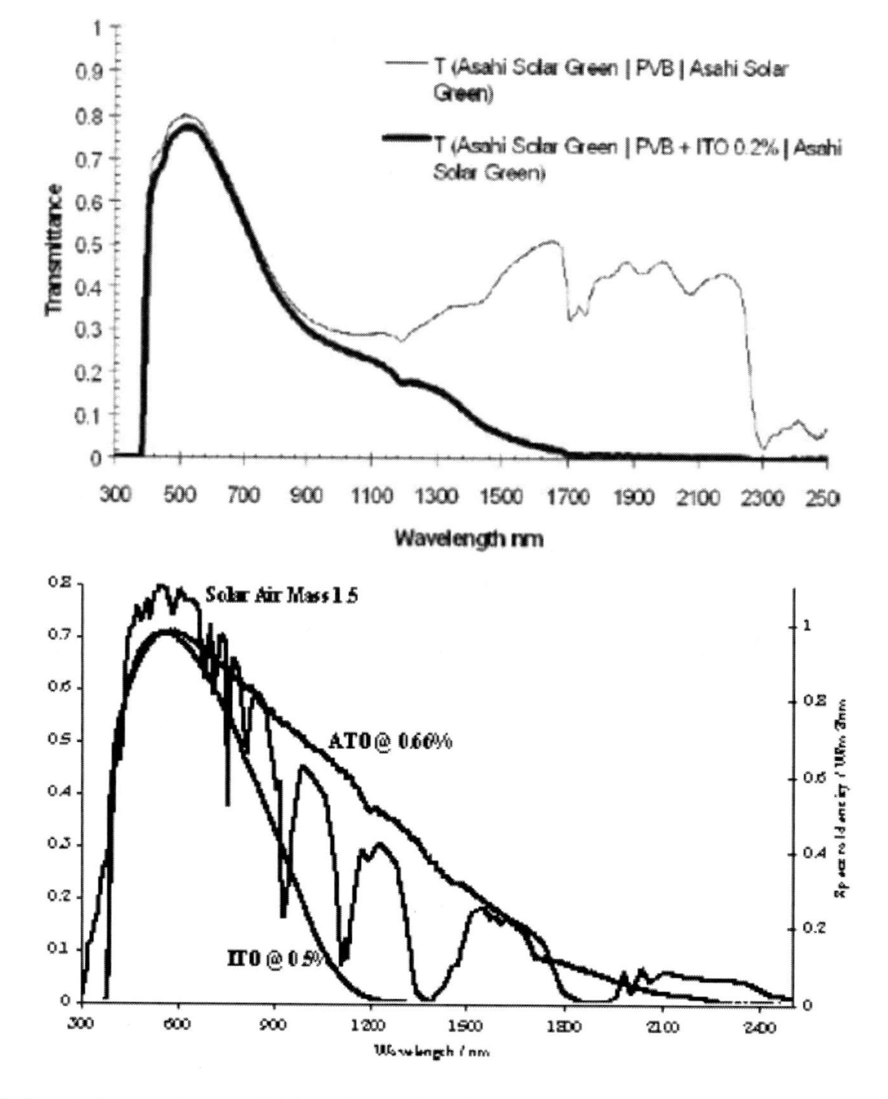

Figure 3. Spectral transmittance of thin polymer sheet between glass doped with nanoparticles of ITO and antimony tin oxide [20].

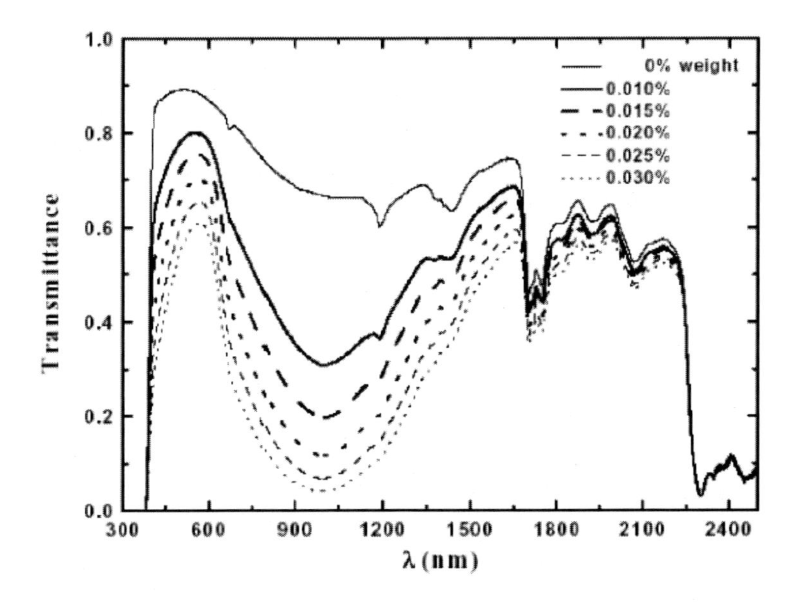

Figure 4. Spectral transmittance of laminated glazing doped with LaB6 nanoparticles [21,22].

3. Nanoparticle Plasmon Resonances for Spectral, Directional and Haze Control in Glazing and Skylights

Most glazing is clear and transparent, though it may be visibly tinted to reduce glare. Admission of daylight, provision of a clear view, and thermal insulation, provide the best combination for energy efficiency in cool climates. For temperate and warm climates, it is also desirable to block the NIR component of solar radiation to reduce cooling loads. Table 1b gives various established approaches to the latter goal using thin films. An alternative approach is now available using nanoparticle doped thin polymer layers which are typically either laminated between glass sheets or stuck onto the glazing surface. The particles in combination with their associated glass must satisfy three criteria, viz. (i) absorb strongly within the desired wavelength (NIR) range, (ii) produce negligible visible haze, and (iii) have an acceptable colour rendering impact on transmitted light. The surface plasmon resonance (SPR), which can occur in conducting nanoparticles when they have a negative real part ε_c' of the dielectric constant of conductor, can fulfil all three criteria provided

1 the material used has a plasma frequency which is not large enough to cause visible SPR resonances,
2 the particle is small enough that Rayleigh scattering in the blue wavelength range is negligible,
3 the particle is small enough that plasmon resonant scattering centred in the NIR is strictly dipolar [22] (multipoles not present),
4 particle shape effect on resonant frequency gives desired resonant frequency [21], and
5 the medium in which the particles are embedded gives a suitable SPR location

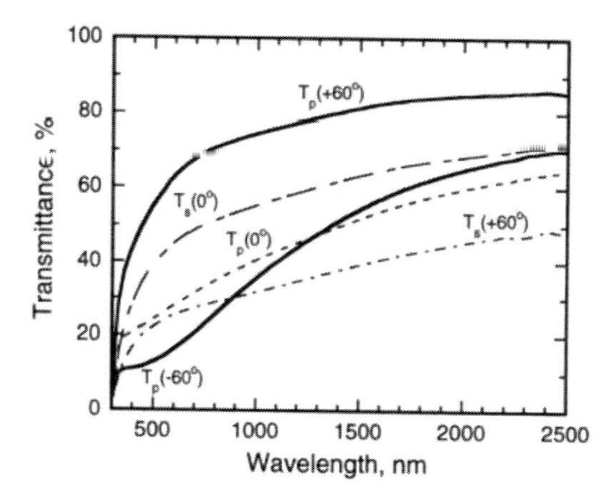

Figure 5. Angle of incidence and polarisation dependence of spectral transmittance for obliquely deposited Ag nano-columns mixed with oblique silicon oxide. Strong angular selectivity is evident.

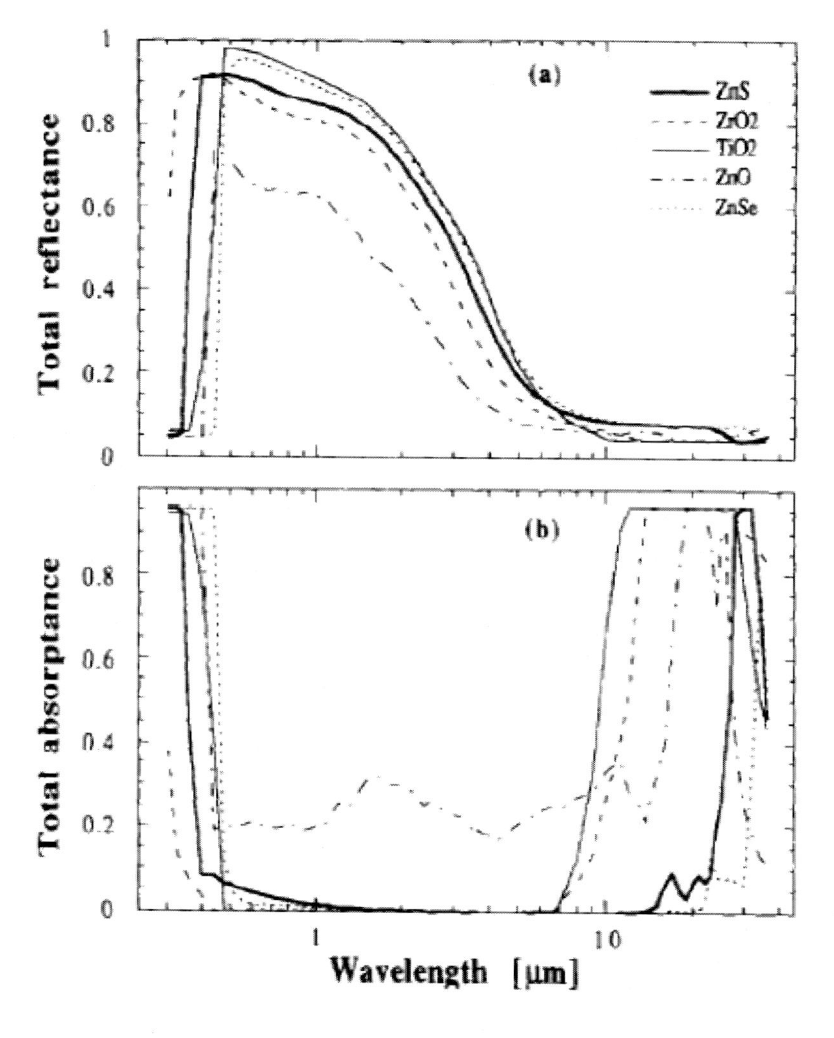

Figure 6. Spectral selectivity for radiative cooling purposes in polymer foils doped with various inorganic pigments [40,41]. Reproduced by permission of Elsevier.

Plasma frequency ω_p determines at what wavelength ε_c' becomes negative. ω_p is given by equatio (5) with n_e as the carrier density and m^* as the carrier effective mass.

$$\omega_p^2 = \frac{n_e e^2}{\varepsilon_o m^*} \tag{5}$$

Gold, silver, aluminium and titanium nitride have carrier densities which cause too strong attenuation in the visible, but lowering n_e below that in metals (such as gold and silver) shifts the plasma frequency down as desired. It is also desirable that inter-band absorption terms are negligible or weak in the range of interest, otherwise plasmon effects may be washed out or not present. This means we can effectively use a Drude model, though in practice, we always start with known dielectric constants. Residual interband effects, while not adding absorption, can cause polarisation, and hence, have an important influence. They are usually included in a modified Drude model (equation (6)) via the constant ε_∞. This is because they have a weak frequency dependence in the range of interest, but do cause some screening of plasmon response, and hence, a lower SPR frequency

$$\varepsilon_c = \varepsilon_\infty - \left[\frac{\omega_p^2}{\omega(\omega + i\omega_\tau)}\right] \tag{6}$$

here ω_τ is the relaxation frequency, and the nanoparticle sizes of interest (< 60 nm) must be modified from bulk values to account for charge collisions with the particle surface [21,22]. The effect of shape is given by equation (7) assuming a spheroidal particle with depolarisation factor L for field along the axis of interest.

$$\varepsilon_c' = -\varepsilon_h\left(\frac{1}{L} - 1\right) \tag{7}$$

here ε_h is the dielectric constant of the host matrix. For spheres, the bracket in equation (7) reduces to 2 since then $L = 1/3$. Substitution of the real part of equation (6) into equation (7) leads to the material (via n_e, ε_∞) and shape (via L) impacts on resonance energy in one expression. The carrier densities and location of inter-band terms in transparent conductors such as indium tin oxide (ITO) and antimony tin oxide (ATO), means they can be used for NIR blocking, but are inefficient because n_e is so low that the sphere resonance peak lies around 2–2.5 μm while we need most absorption (figures 3, 4) from 0.75–1.3 μm. To date, a material identified to bridge the SPR frequencies of metals and the conducting oxide materials, and hence, give good nanoparticle results for solar control, is LaB_6. But LaB_6 also has limitations; shape must be correct and not too variable because of the combined broadening effects of the collisions with nanoparticle surfaces and shape impact on resonance location. LaB_6 spheres, for example, cause too much absorption in the visible red region, but slightly oblate ellipsoids, oriented as in typical extruded sheets, give a shift in the SPR peak to near 1 μm, whereas spheres resonate at 900 nm [21]. Resonance broadening can also arise from a wide spread in shapes as then there is a spread in SPR locations given by equation (7). Thus, LaB_6 is very efficient in attenuating NIR solar energy, typically needing, for example, just 0.02–0.03% by weight in a 0.7 mm thick polymer foil for large NIR attenuation while maintaining good daylight gain. In contrast ITO requires 0.2–0.5% for similar levels of NIR solar attenuation, because it relies on resonance tails well away from the SPR peak to block solar NIR. The problem with LaB_6 is that visible attenuation does increase as concentration increases since the SPR spills over into the visible range (figs. 4, 7b). A detailed generalised analysis [22] indicates an ideal location for the SPR around 1.5 μm to achieve a combination of low particle concentration, no colour impact or visible attenuation, and excellent NIR blocking. If, however, visible tinting for glare reduction with some daylight is not a problem, as in much architectural glazing, then LaB_6 is fine.

Modelling of optical response can be done quite accurately in these systems provided particle shapes are approximately known and particles are well dispersed. Lack of dispersion also cannot be tolerated because clusters cause haze. The concentrations are so small that we can neglect inter-particle interactions in modelling absorption or scattering. For particles under 50 nm diameter we can also neglect scattering in the visible region, and it is also weak in the NIR. The absorption can be found from either calculating the combined dielectric constant ε^* from adding up the quasistatic induced dipole moment of all particles in a unit volume, then finding the complex effective RI $N = (n^* + ik^*) = \sqrt{(\varepsilon^*)}$. Equivalently, the absorption coefficient α is given in the dilute limit by

$$\alpha = f\frac{2\pi\sqrt{\varepsilon_h}}{\lambda}\mathrm{Im}\left[\frac{\varepsilon_c - \varepsilon_h}{\varepsilon_h - L(\varepsilon_c - \varepsilon_h)}\right] \tag{8}$$

where f is the particle volume fraction. This model fits data well, as shown in Figure 7(a). Discrepancies are probably mainly due to some scattering in the NIR.

It should be noted that, for the sizes of interest of up to around 80–100 nm, scattering is actually stronger in the NIR due to the SPR than it is due to normal polarisation effects at blue wavelengths, as shown in Figure 8. The latter gives classical Rayleigh and Rayleigh-Gans scattering. The fact there is a gap between the two scattering zones, as in Figure 8, is technologically important. There are also some important qualitative distinctions between the two types of scattering, which is not often realised. The Rayleigh-Gans term is only isotropic at the very small particle limit (~10 nm) where it is a pure dipole, but becomes mainly forward even by 80 nm diameters and almost all forward scattering by 150 nm. In contrast, the SPR scattering in the NIR remains quite isotropic up to 150 nm [22]. This is because dipole terms dominate in the latter until much larger particles. It is thus the onset of blue haze which limits the particle size for transparent glazing applications. In diffusing skylights, nanoparticle contribution to haze is less important, but one still needs to avoid colour related effects, which can be caused by Rayleigh or SPR scattering.

Another issue is that scattering changes affect the resonance width, and hence, spectral selectivity. This was recently highlighted by Schelm and Smith in the context of dielectric core-metal shell plasmon resonances [53], but it is also important in single material conducting particles.

Once multipoles come into play, the extinction resonance broadens substantially. It is the dipole absorption resonance by itself that gives the narrow spectral response, and makes spectral control manageable.

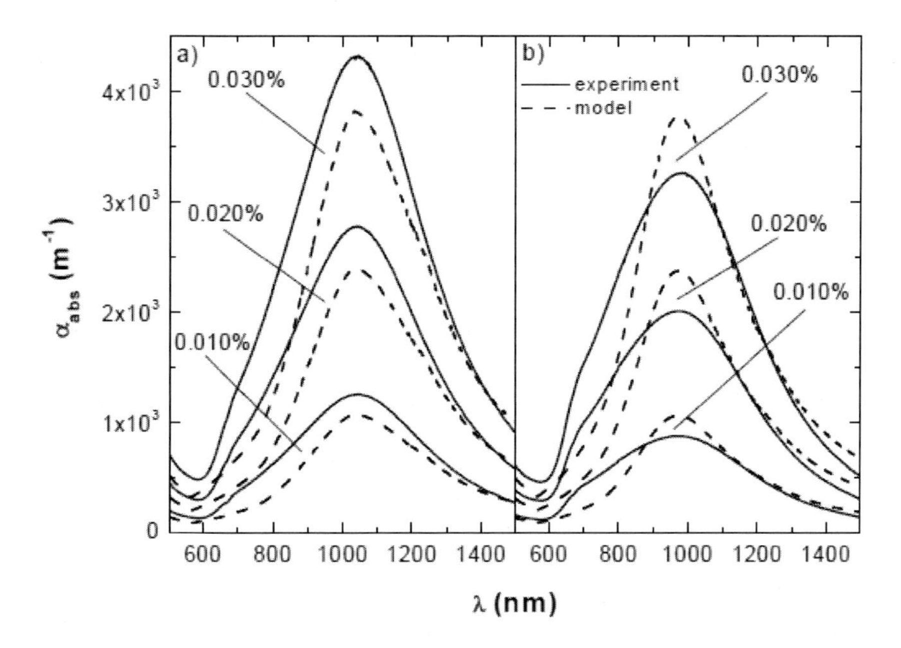

Figure 7. (a) Plasmon resonant absorption spectra in thin polymer sheets doped with LaB6 nanoparticles [21].

Figure 7. (b) LaB6 nanoparticle glass laminated samples with various doping levels against a bright blue sky.

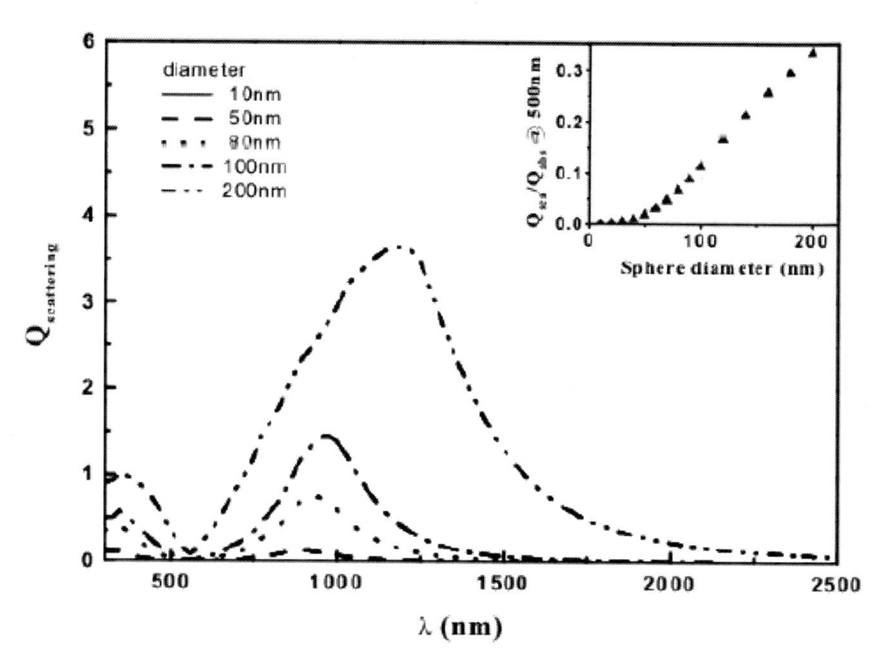

Figure 8. Scattering efficiency (and inset scattering to absorption ratio at 500 nm) as a function of wavelength for different diameter nanospheres in polymer showing scattering due to both NIR plasmon resonance and short wavelength classical Rayleigh scattering [22].

Once nanoparticles start to interact, equation (8) no longer applies. If scattering is still absent, various quasi-static approaches to model the effective dielectric constant ε^* are used, depending on the order if any, in the array, local clustering and percolation exist. The simplest and most widely used approximation assumes an array of dipoles in which each dipole interacts with a homogeneous average field due to all other dipoles. The 3-fold degeneracy of polarisation modes in an isolated sphere is not lifted in this approximation, which yields the Maxwell-Garnett (MG) model [54]. The MG model is inaccurate once any order sets in, or particles touch, as then multipoles also contribute to the dipole moment on each particle and the core assumption fails. Lattice [55] and local structure [56] models can then be useful. For a sphere once such interactions occur, the simple 3-fold degeneracy is lifted, and sets of symmetric and anti-symmetric polarisation modes should be observed for particle arrays. One of the most interest for solar work is the anti-symmetric mode in uniform particles which, relative to the original isolated dipole, moves to longer wavelength or lower energy. In practice, multiple modes are rarely observed as most samples do not have ideal lattice structures, so such effects are blurred out. However, some ordering is recently easily achievable using metal spheres coated with dielectric to form dense arrays of separated plasmon resonant spheres [57]. Mode splitting can also occur in a single inhomogeneous nanosphere [58] as in dielectric core-metal shell nanoparticles where plasmon resonance degeneracy is lifted. Then, however, it is the symmetric mode which is at the lowest energy, so it is one of the potential use in NIR blocking. The symmetric mode there and in other shell systems, such as nanotubes, has like charges at opposite sides of the thin conducting layer. This leads to the lowest energy for the whole sphere. The associated local field distribution is also quite unusual near resonance with the maximum field in the sphere shell at a position normal to the applied field [59] in the symmetric mode, and thus, near field profiles can be quite different to those on a uniform sphere at resonance. This has potential uses for molecular sensing.

In nanoparticle systems of high density, it is now possible to create arrays of conducting nanoparticles which fill most of the composite but remain separated. Thus, it then becomes possible to explore effective medium response and models for closely spaced but separate particles. One way is to coat the particles with an oxide of uniform thickness, and then pack the particles densely [57]. This creates an ordered array of metal particles, and results diverge from MG model predictions at fill factors of around 0.4–0.5, as expected. Another approach suitable for gold is to self assemble the nano-spheres in solution using organic di-thiol linker molecules to both join them up and keep them apart [60]. Spacing depends on linker length, and shorter linker molecules can provide high densities of nanoparticles. The nanoparticles must be small (< 8 nm). In these composites randomness is retained, and it appears that simple dipole response for the gold is retained as a result. In both examples, there are actually three mediums present since voids are also retained in the structures as they assemble. In the linker case with the self assembly occurring in solution before film deposition, the voids tend to phase separate from the linker-nanoparticle composite. A simple effective medium approach for the random three-component system works well only if the model is created in two steps which are consistent with the observed morphology: first a mix of gold in linker (to give the effective medium labelled AuL), then a mixture of the AuL medium with the voids. The AuL phase seems to be sufficiently small not to cause scattering.

The results on this system confirm a prediction of the random unit cell concept [61], namely, provided the randomness is retained, quite high densities of nanoparticles will not display multipolar response.

4. Energy Efficient Scattering Materials: Applications in Attractive Lighting Design Skylights, Display, Light Mixing and Homogenization

4.1. Sheet and Thin Layer Formats

Translucent polymer sheets and other sheets used in light fittings, skylights for glare reduction and more uniform illuminance, and some display work, are either traditionally pigmented with inorganic additives such as $CaCO_3$, TiO_2 and $BaSO_4$.or have special surface treatments. The RI of particles relative to the polymer host, their specific micron scale sizes and the amount added, combine to produce significant spreading of transmitted light. This spreading may be tailored as desired to reduce glare when the transmitting surface is viewed directly, and to provide more uniform lighting when used in skylights or luminaire covers. Traditional Mie scattering theory, modified to include multiple scattering effects, is able to semi-quantitatively describe their spectral transmittance and reflectance, and light spreading [62]. Thus, the emphasis on the use of these materials is on visual comfort, glare control and better use of the available light. Only the latter one has a clear impact on the overall energy efficiency in lighting. Most skylights are diffused to give more uniform illumination. But these traditional diffusers have an intrinsic negative impact on the energy efficiency. Pigment additives at concentrations high enough to achieve a reasonable spread of transmitted light cause a significant degree of backscattering, as is observable from their whiteness or brightness, when viewed in reflection. Consequent reduction of hemispherical transmittance and loss of lumens from lamps or the sun is commonly in the range 20–50%. It should also be noted that backscattering could be used to control light and solar energy throughput in some switchable window types (such as thermo-tropics) which switch from clear to diffuse on heating [63]. Strong backscatter usually requires small (sub-micron) size particles. Light diffusion with surface prism features on a sheet is another approach, but this often causes glare and less overall light spreading.

An easily implemented solution to this problem of eliminating backscatter while achieving controlled light spreading has recently become available in the form of a particular physical class of additive, which we now discuss in detail. This virtual elimination of backscatter comes with the associated optical properties, which provide not only energy efficiency but a whole new range of geometric configurations for lighting systems, new ways of achieving energy efficient "white" light lamps, new approaches to aesthetically pleasing light patterning, and new display and signage options. These will be covered in the next section, after we address the basic optical properties. The key is in the combination of two features: (i) a large enough particle size for which Rayleigh-Gans scattering is weak and geometric optics is a good approximation, and (ii) a particle material whose RI is very close to that of the host material. The host is PMMA or related acrylics in our examples and in many applications. We have given to this class of additives, the generic acronym TRIMM [43, 47] which stands for transparent RI matched micro-particle. The particles used are normally spherical but the basic principles apply to other shapes. The key parameters are thus R (radius) and m (RI ratio) with $m = n_{particle}/n_{host} = 1 + \mu$, for $n_{particle}$ and n_{host} as the respective individual RIs. Thus, in TRIMM, m is close to 1, and μ is close to zero. In the two systems we deal with explicitly the same cross-linked PMMA particles in two different matrices, one solid PMMA and the other a flexible clear acrylic. The respective values of μ are 0.0114 and 0.0182 at 590 nm. At normal incidence, such a material has

$$R \approx \frac{\mu^2}{4} < 10^{-4}.$$

At the median angle (45°) a ray strikes the spheres, and when $\mu = 0.0114$, polarisation averaged reflectance is 6×10^{-5}. Thus, many particles have to be struck before significant backscattering occurs. The reflectance that does occur, which is still weak, is predominantly for the more oblique rays falling onto the sphere surface, so they go forward and are of use. For these large spheres (diameter > 10 μm), the light striking a sphere near its edges is diffracted, but this is a tiny fraction of the total incident radiation.

Thus, the wavelength does not enter explicitly into the light spreading equations in this geometric limit. All wavelength dependence of any consequence comes about through the wavelength dependence of μ. The ray deviation angle δ, and impact parameter h are shown in Figure 9. The deviation angle for a sphere of radius R is given by

$$\delta = 2\left[\sin^{-1}(h/R) - \sin^{-1}\left(h/R(1+\mu)\right)\right] \tag{4}$$

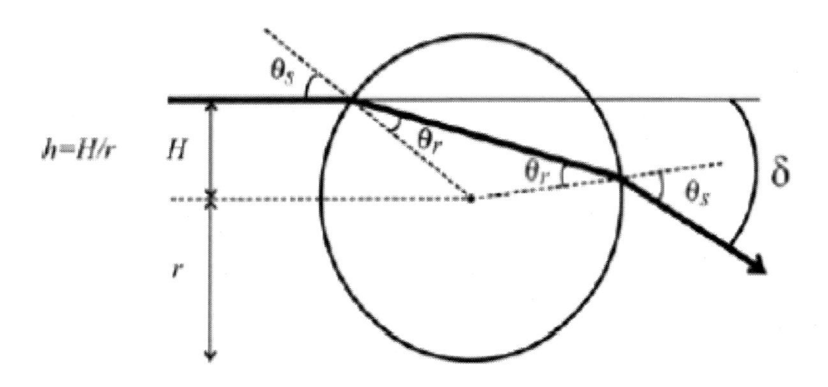

Figure 9. Ray deviation as it passes through a large sphere.

The spread of deviation angles for one sphere and the mean deviation angle are easily calculated [64]. If the probability of a deviation angle between δ and $(\delta+d\delta)$ is $f(\delta)d\delta$, the probability density function $f(\delta)$ can be found analytically first in terms of h, and hence, in terms of δ from equation (4) [64]. For $\mu = 0.0114$, f is plotted versus δ in Figure 10.

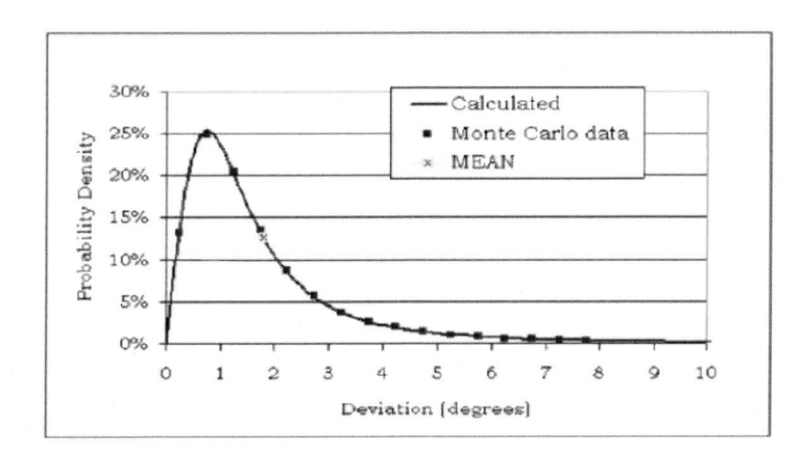

Figure 10. Probability density distribution function for angular deviations of all rays striking a TRIMM sphere in polymer, with □ = 0.0114. [C. Deller, PhD dissertation, UTS, 2005].

Also shown in this figure is the distribution of deviation angles found by Monte Carlo simulation. The mean value of δ, which is quite small at 1.8°, is also plotted on the curve. Note the long but weak high angle tail.

For the flexible matrix with $\mu = 0.0182$, the mean deviation is 2.1°, and for small μ, the mean deviation angle is approximately linear in μ. Such a small but significant linear deviation on average per sphere intersection makes light spreading control relatively easy, since average spread is then controlled through the average number of spheres with which a ray intersects.

This is, in turn, easily controlled through particle concentration or sheet thickness. It is easily shown that the average spacing p between particles of radius r ($2r$ ~17 μm in our samples) is given in terms of volume fraction V_f of spheres in the matrix by

$$p = \frac{4r}{3V_f} \tag{7}$$

Where p is the main geometric parameter needed in simulation of multiple scattering. Thus, we have close to an ideal diffuser from an energy savings viewpoint because we can engineer the degree of light diffusion while losing hardly any light.

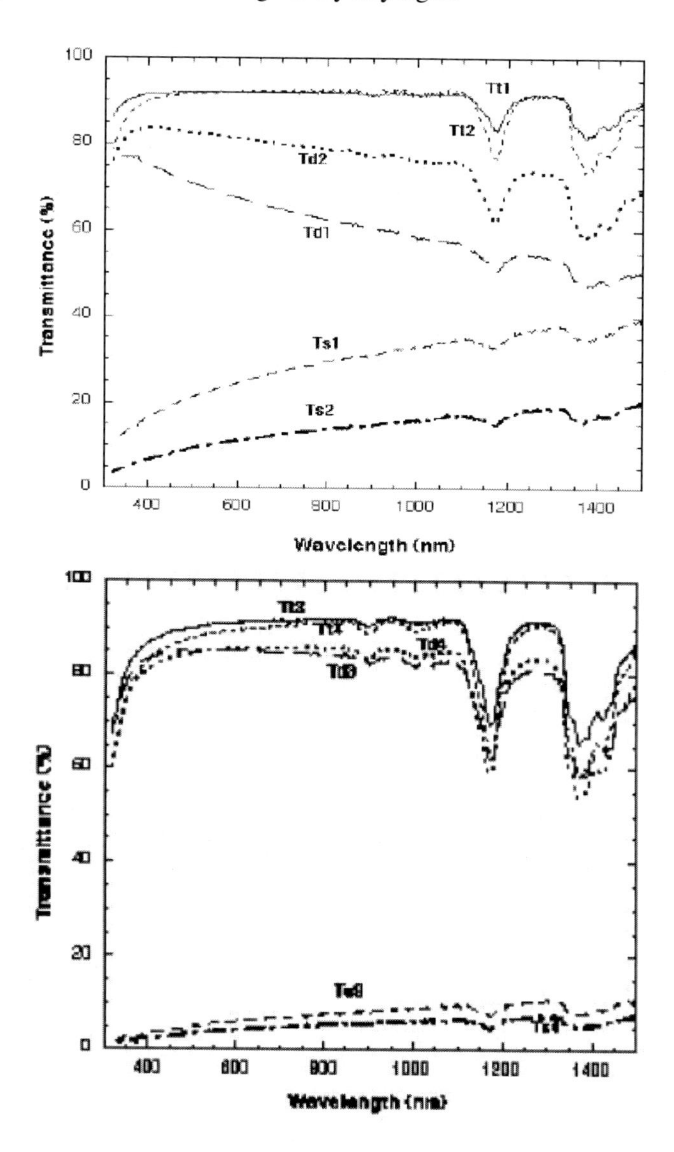

Figure 11. Total hemispherical transmittance, and specular and diffuse components of transmittance, for 1, 2, 3 and 4 mm thick PMMA sheet moderately doped with TRIMM spheres [43]. d = diffuse, s = specular.

The other important aspect is colour shift of light on transmission due to TRIMM scattering. Since this just depends on how μ varies with λ, it is much less wavelength sensitive than Rayleigh scattering. For most sheets that we have studied, the colour rendering index of white light is not changed until quite large thicknesses.

Only if doping and thickness are high, it becomes significant. All of these issues can be seen in the spectral transmittance plots of Figure 11, where total transmittance and the specular and diffuse components are shown together for four different thicknesses, namely, 1 mm, 2 mm, 3 mm, and 4 mm, all at the same composition.

While the specular and diffuse admix changes as thickness, and hence average number of particle-ray intersections increases, the total hemispherical transmittance at visible wavelengths is almost unchanged. In addition, there is little fall off in total T with wavelength λ, so the transmitted light usually appears unchanged in colour.

However, the relative shift in diffuse and specular admix with wavelength indicates that μ is varying with wavelength enough to spread blue more than red, but the shift in μ never gets large enough to cause any significant backscattering. Once the diffuse component dominates, there is apparently a slight fall off in total T at short wavelengths (figure 11).

This is not due to backscattering but rather due to the measurement set up, since light, as shown in Figure 12, can be transported internally in the sheet by total internal reflection away from the integrating sphere aperture. In a large area sheet, total transmittance does not fall off as shown in Figure 11, at blue wavelengths.

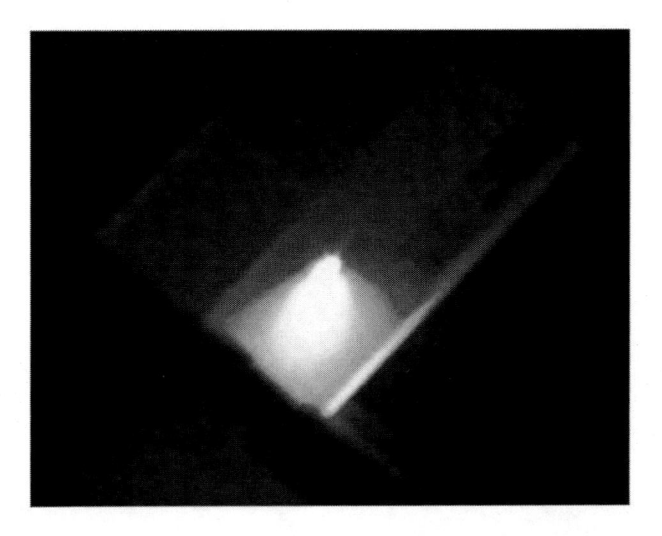

Figure 12. Photograph of exit light dispersion from a small illumination spot on a TRIMM doped sheet due to internal light guiding transport resulting from deviations that occur before emission. Note the light which makes it all the way to the edges [43].

The variation in degree of diffuseness with λ correlates with direct measurements [64] of the dispersion of n_{host} and n_{TRIMM}, and their ratio, as the wavelength decreases. In practice, once rays have undergone a moderate number of interceptions with spheres, the diffuse component dominates. Considering all of these issues, it is clear that the trio; colour engineering, light spread control and low loss, is achievable in combination simply with TRIMM particle concentration control.

4.2. Integration of Light Transport and Controlled Side Emission Using Doped Polymer Fiber Optic Cable

The ability to diffuse light not only by backscatter, but also to diffuse it only a small amount at each interaction, means it is possible to have a diffuser in which some internal light can be transported to long distances, so the fiber is continually emitting at a rate that is consistent with the maintenance of transport. This can be manifested in a fiber optic light guide, which can then be used for continuous illumination or display along its length with light input at one or both ends. The concentration, or more specifically the spacing parameter p, must be adjusted for the length. In contrast, doping a light guide with conventional Mie scattering diffuser pigments leads usually to two outcomes in addition to much backscattering loss. Either light is lost from the guide sides very quickly, so only short lengths are feasible, or if concentration is kept low enough to achieve moderate transport lengths, then side output at any point is quite weak. Interest in fiber optic based lighting has increased significantly with the growth in bright LED availability and performance in recent years, especially for moderate diameter light guides. If small diameter guides are used, fiber bundles are required which are quite expensive, especially if made of glass fibers, or extruded polymers. Single core high transmittance fibers, however, are relatively cost effective, and can be made with good flexibility in their diameters.

These are cast with high optical quality in single lengths up to 200 or more metres long. The transmitting jacket material is low index. Smaller diameters are usually extruded, but with these, it is less easy to couple in a lot of lumens into a single fiber. To extract light from the side of a fiber usually meant treating its surface, for instance, roughening or painting or screen printing one side. This has to be done intimately with the guide surface, so any existing jacket or low index coating has to be either removed where light is needed, or deliberately damaged in some way.

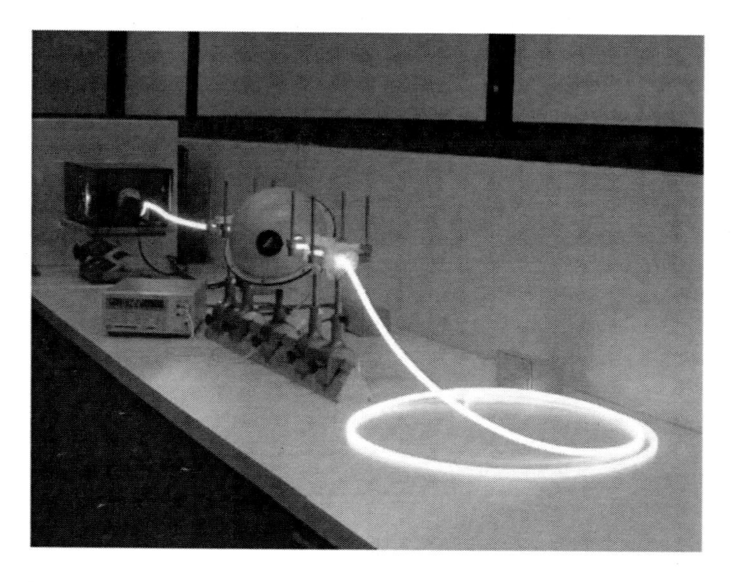

Figure 13. Set up for measuring side-light output as a function of length along a TRIMM doped flexible polymer optical fiber with a single core. Continuous bright emission along the fiber length is obvious.

This is messy and costly. If the guide itself can integrate transport and emission, lighting engineering possibilities are simpler, of lower cost, and more diverse. Thus, TRIMM doped single core light guides have vast potential in all areas of lighting and display. They also have another important capability for end lighting as opposed to side lighting, when lightly doped, which normal undoped single fibers do not. Good quality clear fibers produce caustics or bands of light at their exit end, which are sometimes called "hot spots", the details of which depend on the source beams characteristics [65]. For sources designed to efficiently couple into an optical fiber using focussing or parallel beams, the output caustics tend to be worse and quite visually uncomfortable. The way TRIMM doping can remove this problem without loss of energy is discussed in detail in the next section on end lighting, along with several important applications.

Here we focus on side lighting from single core polymer fibers, and the technology that flowed from this work we called "supersidelight". Though a recent development, it is available from Poly Optics Pty Ltd in Australia (www.fiberopticlight.com) who supported this research, and it is starting to be used by various specialist lighting businesses. Measurements of total output as a function of length have been carried out with an integrating sphere as shown in Figure 13, where the continuous output can also be seen. Fall-off plots based on spot photometer measurements agree with those from the broader distance sampling integrating sphere. The fall off in brightness is approximately exponential, and the exponent's dependence on concentration is complex [64]. The first TRIMM doped long fiber optic cables were made by us in 2000, and first reported openly in 2002 after earlier patents [66]. A French chemical group also reported on a similar system in 2002 [67]. Our models show their use of single event Mie scattering to describe the output is not appropriate to most of these systems. Typically, many scattering events occur before escape. In essence, since $\mu = 0.0182$ at 590 nm for TRIMM in the flexible matrix, the only material parameter to adjust is the particle density in the optical fiber. The axial particle number a in fiber length L (i.e. a = the total number of particles in length L) is the important parameter in terms of both end and side light applications. The related particle spacing parameter is $p = L/a$ with p defined in equation (7) determining either the degree of homogenisation for end light, or the emission rate for side lighting. The smaller p means the more interactions in a given guide length, and hence, larger average deviation. As more rays approach the critical angle, more will exit. The initial light angular distribution profile just inside the fiber is dictated by the source input luminance profile, and the numerical aperture of the fiber. This profile gradually evolves with distance along the fiber. After a critical length L_c dependent on p and guide diameter $2R$, a steady state distribution is reached, and from then on, the rate of emission is a fixed fraction of the remaining light. This has been simulated in a special Monte Carlo ray tracing program using measured profiles on each source in a photogoniometer as input, known p values, and equation (4). Poisson statistics are used to determine whether a particle is struck in terms of p and length l travelled from the last interaction. The ray path is determined by three parameters – r (the radial distance from the axis of the last interaction and the direction defined by (θ, ϕ)), θ (the angle to the axis) and ϕ (the rotational angle about the axis). If a particle is struck, a new ray direction is calculated using spherical geometry [68]. If a wall is struck, the angle of incidence K must first be defined to determine if the ray exits, or if it is totally internally reflected. In the latter case, the new ray direction is simply found, and free path distance l

continues to accrue. For a circular cross-section guide of radius R, in terms of (r, θ, ϕ), K is given by

$$K = \cos^{-1}\left[\sin\theta\sqrt{1 - \frac{r\sin\phi}{R}}\right] \tag{8}$$

A plot showing output profiles for different distances travelled is in Figure 14. The onset of the steady state distribution is clearly seen in this figure. By varying concentration over a wide range, we can achieve energy efficient side lighting over fiber lengths from a few cm up to 20–30 meters depending on the optical quality of the undoped fiber.

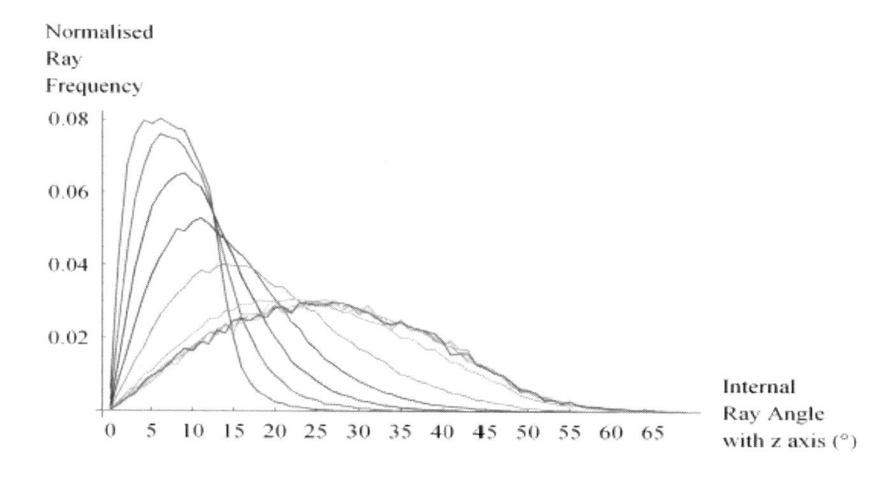

Figure 14. Evolution of internal beam spread with transport distance along a TRIMM doped optical fiber. The first five normalised curves are for transport distances 1, 10, 25, 50, and 100 cm. The remaining curves show a constant beam profile for various distances ranging from 2.5–100 m. Increasing doping simply compresses the distance scale.

The main challenge with the longer lengths is to achieve good colour balance far from the source, but this seems possible as long as these lengths of clear fiber also do not lead to faster loss of blue components. If sources are used at both ends, length limits will be dictated either by the source intensity or by the onset of yellowing of output near the middle.

Because the exit rays are being predominantly forward scattered, the output profile is also forward biased, but the low index jacket helps to make it more uniform. With sources at both ends or a reflector at one end, a more uniform output occurs, but it is still weakest when viewed along any line normal to the guide walls.

Many applications require light only from one side of the guide, so a diffuse reflector is applied to re-direct rear exiting rays. It is applied either to the outside surface of the low index jacket or mounted externally. The net forward output is then usually quite homogeneous.

This ability to easily and efficiently make use of light going away from useful directions is a major advantage with fiber optic side lighting as opposed to the conventional elongated tubular powered source or fluorescent lamps. After reflection, rear-going light passes straight back through the optical fiber. In contrast, in a tube lamp mounted in a reflective luminare, much light can be blocked or lost due to the lamp tube itself. Thus, we again gain on energy efficiency. Related to this is the ability to engineer the light after it exits the guide without any impact on the internal transport performance of the guide. The coating can be on or flush with the outer surface of the low index layer, and still does not affect transport. With undoped fiber, it is difficult to simply control output with a coating directly applied to the bare guide walls. So, complex screen printing patterning is needed, and is only used in practice over limited lengths otherwise the output would appear too "spotty". A continuous coating on bare guide walls would severely limit achievable transport lengths, and hence, applications.

Multiple small lamps such as LEDs can be used simultaneously as fiber input sources, and are rapidly mixed within "supersidelight" or weakly doped end light guides, so that a wide variety of uniform new colours can be output, and also, changed in a dynamic fashion. If different coloured lamps are input from opposite ends or smaller diameters twisted together, even more varieties of colour and colour dynamics as a function of position along the guide can be achieved with very low loss. This mixer approach to arbitrary coloured lighting is clearly more energy efficient than using colour filters. An undoped clear guide using external means to get light out its side, and diffuse it, can easily yield inhomogeneities. Individual caustics, such as those reported by Swift *et al* [65] for clear polymer rods used for mirror skylights (as in Figure 1), can cause quite unpleasant visual effects. Light mixing in doped guides occurs very soon after the input end, and requires very low doping levels. Mixing for end light without side loss is thus possible, and is discussed next.

4.3. Energy Efficient Light Mixing and Homogenisation

The applications here are concerned with using end light from optical fibers or edge light from sheet light guides. There are several challenges when using single core fibers or other light guiding systems such as clear flat panels for illumination, when light is to be used from one side or the fiber end. They include efficient input coupling, removal of caustics, for multiple sources having them fully mixed, keeping mixing distance small and keeping energy loss low or luminous efficacy high. It is often possible to achieve mixing and homogenisation by using strongly doped conventional diffusers, but then much brighter sources are needed to compensate for the large energy losses that occur. So, net efficacy is then very poor. Also, as many applications are in compact, portable and hand held devices, longevity of battery power between re-charges is of major concern. Thus, achieving high brightness with low power is of considerable value. As an example, in surgical applications involving hand held devices, for instance, looking into a complex structure such as the mouth or throat, uniform well directed but dispersed and suitably localised illumination makes the task much easier, and reduces risk of errors. However, residual caustics can distort perception and create false highlights, dark spots and errors in depth perception. A lightly doped TRIMM small diameter guide is ideal for such applications. Inspection lighting used in such manufacture can benefit in the same way.

Mixing and homogenisation is aided by both multiple reflections off the light guide walls (effectively creating multiple sources), and by the spread induced from interaction with TRIMM spheres. The combination leads to much better performance than either alone or with the use of the light guiding reflections, meaning that very few collisions with particles are needed along the guide. This is important for energy efficiency since it means most light can reach the output end or edge without exiting the guide walls. As before, backscattering losses from the particles are not important while Fresnel back reflections from the front and exit ends are the most significant losses, as shown schematically in Figure 15. The Fresnel losses reduce net efficiency to around 88%. These end losses are increased by the extra divergence from the sphere interactions, but can be reduced in various ways and to various degrees, sometimes with quite simple procedures. 92% efficiencies (and maybe even 95%) should be achievable at low cost. In experiments with short mixing rods (~10 cm) backed up by computer simulations [47], it is found that exact profiles of the beams from LEDs are essential to get optimum results. In practice, they often have sharp features though many modellers and much supplier data approximate them as having smoothly varying intensity profiles with exit angle and change in direction. Furthermore, in simulation with doped guides, the extended source area, and precise placement of the source relative to the guide centre, must be used. Significant errors occur if it is treated as a point source, when moderately close to the guide entrance. Earlier mixing work [69] had used clear rods which had to be quite long to remove most caustics, but still could not provide homogeneous colour or brightness output without an additional diffuser.

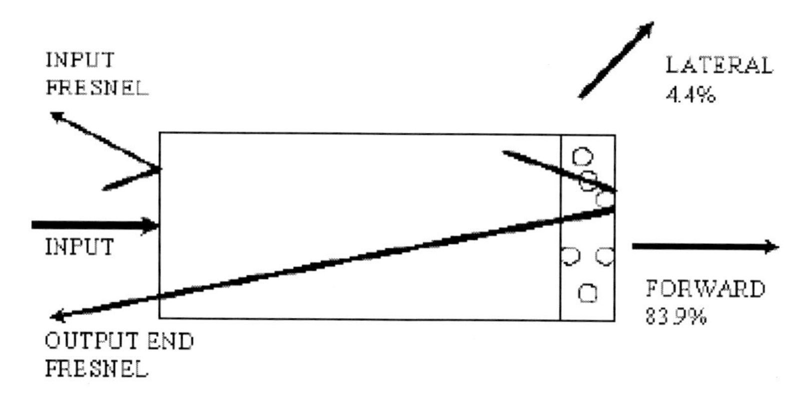

Figure 15. Composition of light mixer losses in a clear light guide terminated with an optically bonded TRIMM sheet. Simple modifications of exit end to eliminate most Fresnel losses can provide > 92% efficiency and combined with LED optical integration at input end up to ~ 96% mixing efficiency [49].

The results for 3×3 mm^2 red, green, and blue LED outputs mixed in a 10 cm long, 12.7 mm diameter PMMA rod, doped with just 0.45% by volume of TRIMM particles show mixing is just complete, and over 93% of lumens, that enter the guide, are output as useful lumens. The average ray has struck only ~ 20 particles. Mixing in even shorter lengths with similar losses is possible with slightly higher concentrations. The central bright output zone is also quite uniform in terms of luminance, then falls off smoothly. In contrast, a clear rod produces both local colour fluctuations, which are easily discerned by eye, and luminance spikes which can be disconcerting.

An alternative combination of guide and TRIMM that works well is that shown schematically in Figure 15. It uses a clear rod and a thin TRIMM doped sheet with concentration high enough to get around 20 collisions across the sheet. It is optically glued to the output end [49]. The side loss from this top sheet will generally contribute to desired illumination. This system or the lightly doped rod thus has the potential to be the nucleus of the compact, energy efficient, low cost white light "lamp" based on RGB LED combinations, that has been sought as an alternative to phosphor based LED white light systems. Other mixing options such as micro-optic arrays on polymer are much more expensive.

The decorative, display and sign options with these types of lamps or the longer doped guides in terms of colour options, dynamic colour shifting, light distribution and visual interest and excitement are numerous, and remain to be exploited.

5. Roof, Car and Wall Cooling Based on Spectrally Selective Coatings and Pigments

5.1. Coloured Solar Reflective Paints

Until recently, solar reflection required white or pale coloured paints on buildings or cars. This is often not attractive to many people, limits decorative variety in warm climates where it is used in most buildings, and can cause substantial exterior glare. However, by using pigments or dyes with narrow absorption bands in combination with appropriate substrates, it is possible to achieve a variety of colours, even black, with strong to medium solar reflection. In actual outdoor tests, we have found that interior spaces can be kept several degrees cooler in some spaces using such paints compared with paints that are visually almost identical, but do not reflect the solar NIR component [5,6].

There are two underlying mechanisms that have been identified and tested, and both rely on using metal to reflect the near IR wavelengths. In one case, the latter is the substrate, and in the other, it is the core of the pigment particle. Figure 16 shows reflection off a binder layer on a copper substrate, with and without various dyes which make it black to the eye. This is compared with traditional black paint in the same figure. This normal black has a flat spectral response, and absorbs 95% of incident solar radiation. The binder is partly transparent in the NIR so that the substrate reflects a lot of incident NIR. An even more NIR transparent binder would allow a black with even higher R_{sol} than the one shown. The best solar reflective "black" in Figure 16 has R_{vis} = 5% and R_{sol} = 30% compared with the standard high quality black with R_{sol} and R_{vis} both about 5%. A solar reflective black reflecting up to 40% of incident solar energy may be possible.

The production of colours on large area metal sheets so as to also reflect the NIR has been known for some time. It relies on simple interference layers, usually oxides, where the thin film absorption maximum is located at appropriate visible wavelengths for the desired colour. The position of this maximum is tuned simply by varying oxide thickness, and with an appropriate metal substrate, it can be a strong NIR reflector. To achieve this capability in the convenience of paint, the pigment itself is a metallic flake, such as aluminium, and is coated with a thin oxide layer for interference on each flake. The layer thickness must be identical on all flakes for a uniform colour. Coatings with such precision on flake batches to over 200 kgs of metal can be achieved on an industrial scale using either sol-gel or hot fluidised bed chemical vapour deposition. Coating thicknesses are typically between 20–150 nm.

A detailed treatment of this type of paint for energy efficiency applications shows a number of attractive colours with good solar reflectance, which are possible together with using iron oxide as the interference layer on an Al flake [5,6]. These papers also include a schematic of the paint layers, showing how the flakes are oriented. The flake layering details have a strong visual and thermal impact. Colours currently available include blue, grey, silver, orange, red, gold, and purple. Solar absorptance varies between 0.7–0.4. These values were higher than that expected from a simple thin film theoretical analysis of individual flakes in the binder where A_{sol} values below 0.4 were predicted for most colours. The production paints still gave much superior thermal rejection than conventional paints of the same colour in outdoor tests. The discrepancy between theory and experiment was traced to the distribution, gaps and orientation in the flakes, and associated multiple reflections and light trapping.

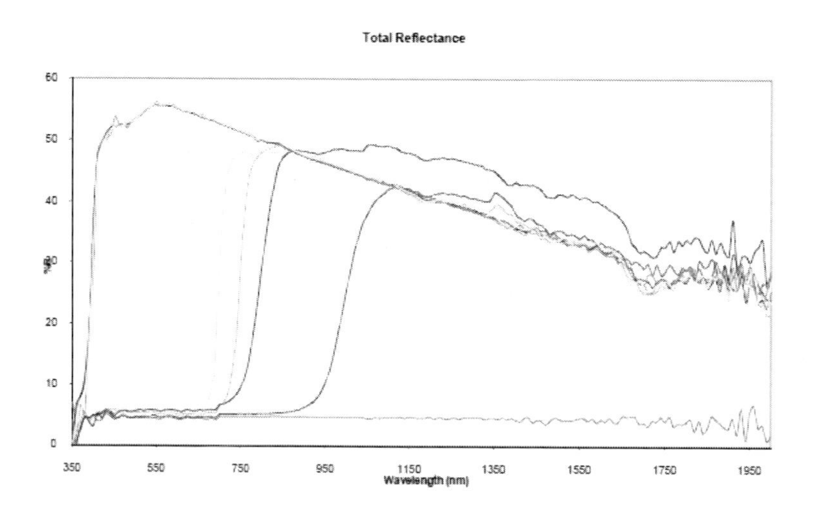

Figure 16. Spectral reflectance of a NIR reflecting painted surface on metal. It is visibly black, and is compared with a standard black (bottom curve) and the binder only with no additive. A binder which is more NIR transmitting would allow the same degree of "blackness" and solar reflectance over 40%.

Improvements may be possible if a simple way of improving paint application to limit the extent of flake miss-orientation could be found, but this slight miss-orientation can add visual appeal via slight texturing.

A final point on these systems concerns the enhanced decorative options which retain solar reflectance when a double layer coating is put on each flake. A study was also presented of a thin SiO_2 film as the first layer, over-coated with a thin Fe_2O_3 layer. Whereas the single layer flakes have little variation in perceived colour with viewing angle, the double layer flakes give a distinct colour shift as viewing angle varies relative to the plane of the coating. Various colour shifts with viewing direction are possible, for instance, from pink to green and aqua. Paints are also available that give such colour shifts, but do not reflect the NIR. These are also flakes with thin film coatings, but since the flakes themselves are not metallic (for instance mica), the desired thermal benefiting NIR effect is lost.

5.2. Radiative Cooling Paint and Pigmented Plastic Layers

The final example of spectral selectivity, we treat in detail, is of interest for cooling of buildings and for harvesting water from the atmosphere. These are surfaces which have high solar reflectivity and low thermal IR reflectivity – the latter one means high thermal emittance. In particular, they must have low reflectivity in the wavelength range of the atmospheric window between 8–13 μm. Moderately low reflectivity across the thermal IR is also usually common in most paint systems, though ideally the black body radiation outside the atmospheric window should be reflected or transmitted by the pigment, and reflected by the substrate [40,41].

A paint, which is specially designed for roofs containing several additives to provide both the special texture (which enhances strong solar reflection) and high thermal emittance, has recently been tested in our laboratories, and used on several Australian buildings. It contains TiO_2, other inorganic particles and binders. Work on related coatings, which contain TiO_2 as the main ingredient, goes back many years [39]. Related work on ZnS pigments for use in polymers has also been reported [40,41] and trialled for atmospheric water collection by radiative cooling. The coating we have studied gave a number of intriguing and useful results in outdoor tests [70] and some quite astounding reductions in HVAC cooling energy requirements, when used on some Australian supermarkets. Such systems work in the day by minimising solar gain through the roof while simultaneously radiating to the sky. They are ideal for locations where clear sky is common, and humidity is not excessive. Heat from inside can build up under the roof even if gains from incident sun are small, and can then be partly radiated away. The paint we studied had a hemispherical solar reflectance of 0.85 and a hemispherical thermal emittance of (0.95±0.02). At night, large cooling is possible with these coatings, the benefits of which can carry over well into the next day, especially if daytime solar gains are low.

The key feature we found was that on a clear sunny day it was still possible to achieve a small but net cooling from the roof, while at night cooling to ~2 °C below ambient was achieved in the space under the small test roof space. Identical spaces covered with more conventional roof integrated whitish coatings, and sold as "helping to cool", performed much worse. Both systems had a low level of insulation under the metal roof (R = 1.5 W^{-1} m^2 °C). The special coating under a solar flux of 1062 Wm^{-2} reflected 894 Wm^{-2}, the roof temperature equilibrated at 32 °C, and the space under the roof at 27.5 °C. In comparison, the standard whitish metal roof reflected 542 Wm^{-2}, the roof equilibrated at around 60 °C and the air under the roof at around 40 °C.

The actual supermarket cooling energy savings exceeded 50% over one year and considerably exceeded the savings predicted by a rigorous computer simulation, which at first made us to think our materials optical and emittance data were wrong. We believe now that the ability of the coating to keep the roof cool also impacted strongly on the performance of the roof mounted heat exchangers, since they were drawing in cooler air. Hence the coefficient of performance of the whole HVAC system was elevated considerably. Another, at first glance surprising result, came when we compared this coating in the daytime on an uninsulated roof in a small test unit, with one with R = 1.5 insulation underneath. In the insulated case, the roof was slightly cooler. Normally, over an insulation layer, solar irradiated metal roofs get much hotter, so they can send more heat up than down. This unexpected result occurred only under a clear sky. Given it is unusual, it was checked a number of times with different sensors, which were also recalibrated. The observation was confirmed. This is only possible when the net flow of heat from the roof is larger than that absorbed in both cases (insulated and not insulated). The uninsulated roof can extract more heat from below, and hence, stays warmer.

At night, the interior space under the roof cooled to a lower temperature than ambient by between 0.5–1 °C with insulation, and between 2–3 °C without uninsulation. This ability can be adapted to collect water from the atmosphere by giving much longer periods for the material below the dew point. Even better cooling performance can be achieved with attention to other thermal flows from the environment. ZnS pigments in polymer have been successfully tested for water collection. Ultimately we need pigments or materials that selectively radiate at the atmospheric window while being transparent or reflective over the rest of the thermal IR [42]. Then very substantial cooling and water collection can be possible.

CONCLUSION

It is becoming increasingly recognised that the visual appearance of energy saving technologies is often just as important as the amount of energy they save. Pleasing aesthetics is essential in achieving major market impacts. The core goal of energy savings follows if that capability co-exists. Daylighting and the popularity of skylights are longstanding examples of the appeal factor, but until recently, most skylights gave thermal energy problems. As pointed out by Kurtz recently [71], off-putting visual impact is a common weakness when new energy saving technologies first come to market, both in lighting and in solar energy. Solar energy technology has often been held back by poor aesthetics, although major improvements in recent years are helping its growth. Paints now exist that provide solar gain or rejection as required, while preserving full decorative options, and in some cases, with the ability to cool or collect water.

Attention to both design and to how new optical science can help to achieve energy savings in an attractive package is now overcoming these obstacles, as demonstrated in the examples given in this discussion. Lighting has additional problems. Glare and better light distribution, without further energy loss and reduced appeal, is the key challenge. A scientific approach to control glare without energy loss has been demonstrated using special TRIMM additives in clear polymer. This, in turn, led to novel decorative options integrated with the energy savings, such as colour control, dynamics, and luminance homogeneity. It also gave a new approach to efficient white light lamps using LEDs. Appearance and light distribution control, combined with energy saving, involves complex optical and materials science. In this chapter, many examples have been given where both issues are helped by new materials and optical developments over recent years. Advances in polymer and nanostructure optics underpin almost all of these advances, and promise many more such opportunities in the future, given the large current activity in both fields.

A final example of optical multi-functionality for energy efficiency and aesthetics: as an example, how far we might go in this quest to make the most of our every day building elements and their inter-play with the resources in the environment, consider a window which we engineer to do all of the following at once: block UV, photocatalytically using solar UV clean intake air or organic deposits, self clean with rain or light spray, provide a view, provide plenty of well distributed and comfortable daylight, give good interior thermal control, appear visually interesting from the outside, and finally use the solar NIR radiation to do one of (i) provide PV electricity, (ii) provide useful heat within a suitable fluid, and (iii) desalinate water. Having an optically clear, clean window is also an important aesthetic issue. This ambitious goal is quite feasible using known physics and chemistry, especially with further developments in nano-systems, to create the requisite materials, surfaces, and systems.

ACKNOWLEDGMENTS

My colleague Jim Franklin and recent research students Christine Deller, Alan Earp, Stefan Schelm, Angus Gentle, and Swedish (Uppsala) student Jacob Jonsson, contributed much to the work described. Skilled technical and imaging support from Geoff McCredie and Ric Wuhrer is much appreciated. A number of companies supported or had input into different aspects of this work including BASF (Ludwigshafen), Poly Optics Pty Ltd (Burleigh Heads, Qld), Skydome Industries Pty Ltd (Sydney), Solutia (Springfield, US), Lehmann Pacific (Sydney) and Evonik (Darmstadt) plus CSIRO Industrial Physics labs. Much of the research described was supported in part by various Australian Research Council grants.

ABOUT THE AUTHOR

Geoff Smith is a Professor of Applied Physics at the University of Technology (Sydney, Australia), and currently leads research programs in Physics at the Institute of Nanoscale Technology. He has worked on solar energy and energy efficiency using advanced optical materials for thirty years. His recent focus has been on windows, skylights, lighting, daylighting and thermal control, using novel polymer composites, polymer light guides, and nanostructured thin films. Recent technologies he has helped to develop include solar driven fluorescence for light piping to interiors, nanoparticle doping for transparent solar control glazing, energy efficient RGB light mixers, and one-step polymer fibers which either integrate efficient transport and controlled side emission, or produce uniform caustic free output from one end. Theoretical and modelling studies of complex composite media have featured strongly in this work. He is a Renewable Energy Pioneer (World Renewable Energy Network), has a Doctoral degree (honoris causa) from the University of Uppsala (Sweden), and is a Fellow of the Australian Institute of Physics. He has over 180 reviewed publications, a book, 4 book chapters and a number of patents.

REFERENCES

[1] Balcomb, D., *Proc. PLEA '98*, 1998, 33–37.
[2] Fontoynont, M. *Solar Energy* 2002, 73, 83–94.
[3] Mardaljevic, J., Rylatt M. *Energy and Buildings* 2003, 35, 27–35.
[4] Ward Larson, G., Shakespeare, R. Rendering with radiance: the art and science of lighting visualisation, Morgan Kaufmann, San Francisco, CA, 1998.
[5] Smith, G. B., Gentle, A., Swift, P., Earp, A., Mronga, N. *J. Solar Energy Materials and Solar Cells* 2003, 79, 163–177.
[6] Smith, G. B., Gentle, A., Swift, P., Earp, A., Mronga, N. *J. Solar Energy Mat. and Solar Cells* 2003, 79, 179–197.
[7] Granqvist, C. G.; Energy-efficient windows: present and forthcoming technology, In: *Material Science for Solar Energy Conversion Systems,* Pergamon, Oxford, UK, 1991.
[8] Smith, G. B.; Dligatch, S.; Sullivan, R.;. Hutchins, M. G. *Solar Energy* 1998, 62, 229–244.
[9] Weber, M. F.; Stover, C. A.; Gilbert, L. R.; Nevitt, T. J.; Ouderkirk, A. J. *Science* 2000, 287, 2451–2456.
[10] Granqvist, C. G.; *Handbook of Inorganic Electrochromic Materials*; Elsevier, Amsterdam, The Netherlands, 1995.
[11] Wittwer, V.; Datz, M.; Ell, J.; Georg, A.; Graf, W.; Walze, G. *J. Solar Energy Mat. and Solar Cells* 2004, 84, 1–4, 305–314.
[12] Takabayashi, T.; Ozeki, Y.; Tanabe S. *J. Soc. Automotive Engg.* 2003, 2791, 520–525.
[13] Roos, A.; Polato, P.; Van Nijnatten, P. A.; Hutchins, M. G.; Olive, F.; Anderson, C. *Solar Energy* 2000, 69 (Supplement), 15–26.
[14] Karlsson, J.; Rubin, M.; Roos, A. *Solar Energy* 2001, 71, 23–32.
[15] Apian-Bennewitz, P.; *Proc. SPIE* 1999, 2255, 697–706.
[16] Andersen, A.; Rubin, M.; Scartezzini, J.-L. *Solar Energy* 2003, 74, 157–173.
[17] Eriksson, T. S.; Granqvist, C. G. *J. Appl. Phys.* 1986, 60, 2081–2091.
[18] Hamberg, I.; Granqvist, C. G. *J. Appl. Phys.* 1986, 60, R123–159.
[19] Cochrane, G.; Zheng, Z.; Smith, G. B. *Proc. SPIE* 1993, 2017, 68–74.
[20] Smith, G. B.; Deller, C. A.; Swift, P. D.; Gentle, A.; Garrett, P. D.; Fisher, W. K. *J. Nanoparticle Res.* 2002, 4, 157–165.
[21] Schelm, S.; Smith, G. B. *Appl. Phys. Lett.* 2003, 82, 4346–4348.
[22] Schelm, S.; Smith, G. B.; Fisher, W. K.; Garrett, P. D. *J. Appl. Phys.* 2005, 97, 124314-1–8.
[23] Alfrey, T.; Gurnee, E. F.; Schrenk, W. J. *Polymer Engg. Sci.* 1969, 9, 400.
[24] Smith, G. B.; Swift, P. D.; Bendavid, A. *App. Phys. Lett.* 75, 630–632 (1999).
[25] Andersson, K. E.; Wahlstrom, M. K.; Roos, A. *Thin Solid Films* 1992, 214, 2132–218.
[26] Ginley, D. S.; Bright, C. *MRS Bulletin* 2000, 25, 15–21.
[27] Jahan, F.; Smith, G. B. *Thin Solid Films* 1998, 333, 185–190.
[28] Le Bellac, D.; Niklasson, G. A.; Granqvist, C. G. *J. Appl. Phys.* 1995, 77, 6145–6150.
[29] Gombert, A.; Blasi, B.; Buhler, C.; Nitz, P.; Mick, J.; Hossfeld, W.; Niggemann, M. *Opt. Engg.* 2004, 43, 2525–2533.
[30] Blasi, B.; Buhler, C.; Georg, A.; Gombert, A.; Hossfeld, W.; Nitz, P.; Walze, G.; Wittwer, V. *Microstructured surfaces in architectural glazings* (preprint).

[31] Reppel, J.; Edmonds, I. R. *Solar Energy* 1998, 62, 245–253.
[32] Hui, S. C. M.; Muller, H. F. O. *Architectural Sci. Rev.* 2001, 44, 221–226.
[33] Zhang, Q. C.; Mills, D. R. *App. Phys. Lett.* 1992, 60, 545–547.
[34] Zajac, G.; Smith, G. B.; Ignatiev, A. *J. Appl. Phys.* 1980, 51, 5544–5554.
[35] Andersson, A.; Hunderi, O.; Granqvist, C. G. *J. Appl. Phys.* 1980, 51, 754–763.
[36] Smith, G. B.; Sabine, T. M. *J. Australian Ceram. Soc.* 1978, 14, 4–8.
[37] Orel, Z. C.; Gunde, M. K. *Solar Energy Mat. and Solar Cells* 2000, 61, 445–450.
[38] Orel, Z. C.; Gunde, M. K. *Solar Energy Mat. and Solar Cells* 2001, 68 3–4, 337–353.
[39] Addeo, A.; Nicolais, L.; Romeo, G.; Bartoli, B.; Coluzzi, B.; Silverstrini, V. *Solar Energy* 1980, 24, 93–98.
[40] Nilsson, T. M. J.; Niklasson, G. A.; Granqvist, C. G. *Solar Energy Mat. and Solar Cells* 1992, 28, 175–193.
[41] Nilsson T. M. J., Niklasson G. A. *Solar Energy Mat. and Solar Cells* 1995, 37, 93–118.
[42] Eriksson, T. S.; Lushiku, E. M.; Granqvist, C. G. *Proc. SPIE* 1983, 428, 105–111.
[43] Smith, G. B.; Jonsson, J.; Franklin, J. *Appl. Opt.* 2003, 42, 3981–3991.
[44] Jonsson, J. C.; Smith, G. B.; Niklasson, G. A. *Opt. Commun.* 2004, 240, 9–17.
[45] Horibe, A.; Baba, M.; Nihei, E.; Koike, Y. *IEICE Trans. Electron.* 1998, E81-C,1697–702.
[46] Edmonds, I. R.; Moore, G. I.; Smith, G. B.; Swift, P. D. *Lighting Res.Tech.* 1995, 27, 27–35.
[47] Deller, C.; Smith, G.; Franklin, J. *Opt. Exp.* 2004, 15, 3327–3333.
[48] Deller, C.; Smith, G. B.; Franklin, J.; Joseph, E. *Proc. Aust. Institute of Physics National Congress* 2002, 307–309 (CD Causal Productions, Adelaide). Ph.D. dissertation Deller, C. (UTS library Online), 2005.
[49] Deller, C. A.; Smith, G. B.; Franklin, J. B. *Proc. SPIE* 2004, 5530, 231–240.
[50] Earp, A. A.; Smith, G. B.; Swift, P. D.; Franklin, J. *Solar Energy* 2004, 76, 655–667.
[51] Earp, A. A.; Smith, G. B.; Franklin, J.; Swift, P. D. *Solar Energy Mat. and Solar Cells* 2004, 84, 411–426.
[52] Stuart, H. R.; Hall, D. G. *Appl. Phys. Lett.* 1996, 69, 2327–2329.
[53] Schelm, S.; Smith, G. B. *J. Opt. Soc. Am.* A. 2005, 22, 1288–1292.
[54] Maxwell Garnett, J. C. *Philos. Trans. R.. Soc.* 1904, 203, 385–420 and 1906, 205, 237–288.
[55] McPhedran, R. C.; McKenzie, D. R.; *Proc. Roy. Soc. Lon.*, 1978, 359, 45–63.
[56] Reuben, A. J.; Smith, G. B. *Phys. Rev. E* 1998, 58, 1101–1111.
[57] Ung, T.; Liz-Marzan, L. M.; Mulvaney, P. *Colloids and Surfaces A: Physicochemical and Engineering Aspects* 2002, 202, 119–126.
[58] Prodan, E.; Radloff, C.; Halas, N. J.; Nordlander, P. *Science* 2003, 302, 419–422.
[59] Schelm, S.; Smith, G. B. *J. Physical Chem. B* 2005, 109, 1689–1694.
[60] Schelm, S.; Smith, G. B.; Wei, G.; Vella, A.; Wiecorek, L.; Muller, K. H.; Raguse, B. *Nano Lett.* 2004, 4, 335–339.
[61] Smith, G. B. *Appl. Phys. Lett.* 1979, 35, 668–670.
[62] Vargas, W. E. *J. Opt. Soc. Am. A* 1999, 16, 1362–1372.
[63] Nitz, P.; Ferber, J.; Stangl, R.; Wilson, H. R.; Wittwer, V. *Solar Energy Mat. and Solar Cells* 1998, 54, 297–307.
[64] Deller, C. *Ph.D. Dissertation 2005* (Available Online UTS library).
[65] Swift, P. D.; Smith, G. B; Franklin, *J. Lighting Res. and Tech.* 2006, 38, 19–31.

[66] Deller, C.; Smith, G. B.; Franklin, J.; Joseph, E., *Proc. Aust. Institute of Physics National Cong.* 2002, 307–309.

[67] Berthet, R.; Brun, G.; Zerrouchi, A.; Druetta, M. *Proc. POF* 2001, 361–367.

[68] Jonsson, J. C.; Smith, G. B.; Deller, C.; Roos, A. *Appl. Opt.* 2005, 44, 2745–2753.

[69] Zhao, F.; Narendran, N.; Van Derlofske, J. *Proc. SPIE*, 2002, 4776, 206–214.

[70] Bell, J. M.; Smith, G. B.; Lehman, R. *Proc. SASBE*, 2003 (Paper T606).

[71] Kurtz, S.; *Proc. SPIE* 2004, 5530, 316–325.

In: Advances in Laser and Optics Research. Volume 10
Editor: William T. Arkin
ISBN: 978-1-62257-795-8

Chapter 8

HIGH-SPEED OPTICAL MODULATORS AND PHOTONIC SIDEBAND MANAGEMENT

Tetsuya Kawanishi
National Institute of Information and Communications Technology,
Tokyo, Japan

Abstract

This chapter shows mathematical expressions for various types of electro-optic modulators, such as, intensity modulators, single-sideband modulators, frequency-shift-keying modulators, phase-shift-keying modulators, etc, based on optical phase modulation. We also discuss electrode structures for high-speed operation. Resonant electrodes can enhance modulation efficiency in a particular band. Design schemes for two types of resonant electrodes: asymmetric resonant structure and double-stub structure, are given. Most important application of optical modulators is conversion of electric signals into lightwave signals. This is indispensable to high-speed optical links. However, recently, we need new functions such as optical label processing at nodes, signal timing control to prevent packet collision, etc. In addition to basic function of optical modulators, this chapter describes the photonic sideband management techniques which can manipulate optical or electrical signals in time or frequency domain, by using sideband generation at the modulators.

1. Introduction

In lightwave telecommunication systems, we have to use some interactions with unmodulated lightwave, in order to obtain optical signals having some information. There are two types of interactions. One is the interaction between electric and optical signals. The other is between two or more lightwaves. Recently, many types of all-optical systems using interaction among lightwaves were proposed to enhance processing performance at nodes. However, it does not decrease the importance of the interaction between electric and optical signals, because most of information which can be accessed by human beings is described by electric signals, at this moment. We have to use the interaction between electrical and optical signals at terminals of the lightwave telecommunications system, while all-optical interaction can be used at the nodes. This chapter describes recent results in high-speed optical modulation and its applications. Optical intensity modulators, which are most commonly used in commercial systems, can be constructed by using electro-optic (EO) or electro-absorption (EA) effect. However, recently, other modulation formats, for example, differential phase-shift-keying (DPSK) or frequency-shift-keying (FSK), are used for long-haul system, optical labeling technique, etc. In this chapter, we focus on the EO modulator, which can control the phase and frequency of lightwave, as well as the intensity. In a material having the EO effect, the refractive index (RI) depends on the voltage applied on the material. Thus, in photonic sideband management (PSBM) techniques, we can obtain an optical phase modulator comprised of an optical waveguide and an electrode. The other types of modulators, such as intensity modulator, FSK modulator, can be constructed by optical phase modulators.

Firstly, some mathematical expressions for optical modulators using the EO effect are given. We also discuss performance of various types of modulators. Secondly, we describe the electrode structures for high-frequency operations [1, 2, 3, 4], and the design scheme for resonant electrodes [5, 6, 7]. In this chapter, we also review a new concept of optical sideband modulation techniques, i.e., PSBM, where optical or electrical signals can be manipulated in time or frequency domain by using sideband generation in optical modulators. A couple of PSBM techniques, such as, optical label swapping (OLS) using double-sideband (DSB) modulation [8], reciprocating optical modulation (ROM) for high-order harmonic generation [9, 10, 11, 12, 13, 14], and optical tunable delay [15], are described, where the frequency and time domain profile of the input lightwave is controlled by an electric signal. PSBM techniques enable us to manipulate optical signals by electric control signals, while most of modulators are just to convert the electric signal to the optical one. In addition, basically, PSBM does not use optical oscillation or non-linearities, so that there are no chaotic phenomena, such as mode-hopping, mode-competitions. Thus, we can obtain stable and agile manipulation of the optical signals in frequency and time domains.

2. Optical Modulators Based on Phase Modulation

Mathematical expressions for optical phase modulators are given to describe the basic principle of modulators using the EO effect. We also describe various types of optical modulators consisting of phase modulators: 1) intensity modulator, 2) single-sideband (SSB) modulator [16, 17, 18], 3) FSK modulator [19, 20], and 4) PSK modulator.

2.1. Phase Modulator

We can electronically control optical phase by using EO materials, such as lithium niobate (LN), lithium tantalate (LT), gallium arsenide (GaAs), etc. Fig. 1 shows a schematic of an optical modulator using EO effect. An electrode is placed along with an optical waveguide. By applying electric voltage on the electrode, we can change the RI in the waveguide, so that the phase of the lightwave can be controlled. For simplicity, the input lightwave is assumed to be monochromatic, and be described by $\mathrm{e}^{2\pi \mathrm{i} f_0 t}$. The output lightwave R can be expressed by

$$R = A^{\mathrm{LW}} \mathrm{e}^{2\pi \mathrm{i} f_0 t + \mathrm{i} f(t)} \quad (1)$$

$$f(t) \equiv KV(t) \tag{2}$$

where $V(t)$ and K, respectively, denote the electric voltage on the electrode and the coupling coefficient between the electric signal and the lightwave signal. A^{LW} is the optical transmittance in the waveguide. When $V(t)$ has high-frequency components, the frequency response of K should be taken into account. The coefficient K can be assumed to be a constant if the wavelengths of the high-frequency components on the electrode are much longer than the length of the electrode. The response of EO materials is much faster than the frequency response of K, which is dominated by that of the electrode and by the phase mismatch between the electric and lightwave signals.

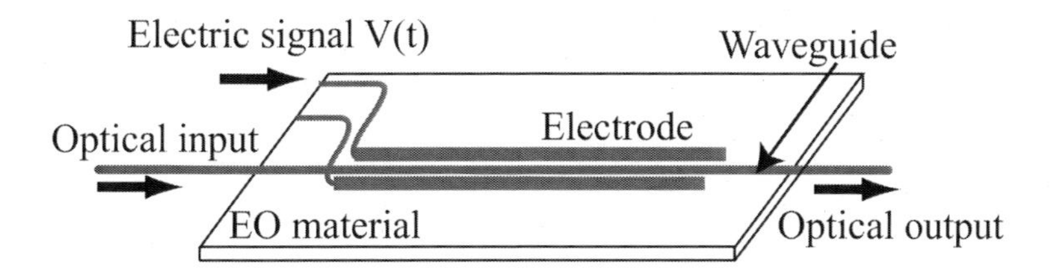

Figure 1. Optical phase modulator.

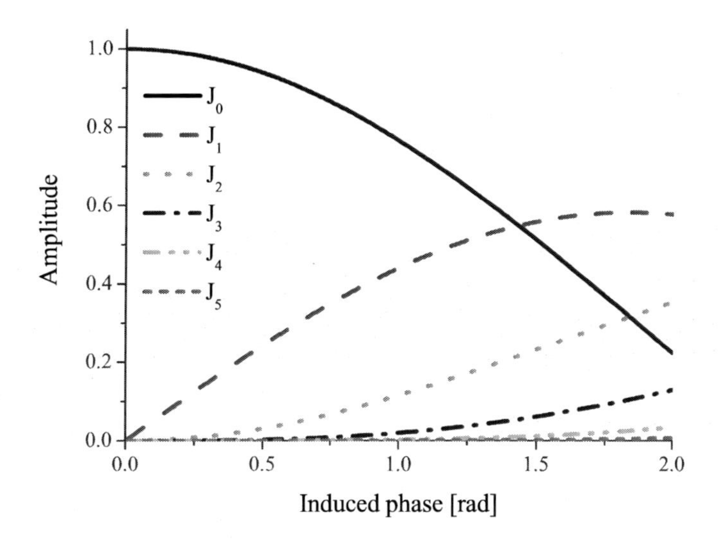

Figure 2. Bessel functions.

Here, we consider that $f(t)$ is a sinusoidal signal described by $A^{\mathrm{RF}} \sin 2\pi f_m t$. The optical output can be expressed by

$$\begin{aligned} R &= A^{\mathrm{LW}} \exp \mathrm{i}[f_0 t + A^{\mathrm{RF}} \sin f_m t] \\ &= A^{\mathrm{LW}} \sum_{n=-\infty}^{\infty} J_n(A^{\mathrm{RF}}) \exp 2\pi \mathrm{i}[f_0 t + n f_m t], \end{aligned} \tag{3}$$

where $J_n(A)$ is the first-kind n-th order Bessel function. $A^{\rm RF}$ is an index for optical phase deviation induced by the electric signal, so that it is called induced phase. When $f(t) = A^{\rm RF}\cos f_m t$, the output is expressed by

$$\begin{aligned} R &= A^{\rm LW}\exp{\rm i}[2\pi f_0 t + A^{\rm RF}\cos 2\pi f_m t] \\ &= A^{\rm LW}\sum_{n=-\infty}^{\infty}{\rm i}^n J_n(A^{\rm RF})\exp 2\pi{\rm i}[f_0 t + n f_m t]. \end{aligned} \tag{4}$$

The spectrum component of $e^{2\pi i(f_0+nf_m)t}$, so-called the n-th order sideband, can be expressed by the n-th order Bessel function, which are shown in fig. 2. The first order function $J_1(A)$ has the maximum point which can be defined by

$$J_1'(A_m) = 0, \quad \left(J_1'(A) \equiv \frac{dJ_1(A)}{dA}, \quad 0 \le A \le \pi\right), \tag{5}$$

where $J_1(A_m) = 0.583, \quad A_m = 1.841$. In most applications of optical modulation techniques, the desired component is the first order sidebands, so that the induced phase A should be less than A_m to achieve effective modulation without generation of undesired high-order sideband components.

2.2. Intensity Modulator

Mach-Zehnder (MZ) structures are commonly used to achieve intensity modulation by EO effects. As shown in fig. 3 the MZ structure consists of two optical phase modulators. We can control the output optical intensity by applying a pair of electric signals to the two phase modulators. Balanced push-pull scheme is ideal for intensity modulation, where the induced phase of the modulator A is $g(t)/2$, and that of the modulator B is $-g(t)/2$. $g(t)$ describes the optical phase difference between the two arms of the MZ structure. We can obtain the balanced push-pull operation by using an x-cut LN substrate and a coplanar waveguide (CPW) structure for electric signal guiding, as shown in fig. 4, where an electric signal is fed to the CPW. Fig. 5 shows the principle of operation. When the lightwaves in the two arms are in phase $g(t) = 2n\pi \quad n = \cdots, -1, 0, +1, \cdots$, the modulator is in "on" state. At the junction of the output, lightwaves interfere constructively. On the other hand, when lightwaves are in opposite phase $g(t) = (2n+1)\pi$, they are converted into high-order radiative mode waves which cannot propagate through the waveguide for output. Thus, we turn the modulator "on" and "off" by changing the voltage on the electrode. The intensity can be expressed by

$$|R| = \frac{1}{2}\left|\left(e^{ig(t)/2} + e^{-ig(t)/2}\right)e^{2\pi i f_0 t}\right| \tag{6}$$

$$= \left|\cos\left[g(t)/2\right]\right|. \tag{7}$$

We call the MZ structure for intensity modulation as the MZ modulator (MZM). The output intensity ratio between the "on"-state and the "off"-state, called the extinction ratio, is an important index of the MZM. In the ideal case, described by eq. (7), the extinction ratio goes to infinity. However, there is some residual lightwave even in the "off" state, due to amplitude imbalance in the arms, and crosstalk between the high-order and waveguide modes, as shown in fig. 6. Typical extinction ratios of commercially available LN modulators are 20-30 dB. Recently, we proposed a novel technique for compensation of the amplitude imbalance due to fabrication errors [21]. By using the trimmers, we can compensate the amplitude imbalance due to fabrication errors. Fig. 7 shows the on-off switching performance of the LN MZM with trimmers. The extinction ratio was 50 dB, while that of conventional technique was 37 dB.

When $g(t)$ is a sinusoidal signal described by $2A^{\mathrm{RF}}\sin 2\pi f_m t + \phi_{\mathrm{B}}$, the optical output can be expressed by

$$R = \frac{1}{2}A^{\mathrm{LW}}\mathrm{e}^{2\pi\mathrm{i}f_0 t}\left[\mathrm{e}^{\mathrm{i}(A^{\mathrm{RF}}\sin 2\pi f_m t+\phi_{\mathrm{B}}/2)} + \mathrm{e}^{-\mathrm{i}(A^{\mathrm{RF}}\sin 2\pi f_m t+\phi_{\mathrm{B}}/2)}\right] \quad (8)$$

$$= \frac{1}{2}A^{\mathrm{LW}}\mathrm{e}^{2\pi\mathrm{i}f_0 t}\sum_{n=-\infty}^{\infty} J_n(A^{\mathrm{RF}})\mathrm{e}^{2\pi\mathrm{i}[f_0 t+nf_m t]}\left[\mathrm{e}^{\mathrm{i}\phi_{\mathrm{B}}/2} + (-1)^n\mathrm{e}^{-\mathrm{i}\phi_{\mathrm{B}}/2}\right] \quad (9)$$

$$= A^{\mathrm{LW}}\mathrm{e}^{2\pi\mathrm{i}f_0 t}\Big[\cos\frac{\phi_{\mathrm{B}}}{2}\sum_{n=-\infty}^{\infty} J_{2n}(A^{\mathrm{RF}})\mathrm{e}^{2\pi\mathrm{i}[f_0 t+2nf_m t]}$$
$$+\mathrm{i}\sin\frac{\phi_{\mathrm{B}}}{2}\sum_{n=-\infty}^{\infty} J_{2n+1}(A^{\mathrm{RF}})\mathrm{e}^{2\pi\mathrm{i}[f_0 t+(2n+1)f_m t]}\Big]. \quad (10)$$

The output optical intensity $|R|^2$, which can be detected by a high-speed photodetector, is expressed by

$$|R|^2 \simeq |A^{\mathrm{LW}}|^2\Big[J_0^2(A^{\mathrm{RF}})\cos^2\frac{\phi_{\mathrm{B}}}{2} + 2J_1^2(A^{\mathrm{RF}})\sin^2\frac{\phi_{\mathrm{B}}}{2}$$
$$-4J_0(A^{\mathrm{RF}})J_1(A^{\mathrm{RF}})\sin\frac{\phi_{\mathrm{B}}}{2}\cos\frac{\phi_{\mathrm{B}}}{2}\sin 2\pi f_m t$$
$$+2\left\{2J_0(A^{\mathrm{RF}})J_2(A^{\mathrm{RF}})\cos^2\frac{\phi_{\mathrm{B}}}{2} - J_1^2(A^{\mathrm{RF}})\sin^2\frac{\phi_{\mathrm{B}}}{2}\right\}\cos(2\cdot 2\pi f_m t)\Big], \quad (11)$$

where we assumed that $A^{\mathrm{RF}} << 1$, and the high-order components are neglected. By using Taylor's expansion of Bessel function, we get a simple equation,

$$|R|^2/|A^{\mathrm{LW}}|^2 = \frac{1}{2} + \frac{1-|A^{\mathrm{RF}}|^2}{2}\cos\phi_{\mathrm{B}} - A^{\mathrm{RF}}\sin\phi_{\mathrm{B}}\sin 2\pi f_m t$$
$$+\frac{1}{2}|A^{\mathrm{RF}}|^2\cos\phi_{\mathrm{B}}\cos(2\cdot 2\pi f_m t) \quad (12)$$

which is useful to discuss performance of optical systems for rf-signal transmission where the photodetector generates rf-signals. The intensities of the fundamental component $\sin(2\pi f_m t)$ and the second order harmonic $\cos(2\cdot 2\pi f_m t)$ can be controlled by the dc-bias ϕ_{B}. The fundamental and second order components are proportional to $\sin\phi_{\mathrm{B}}$ and $\cos\phi_{\mathrm{B}}$, respectively. The ratio between the average power and the rf-signal component dominates conversion efficiency from lightwaves to rf-signals at the photodetector. The ratios for the fundamental and second order components are, respectively,

$$D_1 = \left|\frac{2A^{\mathrm{RF}}\sin\phi_{\mathrm{B}}}{1+(1-|A^{\mathrm{RF}}|^2)\cos\phi_{\mathrm{B}}}\right| \quad (13)$$

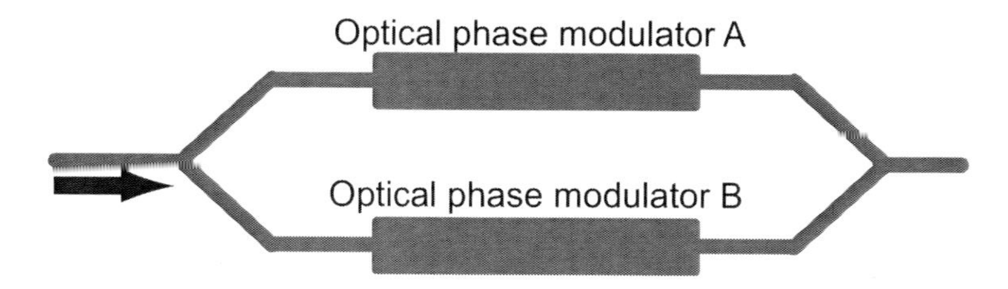

Figure 3. MZ structure

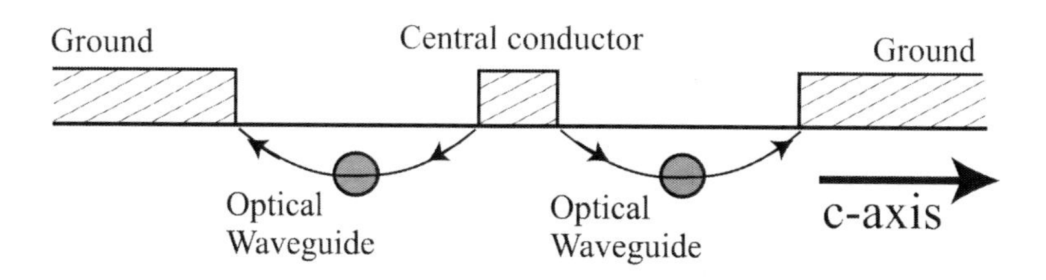

Figure 4. Cross section of balanced MZ structure for push-pull operation.

$$D_2 = \left| \frac{|A^{\mathrm{RF}}|^2 \cos\phi_{\mathrm{B}}}{1 + (1 - |A^{\mathrm{RF}}|^2)\cos\phi_{\mathrm{B}}} \right|. \tag{14}$$

In the case of $\phi_{\mathrm{B}} = \pi$, the even order components in the output R, including the carrier component $\mathrm{e}^{2\pi \mathrm{i} f_0 t}$, become zero and the average power $|R|^2$ goes to a minimum of $\frac{|A^{\mathrm{RF}}|^2}{2}$, where the dominant components are the first order upper and lower sidebands. This modulation scheme, called double-sideband suppressed carrier (DSB-SC) modulation, is suitable for rf-signal transmission. This is because undesired dispersion effects in optical fibers can be suppressed, and the ratio between the average power and the second order rf-signal D_2 becomes a maximum. In addition, DSB-SC has an impotant role in PSBM to manipulate lightwave signals in frequency domain, as described in later part of this chapter. The carrier suppression ratio depends on the extinction ratio of the MZM, so that the carrier can be also suppressed largely by using the intensity trimmers for high-extinction ratio [21]. Fig. 8 shows optical spectra of DSB-SC signals with and without imbalance compensation in the LN MZM with the trimmers, where the modulation frequency f_m was 10.5 GHz.

2.3. SSB Modulator

Fig. 9 shows a schematic of an optical SSB modulator consisting of parallel four optical phase modulators [16, 17]. The SSB modulator has two electrodes $\mathrm{RF_A}$ and $\mathrm{RF_B}$ for high-speed signals, where a pair of rf-signals of the same frequency f_m are fed to the electrodes. The electric field of the output can be expressed by

$$R = \frac{\mathrm{e}^{2\pi \mathrm{i} f_0 t}}{4} \sum_{j=1}^{4} \sum_{n=-\infty}^{\infty} J_n(A_j^{\mathrm{RF}}) \mathrm{e}^{\mathrm{i}n(2\pi f_m t + \phi_j^{\mathrm{RF}})} \times A_j^{\mathrm{LW}} \mathrm{e}^{\mathrm{i}\phi_j^{\mathrm{LW}}} \tag{15}$$

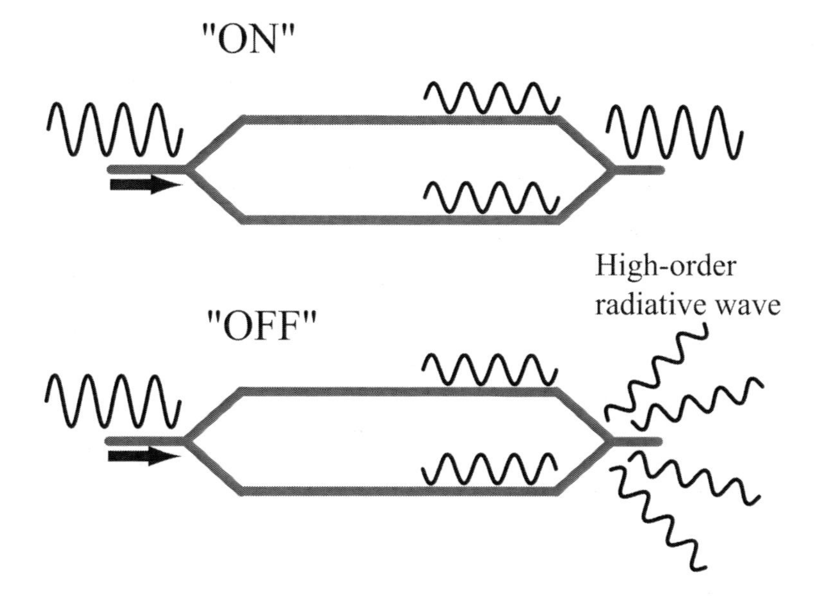

Figure 5. Principle of intensity modulation.

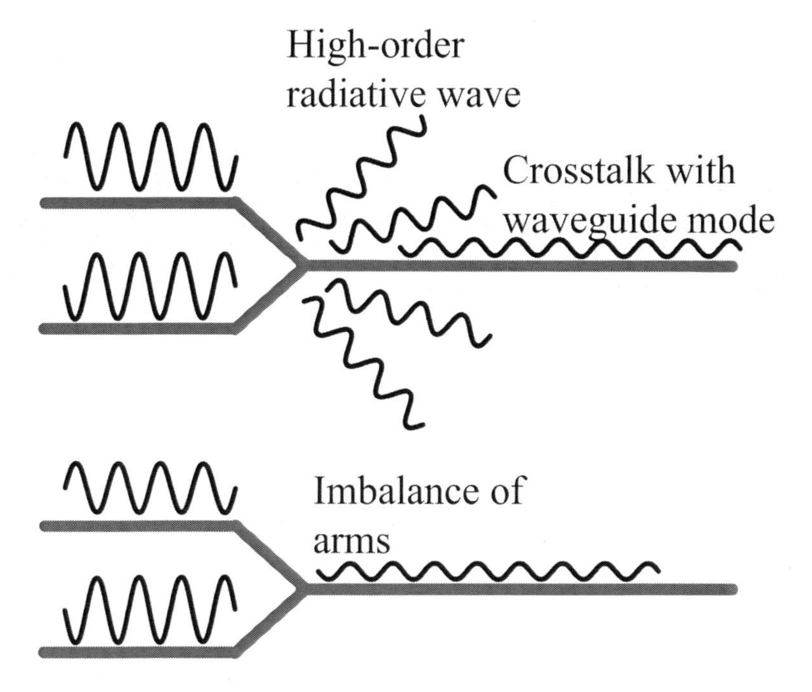

Figure 6. Degradiation of extinction ratio.

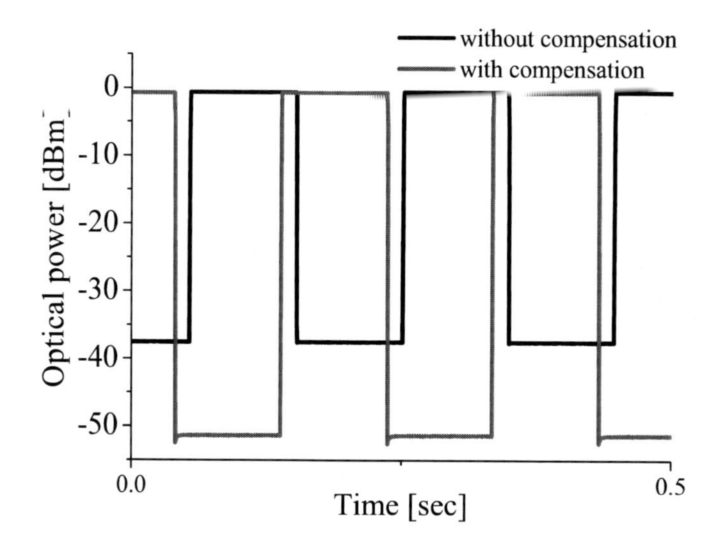

Figure 7. High-extinction ratio on-off switching with compensation of imbalance in MZM using intensity trimmers.

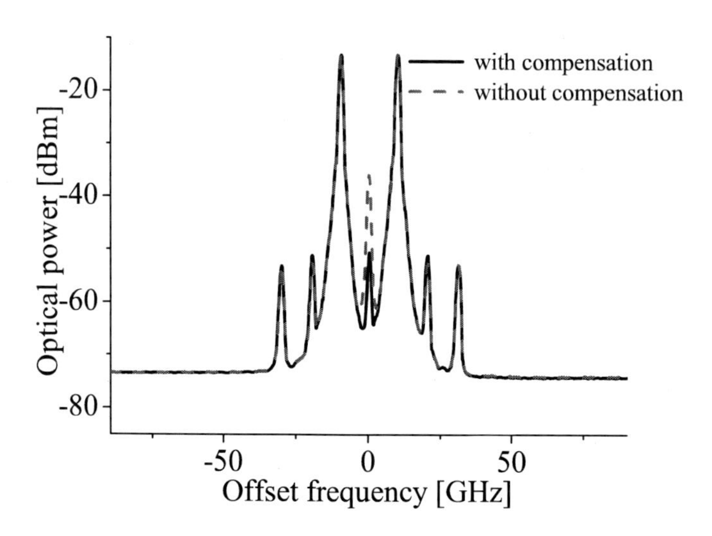

Figure 8. High carrier suppression in DSB-SC using an MZM with intensity trimmers.

$$= \frac{e^{2\pi i f_0 t}}{4} \sum_{n=-\infty}^{\infty} e^{2\pi i n f_m t} \sum_{j=1}^{4} J_n(A_j^{RF}) P_{n,j} A_j^{LW}, \tag{16}$$

where

$$P_{n,j} \equiv \exp i \left[(T - Sn) j \frac{\pi}{2} + \Delta\phi_j^{LW} + n\Delta\phi_j^{RF} \right] \tag{17}$$

$$\phi_j^{LW} = T\frac{j\pi}{2} + \Delta\phi_j^{LW} \tag{18}$$

$$\phi_j^{RF} = -S\frac{j\pi}{2} + \Delta\phi_j^{RF} \tag{19}$$

$$S = \pm 1, \quad T = \pm 1. \tag{20}$$

ϕ_j^{RF} and ϕ_j^{LW}, respectively, denote the phases of rf-signal and lightwave in path j. $\Delta\phi_j^{RF}$ and $\Delta\phi_j^{LW}$ are the deviations of the phases from the ideal condition for the SSB modulation. A_j^{RF} is the induced optical phase due to the rf-signal in path j, and is proportional to the amplitude of the rf-signal applied to the electrode. A_j^{LW} denotes the amplitude of lightwave in path j. f_0 and f_m are the frequencies of input lightwave and rf-signal, respectively. When $\Delta\phi_j^{RF} = \Delta\phi_j^{LW} = 0$, the phases are 0, $\pi/2$, π and $3\pi/2$ (0°, 90°, 180° and 270°). The SSB modulator has a pair of Mach-Zehnder structures on an x-cut LN substrate with a CPW, so that we can apply rf-signals of 0°, 90°, 180° and 270° by feeding a pair of rf-signals with 90° phase difference at two rf-ports (RF_A, RF_B) [17]. The optical phase differences are also set to be 90°, by using dc-bias ports (DC_A, DC_A, DC_A). The output optical spectrum, defined by eq. (16), can be expressed by

$$R = A^{LW} e^{2\pi i f_0 t} \sum_{n=0}^{\infty} J_N \left(A^{RF} \right) e^{2\pi i N f_m t} \tag{21}$$

$$\begin{aligned} &= A^{LW} U e^{2\pi i f_0 t} \Big[J_1(A^{RF}) e^{2\pi i U f_m t} \\ &\quad - J_3(A^{RF}) e^{-3\cdot 2\pi i U f_m t} + J_5(A^{RF}) e^{5\cdot 2\pi i U f_m t} \\ &\quad - J_7(A^{RF}) e^{-7\cdot 2\pi i U f_m t} + \cdots \Big], \end{aligned} \tag{22}$$

where N and U are defined by

$$N \equiv U \times (2n+1) \times (-1)^n \tag{23}$$

$$U \equiv S \times T \tag{24}$$

When the intensity of the electric field is so small that we can neglect high-order harmonic generation at the optical phase modulation, the output for $U = 1$ can be approximately expressed by

$$R \simeq A^{LW} e^{2\pi i f_0 t} \left[J_1(A^{RF}) e^{2\pi i f_m t} - J_3(A^{RF}) e^{-3\cdot 2\pi i f_m t} \right]. \tag{25}$$

Because $J_1 > J_3$, the dominant component in the output is the first order upper sideband (USB), which corresponds to the frequency shifted component. Fig. 10 shows the principle of the SSB modulation. Each sub Mach-Zehnder structure (the pair of paths 1 and 3, or that of paths 2 and 4) is in null-bias point. Thus, the input lightwave component ($e^{2\pi i f_0 t}$) is vanished, and lower and upper sidebands are generated. Due to the 90° phase differences in the rf-signal and the lightwave, the polarity of lower sideband (LSB) component at point "P" is the opposite of "Q". The LSB would be vanished at the output port, so that the SSB modulated signal consisting of USB can be obtained. On the other hand, in the case of $U = -1$, LSB can be generated instead of USB. We can easily switch the polarity of T, by changing the dc-bias voltage applied on the port DC_C. The signal-to-noise-ratio (SNR) and conversion efficiency of the frequency shift by the modulator are given by $J_1(A^{RF})/J_3(A^{RF})$ and $J_1(A^{RF})$, respectively (see fig. 11). The conversion efficiency has a maximum of $0.582 \quad (-5.36\,\mathrm{dB})$ when $A^{RF} = A_m$. As shown in eq. (22), the output lightwave has undesired high-order sidenband components whose orders are $-3, +5, -7, \cdots$ for $U = +1$, and $3, -5, +7, \cdots$ for $U = -1$. Thus, the frequency difference between the generated spectral components equals $4f_m$. We note that the dominant high-order component (J_3) can be suppressed by using a predistortion technique [18].

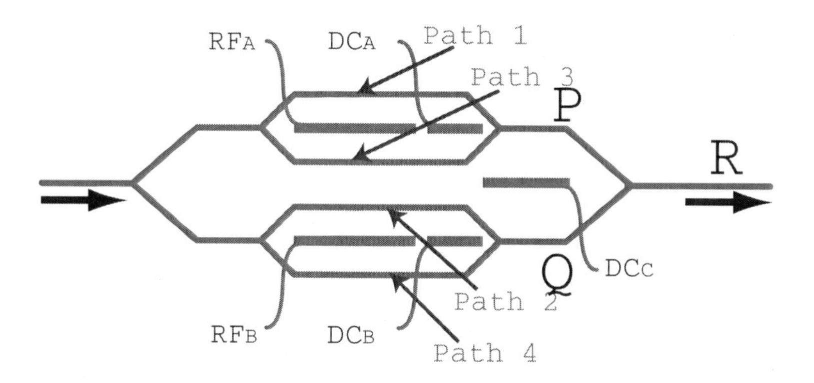

Figure 9. Optical SSB modulator.

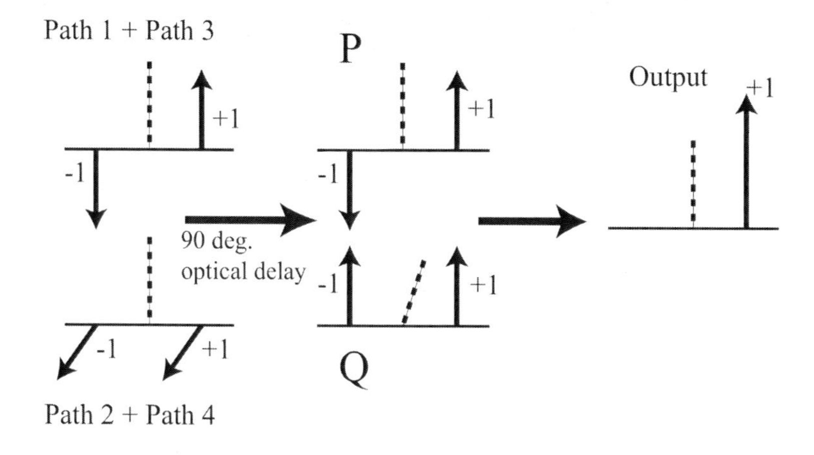

Figure 10. Principle of optical SSB modulation.

We consider the optical frequency shift with the SSB modulator in wavelength-domain-

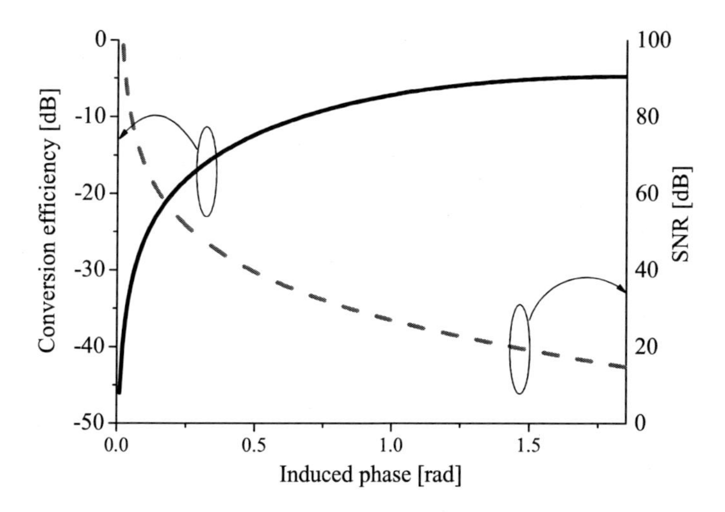

Figure 11. Conversion efficiency and SNR of wavelength shifter using SSB modulation.

multiplexing (WDM) systems with numerical results. Fig. 12 shows a model of a 9-channel WDM system for the simulation with a lightwave network simulator, $Optisystem^{\mathrm{TM}}$ 2.2, where the channel spacing was 25 GHz. The bit rate and the optical frequency of the n-th channel were, respectively, $2.5 \times [(n-5) \times (r/100+1)]$ Gbps and $193.5+(n-5) \times 0.025$ THz, where $r = 2.0$. Thus, the bit rate of the fifth channel was 2.5 Gbps. Those of neighboring channels were 2% higher or lower than that of the fifth, in order to suppress unphysical cross-talks which are due to the finite time windows in this simulation, where the sampling rate and time window in the simulator were 160 Gbps and 40.96 ns, respectively. The non-return-to-zero (NRZ) signals were generated by an LN optical intensity modulator. We used a model for the SSB modulator consisting of four optical phase modulators as shown in fig. 9. The lightwave signals were fed to the SSB modulator via a WDM multiplexer. In order to compensate the loss at the wavelength shift, an EDFA was put at the output lightwave port of the modulator, where the output power was controlled to be 10 dBm by tuning the pump laser power. The noise figure of the EDFA was assumed to be 4 dB. We calculated the optical spectra at the input and the output ports of the SSB modulator with the resolution of 0.01 nm. As shown in fig. 13, optical frequency shift (25 GHz) of WDM channels was successfully demonstrated in this simulation. The undesired third order sideband components were also generated in the output spectra. The frequency difference between the desired frequency shifted component (the first order USB) and the undesired sideband was $4f_m$, so that there were four peaks of them in the spectrum. The other third order components were overlapped with the desired components. In the case of $A^{\mathrm{RF}} = 0.92$, the fifth order sidebands were also generated, in addition to the third order sidebands which were dominant in undesired components. Fig. 14 shows Q-factors of the fifth channel at the electric signal monitor in fig. 12, and the total optical power at the output port of the SSB modulator. The electric output of the photo detector was fed to the monitor via a transimpedance amplifier whose impedance was 600Ω and noise figure was 6 dB. Q-factor of the reference signal, where an attenuator was placed at the output port of the WDM multiplexer instead of the SSB modulator, is also shown in fig. 14. The output power was an increasing function of the induced phase A^{RF}. However, the SNR of the output signal is a decreasing function as shown in fig. 11. Due to this trade-off relations, the Q-factor had a peak around $A^{\mathrm{RF}} = 0.5$.

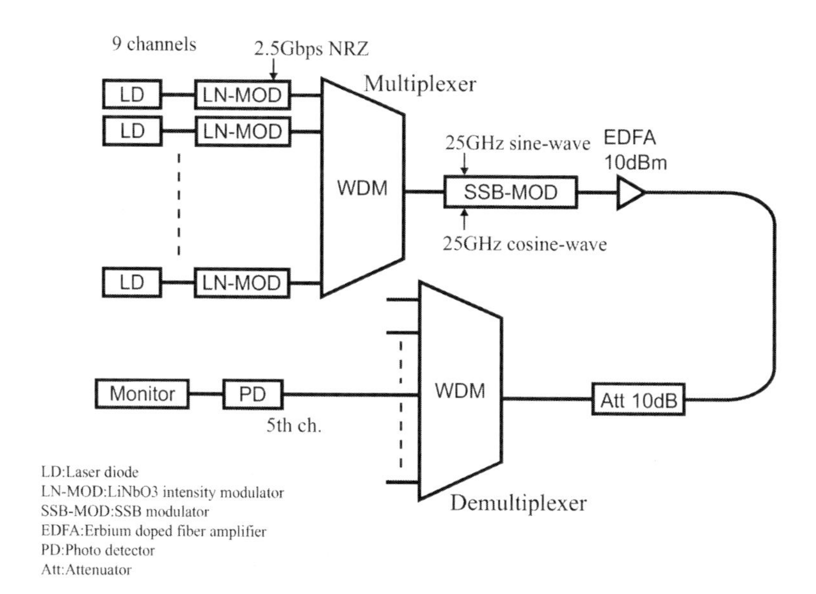

Figure 12. Model for a 2.5 Gbps WDM system with a wavelength shifter using SSB modulation.

2.4. FSK Modulator

The FSK modulator consists of parallel four optical phase modulators as shown in fig. 15 [19, 20]. The device structure is almost the same in the SSB modulator, but the FSK modulator has an electrode ($\mathrm{RF_C}$) for high-speed FSK signal, instead of a dc-bias electrode ($\mathrm{DC_C}$) in the SSB modulator. When we apply a pair of rf-signals, which are of the same frequency f_m and have a 90° phase difference, to the electrodes $\mathrm{RF_A}$ and $\mathrm{RF_B}$, frequency shifted lightwave can be generated at the output port of the modulator. A sub Mach-Zehnder structure of paths 1 and 3 should be in null-bias point (lightwave signals in the paths have 180° phase difference), where the dc-bias can be controlled by $\mathrm{RF_A}$. The other sub Mach-Zehnder structure of paths 2 and 4 are also set to be in null-bias point by using $\mathrm{RF_B}$. To eliminate USB or LSB, the lightwave signal in each path also should have 90° phase difference with each other, as described in Section 2.3.. When the phase difference induced by $\mathrm{RF_C}$ is ±90°, we can get carrier-suppressed SSB modulation comprising one of the sideband components (USB or LSB). Thus, the optical frequency of output lightwave can be switched by changing the induced phase at $\mathrm{RF_C}$. In the SSB modulator, the electrode for optical phase control $\mathrm{DC_C}$ was not designed for high-speed operation, so that the switching time was limited by the response of the electrode. On the other hand, the FSK modulator has the electrode $\mathrm{RF_C}$ for high-speed optical phase switch at the junction of a pair of sub Mach-Zehnder structures. The amplitudes of USB and LSB are, respectively, described by

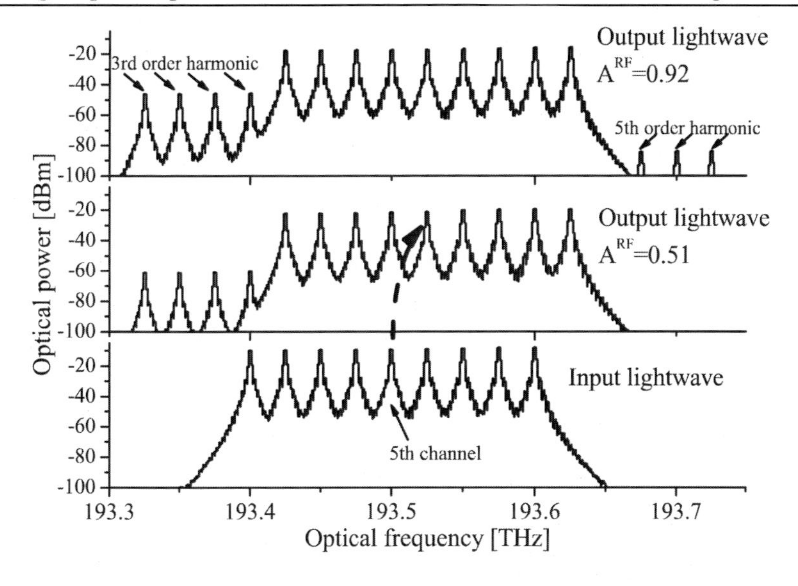

Figure 13. Input and output optical spectra of the SSB modulator.

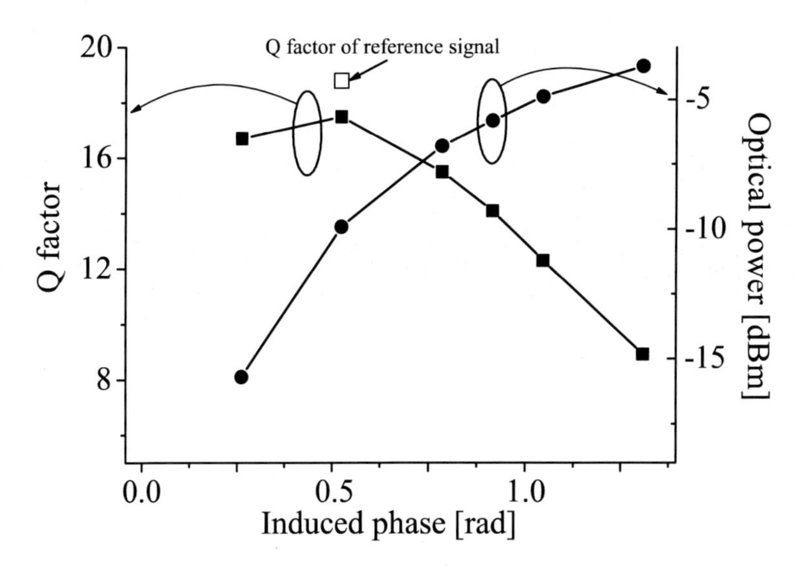

Figure 14. Q-factor of the fifth channel measured by an electric signal monitor, and the total optical power at the output port of the SSB modulator. Q-factor of the reference signal is 18.8, where an attenuator is placed at the output port of the WDM multiplexer instead of the SSB modulator.

$[1 + \mathrm{i}\exp(\mathrm{i}\phi_{\mathrm{FSK}})]/2$ and $[-1 + \mathrm{i}\exp(\mathrm{i}\phi_{\mathrm{FSK}})]/2$, where ϕ_{FSK} is the induced phase difference at $\mathrm{RF_C}$, and $\phi_{\mathrm{FSK}} = -90°$ corresponds to an optimal condition for USB generation ($U = 1$). Thus, by feeding an NRZ signal, whose zero and mark levels correspond to $\phi_{\mathrm{FSK}} = -90, +90°$, to $\mathrm{RF_C}$, we can generate an optical FSK signal, without any parasitic intensity modulation. The bandwidth for the FSK signal should be smaller than the rf-frequency f_m, when the FSK signal is demodulated by optical filters.

An optical FSK modulator having an LN lightwave circuit with traveling wave electrodes to obtain high-speed response, was reported in Ref. [8], where the rise time was about 100 ps, which was limited by driver and receiver circuits, and 3 dB bandwidths of the electrodes ($\mathrm{RF_A}$, $\mathrm{RF_B}$ and $\mathrm{RF_C}$) were 18 GHz, approximately. Thus, the frequency switching time of the optical FSK modulator would be less than 100 ps. We also measured the FSK response by feeding a sinusoidal signal to $\mathrm{RF_C}$. The FSK signal was demodulated into an intensity modulated (IM) signal by an optical filter. We also measured bit-error-ratio (BER) performance of FSK transmission. As shown in fig. 2.4., a 9.95 Gbps NRZ ($2^{31} - 1$ pseudorandom binary sequence:PRBS) signal was applied to $\mathrm{RF_C}$ of the FSK modulator, where a fiber Bragg grating (FBG) was used to convert an FSK signal into an IM signal for demodulation. Fig. 17 shows BER curves and eye-diagrams of 9.95 Gbps FSK for back-to-back, and after transmission through a 60 km single-mode fiber (SMF). The rf frequency f_m was 12.5 GHz. The results show that the eyes are clearly open and that error-free transmission of 10 Gbps FSK-60 km SMF is possible. Power penalties of 60 km transmission with reference to back-to-back at $-\log(BER) = 9$ was -0.8 dB. We deduce that the negative penalty was due to the dispersion of the FBG. In addition, by using balanced detection scheme, we can achieve long-distance transmission and enhanced receiver sensitivity. 10 Gbps FSK transmission over a 130 km SMF was demonstrated by balanced receiver with group delay compensation between USB and LSB components [22].

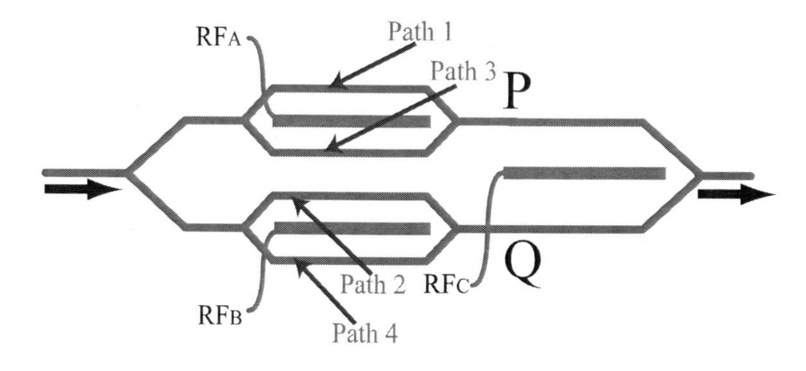

Figure 15. Optical FSK modulator.

2.5. PSK Modulator

By using the SSB modulator designed for optical frequency shift, we can control both phase and amplitude of the output lightwave simultaneously. We apply a pair of baseband signals $g_1(t)$ and $g_2(t)$ to the electrodes $\mathrm{RF_A}$ and $\mathrm{RF_B}$, respectively, where the dc-bias condition

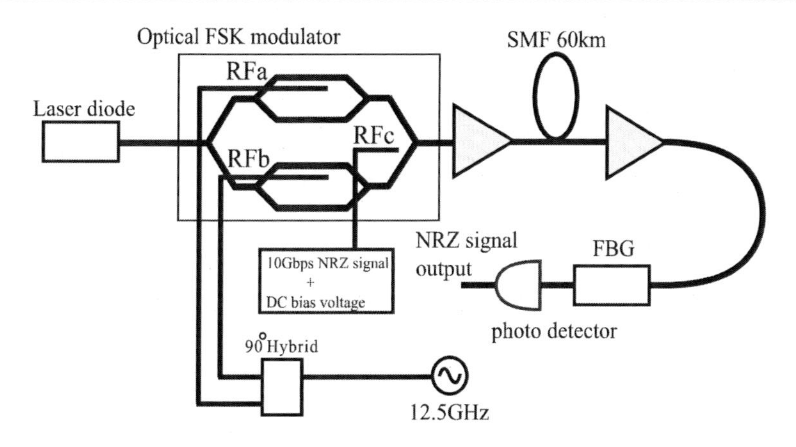

Figure 16. Setup for FSK transmission.

is identical to that of SSB modulation. The output can be expressed by

$$R = \frac{A^{\mathrm{LW}} \mathrm{e}^{\mathrm{i}\pi/4}}{2} \mathrm{e}^{2\pi \mathrm{i} f_0 t} \left[\cos[g_1(t)/2] + \mathrm{i}\cos[g_2(t)/2]\right] \tag{26}$$

$$= \frac{A^{\mathrm{LW}} \mathrm{e}^{\mathrm{i}\pi/4}}{2} \sqrt{\cos^2[g_1(t)/2] + \cos^2[g_2(t)/2]} \times \exp \mathrm{i}\left[2\pi f_0 t + \tan^{-1}(\cos[g_2(t)/2]/\cos[g_1(t)/2])\right] \tag{27}$$

where U is assumed to be -1 ($\phi_{\mathrm{FSK}} = \pi/2$). By using four symbols of $(g_1, g_2) = (0, 0)$, $(2\pi, 0)$, $(0, 2\pi)$, $(2\pi, 2\pi)$, we can generate a quadrature PSK signal, where the phases of the output, $\tan^{-1}(\cos[g_2(t)/2]/\cos[g_1(t)/2])$, are $0, \pi/2, \pi, 3\pi/2$. Similarly, binary PSK can be achieved by an MZM with symbols of $g = 0, 2\pi$.

3. Electrode Design for High-Speed Signals

The EO effects of LN, LT, GaAs, etc. are so small that the length of the modulator should be a few centimeters to obtain effective modulation. When the frequency of the electric signal on the electrode is high, the wavelength of the electric signal would be less than the electrode length. In addition, there is phase delay due to lightwave propagation along the waveguide. Thus, we have to use a transmission line theory to design the electrode for high-frequency operation. The moving coordinate system also should be used to investigate the effect of the phase delay due to the propagation. There are two types of electrode for high-frequency operation: traveling-wave and resonant electrodes. The traveling-wave electrode gives broadband response from dc to 50 GHz. Thus, the modulator with the traveling-electrode is suitable for high-speed digital transmission system. On the other hand, the resonant electrode can be used for band operation, for example, 25-26 GHz band [23, 24]. The merits of the resonant electrodes are 1) the length of the electrode is much shorter that that of the traveling-wave, and 2) the modulation efficiency is large at the particular frequency. The modulator cannot be used for broadband digital systems, but it is useful for some special applications, such as, clock generation, sideband generation with a fixed

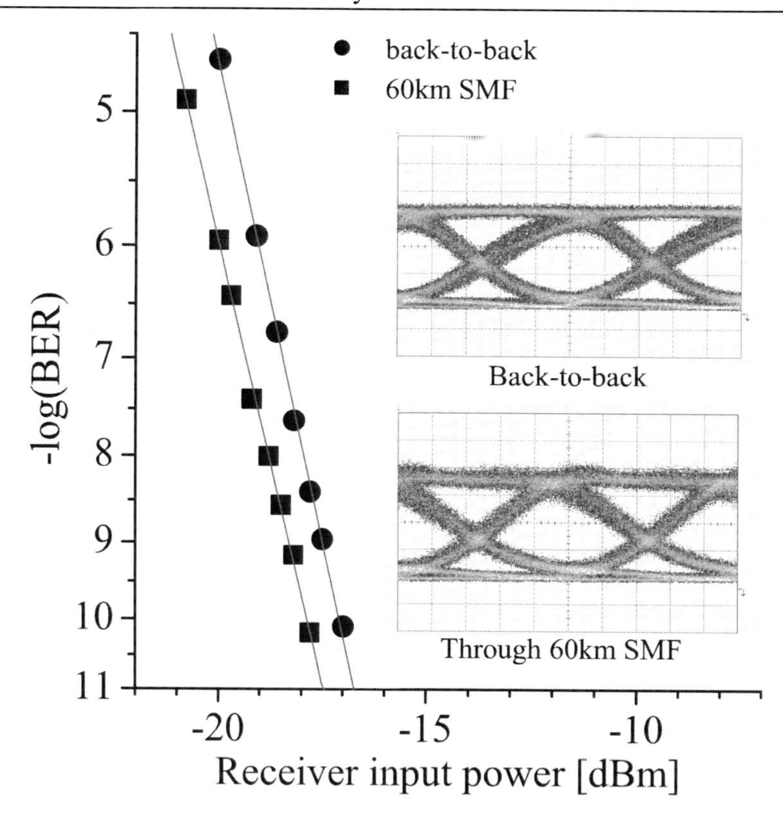

Figure 17. Bit-error-ratio curves and eye-diagrams of 9.95 Gbps FSK.

frequency separation and radio-on-fiber systems. The modulator size is much smaller than the conventional one, so that this is suitable for integrated modulators which have several functions. In this section, we investigated two types of resonant structures: an asymmetric resonant structure and a double-stub structure, to obtain compact and effective optical modulators for band-type operation. Combination of series and parallel resonance was used in these structures. The impedance at the feeding point can be increased by parallel resonance between a modulating electrode and a stub.

3.1. Asymmetric Resonant Structure [5]

As shown in fig. 18, the modulating electrode has two arms that are asymmetric coplanar waveguides (ACPWs). The ground plane, modulating electrode, and feeding line consist of gold whose thickness is t_{EL}, where s is the width of the gap between the ground plane and the modulating electrode, and w is the width of the modulating electrode. The feeding line, which is connected at the junction of the two arms, is a CPW. A buffer layer, whose thickness is t_{BUF}, is there to reduce the loss of the lightwave propagating in the optical waveguides under the metal electrodes. The short arm (arm I), with length denoted by L_1, is short-ended, while the long arm (arm II), with length denoted by L_2, is open-ended. The two arms differ in length and termination, so that we termed this configuration the asymmetric resonant structure. The equivalent circuit is shown in fig. 18 (c), where the

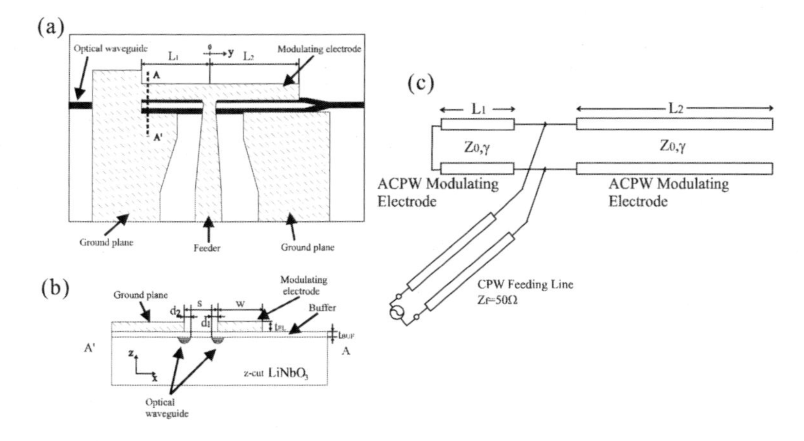

Figure 18. Asymmetric resonant structure. (a) top-view, (b) cross-section, (c) equivalent circuit.

impedance of arm I (Z_1) and that of arm II (Z_2) are expressed by

$$Z_1 = Z_0 \tanh \gamma L_1 \tag{28}$$

$$Z_2 = Z_0 \coth \gamma L_2. \tag{29}$$

Z_0 is the characteristic impedance of ACPW. $\gamma(= \alpha + \mathrm{i}\beta)$ denotes the propagation coefficient. The total impedance of the asymmetric resonant structure is given by

$$Z_L = Z_0 \frac{\tanh \gamma L_1 \coth \gamma L_2}{\tanh \gamma L_1 + \coth \gamma L_2}. \tag{30}$$

The voltage on the modulating electrode is given by

$$V(y,t) = T_r G(y) V_{in} \mathrm{e}^{2\mathrm{i}\pi f_m (t-t_0)}, \tag{31}$$

where $V_{in}\mathrm{e}^{2\mathrm{i}\pi f_m (t-t_0)}$ is the input voltage whose frequency is f_m. The voltage transmittance at the junction is defined by

$$T_r \equiv \frac{2Z_L}{Z_L + Z_f}, \tag{32}$$

where Z_f is the characteristic impedance of the feeding line. $|T_r|$ is an increasing function of $|Z_L|$, and the range is from 0 ($|Z_L| = 0$) to 2 ($|Z_L| = \infty$). $G(y)$ expresses the distribution of the standing voltage wave on the electrode, and is defined by

$$G(y) \equiv \begin{cases} \frac{\cosh \gamma (L_2 - y)}{\cosh \gamma L_2} & (y > 0) \\ \frac{\sinh \gamma (L_1 + y)}{\sinh \gamma L_1} & (y < 0) \end{cases}. \tag{33}$$

The induced phase at each optical waveguide is the sum of Pockels effect with respect to the coordinate system moving along with the lightwave propagating in the optical waveguide. Thus, the difference of the induced phases of the two optical waveguides of the Mach-Zehnder structure can be expressed by

$$\psi = \frac{\pi}{\lambda_0} n_0^3 r_{33} \frac{\Gamma}{s} L \Phi V_{\text{in}} \quad (34)$$

$$\Phi \equiv \frac{1}{L V_{\text{in}}} \int_{-L_1}^{L_2} V(y, \frac{y}{c} n_0 + t) \mathrm{d}y \quad (35)$$

$$\Gamma = \Gamma_1 + \Gamma_2 \quad (36)$$

where c is the speed of light. λ_0 and n_0 are the wavelength and RI of the lightwave, respectively. $L \equiv L_1 + L_2$ is the total length of the modulating electrodes. Γ_1 is the overlap integral between the field induced by the electrode and the field of the lightwave in the waveguide under the edge of the modulating electrode. Γ_2 is that of the ground plane. We consider a z-cut LN substrate where the c-axis is in z-direction. Pockels effect between the z-direction component of the electric field induced by the modulating electrode and lightwave whose electric field is polarized in the z-direction is considered, because the electrooptic coefficient of the effect mentioned above (r_{33}) is the largest one. The factor of $\frac{y}{c} n_0 + t$ means the phase difference due to the propagation delay of the lightwave. The integral of the voltage on the electrode defined by eq. (35), Φ is termed normalized induced phase, and shows the effect of the resonant structure. The normalized induced phase for an optical modulator with a resonant structure can be larger than unity, while the normalized induced phase for a lossless perfectly velocity-matched traveling wave modulator equals unity. The half-wave voltage of the modulator V_π is given by $\pi/|\phi|$. Similarly, $V_\pi L$, defined by $\pi L/|\phi|$ ($\propto |\Phi|^{-1}$), is the voltage-length product which shows the degree of the modulation efficiency and the scale of the modulator.

To obtain compact and effective optical modulators for band-operation, we designed them to have large normalized induced phases. We consider an ACPW with the following: $t_{BUF} = 0.55$ μm, $s = 27$ μm, $w = 50$ μm, and $t_{EL} = 2$ μm. γ and Z_0 at 10 GHz were $11.24 + \mathrm{i}891.8\ m^{-1}$ and 34.71 Ω, which were obtained by using an electro-magnetic field calculator with the finite element method (HP-HFSS5.4). Thus, the wavelength of a 10 GHz voltage wave on the ACPW was 7.03 mm. The following parameters were used for numerical calculations: $\varepsilon_{\text{SiO}_2} = 4.0$, $\varepsilon_{\text{LN[xy]}} = 43.0$, $\varepsilon_{\text{LN[z]}} = 28.0$ and $\sigma_{Au} = 4.3 \times 10^7 (\Omega\text{m})^{-1}$. $\varepsilon_{\text{LN[xy]}}$ and $\varepsilon_{\text{LN[z]}}$ denote permittivities of LN in the xy-plane and the z-direction, respectively. $\varepsilon_{\text{SiO}_2}$ and σ_{Au} are permittivity of the buffer layer (SiO_2) and conductivity of the electrodes (Au), respectively. The central inductor of the feeding line was 50 μm wide and the characteristic impedance was 50 Ω. By using eqs. (34) and (35), we obtained a combination of L_1 and L_2, which gave a maximum of Φ. When $L_1 = 0.03\lambda$ and $L_2 = 0.22\lambda$, Φ had a maximum of 3.7 at 10 GHz. The total length of the modulating electrodes L was 1.76 mm. Fig. 19 shows the normalized induced phase as a function of the frequency f_m, and the voltage wave distribution on the modulating electrode. In addition to a peak at 10 GHz, there are peaks due to high-order resonance at 30 and 50 GHz. In fig. 20, the voltage reflectivity at the junction and the total impedance of these two modulating electrodes were shown as functions of f_m. The denominator of eq. (30) goes to a minimum at resonance, which corresponds to parallel resonance of the two arms, so that the impedance has peaks at the resonant frequencies (10, 30, 50, and 70 GHz). As shown in fig. 20(a), the reflectivity has dips, where the impedance has peaks. The peaks of the optical response were caused

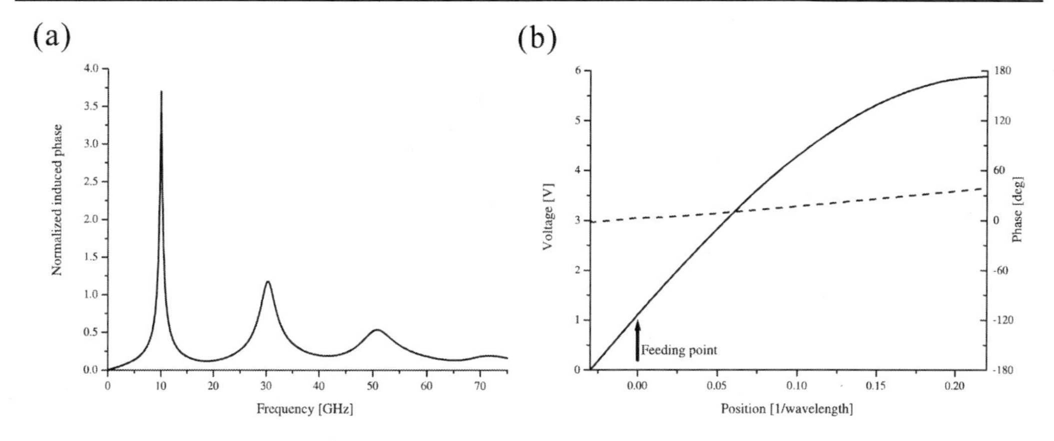

Figure 19. (a) Optical response of the asymmetric resonant structure, and (b) voltage wave distribution on modulating electrodes, with respect to the coordinate system moving along with the lightwave propagating in the optical waveguide $V(y, \frac{y}{c}n_0 + t)$. Solid line and dashed line denote the amplitude and the phase of the voltage wave, respectively.

by enhancement of T_r. Actually, the impedance Z_L is not equal to Z_f at the first-order resonance (10 GHz), as shown in fig. 20(b). The impedance matching performed at the feeding point is not necessary to obtain a large optical response, because Pockels effect is caused by the electric field induced by the electrodes regardless of the power input from the feeding line.

If both arms were terminated by the same impedance (e.g. open or short), the polarity of the standing-wave voltage profile would change on the electrodes and the length L_1 or L_2 should be longer to obtain the resonance. In an asymmetric resonant structure, however, both L_1 and L_2 are smaller than a quarter of the wavelength λ. As shown in fig. 19(b), the polarity does not change, which is why an asymmetric structure can reduce the half-wave voltage and the electrode length. At the end of arm II, the voltage amplitude is approximately six times as large as the input voltage. The length of arm II is slightly shorter than $\lambda/4$, so that arm II is nearly in resonance and that the impedance of arm II is small. But, the total impedance Z_L is large owing to arm I which has two functions: as a stub to enhance the impedance and as a modulating electrode. Arms I and II are in parallel resonance, where arm I is capacitive and arm II is inductive.

3.2. Dual-Stub Structure [6]

As shown in fig. 21 (a), the electrode structure consists of a modulating electrode and two stubs. The modulating electrode has two arms of the same length, and is an ACPW. A feeding line, which is a CPW, is connected at the center of the modulating electrode. Two stubs, which are also CPWs of the same length, are also connected at the junction of the modulating electrode and the feeding line, so that we call it the double-stub structure. The equivalent circuit is shown in fig. 21 (b). The total impedance is given by

$$Z_L = \frac{Z_{01} Z_{02} \coth \gamma_1 L_1 \tanh \gamma_2 L_2}{2(Z_{01} \coth \gamma_1 L_1 + \tanh \gamma_2 L_2)}, \tag{37}$$

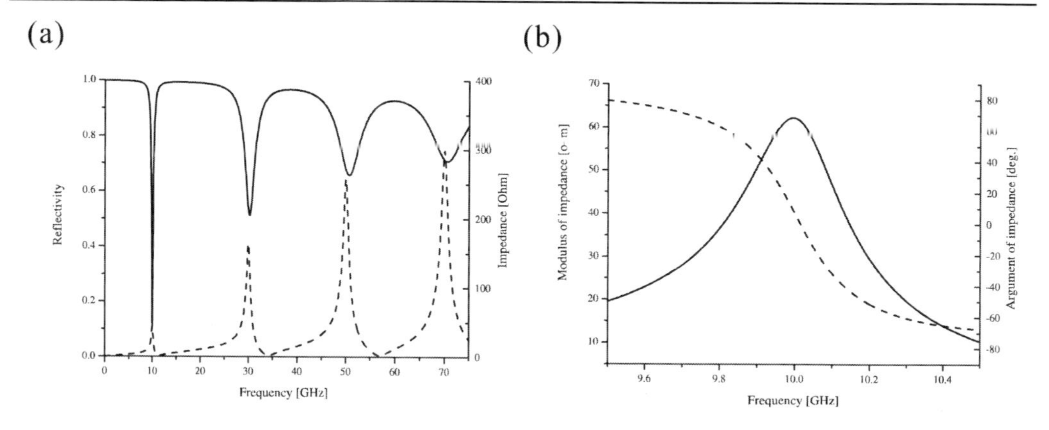

Figure 20. Electric response of the asymmetric resonant structure (a) solid line and dashed line denote voltage reflectivety at the junction and impedance of the optimized structure, respectively. (b) impedance of the modulator near the peak at 10 GHz. Solid line and dashed line denote modulus and argument of the impedance.

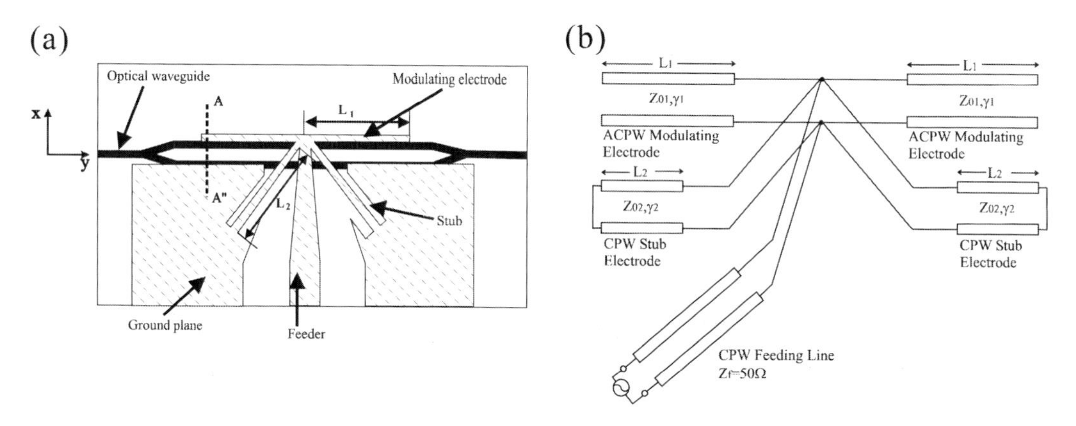

Figure 21. Double-stub structure (a) top-view, (b) equivalent circuit.

when we consider the case in which the modulating electrode is open-ended and the stub is short-ended. L_i, γ_i and Z_{0i} denote the lengths, the propagation coefficients and the characteristic impedances of the modulating electrode ($i = 1$), and of the stub ($i = 2$). The voltage on the modulating electrode is given by eq. (31), where the distribution of the standing wave voltage wave on the modulating electrode of the double-stub structure is expressed by

$$G(y) = \frac{\cosh \gamma_1 (L_1 - |y|)}{\cosh \gamma_1 L_1}. \tag{38}$$

The difference of the induced phases of the two optical waveguides of the Mach-Zehnder structure can be expressed by eq. (34), but the normalized induced phase for the double-stub structure is defined by

$$\Phi \equiv \frac{1}{V_{\rm in} L} \int_{-L_1}^{L_1} V(y, \frac{y}{c} n_0 + t) \mathrm{d}y, \tag{39}$$

where $L \equiv 2L_1$ is the total length of the modulating electrodes.

We considered the electrode structure with the following parameters. The gap between the electrodes of ACPWs and CPWs is 27 μm. The width of the modulating electrode is 5 μm, while that of the stub is 50 μm to reduce loss at the stubs. The characteristic coefficients of the modulating electrode (ACPW) are $\gamma_1 = 25.92 + i735.2\, m^{-1}$, $Z_{01} = 70.1\Omega$. Those for the stub (CPW) are $\gamma_2 = 14.29 + i863.5\, m^{-1}$, $Z_{02} = 28.7\Omega$. By using eq. (35), we obtained a combination of L_1 and L_2, which gave a maximum of Φ. When $L_1 = 0.19\lambda_1$ and $L_2 = 0.12\lambda_2$, Φ had a maximum of 2.44 at 10 GHz, where λ_1 and λ_2 are, respectively, the wavelength on the modulating electrode and the stub. The total length of the modulating electrodes L was 3.25 mm. As shown in fig. 22, the optical response and

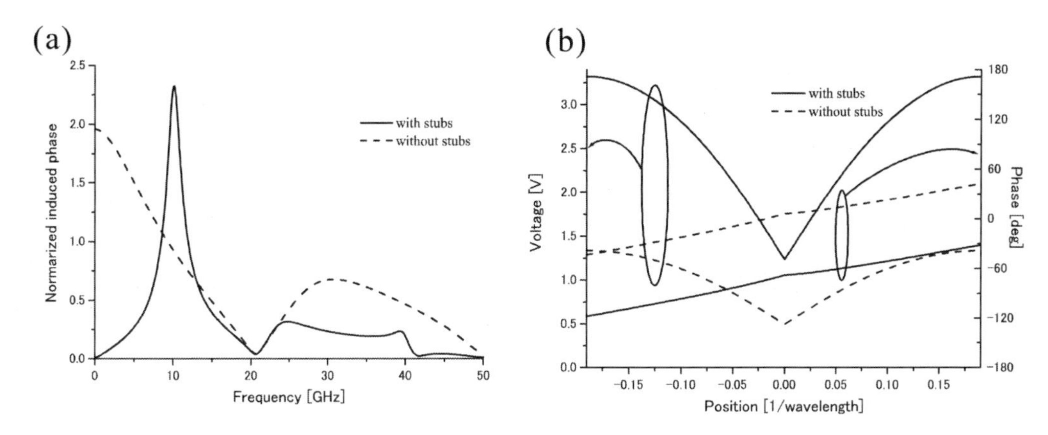

Figure 22. (a) Optical response of the double-stub structure, and (b) voltage wave distribution on modulating electrodes, with respect to the coordinate system moving along with the lightwave propagating in the optical waveguide $V(y, \frac{y}{c}n_0 + t)$. Solid line and dashed line correspond to the structures with stubs and without stubs, respectively.

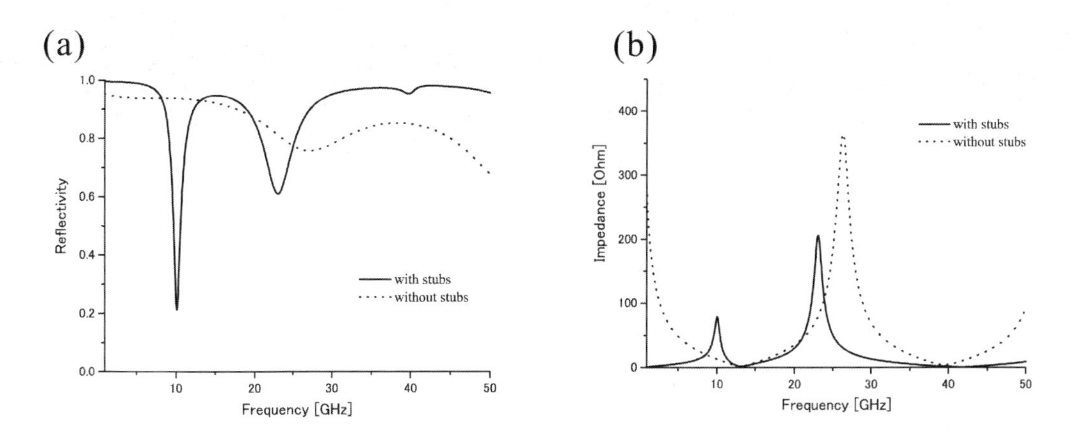

Figure 23. Electric response of the double-stub structure. (a) voltage reflectivity at the junction. (b) total impedance of the resonant structure. Solid line and dashed line correspond to the structures with stubs and without stubs, respectively.

the voltage profile at 10 GHz can be enhanced by using the stubs. Fig. 23 shows the electric response of the designed structure. The impedance had a peak at the resonant frequency, due to the parallel resonance between the modulating electrode and the stub, in the same way as the asymmetric resonant structure. Even in the case of the structure without stubs, there was a peak in the impedance, but it does not correspond to the enhancement of the optical response.

3.3. Numerical and Experimental Results

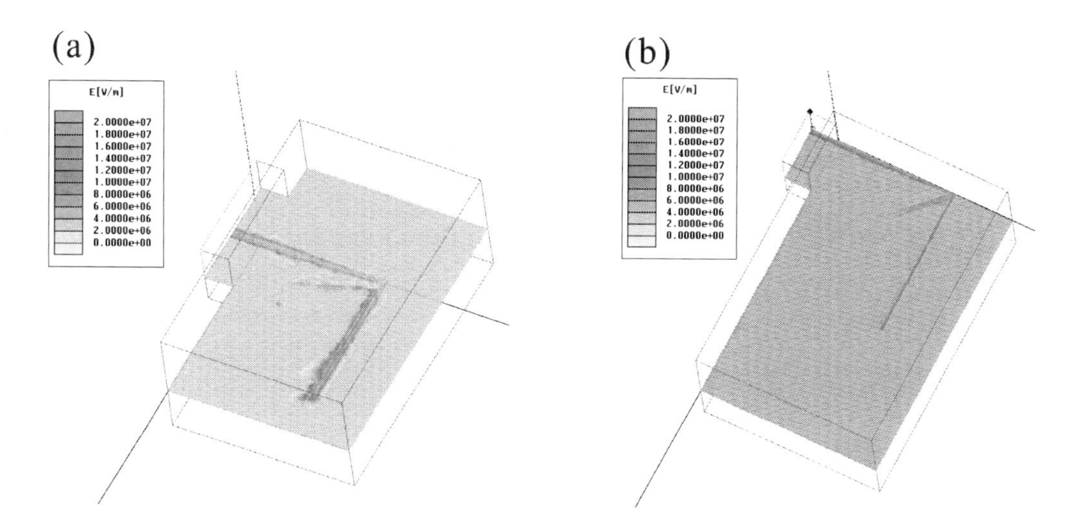

Figure 24. Electric field distribution on the resonant structures, calculated by FEM. Input power and frequency were 1W and 10GHz, respectively. (a) asymmetric resonant structure. (b) double-stub structure.

By using the finite element method (FEM), we investigated electro-magnetic field in the resonant structures. Fig. 24 shows the electric field distribution on the resonant structures. The double-stub structure is symmetric with respect to xz-plane, so that a half of the structure $(y > 0)$ was used in FEM, as shown in fig. 24 (b). The electric field on the modulating electrodes were enhanced both in the asymmetric resonant structure and in the double-stub structure.

We also measured optical and electric responses of fabricated modulators. The Mach-Zehnder interferometer consisted of two optical waveguides formed by the titanium diffusion technique on a z-cut LN substrate. The RI n_0 was assumed to be 2.2 for $\lambda_0 = 1.55\ \mu$m. The effective width and depth of the optical waveguides were 10 μm and 8 μm, respectively. The positions of the waveguides were as follows: $d_1 = 2\ \mu$m and $d_2 = 3\ \mu$m. The overlap integral Γ, calculated by TEM approximation [25], was 0.92, where the thickness of the electrode t_{EL} was neglected. The profile of the mode guided in the optical waveguide was elliptic and the field distribution was assumed to be uniform. The modulator also had an open-ended dc-bias electrode to set the Mach-Zehnder switch function to the quadrature point. The normalized induced phase can be measured as the ratio of the optical response of a resonant electrode at a particular frequency to that of a non-resonant electrode at dc, where both electrodes were of the same length. Γ obtained by the response measured at dc was 0.82, which was 89% of the calculated one. This discrepancy was due to the effect of the finite electrode thickness or deformation of the lightwave field pattern. Fig. 25 (a)

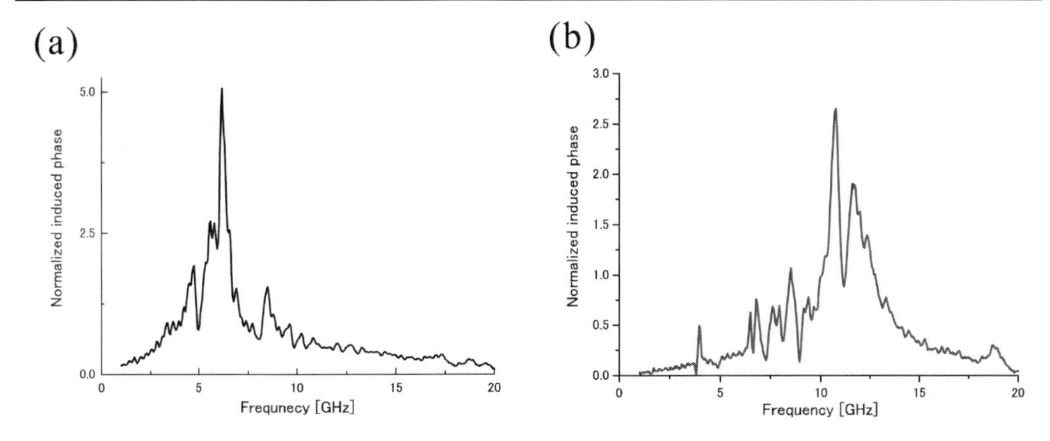

Figure 25. Measured optical responses (normalized induced phase) of fabricated modulators. (a) asymmetric resonant structure. (b) double-stub structure.

shows the optical responses of the asymmetric resonant structure for the 1.55-μm region, obtained from the responses of the resonant and dc-bias electrodes. The responses had a peak of 4.94 at 6.2 GHz, while the designed resonant frequency was 10 GHz. We deduce that this difference is due to the asymmetry of the resonant structure. The electric field of the fundamental CPW mode is symmetric, so that the mismatch of the field pattern occurs at the junction of the modulating electrode, as shown in fig. 24 (a). On the other hand, the double-stub structure is symmetric with respect to the central electrode of the feeding line, so that the electric field is also symmetric. The optical response of the double-stub structure had a peak of 2.65 at 10.75 GHz, which is in good agreement with the result of the equivalent circuit. We also measured the responses of the double-stub structures whose electrode lengths L_1 and L_2 were slightly changed from the designed values [6]. V_π and Φ were sensitive to the stub length L_2. The normalized induced phase of a modulator whose stub length was 90% of the designed one, and had a peak of 3.41 at 10.55 GHz. The half-wave voltage was 13.7 V at the peak, while the total length L was 3.25 mm.

Two types of resonant structures for band-operation optical modulators, the asymmetric resonant structure and the double-stub structure, were investigated, by using the equivalent circuits, the finite element method technique, and conventional measurements for electric and optical responses. The normalized induced phase of the fabricated modulator using the asymmetric resonant structure was 4.94. The half-wave voltage was 17.9 V in the 1.55 μm wavelength range, in spite of the short electrode length of 1.76 mm. However, due to the asymmetry of the structure, the resonant frequency was shifted from the frequency designed by the equivalent circuit. On the other hand, the optical response of the double-stub structure had a resonant peak near the designed frequency, and the half-wave voltage was 13.7V with the electrode length of 3.25 mm.

Resonant-type modulators can acts both on forward and backward lighwaves, while traveling-wave electrodes cannot on backward lightwaves. Thus, the lightwave reflection in the resonant-type modulator can double the modulation efficiency [7]. The halfwave voltages of a double-stub structure with lightwave reflection was 2.9 V at 7.0 GHz.

4. Photonic Sideband Management (PSBM) Techniques

PSBM techniques, can manipulate optical or electrical signals in time or frequency domain by using sideband generation at optical modulators. A couple of PSBM techniques, such as, reciprocating optical modulation (ROM), optical label swapping (OLS) using double-sideband (DSB) modulation and optical tunable delay, are described, where the frequency and time domain profiles of the input lightwave are controlled by an electric signal. ROM can generate lightwave modulated by a high-frequency signal whose frequency is an integer multiple of the electric signal fed to the modulator. Thus, we can easily obtain a lightwave signal having a millimeter-wave from a microwave signal. OLS with DSB modulation can be used for optical packet switching systems. This technique can be applied for a bundle of wavelength-domain-multiplexing (WDM) channels, so that we can construct simple nodes. The tunable delay line can also be used in optical packet systems, to prevent the packets from colliding at the nodes.

4.1. IM/FSK Modulation and OLS Using Double-Sideband Modulation [8]

A combination of IM and FSK is a promising technique for optical label switching in optical packet systems [26, 27]. In this technique, payload signals are in IM format, while label information is written by FSK signal. The merit of this FSK labeling is that an FSK transmitter generates the label information on the optical carrier frequency without affecting its intensity. Simultaneously modulated IM and FSK signals are independently demodulated by using an optical filter, so that the label information can be extracted without affecting the payload signal. OLS is a promising technique for implementing packet routing and forwarding functions over optical packet networks. Previously reported OLS systems used cross modulations between optical spectral components, such as cross phase modulation in an optical amplifier, in order to generate a new label signal. Thus, the OLS system should have additional light sources for the cross modulation. In an OLS system having wavelength-domain-multiplexing (WDM) channels, we have to use a pumping light source and a cross modulation device for each WDM channel, so that the setup for the OLS with WDM becomes extremely complicated. In this section, we describe an OLS technique using DSB modulation technique. As shown in fig. 26, the setup is very simple where pumping light sources are not necessary. IM/FSK signal has a pair of optical carriers whose frequencies are $f_0 - f_m$ and $f_0 + f_m$, where f_0 is the optical carrier frequency at the output port of the light source. f_m denotes the frequency of the modulating signal fed to the electric input ports $\mathrm{RF_A}$ and $\mathrm{RF_B}$ of the FSK modulator. When the FSK switching signal fed to $\mathrm{RF_C}$ is in "1" state where the modulator generates USB ($\phi_{\mathrm{FSK}} = -90°$), the output carrier frequency is $f_0 + f_m$. That of "0" state, where the modulator generates LSB ($\phi_{\mathrm{FSK}} = +90°$), is $f_0 - f_m$. The IM/FSK signal is fed to an intensity modulator which is in null-bias point to obtain DSB-SC modulation. The modulating signal is a sinusoidal wave whose frequency is f_m. The output has three optical spectral components whose carriers are $f_0 - 2f_m$, f_m and $f_0 + 2f_m$. When the FSK switching signal is in "1" state, the output of the intensity modulator has two carriers of f_0 and $f_0 + 2f_m$. On the other hand, the output of "0" has two carriers of f_0 and $f_0 - 2f_m$. The output always has the f_0 component regardless of the FSK switching signal. Thus, by using an optical bandpass filter whose center frequency is f_0, we can remove the FSK signal from the IM/FSK signal, and restore a pure IM signal which does not have an FSK signal. The output of the DSB-SC modulation has only one optical carrier of f_0, so that we can add a new FSK label by using another FSK modulator. As shown in fig. 27, the FSK modulator can be used for a WDM signal which has many IM channels. We can simultaneously shift the carrier frequencies of the channels by one FSK modulator, as described in Section 2.3.. For an FSK-WDM signal, an optical filter which has a periodic optical frequency response, such as an interleaver, an arrayed-waveguide, and so on, can be used to demodulate the FSK signal. The proposed label swapping technique is also applicable to WDM systems by using an optical filter having the periodic response.

An experimental setup for IM/FSK modulation and our proposed OLS system is shown in fig. 28, where the bit rates of FSK and IM were, respectively, 1 Gbps and 9.95 Gbps (NRZ, PRBS $2^{31}-1$). The frequency deviation of FSK (f_m) was 12.5 GHz. The extinction ratio of the IM modulation was 3.9 dB. Residual IM in the FSK modulation was so small that the IM output which can be generated just by feeding the IM/FSK signal to a high-speed photodetector has a clearly opened eye-diagram. The FSK signal was demodulated by using an optical filter which can discriminate between USB and LSB. Fig. 29 shows BER curves and eye-diagrams of the FSK and IM signals, where error-free transmissions for FSK and IM were obtained. By using DSB-SC modulation, we demonstrated the restoration of a pure IM signal which has one optical carrier. Fig. 30 shows the optical spectra at the output port of the DSB modulator and at the output port of the optical bandpass filter. The spectral components of $f_0 - 2f_m$, f_0 and $f_0 + 2f_m$ were generated at the DSB modulator. The restored pure IM signal having the f_0 component was extracted by the bandpass filter. We also obtained clear eye-opening and error-free transmissions for the restored IM, as shown in fig. 31. An IM/FSK signal with a new FSK label can be generated by another FSK modulator placed at the output port of the optical filter, so that we can construct a simple OLS system consisting of a DSB-SC and an FSK modulators.

4.2. Reciprocating Optical Modulation [14]

An ROM, consisting of a pair of optical filters and an optical modulator, can generate high-order sideband components of EO modulation, as shown in fig. 32, where the FBGs were fixed in V-grooves on SiO_2 substrates and directly attached to the LN modulator chip. One of the optical filters is placed at the optical input port (input filter), and the other is at the output port (output filter). By using an ROM, we can generate a lightwave modulated by an rf signal whose frequency is an integer multiple of an electric rf signal applied to the modulator. The phase-shifted FBG has a narrow pass band of 1.5 GHz in the reflection band. The output filter was a conventional FBG with a heater to let the center wavelength equal to that of the input filter. When the input wavelength was in the pass band of the input filter, both USB and LSB reciprocate between the input and output filters, so that high-order DSB components can be effectively generated. The resonant electrode enhances modulation efficiency for forward and backward lightwaves, while a traveling-wave electrodes acts only on forward propagating lighwaves. The resonant frequency of the electrode was designed to be equal to inverse of delay in a reciprocation process, so that the successive modulation process can be in phase. Fig. 33 shows the output spectrum of the fabricated ROM and that

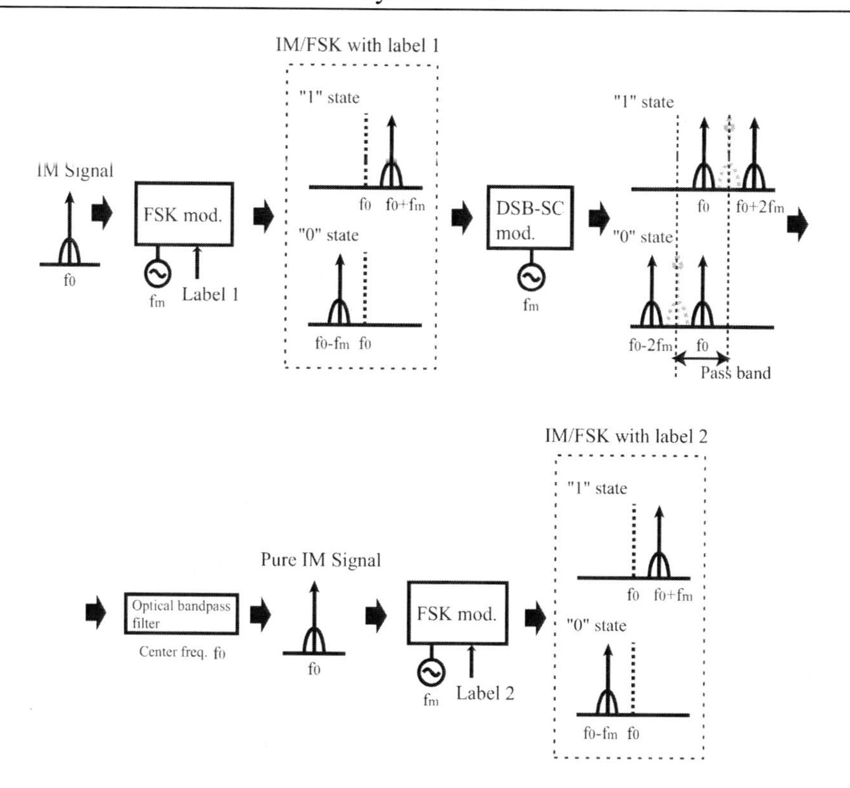

Figure 26. Principle of OLS of IM/FSK signal.

of the conventional optical phase modulation, where the input rf-power and frequency were, respectively, 19.8 dBm and 5 GHz. Optical power and frequency of the input lightwave were, respectively, 4.2 dBm and 192.956 THz. Sideband components lower than 5th-order were in the reflection band, so that 5th or 6th order harmonic components were generated effectively.

4.3. Optical Tunable Delay Line Using SSB Modulation [15]

Optical buffers that are composed of delay lines are key components in optical packet switching systems. To prevent packets from colliding, it is necessary to vary the delay at the buffer. There are two types of configurations for tunable optical delay lines. One is comprised of an optical switch and a fiber loop, where the operation of the switch must be synchronized with the propagation cycle of the packet in the loop. The other consists of an array of several delay lines of various lengths connected to an optical switch by which we can select one from the array delay line. Each line has an optical amplifier and an optical switch, which results in a complicated setup. In this section, we describe an optical tunable delay line consisting of an optical single-sideband (SSB) modulator, a fiber Bragg grating, and an optical fiber loop. The number of times, a lightwave circulates in the loop, depends on the frequency of the electric signal applied to the modulator. Thus, we can control the delay by switching the frequency without using delay lines of various lengths. In addition, it is not necessary to synchronize the switching timing with the cycle of the packet propagation.

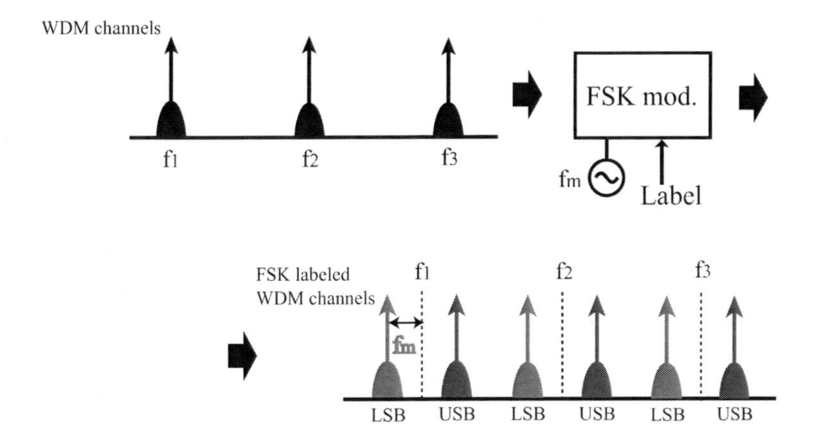

Figure 27. FSK labeling on bundled WDM channels.

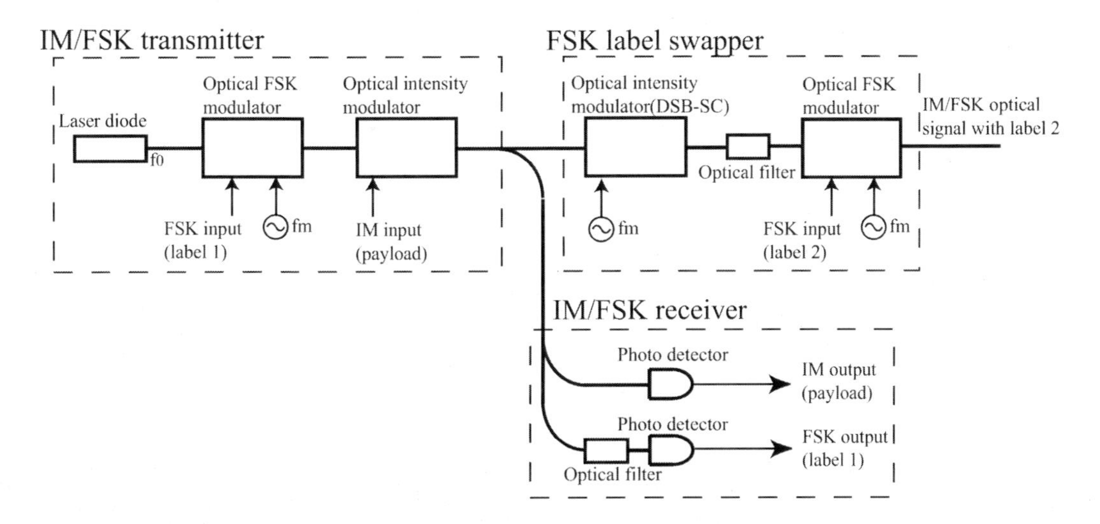

Figure 28. Setup for IM/FSK modulation and OLS using DSB-SC.

As shown in fig. 34(a), the tunable delay line has an optical port consisting of two optical circulators and an FBG placed between them. A lightwave whose optical frequency is out of the reflection band of the FBG can pass through the port, while a lightwave in the band cannot. Thus, an input lightwave whose frequency is out of the reflection band comes into the loop. When the modulator is not in operation, the lightwave goes through the loop once, and then it exits from the port. Consider that the SSB modulator is in operation, and that the input optical frequency (f_0) is slightly lower than the edge of the reflection band, as shown in fig. 34(b), where f_m denotes the frequency of the electric signal fed to the modulator, and f_r is the reflection band width of the FBG. The input optical frequency can be shifted at the SSB modulator. We assume that the dc-bias voltages for the modulator are tuned to generate the upper sideband. Of course, we can use the lower sideband instead of the upper sideband when f_0 is slightly higher than the upper edge of the reflection band. The f_0 lightwave, coming into the loop through the port, is input to the SSB modulator, and then its frequency is shifted into the reflection band (f_0+f_m). Thus, the lightwave is reflected by the FBG, and circulates in the loop again. During the successive circulation steps, the optical frequency is

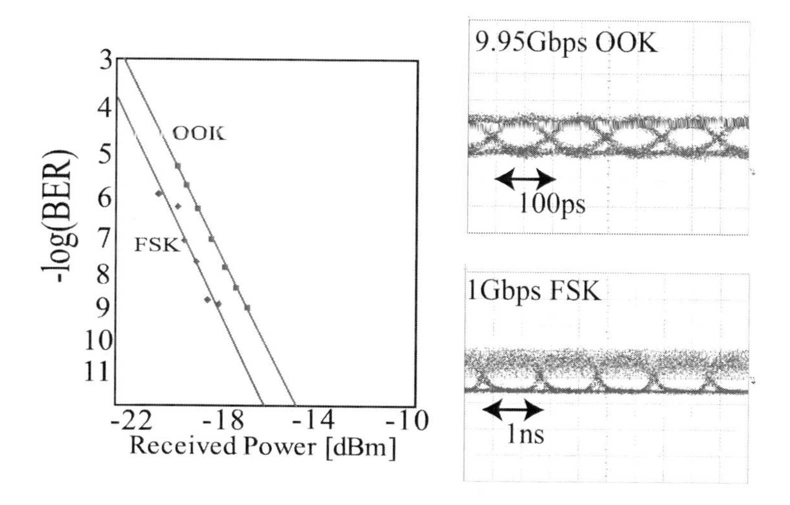

Figure 29. Bit-error-ratio curves and eye-diagrams for FSK and IM signals.

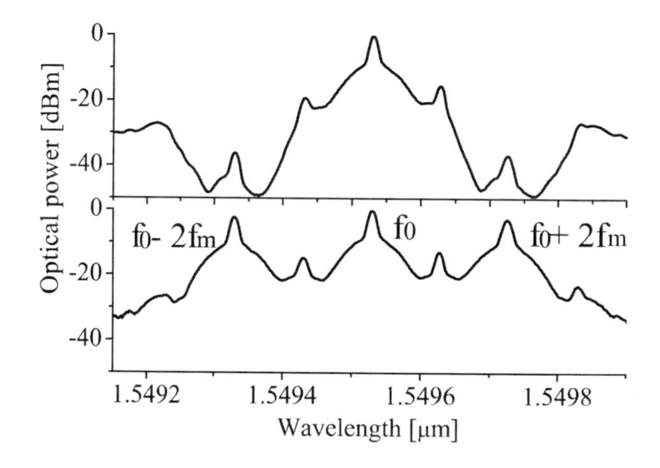

Figure 30. Optical spectra at the output port of the DSB modulator (lower) and at the output port of the optical bandpass filter (upper).

shifted at the modulator. After n times circulation, the frequency of the lightwave becomes $f_0 + nf_m$. We call such a spectral component an n-th order channel. After circulating several times, the lightwave exits the reflection band, where the number of circulation steps p is given by $pf_m > f_r > (p-1)f_m$. In this section, we call the spectral component of $f_0 + pf_m$ the prime channel, which is the lowest order channel in the channels whose frequency is higher than the reflection band. The output lightwave contains some other components resulting from the nonlinearity of the modulator and residual transmittance in the reflection band. However, the prime channel would be dominant when we can consider that f_m is larger than the edge slope width of the reflection band, and that any high-order harmonic generation at the modulator is negligible. The propagation delay of a circulation step is given by $\tau = l/c$, where l denotes the effective optical length of the loop, and c is the speed of light. The delay of the prime channel at the loop is dependent on f_m, as follows: $p\tau = \tau \times Int(f_r/f_m)$, where $Int(x)$ gives the smallest integer larger than x. Thus, we can

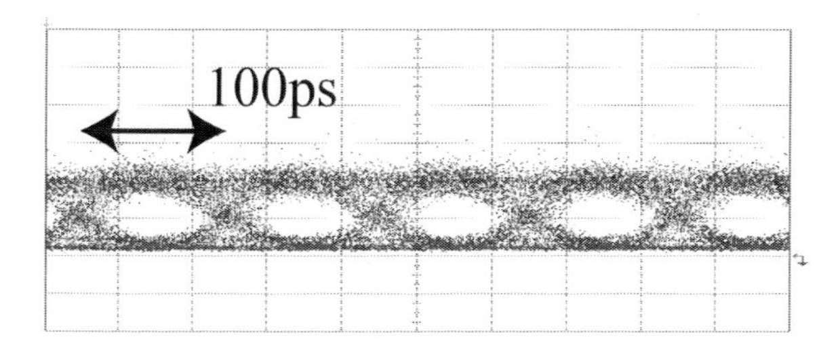

Figure 31. Eye-diagram of restored pure IM signal at the output port of the optical band pass filter.

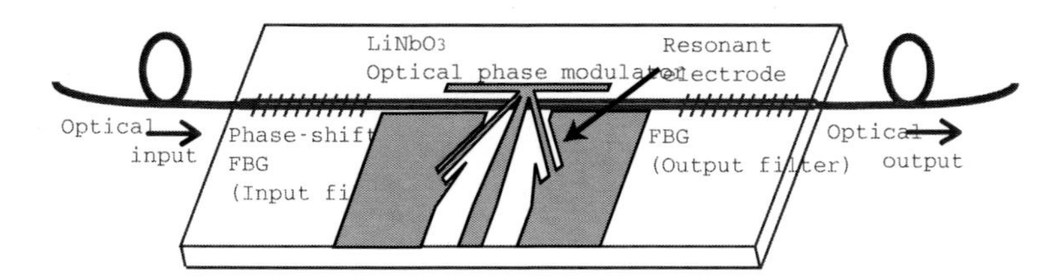

Figure 32. Reciprocating optical modulator with resonant electrode.

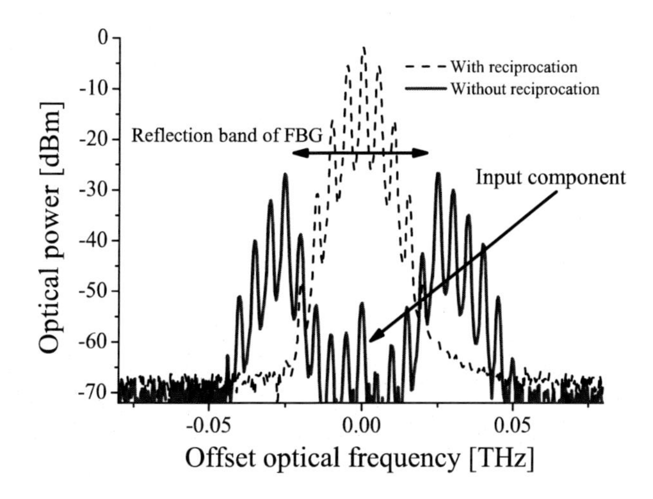

Figure 33. Output spectra modulated by ROM and by conventional phase modulation.

change the delay $p\tau$ by switching the frequency of the electric signal f_m. The frequency can be switched at any time except when a packet is passing through the SSB modulator. Thus, it is not necessary to synchronize the switching timing with the propagation cycle of the packet. In addition, we note that pulses longer than the optical length l can pass through the delay loop without any collision owing to frequency shift at the modulator.

Fig. 35 shows the time domain envelopes and the frequency domain spectra of the output lightwaves. The input lightwave was intensity modulated by a train of pulses whose period and duty ratio were, respectively, 4 μs and 20%, as shown in the plot at the bottom of fig. 35 (b). An FBG was placed at the optical output port to eliminate the spectral component of the input lightwave reflected at the FBG located between the two circulators. The reflection band width f_r was 22.5 GHz (FWHM), and the edge slope width Δf_r

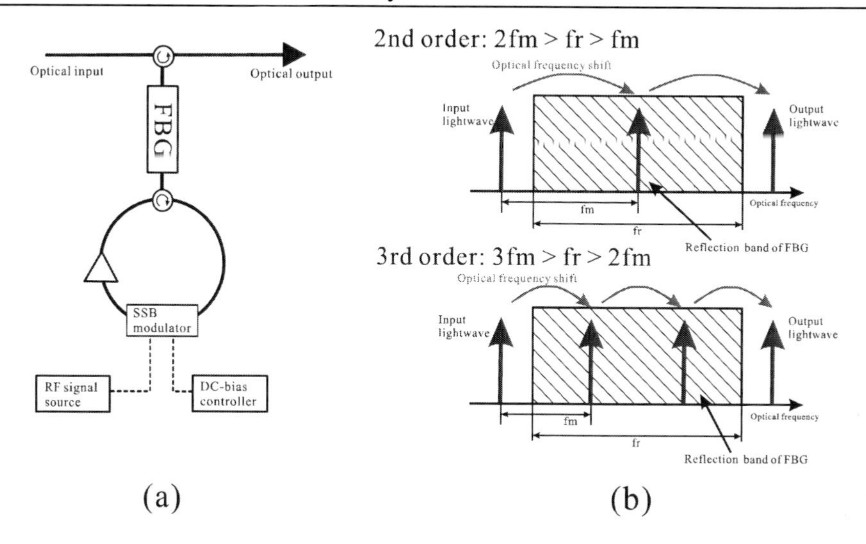

Figure 34. Tunable delay using SSB modulation.

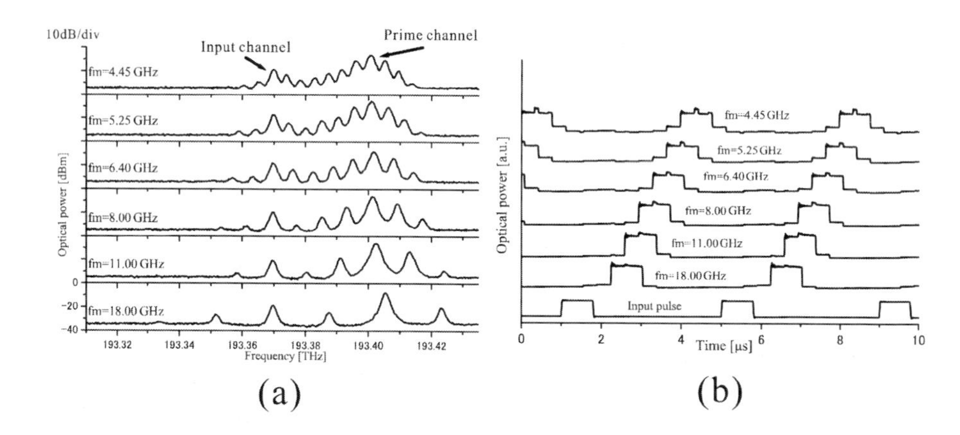

Figure 35. Output lightwaves: (a) Optical spectra (b) Time domain envelopes.

was 6.0 GHz. Δf_r was defined by the width of the slope where the reflectivety is from 10% to 90% of the maximum. We used an x-cut LN optical SSB modulator, as the optical frequency shifter, where the carrier-suppression ratio was over 25 dB, and the frequency conversion loss was 23 dB. The dc-bias voltages were adjusted to maximize the first order upper sideband in the output of the SSB modulator. An optical amplifier was used, as shown in Fig. 34(a), to compensate for loss at the modulator. When $f_m = 18.00$ GHz, the 2nd-order channel is the primary channel in the output. Similarly, when $f_m = 11.00, 8.00,$ 6.40, 5.25, and 4.45 GHz, the primary channels were the 3rd, 4th, 5th, 6th, and 7th channel, respectively. In the case of $f_m = 18.00$ GHz, the peak of the primary channel exists at 193.406 THz, where the optical frequency of the input lightwave is 193.370 THz. As shown in fig. 35(b), the delay can be controlled by changing the frequency of the electric signal fed to the SSB modulator. The delay due to one step of the circulation (τ) corresponding to the time difference between adjoining channels, was 350 ns, while the pulse width is 800 ns. In the cases of $f_m = 5.25$ GHz and 4.45 GHz, undesired spectral components in the outputs were not negligible, as shown in fig. 35(a), because f_m is not larger than Δf_r. The time domain profiles shown in the figure are deformed due to overlapping of the adjacent channels onto the primary one.

4.4. About the Author

Tetsuya Kawanishi received the B.E., M.E. and Ph.D. degrees in electronics from Kyoto University, in 1992, 1994 and 1997, respectively. From 1994 to 1995, he worked for Production Engineering Laboratory of Matsushita Electric Industrial (Panasonic) Co., Ltd. In 1997, he was with Venture Business Laboratory of Kyoto University, where he had been engaged in research on electromagnetic scattering and on near-field optics. He joined the Communications Research Laboratory, Ministry of Posts and Telecommunications (from April 1, 2004, National Institute of Information and Communications Technology), Koganei, Tokyo, in 1998. He was a Visiting Scholar at the Department of Electrical & Computer Engineering, University of California at San Diego, USA, in 2004. He is now a Senior Researcher of National Institute of Information and Communications Technology, and is currently working on high-speed optical modulators and on RF photonics. He received URSI Young Scientists Award in 1999.

References

[1] Noguchi, K.; Mitomi, O.; Miyazawa, H. *J. Lightwave Technol.* 1998 16, 615–619.

[2] Zhou, Y.; Izutsu, M.; Sueta, T. *J. Lightwave Technol.* 1991 9, 750–753.

[3] Yoshida, K.; Nomura, A.; Kanda, Y. *IEICE Trans.* 1994 E77-C, 1181–1184.

[4] Izutsu, M.; Mizuochi, Y.; Sueta, T. *IEICE Trans.* 1995 E78-C, 55–60.

[5] Kawanishi, T.; Oikawa, S.; Higuma, K.; Sasaki, M.; Izutsu, M. *IEICE Trans. Electron.* 2002 E85-C, 150–155.

[6] Kawanishi, T.; Oikawa, S.; Higuma, K.; Matsuo, Y.; Izutsu, M. *Electron. Lett.* 2001 37, 1244–1246.

[7] Kawanishi, T.; Oikawa, S.; Higuma, K.; Matsuo,Y.; Izutsu, M. *Electron. Lett.* 2002 38, 1204–1205.

[8] Kawanishi, T.; Higuma, K.; Fujita, T.; Ichikawa, J.; Sakamoto, T.; Shinada, S.; Izutsu, M. *J. Lightwave Technol.* 2005 23, 87–94.

[9] Kawanishi, T.; Sasaki, M.; Shimotsu, S.; Oikawa,S.; Izutsu, M. *IEEE Photonics Tech. Lett* 2001 13, 854–856.

[10] Kawanishi, T.; Oikawa, S.; Izutsu, M. *J. Lightwave Technol.* 2002 20, 1408–1415.

[11] Kawanishi, T.; Yoshiara, K.; Oikawa, S.; Shinada, S.; Sakamoto, T.; Izutsu, M. *Jpn. J. Appl. Phys.* 2004 43, 5791–5794.

[12] Kawanishi, T.; Sakamoto, T.; Shinada, S.; Izutsu, M. *IEEE Microwave and Wireless Components Lett.* 2004 14, 566–568.

[13] Kawanishi, T.; Oikawa, S.; Yoshiara, K.; Sakamoto, T.; Shinada, S.; Izutsu, M. *IEEE Photon. Tech. Lett.* 2005 17, 669–671.

[14] Kawanishi, T.; Shinada, S.; Sakamoto, T.; Oikawa, S.; Yoshiara, K.; Izutsu, M. *Electron. Lett.* 2005 41, 69–70.

[15] Kawanishi, T.; Oikawa, S.; Higuma, K.; Izutsu, M. *IEEE Photon. Tech. Lett.* 2002 14, 1454–1456.

[16] Shimotsu, S.; Oikawa, S.; Saitou, T.; Mitsugi, N.; Kubodera, K.; Kawanishi, T.; Izutsu, M. *IEEE Photon. Tech. Lett.* 2001 13, 364–366.

[17] Higuma, K.; Oikawa, S.; Hashimoto, Y.; Nagata, H.; Izutsu, M. *Electron. Lett.* 2001 37, 515–516.

[18] Kawanishi, T.; Izutsu, M. *IEEE Photon. Tech. Lett.* 2004 16, 1534–1536.

[19] Kawanishi, T.; Higuma, K.; Fujita, T.; Ichikawa, J.; Shinada, S.; Sakamoto, T.; Izutsu, M. *IEICE Electron. Express* 2004 1, 69–72.

[20] Kawanishi, T.; Higuma, K.; Fujita, T.; Ichikawa, J.; Sakamoto, T.; Shinada, S.; Izutsu, M. *Electron. Lett.* 2004 40, 691–692.

[21] Kawanishi, T.; Sakamoto, T.; Tsuchiya, M.; Izutsu, M., "High Carrier Suppression Double Sideband Modulation Using an Integrated $LiNbO_3$ Optical Modulator," *Microwave Photonics Conference 2005*

[22] Kawanishi, T.; Higuma, K.; Fujita, T.; Ichikawa, J.; Sakamoto, T.; Izutsu, M., *IEICE Electron. Express* 2005 2, 333–337

[23] Lim, C.; Nirmalathas, A.; Novak, D.; Waterhouse, R. *Electron. Lett.* 2000 36, 442–443.

[24] Kuri, T.; Kitayama, K.; Takahashi, Y. *IEEE Photonics Tech. Lett.* 2000 12, 419–421.

[25] Collin, E. R., Field Theory of Guided Waves, IEEE Press, NY 1991, 247–328.

[26] Vegas Olmos, J.J.; Tafur Monroy, I.; Koon, J. M. A. *Optics Express* 2003 11, 3136–3140

[27] Vlachos, K.; Zhang, J.; Cheyns, J.; Sulur; Chi, N.; Van Breusegem, E.; Monroy, T. I.; Jennen, L. G. J.; Holm-Nielsen, V. P.; Peucheret, C.; O'Dowd, R.; Demeester, P.; Koonen, J. M. A. *J. Lightwave Technol.* 2004 21, 2617–2628.

In: Advances in Laser and Optics Research. Volume 10 ISBN: 978-1-62257-795-8
Editor: William T. Arkin

Chapter 9

AN OVERVIEW OF OPTICAL SENSORS AND THEIR APPLICATIONS

P. K. Choudhury[1] and O. N. Singh[2]

[1]Faculty of Engineering, Multimedia University, Cyberjaya, Malaysia
[2]Department of Applied Physics, Institute of Technology, Banaras Hindu University, Varanasi, India

ABSTRACT

In the developed nations of the world, optical sensor industry is a multimillion-dollar business. This chapter is devoted to the review of the fundamentals of fiber optic sensing techniques, and describes the different types of optical sensors and their applications in different industries. Descriptions are made of intensity based sensors, distributed sensors, and interferometric sensors, their different types and specific applications. Examples of some more forms of FOSs, and their use in the context of environment are also reviewed. In addition, discussions are made of the role of integrated optics technology in sensing applications. It is noteworthy that photonic crystals and STFs have been the emerging areas of research in the present day; the chapter describes the use of such materials and/or mediums in the area of optical sensing.

1. INTRODUCTION

Fiber optic sensors (FOSs) have been the focus of research and development for over twenty five years. In this period, varieties of sensors have been devised by the investigators for multifarious range of applications. Initial devices incorporating such sensors have been designed for very specific applications. However, because of peak research activity, situation has now changed very rapidly, and a variety of such sensors now appeared in the market. The basic conceptual building blocks of FOSs are now well-known, and presently, the principal areas of application of these sensors appear to be in medical systems, aerospace and hydrospace sciences, electrical power supply industries, and petrochemical, oil and gas industries [1–10]. In the developed nations of the world, optical sensor industry is a multimillion-dollar business and, by the end of this century, it is projected to grow over 600 million-dollar industry [11].

Optical technologies have long played a vital role in instrumentation and sensors, especially for non-contact measurements. Although various forms of chemical and medical sensors were invented in the past, such sensors implementing the optical fiber technology offer significant advantages over conventional systems. The basic reason is that the energy source and the information processing involve the transmission of lightwave signals through optical fibers which are made of chemically and electrically inert mediums. To date highly sophisticated instruments for sensing measurements have been investigated which include remote measurement capabilities, multiple sensing, miniaturized spectrometers, miniaturized optical probes to adapt into hypodermic models for clinical and biomedical applications etc. Although, so far as the sensors for chemical and medical applications are concerned, there are still some problems related to poor long-term stability, slow response time, mechanical design of the probe-to-fiber coupling etc., yet the optical fiber sensing technique has been recognized to have enormous potentials. Apart from this, the application of physical sensors to the biomedical area also appears to be extremely promising.

The key factors in the development of FOSs include the fabrication of sensors that are small and trendy, lightweight, immune to electromagnetic (EM) interference and require no electrical power at the point of sensing. Fiber optic sensing technology always faced the problem of competing with the established low cost and highly reliable conventional sensor technologies that provide adequate performance characteristics. In fact, the performance characteristics of FOSs are equally well as compared to their conventional sensor counterparts, but the overall technology becomes more expensive owing to their small market volume. As a result, FOSs are realized only when there are some specialized application needs, e.g. trendy and non-consumption of electrical power etc. Such fiber optic sensing technology is also implemented when there are requirements of multiplexing the sensors (optical fibers can be multiplexed efficiently).

A large number of components depending on the physical principles have been developed over the decade. Today some of the commercially available components are fused and polished couplers, absorption filters, amplifiers and sources made of rare earth doped fibers etc., which are greatly used in all-fiber sensing devices. Efforts are also put for the development of active fiber components such as sandwitchable couplers or fast light modulators. Several attempts have been made in this regard, but they have been inherently limited by the passive nature of fiber material. Factors related to materials are of high importance for both the sensing means and the practical implementation of FOS systems. A great many varieties of optical fiber materials and structures have been introduced leading to various forms of applications in sensing technology. Polished fiber substrate technology has emerged as a powerful tool to perform a large number of basic functions within the fiber itself by providing controlled access to its core, and thereby guiding the wave.

It is noteworthy that majority of the FOS products are based on multimode fiber systems, which have large core diameter and high numerical aperture, resulting thereby high light-collecting efficiency of fibers, and therefore, low cost light sources such as LEDs can be implemented. The output monitored by such FOS systems is typically related to the measurement of the intensity of light either as a direct intensity measurement or via the polarization properties of light. Monomode fiber systems are also developed which are normally based on interferometric techniques. But, for such FOS systems, laser diodes are used as the light source, which are more expansive than LEDs. Also, such systems incorporate costlier components such as fiber couplers, integrated optic (IO) modulators etc., and the measurements are based on the phase delay of the light in the sensor arrangement. A new class of sensors has been investigated by the introduction of monomode fiber components into optical devices. In its simplest form, the fiber itself serves basically as a flexible link for the transmission of optical signals to and from the optical arrangement that constitutes the sensor. In more complex arrangements, optical fiber itself serves not only as the medium of transmission, but also as the sensor. FOSs can be categorized on the basis of these two forms of sensing devices – i.e. intensity or interferometric techniques.

Major advancements in this new exciting field are expected because novel fiber optic components are developed specifically for sensor applications.

FOSs are classified into the categories of optrodes, fiber coupled sensors and distributed FOSs. In all these configurations, optical fibers themselves serve as a passive pipe of light in spectroscopic analyses, or can be coated with a substance to produce an electrochemical response. Optrodes are optical fiber based analogues of electrodes made of an optical fiber, and the tip of the fiber is sensitised via chemical indicators that respond to pH or a reflective diaphragm to change the amount of light reflected at the tip due to changes in chemical properties or pressure. In general, in such sensors, the sensitised tip reacts with the measurand. Several kinds of sensor configurations based on the single-ended transducer type, as shown in Figure 1, have been realized. Such kinds of sensors are generally applied to biomedical and chemical sensing applications, e.g. for measurands including temperature, pressure and chemical parameters. In fiber-coupled sensors, fibers themselves act as passive waveguides. These waveguides take light to and from the points of measurement, and in this way, offer the best ability to perform quasi-remote measurements. In such sensors, the parameters of interest are measured at a specific location. In the case when a longer length of optical fiber is used for sensing measurements, the sensor can be configured as an extended path-integrating device (Figure 2). Alternatively, long sensing fiber can be wrapped on suitable transducers, and such techniques are used in interferometric sensors to provide increased sensitivity to weak physical fields, e.g. in acoustic sensing. In distributed FOSs, all or a part of the fiber is sensitised that reacts with the measurand, and subsequently offer a special distributed measuring capability.

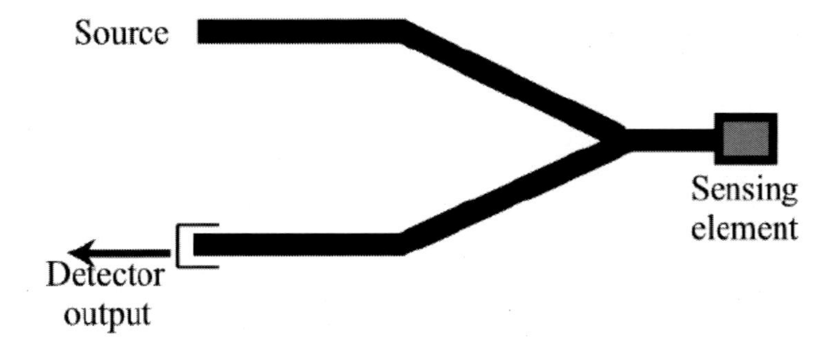

Figure 1. Single-ended sensor transducer.

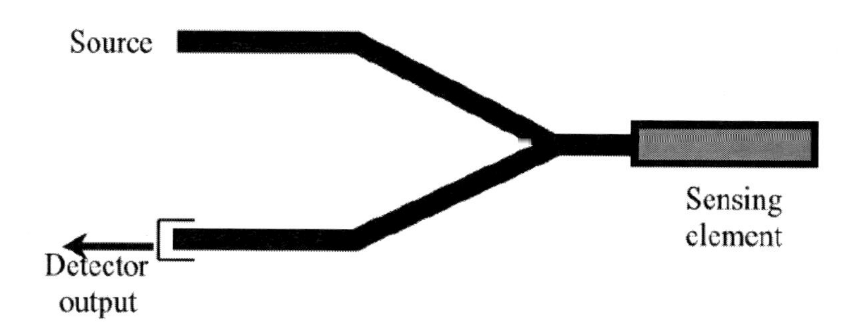

Figure 2. Extended path integrating device.

That is, in such FOS systems, the measurand parameter of interest is monitored at many locations along a fiber, and this technique opens up new potential applications where a spatial mapping of the parameters are required.

2. Intensity Based Sensors

A variety of intensity based FOSs implementing numerous sensing techniques viz. absorption, emission, luminescence, fluorescence, etc., have been devised so far in order to monitor various physical parameters such as temperature, pressure, displacement etc. The techniques involved may be, in general, point sensing or distributed sensing leading thereby to different degrees of sensing capabilities. The most basic configuration of FOS systems is shown in Figure 1 where the sensing element is designed to some change in some physical characteristics of light, e.g. wavelength, intensity, polarization etc., which are then monitored at the optical detector. There are several varieties of probes or transducers used in FOS systems, some of which are coupled fibers, birefringent probes and microbend probes [12–25].

In sensors employing birefringent probes, light from an input fiber is passed through a birefringent element which acts simply as a phase plate that is sensitive to lateral displacements. This birefringent element is placed between two polarizers, and the light signal coming out of the second polarizer is coupled to a detector (Figure 3). The birefringent material changes the state of polarization of the light passing through it, causing thereby a change in the optical power passing through the second polarizer.

In the case of microbend sensors, the transmitted light in the fiber modulates periodic bends in the fiber which are produced by placing it between two corrugated plates (Figure 4). This results in a radiative out-coupling of light from the fiber, and the degree of coupling depends on the degree of bending.

Thus, a kind of attenuation is introduced at a given location in the fiber, and the degree of attenuation depends on the pressure applied to the probe. Such FOSs can be classified into two types, viz., intrinsic and extrinsic. Intrinsic sensors are those where there occurs an interaction between light and measurand, and the light remains guided within the fiber, whereas in extrinsic sensors, the light is coupled from the fiber through a sensing region and back into fiber. The mechanisms of a few forms of intensity based sensors are discussed below.

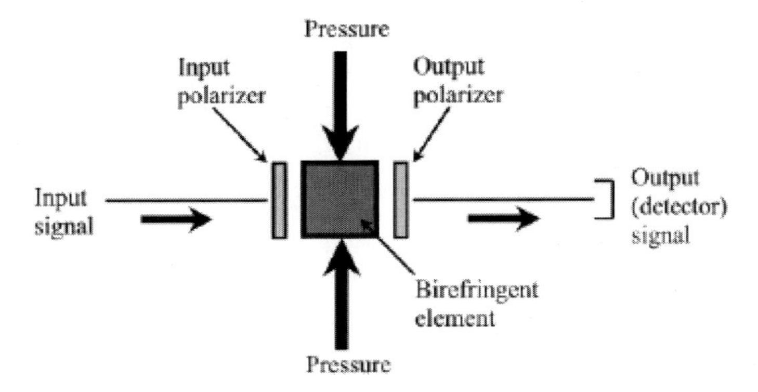

Figure 3. Arrangement implementing birefringent probes.

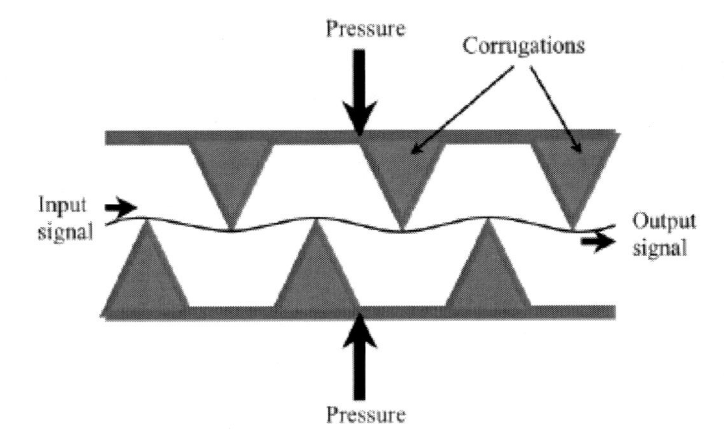

Figure 4. Microbend arrangement used in sensors.

2.1. Fabry-Perot Type Sensors

Several forms of Fabry-Perot sensors are described in the literature [26–29]. A rough sketch of the kind of sensor probe, that can be used with multimode fibers with a simple intensity based signal processing scheme, is demonstrated in Figure 5, where the light from an optical source is coupled to a micro dimensional Fabry-Perot element (consisting of two partially reflecting surfaces separated by a suitable gap of either air or some transparent material) located at the tip of the fiber.

In the present configuration, light reflecting off the two surfaces is coupled back into the fiber, and therefore, the net reflectivity of light is essentially a function of the phase difference between the two paths. The interference may be either constructive or destructive depending on whether the relative phase difference between the reflected optical beams is in phase or out of phase. The reflected power P_R is given as

$$P_R = \frac{1}{2}\xi P_0\left[1+\cos\left(\frac{4\pi n d}{\lambda}\right)\right]$$

where ξ, P_0, n and d are coupling coefficient, input power, refractive index (RI) of the material between the two reflecting surfaces and the thickness of the gap, respectively. Thus, the effective reflection coefficient $P_R/(\xi P_0)$ is a function of wavelength λ, and therefore, on illumination, a spectral response (depending on d, n and λ) is obtained which changes with the applied pressure owing to the change in the thickness of the gap. Shift in the spectral response is a measure of the applied pressure.

2.2. Emission Based Sensors

Functioning of emission based sensors primarily depends on the spectral shift in the blackbody radiation emitted by a body at a certain temperature. Such sensors use a kind of metallic coating at the tip of the fiber [30–32].

The blackbody radiation follows Planck's law (i.e. the total radiation W emitted by a body is proportional to the fourth power of the temperature T at which the radiation is emitted; $W \propto T^4$), and the temperature of the emitting surface can be obtained by the ratio of radiance detected in two band-limited sections of the spectrum. This kind of sensor system works particularly well at higher tepmeratures, i.e. more than 100 ^{0}C.

2.3. Differential Absorption Based Sensors

A simple fiber optic temperature sensor, based on differential absorption phenomenon, is shown in Figure 6, where the light emitting from two optical sources (wavelengths λ_1 and λ_2) is injected into the fiber. The injected light falls on the tip of the fiber where it passes through an absorptive thermochroic material, reflected by a reflective surface, and then returns into the fiber. The amount of light returned to the input end of the fiber is detected by a fiber coupler.

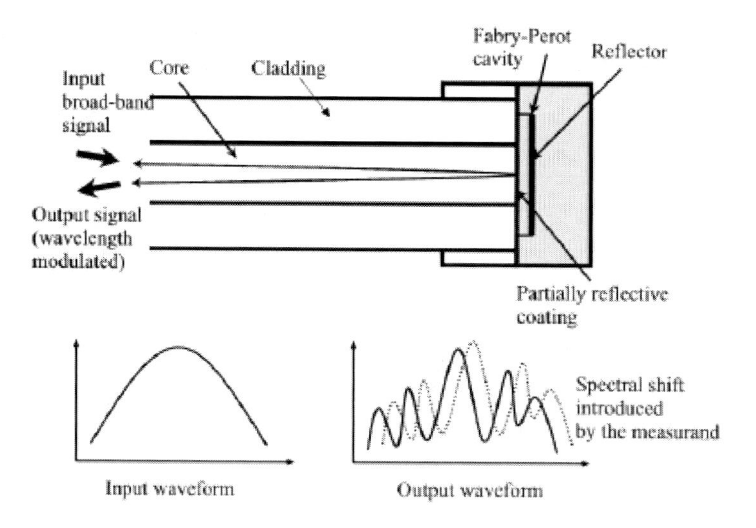

Figure 5. Fabry-Perot type sensor.

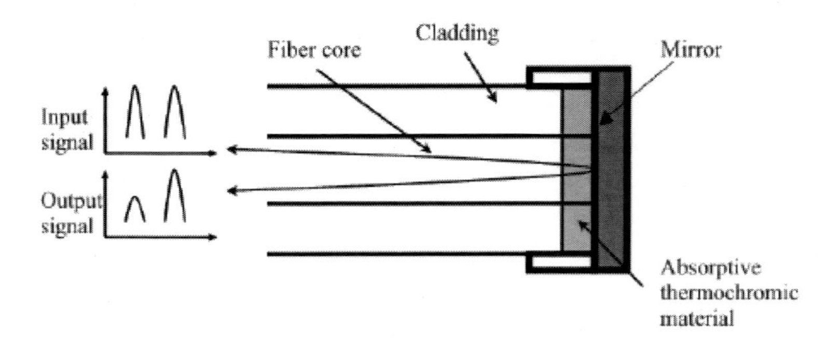

Figure 6. Sensor arrangement based on differential absorption.

The thermochroic material introduces a change in the spectral absorption characteristics with temperature producing thereby a spectral shift. In this kind of sensor system, in general, a solution is encapsulated at the tip of the fiber. However, semiconductor materials such as GaAs or heavily doped glass, that exhibit a shift in the absorption band, can also be used.

Varieties of systems employing optical absorption and differential optical absorption methods are reported in the literature [33–40]. These generally require an open cell configuration, and the light propagating through it acts as the detection system. But these systems meet the disadvantage that a very precise alignment of light is required and, in addition, the light in the near infrared region is transmitted causing thereby to monitor the weaker overtone bands only. However, the low-loss optical fibers are advantageous in the sense that the light can be transmitted to a longer distance, and the sensor head can be remotely located from the monitoring station.

2.4. Flourescence Luminsecence Type Sensors

The fluorescence and luminescence phenomena may also be implemented for sensing [41–48]. Such optical signals of many materials exhibit spectral and temporal changes with temperature. Principle of spectral based sensors depends on the detection of shifts in the fluorescence spectrum introduced by the measurand. The temporal based sensors involve the measurement of the lifetime of a fluorescence signal with the strength of the measurand. Figure 7 represents a typical example of temporal based such kind of FOS where a fiber-tip is sensitized by a fluorescent material (to be sensed), which is excited by a laser light source. In this process, a portion of the emitted light is collected by the fiber, which is then passed through a detector. The emitted fluorescence signal decays with a characteristic time constant that depends on the material and the host substrate. As a result, for a given material, the decay of the fluorescence signal can be measured with time, which ultimately yields the measurement of the change in temperature.

2.5. Evanescent Field Sensors

Sensors employing modified cladding elements are primarily based on the study of the evanescent wave spectroscopy, which play a very important role in the area of sensing technology. The technique involved in this kind of sensor incorporates the study of evanescent wave spectroscopy where the evanescent waves, penetrating the cladding region, are absorbed by the fluid (under testing) which is placed on the uncladded region of an optical fiber. Thus, a part of the fiber is sensitised by the fluid, and the study of the spectral composition of these evanescent waves, after getting absorbed by the fluid, give the estimation of their concentration.

Normally He-Ne laser is used as the optical source, and the uncladed portion of the fiber is sealed in a glass tube, which has facilities for continuous filling and drawing with the fluids of various concentrations. However, the selection of light source essentially depends on the peak absorption wavelength of the measurand. The output end of the fiber is connected to a power meter, and the determination of the transmitted power gives the concentration of the absorbing fluid. In this way a continuous monitoring of the absorbing fluid is done. The sensitivity of this kind of sensor depends on the type of the source used, numerical aperture (NA) of the fiber and the RI of the fluid. This also depends on the launching condition of the beam, and on the geometry of the uncladed portion of the fiber. However, if the uncladed region is tapered and only the selected rays are allowed to launch into the fiber, there occurs a many fold increase in sensitivity. Some particular forms of evanescent field sensors are discussed in the later sections of this review article.

2.6. Photometric Sensors

Types of sensors described so far involves the use of probe by which the measuring parameter is sensed. Photometric sensors are those where no any probe or transducer is used at the fiber end, and are useful for the measurement of several biological parameters in human body.

As for example, such sensors can be used to measure the relative velocity of the blood flow in vessels which is desirable for a wide range of medical purposes [49,50]. The velocity of blood flow in vessels can be determined by using a fiber coupled laser-Doppler velocimeter. The method is based on the Doppler frequency shift of laser light due to scattering from moving red cells. In this method, a graded-index optical fiber is inserted into a blood vessel through a catheter at a fixed angle to the axis of the vessel, and a laser light (of low intensity) is propagated through the fiber, which gets scattered by the blood cells at the outer end.

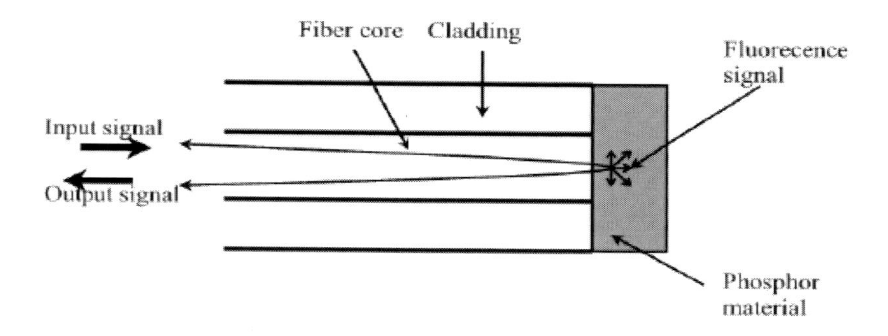

Figure 7. Temporal based fluorescence/luminescence sensor.

Now the blood cells, moving in the artery, will introduce a change in the frequency of the scattered light from that of the original laser light, and this change in frequency will be higher for the cells moving with higher velocity towards the fiber tip. Some of the light is back-scattered, and is collected by the same fiber. The Doppler shift in frequency is determined by the relationship

$$\Delta\nu = \frac{2nv\cos\theta}{\lambda}$$

where n (= 1.33) is the RI of the blood, λ is the free-space wavelength of laser light, v is the velocity of the blood flow and θ is the angle between the direction of the flow of blood and the fiber axis. Now, using this equation, velocity of the flow of blood can be evaluated by measuring the shift in frequency.

Photometric sensors are also used to determine the oxygen content of the blood by the measurement of the oxyhaemoglobin concentration [51–53], which is of high importance for human life. Blood contains two kinds of haemoglobins among which, one carries oxygen and the other does not. Both these kinds of haemoglobins reflect IR light equally, but the former kind reflects red light more than the latter one.

Therefore, two light sources, one emitting IR (805 nm) and the other emitting red (660 nm), are used [54]. A fiber coupler is used where the 660 nm wavelength light is coupled to one fiber while 805 nm wavelength light is coupled to the other, and the lights exiting from the other end of the coupler are reflected by the haemoglobins present in the blood (Figure 8).

The reflected light is collected by another fiber, the other end of which is connected to a detector. Now the intensity of the reflected red light gives the amount of oxyhaemoglobin while the intensity of the reflected IR light gives the total quantity of haemoglobin present in the blood. The concentration of oxyhaemoglobin, $HbNO_2$, is given by

$$\%\ HbNO_2 = A - B \times (R_1/R_2)$$

where A and B are constants that depend on the geometry of the probe, and R_1 and R_2 are, respectively, the reflectance at the first and the second wavelengths [54].

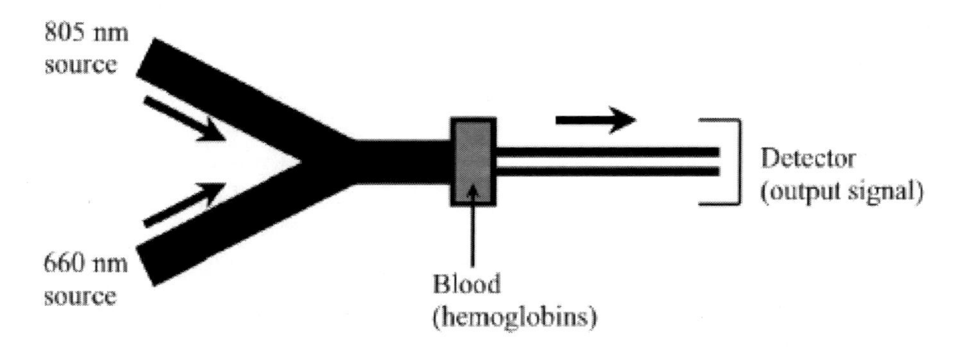

Figure 8. A fiber coupler arrangement.

The similar technique can also be employed for the determination of the cardiac output, which is of high importance for the treatment of cardiac surgery patients because, by this method, one can estimate the ability of the heart and lungs to supply oxygen.

3. Distributed Sensors

As stated earlier, distributed FOSs use all or a part of the sensitised fiber that reacts with the measurand, i.e. the measurand under investigation is monitored at many locations along the length of the fiber. Intrinsic distributed sensing and quasi-distributed sensing are the two techniques that fall under this category. In the former method, the measurand can be determined at any point along the fiber length, i.e. the fiber is sensitized along its entire length, whereas in the latter system, the measurand is determined at a number of predefined locations along the fiber length, i.e. the fiber is sensitive to the measurand at certain numbers of locations. Nowadays a rapid increase is seen in the number of techniques to achieve distributed sensing along optical fibers [55–62].

However, it is rather difficult to generalize on preferred approaches, but the optical time domain reflectometry (OTDR) method, that includes Raman and fluorescent variants, appear to present major advantages for distributed measurements. The achievement of more sensing range-cells per monitoring station, and greater ease of making compressions of the measurand value at each position, are the factors that increase the practical suitability of optical sensor systems. Some of the sensors based on this principle are described below.

3.1. Rayleigh Scattering Based Sensors

In optical fibers, Rayleigh scattering is the process by which light is scattered form the microscopic density fluctuations in the fiber material [63]. The loss due to this kind of scattering represents the fundamental limitation to loss in fibers at around 1.3 μm and 1.5 μm in silica glass fibers. As the loss in fiber at these wavelengths is very low, the amount of (back-) scattered light is also very low, which can be detected back at the input end of the fiber [64–66]. If light pulses are injected into an optical fiber having uniform loss and scattering characteristics, the power of the back-scattered light falls down exponentially. However, this decay appears as a straight line when viewed on a logarithmic scale; the slope of the line being proportional to the product of the loss and scattering coefficient of the glass. Thus, the spatial variation in these parameters gives rise to a change in the slope of this curve. This technique is termed as OTDR, and is widely implemented for the analysis of loss in optical fiber communications [67–71]. Intrinsic distributed fiber sensors typically utilize either basic loss or the loss due to scattering in the sensing part of the fiber. The change in the scattering coefficient has been used in distributed temperature sensing. However, this kind of dependence is very weak in solid core fibers; liquid core fibers are more suitable to perform this kind of sensing measurement. There are several other mechanisms related to loss which can also be implemented in distributed temperature sensing, e.g. the temperature dependence of the radiative loss due to external RI changes, bending loss in plastic clad fibers, absorption in appropriately rare-earth elements doped fibers.

3.2. Raman Scattering Based Sensors

Apart from Rayleigh scattering, Raman scattering has also been used for the demonstration of distributed fiber optic temperature sensors [72–77]. In this process, components consisting of Stokes and anti-Stokes lines (corresponding to the lower and the higher photon energies, respectively) are produced in a broad band about the exciting wavelength, and the amount of the backscattered light is analyzed by a modified OTDR method in which the ratio of the Stokes to anti-Stokes lines are measured. It is to be noted down that the magnitude of the coefficient of Raman scattering [78] is far less than that of Rayleigh scattering, and that is why, significantly high input power and signal with longer times are required; this is the major drawback of this kind of measurement technique.

3.3. Brillouin Scattering Based Sensors

Brillouin scattering is another phenomenon implemented for sensing applications [79–81]. As the light propagates through an optical fiber, there occurs an interaction between photons and phonons (generated by the photorestrictive processes in the fiber), and this causes a kind of scattering called Brillouin scattering of light. Light scattered by phonons presents a 'shift' by an amount determined by the velocity of the acoustic waves, which in turn, depends on the density of the glass, and hence the temperature of the material. The Brillouin shift in a fiber is given by the expression

$$\nu_B = \frac{2n_{eff}\nu_m}{\lambda}$$

where n_{eff} is the effective RI of the fiber, v_m is the velocity of the acoustic waves in the medium, and λ is the operating wavelength. The Brillouin shift in the frequency depends on two parameters, i.e. temperature and strain in the fiber, which is because the effective RI of the fiber and the velocity of phonons (in the fiber) both depend on these two parameters. It has been observed that the efficiency of spontaneous Brillouin scattering is approximately 20 dB weaker than that of Rayleigh scattering. Apart from these, several other forms of distributed fiber optic sensors have been demonstrated by the investigators that include phenomena like non-linear effects, mode coupling etc., e.g. Kerr effect, mode coupling in birefringent fibers etc.

3.4. Fiber Bragg Grating Sensors

The phenomenon of RI change in the core (of an optical fiber) due to the photosensitivity of optical fiber was first discovered in 1978 [82]. Fiber Bragg grating (FBG) is a longitudinal periodic variation of RI in the core of a monomode optical fiber, and fiber Bragg reflectors are written onto the core using a high intensity UV laser source. Optical fibers can be made photosensitive by doping the core of the fiber with materials like germanium (Ge), potassium (K) or boron (B) during the fabrication of preform. It has been found that conventional Ge-doped fibers, when illuminated with UV light in the region of 248 nm, exhibit a relatively strong photosensitive response [83]. There are several techniques to fabricate FBGs of which holographic and phase mask methods are generally implemented [84–88]. In both the cases, a quasi-permanent kind of grating structure is developed in the fiber core with the Bragg wavelength given as $\lambda_B = 2n_{eff}D$, n_{eff} being the effective RI of the fiber core and D being the grating pitch. Such a fiber with grating, when illuminated with a broadband source of light, reflects a narrowband component at the Bragg wavelength, and a perturbation of grating results in a shift in the Bragg wavelength (Figure 9) The use of FBG as sensing elements has been very well established, and by exploiting their inherent properties, these are used for a variety of sensing applications [89–96]. FBG sensors are the class of sensing devices based on the photosensitivity of optical fibers, and are widely used for the measurement of strain [97–100] because it (the strain) causes both the physical elongation of the sensors as well as change in the effective RI of the fiber [92]. The physical elongation results in fractional change in the grating pitch, and the change in RI is caused due to the photoelastic effect. The shift in Bragg wavelength is determined by the expression

$$\delta\lambda_B = (1 - p_e)\lambda_B\varepsilon$$

where p_e is effective photoelastic coefficient ($\cong$ 0.22 for fused silica) and ε is the applied strain; p_e is given by

$$p_e = \frac{n^2}{2}\{P_{12} - \eta(P_{11} + P_{12})\}$$

where η is Poisson's ratio and P_{ij} are Pockel's coefficients of strain optic sensor.

FBG sensors are also used for the measurement of temperature. Bragg wavelength depends on temperature as follows:

$$\delta\lambda_B = \left\{\left(\frac{dD}{dT}\right)D + \left(\frac{dn}{dT}\right)n\right\}\lambda_B\delta T$$

where $(dD/dT)/D$ represents the thermal expansion whereas the term $(dn/dT)/n$ relates to the thermal dependence of RI. It has been observed that in silica fibers that the latter part has more dominant effects on the observed shift. FBG sensors have several advantages over the other sensing techniques, e.g. the output doesn't depend directly on the total light, input power of the source, attenuation, etc. In addition, wavelength division multiplexing (WDM) can also be achieved by assigning each sensor to a different part of the available source spectrum. This is illustrated in Figure 10 where two (or many) grating structures are shown that can be used for a large number of devices; the maximum number of gratings that can be addressed in this way, however, depends on the width of the source profile [90,95,96]. Sensing devices implementing FBGs are based on passive broadband illumination, where light with a broadband spectrum is input to the system, and the narrowband component of the light, reflected by the FBG, undergoes detection.

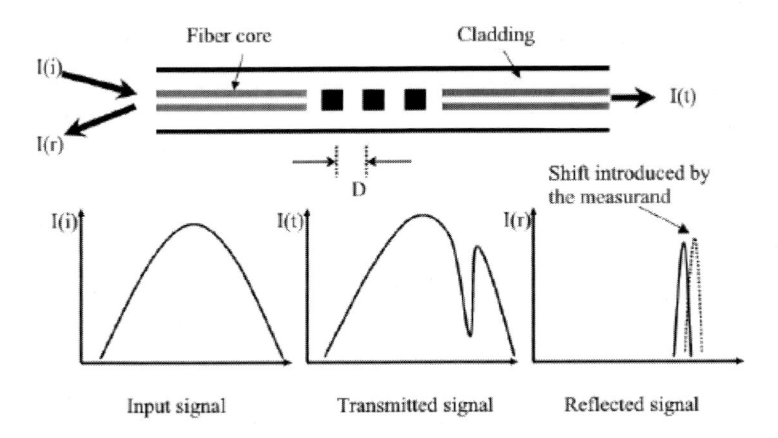

Figure 9. Fiber Bragg grating sensor.

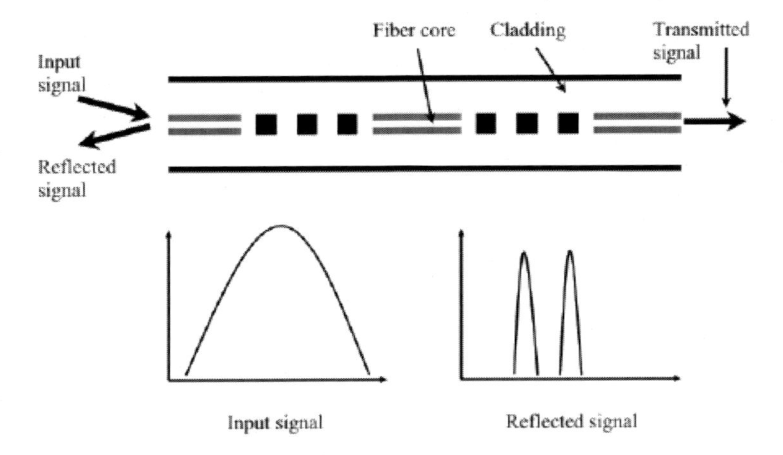

Figure 10. Multiple Bragg-grating fiber sensor.

There are several techniques to measure such Bragg reflected wavelength, e.g. interferometric detection, passive optical filtering etc. [99,100] The general principle of the interferometric method is shown in Figure 11.

An imbalanced interferometer has always a term related to the phase that depends on the wavelength of the input light signal. Light reflected by the grating is directed through the interferometer having unequal paths, and since the phase of the unbalanced interferometer depends on the input wavelength, shifts in Bragg wavelength are naturally converted into phase shifts. The phase shift of the output signal is related to the Bragg wavelength shift as

$$\delta\varphi = \left(\frac{2\pi nD}{\lambda^2}\right)\frac{1}{\delta\lambda}$$

where nD is the optical path difference. Proper choice of the factor nD can make the system sensitive even to smaller shifts in Bragg wavelength.

Another useful area of FBGs has been the fiber lasers where FBGs are used as spectrally narrowband reflectors for certain in-fiber cavities.

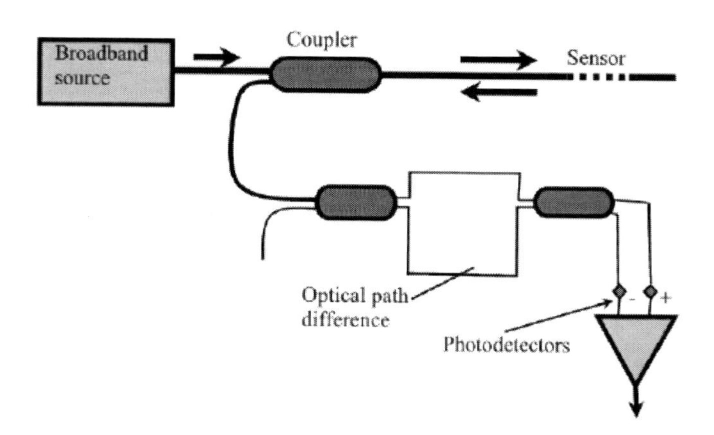

Figure 11. Fiber based interferometric sensor.

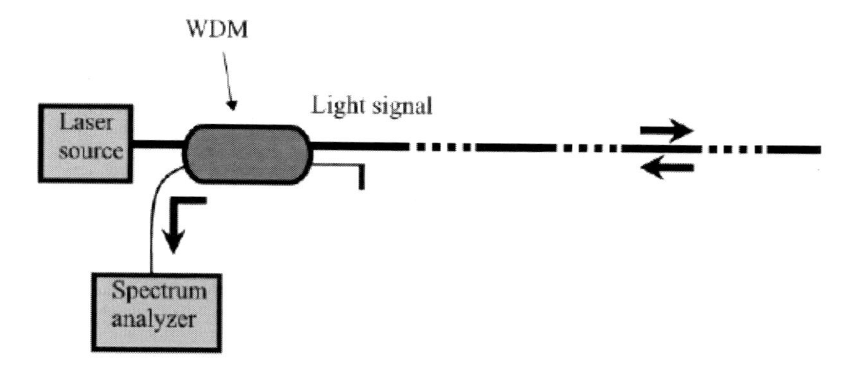

Figure 12. WDM based FBG laser sensor.

Such fiber lasers find useful applications for tunable single frequency devices for WDM networks. These are also used for strain and temperature monitoring [95,96,101–104]. FBG lasers basically utilize two gratings of properly matched Bragg wavelength, which creates an in-fiber cavity. The doped section of the fiber between the gratings allows the optical pumping of the system which ultimately results in the cavity, and therefore, the lasing action starts. Figure 12 represents a typical WDM array of FBG laser sensors. However, several other configurations employing FBG laser sensors have been presented by the investigators.

4. Interferometric Sensors

Interferometric sensors are of great importance because these are able to provide the sensitivity to a wider range of weak fields. These sensors find significant applications in detecting acoustic perturbations. Much work in the fields of interferometric demodulation, multiplexing etc., has appeared in the literature which motivated the investigators to implement this technology for sensing [105–114].

Several forms of interferometer configurations have been proposed by the investigators that include Mach-Zehender, Michelson, Fabry-Perot, and ring resonator types, among which the Mach-Zehender type configuration is widely implemented for sensing measurements. It is to be noted that Mach-Zehender and Michelson type configurations are two-beam systems, whereas Fabry-Perot- and ring resonator-type configurations are multiple-beam systems. Such interferometric sensors yield a non-linear output described by a cosine interference function, and the phase shift introduced by the interferometer (proportional to the strength of the measurand) is encoded non-linearly into an intensity change at the detector stage. Techniques based on the modulation of laser frequency were in use to introduce carrier phase shifts in an unbalanced interferometer, and such methods were found much more useful in the technological point of view. Earlier investigators used He-Ne lasers or 0.8 μm semiconductor lasers for interferometric sensing applications [115]. Subsequent invention of a number of lasers with narrower linewidth made the interferometric systems much more sensitive to the phase detection [116]. Typically, for example, diode-pumped ring cavity Nd-YAG laser operates at 1.319 μm, and exhibits linewidths of the order less than 10 KHz. Such a laser system allows the use of interferometers with long path differences between the arms without affecting the phase sensitivity. Fiber lasers are of another class which are also proved to be suitable optical sources for certain sensing applications implementing interferometric configurations, e.g. short-cavity Bragg grating based erbium-doped fiber lasers exhibit linewidths of the order less than 10 KHz [117, 118].

In connection with the techniques to introduce internal mirrors in a fiber by fusion splicing, end-coated fibers were reported [119,120]. Earlier work in the area of multiple beam interferometer used coatings of monolayer dielectric materials which, after splicing, showed reflectance less than 10%. Fabry-Perot interferometers fabricated with such reflectors generated interference fringes. This is shown in Figure 13. Later multilayer deposited films on the ends of fibers were also tested, and it was found that the fusion splicing of multilayer coated fibers to uncoated fibers produced internal mirrors with much improved reflectance. One of the most widely recognized multiple beam interferometric configurations is the extrinsic Fabry-Perot interferometer (EFPI) in which two fibers, one carrying light from optical source and the other acting simply as a reflector, are inserted together into an alignment tube (Figure 14) [121].

In this system, Fabry-Perot cavity is formed by the air gap between the two uncoated fiber end faces; the interference signal is detected in the reflected light back at the source end of the input fiber. These sensors are used to monitor strain. In such systems, in general, both the fibers are epoxied into the alignment tube.

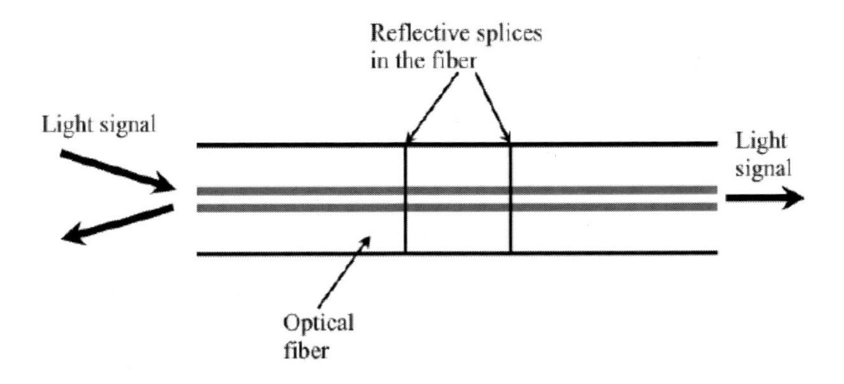

Figure 13. Fiber Fabry-Perot sensor.

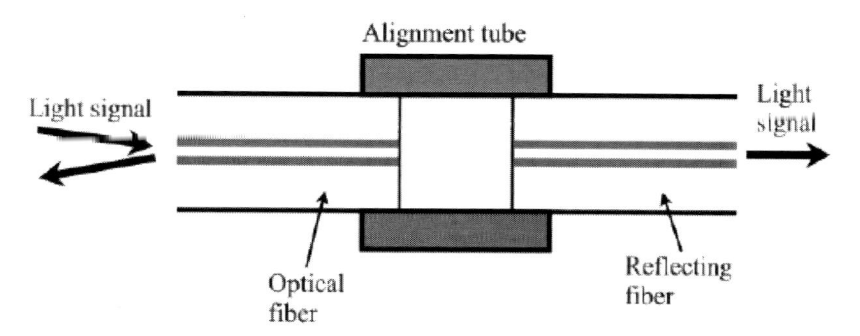

Figure 14. Extrinsic fiber Fabry-Perot sensor.

Application of strain results in a change in the air gap between the fiber end facets which are then detected in terms of the shifts in the signal. An almost similar kind of sensor configuration can also be formed by allowing one of the fibers to move freely within the alignment tube.

5. Some Other Forms of FOSs and Their Applications

The design of a simple and low cost liquid level sensor has been presented that can be fabricated by using an optical source emitting a narrow beam and a photodetector, and a tube containing some liquid. The light beam from the source may travel either through the tube (containing the liquid) or reflected by the surface of the liquid [122]. In this kind of sensor, the intensity of light is changed due to its reflection/refraction from the liquid surface and/or absorption in the liquid, and the intensity (detected by the photodetector) changes as a function of the liquid level. Investigators claim that the sensor can be used in automobile and aeronautical industries, and in the measurement of the level of oil in electrical transformers.

The operation of Bragg displacement sensor is based on the principle of distributed feedback dye lasers. Using Nd:YAG laser as the optical source, experimental demonstration of microdisplacement sensors based on the periodicity of Bragg grating and interference overlap length modulation has been presented by K. Nasrullah [123]. It was found that the displacement length varies linearly with the change of wavelength as well as the optical pulse length. This kind of ultrasmall displacement sensor can be used to determine the frequency of vibration or simple displacements in ultraprecise applications.

Faraday effect current sensors and Sagnac effect fiber optic gyroscopes are also of practical applications [124]. Both the Faraday and Sagnac effects can be detected in terms of phase changes between two counter-propagating lights in ring interferometers [125,126]. Investigators have presented fiber ring lasers for current or rotation sensing in which the non reciprocal phase shift due to the magnetically induced circular birefringence or the Sagnac effect is detected in terms of the beat frequency between the counter-travelling laser beams [127–132]. For such devices, fiber Brillouin lasers or erbium-doped fibers lasers were used.

Low birefringence optical fibers are very useful for Faraday effect current sensors because stable sensitivity can be achieved. Lead glass monomode fibers present six-times larger Verdet constant than a silica fiber with very low photoelastic constant as well as low intrinsic birefringence. As such, these fibers are able to make the polarization of light (propagating in itself) stable against external mechanical perturbations and the effects of temperature, and therefore, such fibers are used for polarization modulation current sensor [133,134].

Fiber ring lasers using laser diode amplifiers and flint glass fibers are also studied where the use of a laser diode amplifier (instead of erbium-doped fiber amplifier) for a gain medium makes the set up economic as well as provides the opportunity to use the device to sense various wavelengths and the integration with other semiconductor based devices [135–137]. The application of the device was demonstrated as frequency-domain fiber current sensor and fiber gyroscope. It was found that, because of the involvement of a Faraday bias element in the cavity, the laser was substantially free from frequency locking, and was able to detect the polarity of current as well as the direction of rotation.

It is rather important to detect the liquid-phase water in many applications. It has strong absorption bands in the near IR region of the EM spectrum, and therefore, near IR spectroscopy is certainly of great importance in the context of water sensing [138–141]. Demonstrations have been done for the application of FTIR absorption spectroscopy in sensing water concentration in milk [142]. Plastic fibers with fluorescent-dye-doped cladding have been reported to monitor breathing conditions and humidity [143]. Although several kinds of optical sources have been proposed for such purposes [144–147], a Tm^{3+} doped YAG crystal seems to be a more suitable one as its small crystal can even yield high pumping efficiency. Investigation of optical fiber water-droplet sensor has been reported which is based on the principle of absorption spectroscopy, and uses the fluorescence from a laser diode pumped Tm^{3+}:YAG crystal as a compact and high power fluorescent light source [148,149]. In the experiment, the fluorescent light was launched into an optical fiber, which acts for both sending the fluorescent light and picking up the perturbed optical signal. It is found that such sensors can be used for the estimation of both the size and the position of water droplets. Also, it can be used to distinguish clearly water from other perturbations such as oil and surface irregularities.

The suitability of a fiber optic portable Raman spectrometer for in-situ measurement of the percentage of water by volume in methanol has been demonstrated by Asundi *et al.* [150]. The basic measurements involved are the measurements of intensities of scattered radiation, from the material under investigation, as a function of wavelength or frequency. Also, it is possible to represent the measured intensity as a function of Raman shift. By exciting the test sample with a suitable laser beam, the scattered radiation was detected with an array of detectors. The intensity of the scattered radiation was represented as a function of Raman shift, which gives the signature of each material (that is unique to that particular material). Raman spectra diagnostic tests show the obtained spectra on par with those obtained by conventional spectrometers. It has been found that the measured intensities at the characteristic Raman shift corresponding to methanol are inversely proportional to the volume of water present in methanol. The experimental procedure can even be applied to measure and quantify the methanol content in gasoline, and can be directly used in the quality control where there are the possibilities of adulteration of gasoline with methanol.

Utilizing the thermal and optical properties of Ta_2O_5 (tantalum pentoxide), a robust, compact and reliable fiber optic low-coherence wavelength modulated thermometric sensor has been presented by Inci *et al.* [151]. The principle of operation of the sensor is based on the temperature dependence characteristics of wavelength, and the sensor is free from vibrational effects and intensity fluctuations, and also, has the potential for multiplexing. Their design was based on the spectral shifts introduced by the thin films of Ta_2O_5. A monolayer film of Ta_2O_5 was deposited on the cleaved end face of a monomode optical fiber, and the film was illuminated with a broad-band super luminescent diode source through a directional coupler. Interference fringes were obtained on reflection from the film, which show shifts with the rise in temperature. This kind of sensor can be used to measure higher temperature ranges up to 650 °C. For the measurement of such a high temperature, Ta_2O_5 is found to be a good candidate as it has a very high melting point, strong adhesion to glass and most other oxides, high RI (typically 2.15 at 1.55 μm), high thermal optical coefficient (about ten times greater than that of fused silica) and low absorption in the near UV (350 nm) to IR (8 μm) region of the EM spectrum.

A fiber microbend sensor (MBS) is an inexpensive intensity based sensor used to detect deformation and several other related quantities such as strain, temperature, pressure etc. [152–156]. However, its sensing performances are rather complicated, and are the function of fiber parameters and deformation. This feature limits their applications up to simple monitoring or on-off sensors. Investigators have reported MBSs based on non-uniform deformations of fiber caused by some means like corrugated plates etc. [157,158]. Yoshino *et al.* [159,160] reported a spiral microbend-loss sensor, and it was shown that, by means of monomode fiber embedded metal springs, strain and deformation can be measured with good linearity and reproducibility, and also, with large dynamic range in strain/deformation measurements. Their spiral-type MBS was based on the uniform change in the bend radius of the fiber so that the characteristics of the sensor can be specified simply from the circular bending loss of the used monomode fiber. Analyses were also made of the effect of spring parameters (e.g. radius, pitch and length) on the sensing performances.

Optical gas sensing methods have proved several advantages (over conventional chemical methods) such as the possibility of fast, remote, selective, nondestructive and intrinsically safe detection [161–168]. An optical method to sense toxic gases (like methane etc.) utilizing evanescent waves of optical fibers has been demonstrated by Tai *et al.* [169]. In their experiment, a single multimode step-index silica optical fiber having 50 μm and 125 μm as core and cladding diameters, respectively, was used as both a sensor and an optical transmission medium, and therefore, the detection of gas could be performed in a simple, flexible, and possibly distributed way. They presented an experimental demonstration for the detection of methane gas by its strong optical absorption at 3.392 μm of a He-Ne laser light. The sensing part of the fiber was made by heating and expanding some short section of the fiber, and the length and the diameter of the sensing part ranged from 5–10 mm and 1.8–7 μm, respectively. Optical transmittance of the fiber for the 3.392 μm-line was measured for various lengths of the fiber, and for various concentrations of methane gas. It was observed that an evanescent wave of 5–40% of the total propagating power was generated outside the fiber, and the use of 1.8 μm diameter and 10 mm length of the sensor fiber showed that the minimum detectable concentration of methane gas was less than the lowest explosive limit.

A variety of FOSs based on immuno-fluorescence spectroscopy has been developed which can detect very low concentration of specific chemicals in complex samples [170]. The chemical selectivity of these sensors depends on the antibody-antigen reaction. A fiber optic microprobe is located at the end of a fiber; the surface of the microprobe contains the antibody. If the microprobe is inserted into the sample containing antigens, these will be captured by the antibodies. When the original fiber is excited by a laser source, antigens fluoresce; the intensity of fluorescence is used for the detection of several viral infections, e.g. screening of cancers etc.

pH sensor is also a kind of FOS which is based on the principle that the absorbance of the phenol red dye depends on its pH value. In such sensors, an optical fiber retroscatter monitor interrogates the indicator at two different wavelengths; one measures the change in reflectance as the dye responds to pH changes, whereas the other provides a normalization signal at a wavelength that is insensitive to pH.

In general, the probe contains polyacrylamide microspheres which are covalently bound to the red dye in order to prevent these from diffusing out of the probe [171]. Typically, the probe diameter is kept about 5–10 μm. These polyacrylamide spheres are sealed in a cellulosic tube that is attached at the ends of two fibers, and permeable to hydrogen ions. Now two lights having their wavelengths 560 nm and 600 nm are passed through one fiber. The 560 nm wavelength light is absorbed by the dye, and the amount of absorption is a function of the pH value of the solution. The 600 nm wavelength light acts as a reference beam, and its absorption does not depend on the pH of the solution. Both the light beams, after multiple scattering and absorption in the cellulosic tube, ultimately return to the other fiber connected to the detector. The ratio of the intensities of two lights reveals the pH of the solution.

The response of fiber optic evanescent field sensors to pH variations has also been presented by Choudhury and Yoshino [172].

In the experiment, a bent probe was used because the sensitivity is remarkably improved in this case. The bend was particularly chosen to be of the U-shape. At this point, it is of worth to mention that Takeo and Hattori [173] reported sensors with a U-shaped probe for the measurement of RIs of liquids. In order to study the sensitivity, an experimental investigation has also been carried out with fiber probes of different values of the radius of bend [172].

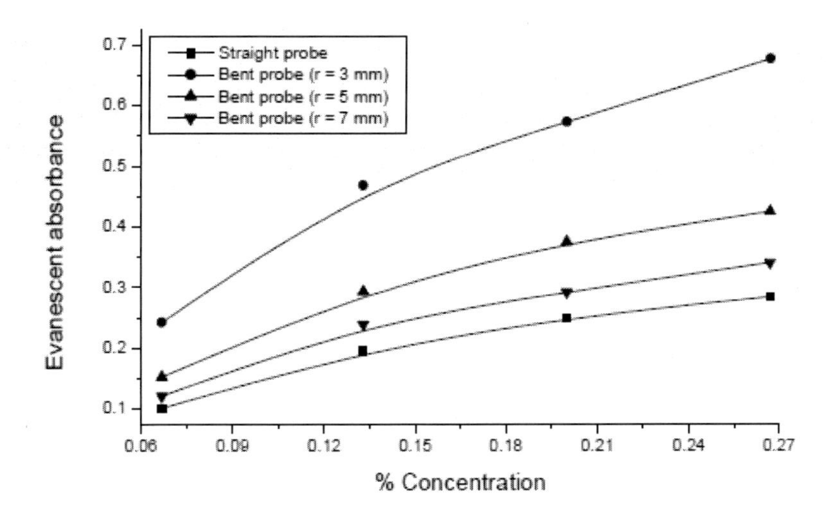

Figure 15. Variation of the evanescent absorption of straight and bent probes.

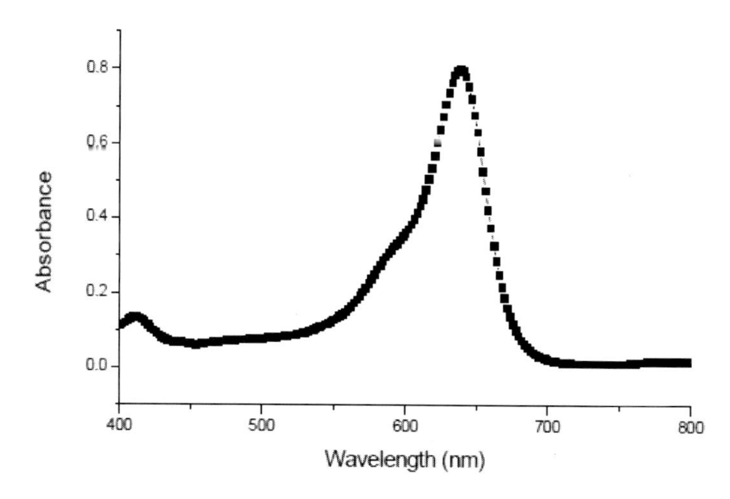

Figure 16. Absorption spectrum of aqueous patent blue VF dye.

Figure 15 shows the evanescent absorption measured for straight probe as well as bent probes. Aqueous solutions of patent blue VF dye with different concentrations serve the purpose of measurand. The RI of these samples is found to be 1.332 (as measured by KEM RA-130 refractometer). A typical absorption spectrum of the aqueous solution of patent blue VF dye is shown in Figure 16, which is obtained by a Hitachi U-3000 spectrophotometer. We observe the peak absorption wavelength near 635 nm, which is very close to the He-Ne laser wavelength. As such, a He-Ne laser can be used as the light source.

Figure 15 shows that the absorption increases with the decrease in bending radius. Compared to the straight probe, we observe that the absorption is increased roughly by 1.2 times for a bent probe of radius of curvature $r = 7$ mm. For $r = 5$ mm, the absorption increases by 1.6 times approximately, and for $r = 3$ mm, the absorption is increased roughly by 2.4 times. Therefore, with the decrease in radius of curvature, the absorption, and hence the sensitivity of the probe increases.

However, the value of optimal bending (in order to get the maximum sensitivity) will be a function of several other factors, viz. the measurand RI, the penetration depth and loss. Thus, to maximize the sensitivity, a trade-off among these factors will have to be considered. Also, the maximum sensitivity will be limited to the value of bend (of the probe) beyond which the process of scattering starts taking place.

Study of the variation of evanescent absorbance for a fixed concentration of patent blue VF dye is shown in Figure 17 for different values of the uncladded fiber length. A linear relationship between the absorbance and the interaction length is observed, which predicts that for a fluid, obeying the Beer-Lambert law of absorption, the evanescent absorbance is linearly related to both the exposed fiber length and the fluid concentration. Paul and Kychakoff [174], however, presented earlier the study of the absorption behavior of rhodamine 6G laser dye, and reported a linear kind of dependence of the absorption coefficient on the measurand concentration. In their experiment, they used organic solvents.

Figure 18 represents the plot of the absorption coefficient with the percentage concentration of aqueous dye solution corresponding to two individual cases, viz. free-beam as well as the method of fiber optics; the latter one implements silica fiber with fixed uncladded length (5 cm).

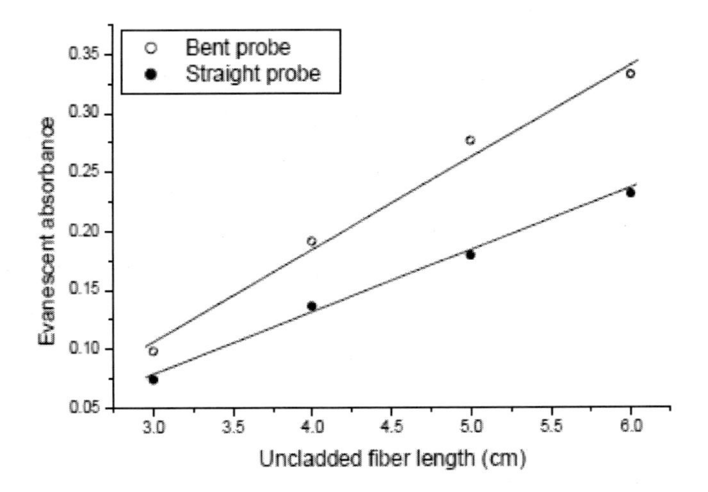

Figure 17. Variation of the evanescent absorption with different probe length.

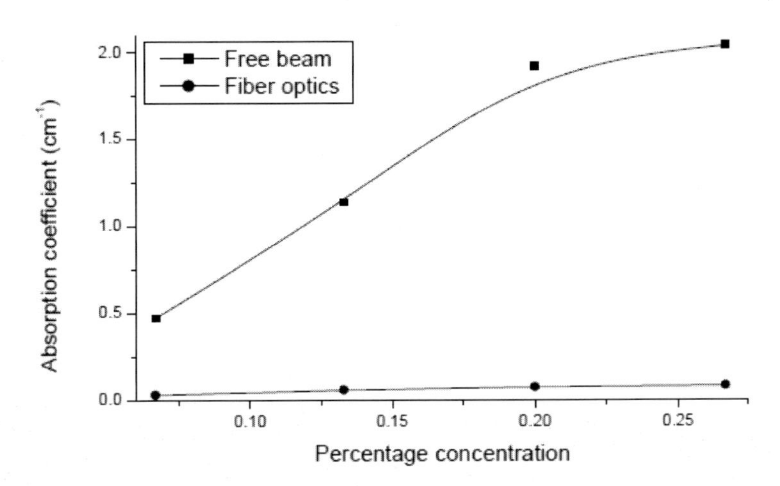

Figure 18. Plot of absorption coefficient with percentage dye concentration.

It is seen that, in the case of free-beam method, the absorption coefficient depends nonlinearly on the measurand concentration. This kind of behaviour cannot be easily seen for the case of fiber optics method, which is because of compact vertical scale in Figure 18. However, the same curve in Figure 15, corresponding to the case of r = 5 mm, shows the nonlinear dependence of the evanescent absorbance on the measurand concentration. A similar kind of nonlinear dependence of the evanescent absorbance on the measurand concentration was reported earlier by Ruddy *et al.*. [175], which is in contrast with the Beer-Lambert law.

Choudhury and Yoshino [176] reported the nonlinear kind of dependence of absorbance on the measurand concentration which is in contrast with the Beer-Lambert type linear dependence. This is attributed to the accumulation of high density of dye molecules on the bare fiber surface, which is due to the electrostatic attraction between polar silica surface and the ionic measurand solution, i.e. the positively charged chromophores. This is justified through the experiments performed under the free-beam condition too [176]. It is reported that the pH has a significant role in determining the nonlinearty of the absorbance response. The degree of nonlinearity decreases with the decrease in pH value, which is owing to the less accumulation of dye molecules on the fiber surface as the value of measurand pH decreases.

Evanescent field sensors can also be used for detecting rate coefficients [176,177] of time dependent chemical reactions. Industrial applications of chemical kinetics require the knowledge of chemical reaction rates, which is essential to establish the optimum physical conditions to perform a reaction. The conventional methods for finding rate coefficients are volumetric analysis [178,179] and the optical transmission [180]. The method of volumetric analysis is much complicated whereas the method of optical transmission is quite expensive. The reported method of evanescent fields would be potentially useful for the study of different chemical kinetics owing to the low cost and simple design. As an illustrative case, typically the reaction between potassium iodide (KI) and potassium peroxdisulfate ($K_2S_2O_8$), which is a time dependent reaction, is reported [176,177]. Equimolar aqueous solutions (0.052 M) of both of these reagents are used. The reaction results in the evolution of iodine, and the quantity of evolution is time dependent. A violet colour solution is formed, and the colour deepens with the increase in time, indicating the increase in the concentration of iodine with time. The absorption spectrum of iodine is presented in ref. [181] and [182]. Both straight as well as bent probes are used to compare the sensitivity (Figure 19). It is found that the result obtained with the U-shaped probe (rate coefficient = 0.0145 min^{-1}) is better than that obtained with a straight probe (rate coefficient = 0.0114 min^{-1}), as confirmed by using a Hitachi U-3200 spectrophotometer (Figure 20; rate coefficient = 0.015 min^{-1}), which is attributed to the increased sensitivity of the U-shaped probe.

In this context, it is also of worth to observe if there is any significant variation in the measurand RI during the chemical reaction, because the studies related to evanescent wave spectroscopy are based on constant RI of the measurand. During the chemical reaction, the measurand RI is monitored by a refractometer (KEM RA-130), and it is seen that the RI remains almost unchanged (Figure 21). This shows the validity of the results obtained by the absorption of evanescent waves. The concept of such an arrangement can be implemented for detecting a variety of other time dependent chemical reactions, and in general, for the study of chemical kinetics.

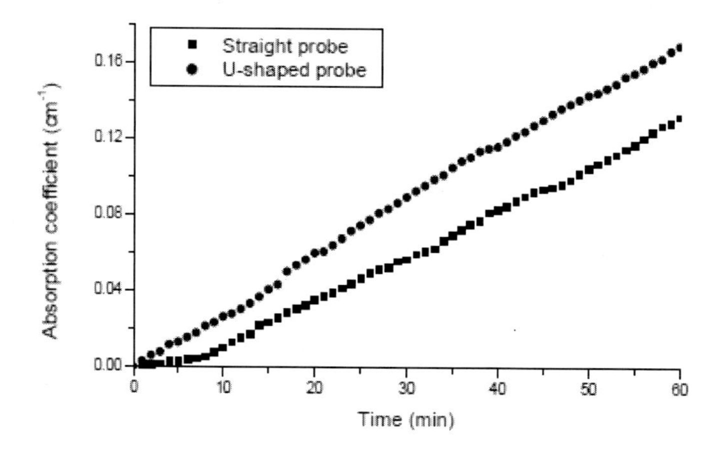

Figure 19. Variation of the evanescent absorption with time for straight and bent probes.

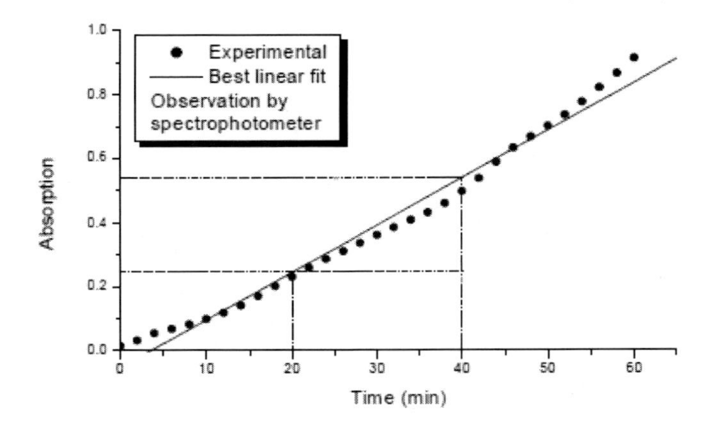

Figure 20. Variation of absorption as observed by spectrophotometer.

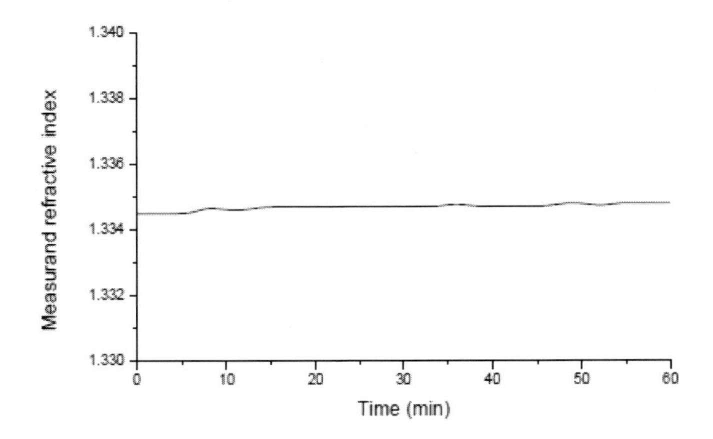

Figure 21. Variation of measurand RI with time as recorded by refractometer.

Further, it is only an illustrative case of chemical reaction presented in the foregoing discussion. However, the process is very general, and can be applicable to monitor any time dependent reaction. The simplicity of the system and the low cost are the additional features of this method for the determination of rate coefficients.

Another important potential market for fiber optic sensing technology is in the areas of aerospace and marine applications [183,184], which is principally because optical fibers are immune to electromagnetic interference. Current technologies are advanced to implement the transfer between the measurand and the modulation function applied to the fiber, and to retrieve the information from all the transducer elements in an array.

The main potential applications of optical fibers in the area of aerospace sensing are the displacement transducers to monitor the position of control surfaces, embedded FOSs for monitoring the performance and fabrication of carbon fiber-reinforced composite materials, and in the advanced testing of gas turbine engines. FOSs find similar applications in respect of marine technology, and high sensitivity interferometric fiber sensors.

Rotationally nonhomogeneous thin films comprised of parallel columns with nano-engineered shapes are a new class of materials addressed as sculptured thin films (STFs) [185–187]. Many new applications of STFs in the area of optical technology are coming up in the horizon [188–190]. The porous characteristic of STFs can be exploited for the purpose of optical sensing. STFs can be morphologically classified into a type called thin-film helicoidal bianisotropic mediums (TFHBMs), which are STFs having cholesteric morphology [191,192]. The other type of STF morphology is nematic [193]. Dielectric TFHBMs can be used for the determination of gas concentrations, and the relevant theory has been developed by A. Lakhtakia [194].

It was reported that bilayers comprised of two matched layers of dielectric TFHBMs with opposite structural handedness have the property to reflect arbitrarily polarized, normally incident, plane waves within a specific frequency range. Porous characteristic of TFHBMs can be exploited to devise a model such that the infiltration of a TFHBM bilayer by a gas would shift as well as reduce the frequency range. Thus, a properly designed TFHBM bilayer can serve as the active element of an optical device for sensing gas concentrations.

The mechanism may be understood in a way that the tunability of the notch-filter response of the TFHBM bilayers [195] can be used for the detection of gas concentration. If the microcolumns of the TFHBM bilayers, which are supposed to be made of chemically inert isotropic materials, the bilayer itself, on the immersion in the gaseous environment, would cause gas molecules (to be sensed) to diffuse in the intermicrocolumnar void regions. This would ultimately result in a change in the permittivity dyadic, resulting thereby a shift in the notch in the transmission spectrum. The amount of shift may be used to estimate the concentration of gas. Thus, in this kind of sensor, nanoscopic-to-continuum models are devised to predict the electromagnetic response of TFHBM bilayers. For sensing some specific gases, such sensors may not be necessarily superior to the existing ones.

However, these may be implemented for gases whose sensing is currently infeasible or problematic. TFHBM bilayers may provide the capabilities of real-time monitoring for sensing several toxic as well as non-toxic gases. Patches of such bilayers may be compactly mounted on the same platform in order to sense different gases simultaneously.

Further, it is expected that TFHBM bilayer sensors are reusable, which is essentially a key feature toward the logical development of such sensors. STF Šolc filter has also been proposed for sensing gas concentrations [196]. The gas, to be sensed, is allowed to enter the filter (which is porous); the electromagnetic response of the filter is modified by the gas itself. Thus, in such sensors again, the transmissivity is found to be strongly dependent on the concentration of the gas present in the inter microcolumnar void region of the thin film. Ultra-narrow band-pass features in the transmission spectrum of the Šolc filter shows a shift with the change in gas concentration.

As such, by using STF Šolc filters, concentration of gas can be determined. It has been shown by Lakhtakia *et al.* [197] theoretically as well as experimentally that, by cascading two identical chiral STF sections with one section twisted with respect to the other by 90° about their common helicoidal axis, a special reflection hole may be realized. Such an arrangement can be used for optical sensing of fluids [198].

In such a device, a chiral STF is treated as a two-phase composit material with locally biaxial dielectric properties, and it has been predicted that the presence of a fluid in the porous film results in a red-shift of the spectral holes. Chiral STFs have been proposed for biosensing applications, which are essentially based on their emission characteristics [199–202].

The reference [202] particularly presents the theory of second-harmonic-generated radiation from a chiral STF, which can be applied for bio-sensing. In such a system, a linear emission is supposedly due to luminescence-producing bio reactions occurring in the void region of the chiral STF platform. In conclusion, theoretical and experimental investigations indicate that chiral STFs have established a strong platform in the area of bio-sensing, and their commercialization will be undoubtedly realized in the near future.

6. FOSs in the Environmental Context

The current technology related to FOSs involves their use in process control to monitor parameters like temperature, liquid level, gas density and image. Advantages of these sensors have been confirmed through their practical use in industrial plants. Such sensors are found to be expensive, and the efforts are underway to improve the fabrication techniques, reliability, and cost reduction. The development of sensor networks plays very important role in cost reduction. Multiplexed and distributed sensing systems enable simultaneous measurements of several points, and these are the key technology to expand the use of such sensors in process control technology.

Nowadays environmental assessment and control market has come in the limelite because it has become a global issue how technology can solve the rapidly growing problems in the environmental context, in order to maintain and improve the quality of life of the nation. Scientists have found that the hottest application of FOSs could be in the emerging area of environmental safety monitoring where a variety of chemical reactions are essentially required. Fiber optic chemical sensors are extremely used in monitoring various chemical processes taking place in the environment. The environmental sensor industry is aware of this major issue, and in the present world of technological improvements, it is rather difficult to classify various environmental sensors by one single technology because the recent trend is now to fabricate hybrid devices, i.e. the entire system involves a combination of various technologies [170,203].

Scientists have invented a variety of gas sensors for the health and safety monitoring purposes. These sensors are used for the detection of various toxic gases and air pollutants like methane and other hydrocarbons, which are deemed to be environmentally threatening [161–169,204]. These toxic gases exist in the air at ppm level, and are hazardous for human beings. For such sensing applications, different methods like evanescent wave spectroscopy, optical absorption, differential optical absorption, infrared fiber optics etc., are usually employed. Several other methods have been reported for the detection of toxic gases like methane which, with the mixture of air, is highly explosive. One of the major advantages of the optical gas detection systems is that these are intrinsically safe. Different methods have been employed for the detection of methane gas using the absorbance of optical power at various wavelengths viz. 1.33 μm, 1.66 μm, 2.25 μm, 3.33 μm, 3.39 μm and 7.7 μm. Systems like fiber optic multipoint sensors are also investigated [205].

The presence of chlorine in drinking water is highly hazardous for health. Using a U-shaped probe, Choudhury and Yoshino [206,207] reported the use of evanescent field FOSs for the detection of chlorine content in drinking water. For this, a green LED with its peak emission wavelength around 530 nm is used as the light source.

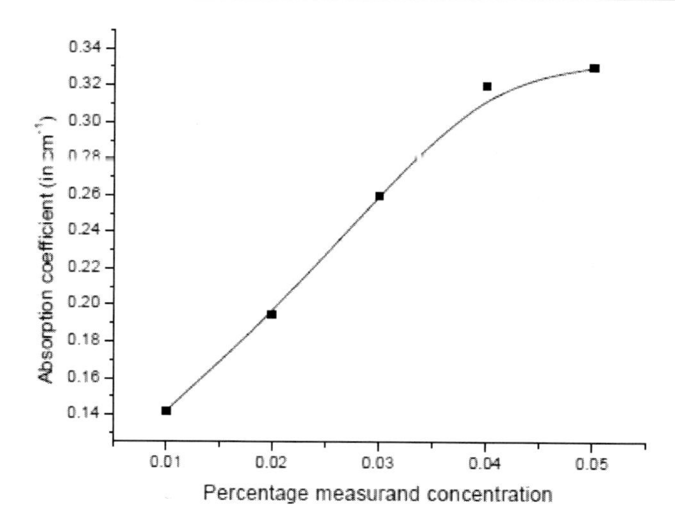

Figure 22. Variation of absorption coefficient against percentage concentrations of chlorine water.

Different concentrations of chlorine water are used as measurands, in which a little amount of diethyl phenylene diamine (DPD) organic compound is added. DPD reacts with chlorine forming an azo compound.

For illustration, the measurements are made for the concentration range from 0.01% – 0.05%, and it is found that, during the experiment, the RI values of the measurands remain almost constant at 1.335 (as measured by KEM RA-130 refractometer). Figure 22 shows the variation of the evanescent absorption coefficient against the different percentage concentrations of chlorine water. It is found that, corresponding to the lower concentration values, the absorption coefficient varies approximately linearly with the percentage concentration. However, for higher concentration values, the dependence is nearly nonlinear, which is attributed to the higher deposition of azo dye over the silica surface. At lower concentrations, less azo dye molecules are deposited on the bare fiber core.

However, for higher concentration values, more azo dye molecules are formed which get deposited on the silica surface of the bare fiber core, reducing thereby the absorption of light. Thus, the characteristic of the sensor is greatly affected due to the loading of azo dye molecules on the fiber surface. Figure 23 shows the reproducibility of the results, which shows almost similar evanescent response curves, though there exists a little deviation owing to the possible presence of azo molecules over the bare fiber surface even after rinse. Thus, the chlorine content in drinking water can be detected by colour formation.

Sensor systems employing evanescent wave spectroscopy has the advantage that it can be coupled (optically and mechanically) directly to an optical fiber. D-shaped fibers are used as evanescent wave methane gas sensor, and for this, attenuation measurements on guided waves are done at a wavelength corresponding to 1.33 μm and 1.66 μm.

Such systems are more rugged and less expansive arrangements [208–210]. Parallel and sequential measurements are also made by employing a number of pairs of optical fibers connected to the individual sample cell for monitoring methane and carbon mono-oxide in respect of their concentrations.

Electrochemical sensors are those that produce electrical signals induced by chemical reactions [180,211]. Bio-sensors are also a kind of electrochemical sensors used in the context of environmental monitoring [212,213]. The detection of gases is achieved by measuring changes in electrical parameters (e.g. resistance or capacitance) that happens due to the absorption of gas molecules by organic polymers like polyanilene mixed with polypyrrole.

Distributed FOSs are used for the detection of methane, which exploit the interaction of the evanescent light wave in a distributed fiber with a methane sensitive coating. Thallium-doped fiber laser operating at a wavelength of 1.648 μm is also a kind of methane sensor, and these can also be used for several gases lying in the wavelength range from 1.65 μm to 2.05 μm. Infrared methanometer is another methane sensor that uses a Febry-Perot etalon filter. One of the most commercially available environmental sensors involves the use of smart skin technology, and is used for leak detection of underground storage tanks [203]. Varieties of sensors, using novel membrane technology, have also been invented for environmental measurements and analyses. However, the use of novel chiral STFs in the area of biosensing is described in Sec.6 above [200–202].

7. The Role of Integrated Optic (IO) Technology in Sensing Applications

There has been a promising role of IO too in the FOS research [214–220]. However, still many problems like fiber coupling, production of suitable masks, reduction in loss, packaging etc., are to be resolved before the IO technology can find commercial implementation to its fullest extent. Silicon IO appears to be highly promising in context of sensors because it offers fairly simple and inexpensive means for precise fabrication of waveguides with much more flexibility than that achievable by glass waveguides formed by the ion exchange method. Apart from this, silicon technology offers the evolution of high-speed modulators, sources as well as detectors too. However, direct combination of silicon microcircuits with silicon optics is of high significance giving substantial rewards.

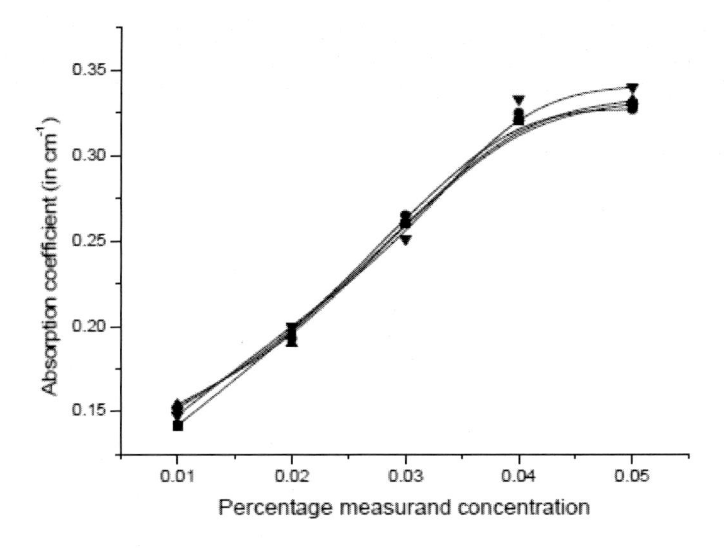

Figure 23. Reproducibility of the results of Figure 22.

Truly speaking, silicon has to play several significant roles in sensing technology because its mechanical properties are also the best among other semiconductor materials. Consequently, it can be implemented wherever both the mechanical integrity as well as the ability to guide light are required; e.g. optical modulators. Just for an example, as a resonator, it can measure slowly varying quantities through the mechanical resonant frequency [216]. It can also be used in accurately defined alignment structures for assembling FOSs and other related precision assemblies [217]. Undoubtedly silicon offers its remarkable scope as optical, mechanical and electronic material, and the RandD community has yet to fully appreciate its true significance.

IO technology based sensors have also their applications in the environmental sensor context. Integrated optical acoustically tunable filters and optochemical microsensors also find applications in environmental monitoring, and are used for the detection of ammonium nitrate and heavy metals. Planar gas and silicon sensors are the other classes that also find extensive use. Semiconductor gas sensors, that use semiconducting oxides of zinc and tin with a small percentage of promoter like platinum, antimony, aluminium etc., find wide applications in the detection of reducing gases like H_2, CO etc. IR gas sensors, that involve specific IR absorption by the rotational-vibrational transition of gas molecules, are used for meteorological measurements, ambient air monitoring etc. The sensitivity of such sensors depends upon the choice of the spectral range and the optical path in closed or open configuration that enhances the limit of detection up to the ppm range. The open configuration system is a kind of compact arrangement of multiple reflections, the length of which may vary from a few centimetres to several meters. Modern technology introduced systems having programmable controls to maintain various parameters like temperature, pollutant concentrations, relative humidity etc., constant even for a period of one year or more, and can be used for simultaneous testing of four gases at a time.

CONCLUSION

Optical fibers are immune to electromagnetic interference, chemically inert and mechanically flexible which enable FOSs to have the unique capability to monitor several parameters as functions of position along the length of fibers even in the worst environmental conditions. These important features of optical fibers combine to provide an attractive baseline technology for several kinds of structural monitoring, and as a result, FOS area has grown to become very broad over the past two decades. Research in the basic sensing techniques and their applications reflects that this technology can be exploited in active structures as the input to a localized control system. These sensors are useful for a diverse range of applications from biomedical science, underwater acoustics to structural health monitoring, safety assessment, and wide range of other related operations.

In this review article, attempt has been made to describe some of the most common methods implemented in FOS technology, and some kinds of FOSs are also presented, which find applications in environmental and biomedical sciences. However, the future of these sensing systems in medical area appears to lie in the integration of several kinds of sensors into a single catheter that should fit through the smallest vessels in a human body. The field of

FOS technology is experiencing rapid developments in the area of smart structures with several demonstrations of embedded sensor systems.

Research and investigations in this area also reflect that such sensor systems can be undoubtedly of great importance so far as the environmental sector is concerned. To date most of the commercially available sensor systems are fiber-coupled instruments in which fibers themselves play the important passive role by taking light to and from remote measuring locations. During the recent years various organizations manufacturing fiber optic technology based sensor systems have shown their big impact, which is certainly evident from their rapid growth. Undoubtedly, it is needless to say that the current and emerging safety and environmental legislation will continue to drive the need for increasingly versatile monitoring systems, and FOSs will play an important part in the evolution of such structures and prove their major impact on future generations of various environmental products.

Although it is not possible to cover all types of FOSs, it is hoped that a flavour of the fascinating subject has been presented in this review article.

ABOUT THE AUTHORS

Pankaj Kr. Choudhury started his research career in 1989 from the Department of Applied Physics, Institute of Technology, Banaras Hindu University (BHU), Varanasi (India), and received Ph.D. degree in Physics in 1992. From 1992 to 1997, he worked as a Research Associate in the Department of Electronics Engineering at BHU. Also, during 1995–1996, he was a Guest Lecturer in Physics at the Faculty of Science and Technology, Mahatma Gandhi Kashi Vidyapith, Varanasi (India). In 1997 he joined the Department of Physics at Goa University, Goa (India) as a Lecturer, where he was until 1999. In 1999 he joined the position of Researcher at the Center of Optics, Photonics and Lasers (COPL), Laval University, Quebec (Canada), where he worked on the fabrication and characterization of ferroelectric optical waveguides. During 2000–2003 he was a Researcher at the Faculty of Engineering, Gunma University, Kiryu (Japan). Presently he is a Professor at the Faculty of Engineering, Multimedia University (Malaysia). His current research interests lie in the theory of optical waveguides and EM wave propagation, which include fiber and integrated optics, fiber optic devices and sensors. He has authored and co-authored over eighty five research publications (including three book chapters) in these areas. He is a Senior Member of IEEE and a Regular Member of the Optical Society of America.

Onkar N. Singh received his M.Sc. and Ph.D. degrees, respectively, in 1963 and 1967 in Physics (Spectroscopy) from the Banaras Hindu University (BHU), Varanasi (India). Joining the BHU faculty in 1963, he became a Professor of Applied Physics in 1986, and served as the Head of the Department of Applied Physics, Institute of Technology (BHU) from 1991 to 1993. He has worked as a visiting/research scientist at the International Center for Theoretical Physics (Trieste, Italy), FOM Institute for Atomic and Molecular Physics (Amsterdam, The Netherlands), Max-Planck Institute for Aeronomy (Lindau, Germany), Ludwig Maxmilian University (Munich, Germany), and the Technical University of Munich (Germany). He has published over 100 research papers and book chapters on molecular spectroscopy, electromagnetic fields, lower stratospheric pollution and climate change, and was co-editor of Vol. 4/E of the *Handbook of Environmental Chemistry* (Springer, Heidelberg, 1999). He was the representative of India to the Atmospheric Chemistry of Commission of IUPAC (1989–1995) and was also a member of the National Committee of World Climate Research Programme (1997–2000). He also co-edited the book *Electromagnetic fields in unconventional structures and materials* (Wiley, US, 2000). He is Member/Fellow of several learned bodies.

REFERENCES

[1] Peterson, J. I.; Vurek, C. G. *Science* 1984, 224, 123–127.

[2] Brenci, M.; Conforti, G.; Falciai, R; Mignani, A. G.; Scheggi, A. M. *Opt. Sensors* 1986, 1, 163–169.

[3] Conforti, G.; Bacci, M.; Brenci, M.; Falciai, R., Mignani, A. G.; Scheggi, A. M. *Proc. SPIE* 1985, 576, 66.

[4] Nuttul, J. D. *Proc IEE J. Optoelectron.* 1985, 132, 250.

[5] Gardnier, P. T.; Edwards, R. A. In: *Fibre Optic Sensors* 1987, IoP Short Meeting Series, Bristol.

[6] Rashleigh, S. C.; Ulrich, R. *Appl. Phys. Lett.* 1979, 34, 768–770.

[7] Hale, K. F. *Phys. Technol.* 1984, 15, 129–135.

[8] Kuribara, M. *Electron. Lett.* 1983, 19, 133–135.

[9] Berthold, J. W. *Proc. SPIE* 1985, 566, 37–44.

[10] Nagai, M.; Shimizu, M.; Oghi N. *Proc. of the 2nd Int. Conf. on Optical Fiber Sensors*, OFS '84, Stuttgart 1984, 207–210.

[11] Dakin, J.; Culshaw, B. *Optical fiber sensors: applications, analysis and future trends;* Artech House: Boston, 1997, Vol. 4.

[12] Bucaro, J. A.; Darty, H. D.; Caron, E. F. *Appl. Opt.* 1977, 16, 1761–1762.

[13] Budiansky, B.; Drucker, D. C.; Kino, G. S.; Rice, J. R. *Appl. Opt.* 1979, 18, 4085–4088.

[14] Hocker, G. B. *Appl. Opt.* 1979, 18, 3674–3683.

[15] Hocker, G. B. *Appl. Opt.* 1979, 18, 1445–1448.

[16] Cielo, P. G. *Appl. Opt.* 1979, 18, 2933–2937.

[17] Tateda, M.; Tanaka, S.; Sugiyama, Y. *Appl. Opt.* 1980, 19, 770–773.

[18] Lagekos, N.; Bucaro, J. A.; Jarzynski, J. *Appl. Opt.* 1981, 20, 2305–2308.

[19] Lyle, J. H.; Pitt, C. W. *Electron. Lett.* 1981, 17, 244–245.

[20] Dakin, J. P.; Liddicoat, T. J. J. *Meas. Control.* 1982, 15, 176–177.

[21] Dakin, J. P. *Proc SPIE* 1984, 468, 219–226.

[22] Marvin, D. C.; Ives, N. A. *Appl. Opt.* 1984, 23, 4212–4217.

[23] Grattan, K. T. V.; Manwell, J. D.; Sim, S. M. L.; Wilson, C. A. *IEE Proc. J. Optoelectron.* 1987, 134, 291–294.

[24] Grattan, K. T. V.; Palmer, A. W.; Saini, D. P. S. *IEEE J. Light. Tech.* 1987, LT-5, 972–979.

[25] Spooncer, R. C.; Jones, B. E.; Ohba, R. *Proc. SPIE* 1987, 798, 137–141.

[26] Petuchowski, S. J.; Giallorenji, T. G.; Sheem, S. K. *IEEE J. Quantum Electron.* 1981, QE-17, 2168–2170.

[27] Franzen, D. L.; Kim, E. M. *Appl. Opt.* 1981, 20, 3991–3992.

[28] Yoshino, T.; Kurosawa, K.; Itoh, K.; Ose, T. *IEEE J. Quantum Electron.* 1982, QE-18, 1624–1633.

[29] Beheim, G. *Electron. Lett.* 1986, 5, 238–239.

[30] Diles, R. A. *J. Appl. Phys.* 1983, 54, 1198–1021.

[31] Dixon, J. *Meas. and Control* 1987, 20, 11–16.

[32] Grattan, K. T. V. *Meas. and Control* 1987, 20, 32–39.

[33] Inaba, H.; Kobayashi, T.; Hirama, M.; Hamza, M. *Electron. Lett.* 1979, 15, 749–751.

[34] Kobayashi, T.; Hirama, M.; Inaba, H.; *Appl. Opt.* 1981, 20, 3279–3280.

[35] Snell, D.; Pitt, G. D. *Proc. BHRA Int. Conf. on Opt. Tech. in Process Control* 1983, 27–41.

[36] Chan, K.; Ito, H.; Inaba, H. *Appl. Phys. Lett.*1983, 43, 634–636.

[37] Chan, K.; Ito, H.; Inaba, H. *Appl. Opt.* 1983, 22, 3802–3804.

[38] Chan, K.; Ito, H.; Inaba, H. Tech. *Dig. 9th Nat. Laser Radar Symp. Japan* 1983, 74–75 (in Japanese).

[39] Chan, K.; Ito, H.; Inaba, H. *IEEE J. Light. Tech.* 1984, LT-2, 234–237.

[40] Chan, K.; Ito, H.; Inaba, H. *Appl. Opt.* 1984, 23, 3415–3420.

[41] Sholes, R. R.; Small, J. G. *Rev. Sci. Instruments* 1980, 51, 882–884.

[42] Bosselmann, Th.; Reule, A.; Schröder, J. *Proc. SPIE* 1984, 514, 151–154.

[43] Grattan, K. T. V.; Palmer, A. W. *Rev. Sci. Instruments* 1985, 56, 1784–1787.

[44] Augousti, A. T.; Grattan, K. T. V.; Palmer, A. W. *IEEE J. Light. Tech.* 1987, LT-5, 759–762.
[45] Grattan, K. T. V.; Selli, R. K.; Palmer, A. W. *Rev. Sci. Instruments* 1988, 59, 1328–1335.
[46] Grattan, K. T. V.; Palmer, A. W.; Zhang, Z. Y. *Rev. Sci. Instruments* 1991, 62, 1210–1213.
[47] Zhang, Z. Y.; Grattan, K. T. V.; Palmer, A. W. *Rev. Sci. Instruments* 1991, 62, 1735–1742.
[48] Zhang, Z. Y.; Grattan, K. T. V.; Palmer, A. W. *Rev. Sci. Instruments* 1992, 63, 3177–3181.
[49] Tanaka, T.; Benedeck, G. B. *Appl. Opt.* 1975, 14, 189–196.
[50] Kilpatrick, D.; Tyberg, J. V.; Parmley, W. W. *IEEE Trans. Biomed. Eng.* 1982, BME-29, 142–145.
[51] Johnson, C. C.; Palm, R. D.; Stewart, D. C.; Martin, W. E. *J. Assoc. Advance Instrum.* 1971, 5, 77–83.
[52] Woodroff, E. A.; Koorajian, S. *Med. Instrum.* 1973, 7, 287–292.
[53] Landsmann, M. H. J. Fiber optic reflection photometry; Verenidge Reproductive Bedrijve, Groningen: Netherlands, 1975.
[54] Gamble, B. G.; Hugenholtz, P. G.; Monroe, R. G.; Polanyi, M.; Nadas, A. S. *J. Intracardiac Oximetry* 1965, 31, 328–343.
[55] Yoshino, T.; Nara, M. *Proc. of the 13th Cong. of Int. Comm. for Opt., Sapporo* (1984), 324–325.
[56] Bjarklev, A. *IEEE J. Light. Tech.* 1986, LT-4, 342–346.
[57] Dakin, J. P. *Proc. SPIE* 1987, 798, 149–156.
[58] Cotter, D. *Opt. and Quantum Electron.* 1987, 19, 1–6.
[59] Appleyard, A. P.; Scrivener, P. L.; Maton, P. D. *Rev. Sci. Instruments* 1990, 61, 2650–2654.
[60] Hartog, A. H. In: Optical fiber sensor technology; Grattan, K. T. V.; Meggitt, B. T. (eds.); Chapman and Hall: UK, 1995.
[61] Fei, L.; Aidong, M.; Deying, Z. *Proc. SPIE* 1996, 2828, 296–300.
[62] Guangping, X.; Keey, S. L.; Asundi, A. *Proc. SPIE* 1999, 3897, 497–504.
[63] Marcuse, D. *Principles of optical fiber measurements;* Academic Press: NY, 1981.
[64] Vita, P. D.; Rossi, U. *Opt. Quantum Electron.* 1980, 11, 17–22.
[65] Kim, B. Y.; Choi, S. S. *Electron. Lett.* 1981, 17, 193–195.
[66] Conduit, A. J.; Payne, D. N.; Hartog, A. H.; Gold, M. P. *Electron. Lett.* 1981, 17, 308–310.
[67] Everard, A. K. J. *Proc. SPIE* 1987, 798, 42–46.
[68] Hartog, A. H.; Payne, D. N.; Conduit, A. J. *Proc. 6th ECOC* (post-deadline paper) 1980, York: UK.
[69] Rogers, A. J. *Electron. Lett.* 1980, 16, 489–490.
[70] Danielson, B. L. *Appl. Opt.* 1985, 24, 2313–2322.
[71] Healey, P. J. *Phys. E: Sci. Inst.* 1986, 19, 334–341.
[72] Ippen, E. P. *Appl. Phys. Lett.* 1970, 16, 303–306.
[73] Stollen, R. H. *Appl. Phys. Lett.* 1972, 20, 62–64.

[74] Stollen, R. H. in Optical fiber telecommunications; Miller, S. E.; Chymoweth, A. G. (eds.); Academic Press: London, 1979.
[75] Farries, M. C.; Rogers, A. J. *Proc. OFS* 1984, 2, 121–132.
[76] Dakin, J. P.; Pratt, D. J.; Bibby, G. W.; Ross, J. N. *Electron. Lett.* 1985, 21, 569–570.
[77] Dakin, J. P.; Pratt, D. J.; Bibby, G. W.; Ross, J. N. *Proc. OFS* (post-deadline paper) 1985, San Diego.
[78] Ferraro, J. R.; Nakamoto, K. Introductory Raman Spectroscopy; Academic Press: London, 1994.
[79] Ippen, E. P.; Stollen, R. H. *Appl. Phys. Lett.* 1972 21, 539–542.
[80] Cotter, D. *Electron. Lett.* 1982, 18, 495–496.
[81] Rogers, *Proc. SPIE* 1986, 718, 65.
[82] Hill, K. O.; Fuji, Y.; Johnson, D. C.; Kawasaki, B. S. *Appl. Phys. Lett.* 1978 32, 647–649.
[83] Hand, D. P.; Russell, R. S. J. *Opt. Lett.* 1990, 15, 102–105.
[84] Melz, G.; Morey, W. W.; Glenn, W. H. *Opt. Lett.* 1989, 14, 823–825.
[85] Anderson, D. Z.; Mizrahi, V.; Erogan, T.; White, A. E. *Electron. Lett.* 1990, 26, 730–732.
[86] Kashyap, R.; Armitage, J. R.; Campbell, R. J.; Williams, D. L.; Maxwell, G. D.; Ainslie, B. J.; Miller, *Bell Syst. Tech. J.* 1993, 11, 150–159.
[87] Askins, C. G.; Tsai, T.; Williams, G. M.; Friebele, E. J. *Opt. Lett.* 1994, 19, 147–149.
[88] Putnam, M. A.; Askins, C. G.; Williams, G. M.; Friebele, E. J.; Baskansky, M.; Reintjes, J. *Proc. SPIE* 1995, 2444, 403–409.
[89] Morey, W. W.; Melz, G.; Glenn, W. H. *Proc. SPIE* 1989, 1169, 98–107.
[90] Morey, W. W.; Dunphy, J. R.; Melz, G. *Proc. SPIE* 1991, 1586, 216–224.
[91] Measures, R. M.; Melle, S. M.; Lui, K. *Smart. Mater. Struct.* 1992, 1, 36–44.
[92] Crossby, P. A.; Powell, G. R.; Fernado, G. F.; Waters, D. N.; France, C. M.; Spooncer, R. C. *Proc. SPIE* 1997, 3042, 141–153.
[93] Ramaswamy, N. S.; Murukeshan, V. M.; Asundi, A. K. *Proc. SPIE* 1999, 3897, 388–396.
[94] Chan, P. K. C.; Jin, W.; Demokan, M. S. *Proc. SPIE* 1999, 3897, 488–496.
[95] Srinivas, T.; Das, I. S.; Selvarajan, A. *Proc. SPIE* 1999, 3897, 480–487.
[96] Jin, W. *Proc. SPIE* 1999, 3897, 468–479.
[97] Melle, S. M.; Liu, K.; Measures, R. M. *Appl. Opt.* 1993, 32, 3601–3609.
[98] Tang, L.; Tao, X.; Choy, C. L. *Smart Mater. Struct.* 1999, 8, 154–160.
[99] Kersey, A. D.; Berkoff, T. A.; Morey, W. W. *Electron. Lett.* 1992, 28, 236–238.
[100] Kersey, A. D.; Berkoff, T. A.; Morey, W. W. *Opt. Lett.* 1993, 18, 72–74.
[101] Jackson, D. A.; Riberio, A. B. L.; Reekie, L.; Archambult, J. L. *Opt. Lett.* 1993, 15, 1192–1194.
[102] Kersey, A. D.; Berkoff, T. A.; Morey, W. W. *Opt. Lett.* 1993 18, 1370–1372.
[103] Davis M. A.; Kersey, A. D. *Electron. Lett.* 1994 30, 75–77.
[104] Koo, K. P.; Tveten, A. B.; Vohra, S. T. *Electron. Lett.* 1999 35, 165–167.
[105] Shupe, D. M. *Appl. Opt.* 1980, 19, 654–655.
[106] Bosselman, Th.; Ulrich, R. *Proc. of the 2nd Int. Conf. on Opt. Fibre Sensors,* OFS '84 (Stuttgart) 1984, 361–365.

[107] Boheim, G. *Appl. Opt.* 1985, 16, 2335–2340.
[108] Dabikewicz, Ph.; Ulrich, R. *Opt. Lett.* 1986, 1986, 11, 5435–5445.
[109] Kersey, A. D.; Dandridge, A. *Electron. Lett.* 1986, 22, 616–617.
[110] Velluct, M. T.; Graingorge, Ph.; Arditty, H. J.; *Proc. SPIE* 1987, 838, 78–83.
[111] Harl, J. C.; Saaski, E. W.; Mitchell, G. L. *Proc. SPIE* 1987, 838, 257–261.
[112] Boheim, G. *Rev. Sci. Instrum.* 1987, 58, 1655–1659.
[113] Mariller, C.; Lequime, M. *Proc. SPIE* 1987, 798, 121–130.
[114] Kock, A.; Ulrich, R.; *Proc. SPIE* 1990, 1267, 128–133.
[115] Dandridge, A.; Tveten, A. B. *Appl. Phys. Lett.* 1981, 39, 530–532.
[116] Dandridge, A.; Tveten, A. B. *Opt. Lett.* 1982, 7, 279.
[117] Ball, G. A.; Morey, W. W. *Opt. Lett.* 1992, 17, 1992, 420.
[118] Kersey, A. D.; Berkoff, T. A.; Morey, W. W. *Electron. Lett.* 1992, 28, 236–238.
[119] Lee, C. E.; Taylor, H. F. *Electron. Lett.* 1988, 24, 193–194.
[120] Lee, C. E.; Taylor, H. F. *IEEE J. Light. Tech.* 1991 9, 129–134.
[121] Murphy, K. A. *Opt. Lett.* 1991, 16, 273.
[122] Ghosh, A. K.; Bedi, N. S.; Paul, P. *Proc. SPIE* 1999, 3897, 522–533.
[123] Nasrullah, K. *Proc. SPIE* 1999, 3897, 578–581.
[124] Rochford, K. B.; Rose, A. H.; Deeter, M. N.; Day, G. W. *Opt. Lett.* 1994, 19, 1903–1905.
[125] Vali V.; Shorthill, R. W. *Appl. Opt.* 1976, 15, 1099–1100.
[126] Arditty, H. J.; Bourbin, Y.; Mapouchon, M.; Puech, C. *Proc. of the Tech. Dig., 3rd Int. Conf. on Integrated Optics and Opt. Fiber Commun.* (San Francisco) 1981, 128–130.
[127] Kadiwar, R. K.; Giles, I. P. *Electron. Lett.* 1989, 25, 1729–1731.
[128] Zarinetchi, F.; Smith, S. P.; Ezekiel, S. *Opt. Lett.* 1991, 16, 229–231.
[129] Raab, M.; Quasat, T. *Opt. Lett.* 1994, 19, 1492–1494.
[130] Kim, S. K.; Kim, H. K.; Kim, B. Y. *Opt. Lett.* 1994, 19, 1810–1812.
[131] Tanaka, Y.; Yamasaki, S.; Hotate, K. *IEEE Phot. Tech Lett.* 1996, 8, 1376–1369.
[132] Kung, A.; Nicati, P.-A.; Robert, P. A. *IEEE Phot. Tech. Lett.* 1996, 8, 1680–1682.
[133] Böhm, K.; Peterman, K.; Weidel, W. *Opt. Lett.* 1982, 7, 180–182.
[134] Kurusawa, K.; Yoshida, S.; Sakamoto, K.; Masuda, I.; Yamashita, T. Trans. *IEE Japan* 1996, 116-B, 93–103.
[135] Takahashi, Y.; Yoshino, T.; Horie, K. *Opt. Rev.* 1997, 4, 417–422.
[136] Takahashi, Y.; Yoshino, T. *Appl. Opt.* 1997, 36, 6770–6773.
[137] Takahashi Y.; Yoshino, T. *IEEE J. Light. Tech.* 1999, 17, 591–597.
[138] Curcio, J. A.; Petty, C. C. J. *Opt. Soc. Am.* 1951, 41, 302–304.
[139] Libnau, F. O.; Kvalheim, O. M.; Chirsty, A. A.; Toft, J. *Vibrational Spectroscopy* 1994, 7, 243–254.
[140] Lin, J.; Brown, C. W. *Vibrational Spectroscopy* 1994, 7, 117–123.
[141] Sanghera, J. S.; Kung, F. H.; Pureza, P. C.; Nguyen, V. Q.; Miklos, R. E.; Aggarwal, I. D. *Appl. Opt.* 1994, 33, 6315–6322.
[142] Hop, E.; Juinge, H. J.; Hemert, H. V. *Appl Spectroscopy* 1993, 47, 1180–1182.
[143] Muto, S.; Sato, H.; Hosaka, T. *Jpn. J. Appl. Phys.* 1993, 33, 6060–6064.

[144] Fan, T. Y.; Huber, G.; Byer, R. L.; Mitzscherlich, P. *IEEE J. Quantum Electron.* 1988, 24, 924–933.
[145] Stonemann, R. C.; Esterowitz, L. *Opt. Lett.* 1990, 15, 486–488.
[146] McAleavey, F. J.; MacCraith, B. D. *Electron. Lett.* 1995, 31, 1379–1380.
[147] Zhu, X.; Cassidy, D. T. *Appl. Opt.* 1996, 35, 4689–4693.
[148] Yokota, M.; Yoshino, T. *Appl. Opt.* 1998, 37, 2526–2533.
[149] Yokota, M.; Yoshino, T. *Meas Sci. Technol.* 2000, 11, 152–156.
[150] Asundi, A. K.; Veeredhi, V. R. *Proc. SPIE* 1999, 3897, 187–192.
[151] Inci M. N.; Yoshino, T. *Opt. Rev.* 2000, 7, 205–208.
[152] Fields, J. N.; Cole, J. H. *Appl. Opt.* 1980, 19, 3236–3267.
[153] Asawa, C. K.; Yao, S. K.; Streans, R. C.; Mota, N. L.; Downs, J. W. *Electron. Lett.* 1982, 18, 362–364.
[154] Yoshino, T.; Nara, M.; Kurosawa, K. *Proc. of the 13th Cong. of the Int. Commission for Opt.* (Sapporo) 1984, 324–325.
[155] Tomita, S.; Tachino, H.; Kasahara, N. *IEEE J. Light. Tech.*1990, 8, 1829–1832.
[156] Berthold, J. W. *IEEE J. Light. Tech.* 1995, 13, 1193–1199.
[154] Oscroft, G. *Proc. SPIE* 1987, 734, 207–213.
[158] Lagakos, N.; Cole, J. H.; Bucaro, J. A. *Appl. Opt.* 1987, 26, 2171–2180.
[159] Yoshino, T.; Inoue, K.; Kobayashi, Y.; Takahashi, Y. *Proc. of the 11th Opt. Fiber Sensors Conf.* (Sapporo) 1996, 264–267.
[160] Yoshino, T.; Inoue, K.; Kobayashi, Y. *IEE Proc. J. Optoelectron.* 1997, 144, 145–150.
[161] Zhu, D. G.; Petty, M. C. Harris, M. *Sens. and Actuat.* B 1990, 2, 265–269.
[162] Qin, S. J. *Sens. and Actuat. B* 1991 3, 255–260.
[163] Vukusic, P. S.; Sambles, J. R. *Thin Solid Films* 1992, 221, 311–317.
[164] Kim, S. R.; Choi, S. A.; Kim, J. D.; Choi, K. H.; Purk, S. K.; Chang, Y. H. *Synthetic Metals* 1995, 71, 2293–2294.
[165] Patel, N. G.; Makhija, K. K.; Panchal, C. J.; Dave, D. B.; Vaishnav, V. S. *Sens. and Actuators B* 1995, 23, 49–53.
[166] Campbell, D.; Collins, R. A. *Thin Solid Films* 1997, 295, 277–282.
[167] Azim-Araghi M. E.; Krier, A. *Appl. Surf. Sci.* 1997, 119, 260–266.
[168] Kim, S. R.; Kim, J. D.; Choi, K. H.; Chang, Y. H. *Sens. and Actuators* B 1997, 40, 39–45.
[169] Hideo Tai, H.; Tanaka, H.; Yoshino, T. *Opt. Lett.* 1987, 12, 437–439.
[170] Choudhury, P. K. *Curr. Sci.* 1998 74, 723–725.
[171] Peterson, J. I.; Goldstein, S. R.; Fitzgerald, R. V.; Buckhold, D. K. *Anal. Chem.* 1980, 52, 864.
[172] Choudhury P. K.; Yoshino, T. *Optik* 2003, 114, 13–18.
[173] Takeo, T.; Hattori, H. *Jpn. J. Appl. Phys.* 1992, 21, 1509–1512.
[174] Paul, P. H.; Kychakoff, G. *Appl. Phys. Lett.* 1987, 51, 12–14.
[175] Ruddy, V.; MacCraith, B. D.; Murphy, J. A. *J. Appl. Phys.* 1990, 67, 6070–6074.
[176] Choudhury, P. K.; Yoshino, T. *49th Spring Meeting of the Jpn. Soc. of Appl. Phys.* (Tokyo) 2002, Extended Abstracts, No. 3, 1206.
[177] Choudhury, P. K.; Yoshino, T. *Meas. Sci. Technol.* 2002, 13, 1793–1797.

[178] Shoemaker, D. P.; Garland, C. W. *Experiments in Physical Chemistry;* McGraw-Hill Kogakusha Ltd.: Tokyo, 1967.

[179] Cotton, F. A.; Darlington, C. L.; Lynch, L. D. *Chemistry – An Investigative Approach,* Houghton Mifflin Co.: Boston, 1980.

[180] Walker, D.; Prudy, K.; Tarczynski, F. *Proc. SPIE* 1998, 3537, 26–33.

[181] Internet http://www.cudenver.edu//chemistry/classes/chem4538/I2DISS. htm.

[182] Internet http://www.mines.edu/fs_home/dwu/classes/pchem/lab/Wet%20 Lab%201/ WetLab1.htm.

[183] Henning, M. L.; Lamb, C. *Proc. OFS '88* (New Orleans) 1988, 84.

[184] Culshaw, B. Application of fiber optic sensors in the aerospace and marine industries. In: *Optical Fiber Sensors – Systems and Applications,* Vol. 2 (eds. B. Culshaw and J. P. Dakin), Artech House, Inc.: 1989.

[185] Lakhtakia, A. *Optik* 1997, 107, 57–61.

[186] Lakhtakia, A. Messier, R. *Mater. Res. Innov.* 1997, 1, 145–148.

[187] Lakhtakia, A. Messier, R. Sculptured thin films: nanoengineered morphology and optics, *SPIE,* 2005, Vol. PM143.

[188] Lakhtakia, A. *Optik* 1997, 110, 289–293.

[189] McCall, M. W.; Lakhtakia, A. *J. Mod. Opt.* 2000, 47, 743–755.

[190] Venugopal, V. C.; Lakhtakia, A. In: Singh, O. N.; Lakhtakia, A. (Eds.), *Electromagnetic fields in unconventional materials and structures,* Wiley: New York, 2000, Chapter 5.

[191] Young, N. O.; Kowal, J. *Nature* 1959, 183, 104–105.

[192] Robbie, K.; Brett, M. J.; Lakhtakia, A. *Nature* 1996, 384, 616.

[193] Messier, R.; Gehrke, T.; Frankel, C.; Venugopal, V. C.; Otaño, W.; Lakhtakia, A. *J. Vac. Sci. Technol.* 1997, 15, 2148–2152.

[194] Lakhtakia, A. *Sensors and Actuators* B 1998, 52, 243–250.

[195] Lakhtakia, A.; Venugopal, V. C. *Microw. Opt. Tech. Lett.* 1998, 17, 135–140.

[196] Ertekin, E.; Lakhtakia, A. *Eur. Phys. J.: Appl. Phys.* 1999, 5, 45–50.

[197] Hodgkinson, I. J.; Wu, Q. H.; Thorn, K. E.; Lakhtakia, A.; McCall, M. W. *Opt. Commun.* 2000, 184, 57.

[198] Lakhtakia, A.; McCall, M. W., Sherwin, J. A.; Wu, Q. H.; Hodgkinson, I. J. *Opt. Commun.* 2001, 194, 33–46.

[199] Lakhtakia, A. *Opt. Commun.* 2001, 188, 313–320.

[200] Lakhtakia, A. Opt. Commun. 2002, 202, 103–111. *Corrections in* Lakhtakia, A. *Opt. Commun.* 2002, 203, 447.

[201] Lakhtakia, A. *Microw. and Opt. Tech. Lett.* 2003, 37, 37–40.

[202] Stelz, E. E.; Lakhtakia, A. *Opt. Commun.* 2003, 216, 139–150.

[203] Pinchbeck, D.; Kitchen, D. *Proc. Conf. Electronics in Oil and Gas Industries* (London) 1985, 496–501.

[204] Chan, K.; Furuya, T.; Ito, H.; Inaba, H. *Opt. Quantum Electron.* 1985, 17, 153–155.

[205] Arakawa, Y.; Fukunaga, H.; Inaba, H. *Proc. of the 4th Int. Conf. on Opt. Fiber Sensors, OFS '86* (Tokyo) 1986, 135–138.

[206] Choudhury, P. K.; Yoshino, T. *Optik* 2004, 115, 329–333.

[207] Choudhury, P. K.; Yoshino, T. *Proc. SPIE* 2005, 5634, 136–139.

[208] Muhammed, F. A.; Stewart, G. *Int. J. Optoelectron.* 1992, 7, 705.

[209] Muhammed, F. A.; Stewart, G. *Electron. Lett.* 1992, 28, 1205–1026.

[210] Muhammed, F. A.; Stewart, G.; Jin, W. *IEE Proc. J. Optoelectron.* 1993, 140, 115.

[211] Bacci, M.; Balidini, F.; Brenci, M.; Conforti, G.; Falciai, R.; Mignani, A.G. *Proc. of the 4th Int. Conf. on Opt. Fiber Sensors (*Tokyo) 1986, 323–326.

[212] Butler, M. A. *Appl. Phys. Lett.* 1984, 45, 1007–1009.

[213] Andrade, J. D.; Vanwagen, R. A.; Gregonis, D. E.; Newby, K.; Lin, J. N. *IEEE Trans. Electron. Dev.* 1985 ED-32, 1175–1179.

[214] Petersen, K. E. *Proc. IEEE* 1982, 70, 420–457.

[215] Jones, B. E.; Philip, G. S. *Proc. of the Conf. on Sens and their Appl,* Manchester (London, Institute of Physics) 1983, 86–88.

[216] Langden, R. M. *J. Phys. E* 1985, 18, 103–115.

[217] Venkatesh, S.; Culshaw, B. *Electron. Lett.* 1985, 21, 315–317.

[218] Stearns, R. G.; Kino, G. S. *Appl. Phys. Lett.* 1985, 47, 1048–1050.

[219] Soref, R. A.; Lorenzo, J. P. *IEEE J. Quantum Electron.* 1986, QE-22, 873–879.

[220] Canham, L. T., Barraclough, K. G.; Robbins, D. J. *Appl. Phys. Lett.* 1987, 51,1509–1511.

In: Advances in Laser and Optics Research. Volume 10 ISBN: 978-1-62257-795-8
Editor: William T. Arkin

Chapter 10

OPTICAL CURRENT SENSORS

Toshihiko Yoshino[1] and Masayuki Yokota[2]
[1]The Kaisei Academy, Nishi-Nippori, Arakawa-ku, Tokyo, Japan
[2]Department of Electronic Engineering, Faculty of Engineering,
Gunma University, Tenjin-cho, Kiryu, Gunma, Japan

ABSTRACT

This review article describes the present state-of-the-art of fiber optic current sensing technology based on Faraday effect. The importance of the fulfillment of HICOC (homogeneous isotropic closed optical circuit) condition is addressed. The optical current sensors are categorized into three types, viz. all-fiber, bulk-optic and hybrid, and their available HICOC techniques, advantageous features, performances and applications are mentioned.

1. INTRODUCTION

In comparison with electric current transformers, an optical current sensor or transformer (OCT) has the invaluable features of non-contact, highly insulating and high-speed measurements [1,2]. Because the OCT technology is closely related to the optical magnetic-field sensing technology, a lot of optical magnetic-field sensors are potentially applicable to OCT as well. The most reliable OCT, however, should be based on Ampere's circuital law, according to which the line integration of current-induced magnetic fields along a closed loop is equal to the current which intersects the closed loop. In order to perform this integration optically, we use Faraday effect, which states that the plane of linear polarization of a light beam propagating in an isotropic material subject to magnetization is rotated in proportion to the magnetization. So far many works on OCT have been made from engineering as well as scientific interests, and some OCTs developed have reached the stage of practical use and are commercially available. This chapter is devoted to the description of the present state-of-the-art of the OCT technology. The basic principles, types, performances and applications of OCTs are addressed.

2. THE BASIS

2.1. Optical Ampere's Circuital Law

When a steady electric current I flows in 3D space, according to the Ampere's circuital law,

$$I = \oint \vec{H} \cdot d\vec{s} \quad , \qquad (1)$$

where $\vec{H}$ is the current-induced magnetic field vector and $\vec{s}$ is the line vector along which the loop integration is carried out. In terms of Faraday effect, equation (1) becomes

$$n\oint V\vec{H} \cdot d\vec{s} = nVI = \phi \quad , \qquad (2)$$

where V is Verdet constant of the loop material (assumed to be constant along the entire optical path), ϕ is Faraday rotation angle due to one round trip of light along the loop, and n is the number of turns of the optical loop. Thus, in order that the OCT based on equation (2) may function properly, the optical loop has to be made of a homogeneous, isotropic and closed optical circuit (HICOC).

2.2. The HICOC Condition

Unless the HICOC condition is fulfilled, the OCT undergoes the measurement uncertainty such that the output signal of the OCT is affected by external currents of the measuring current, and also, by the relative position of the measuring current within the optical loop. In order to evaluate the measurement error due to imperfect fulfillment of the HICOC condition, we assume that the optical loop of an OCT involves some part which has a different Verdet constant V' from the other part, as shown in Figure 1(a). Faraday output signal then undergoes a measurement error due to the cross-talk with (or isolation from) a surrounding current as:

$$e = \left|(V' - V)/V\right|(\theta/2\pi) \quad , \qquad (3)$$

where θ stands for the subtended angle from the current position to the path AB as shown in Figure1(a) [3,4]. Calculation of equation (3) as a parameter of e is shown in Figure 1(b). We consider the case that the optical path is not closed but terminated at separated points A and B. This corresponds to $V' = 0$ on the AB line so that thereby-caused error due to cross-talk is [from equation (3)] $e = \theta/2\pi$. In a particular case, in which the external current is located at a distance d from the AB line (AB = L), θ is then $\theta_1 = 2\tan^{-1}(L/2d)$. Therefore, referring to Figure 1(b), for $x = \theta_1/2\pi$ and $y = 1$, it follows that when d is, e.g. 10 mm, in order to achieve $e < 0.01$, L must be smaller than 0.63 mm. It is thus recognized that even the presence of such a slightly opened loop can cause a significant cross-talk error for close currents. On the other hand, when the traveling light has an elliptical polarization on the path AB, it corresponds to $V' = V \cos\Delta$ (Δ being the polarization retardation) so that $e < 0.01$ requires $\Delta \leq 12°$. Figure 2 shows the experimental demonstration of the cross-talk error due to an open loop [3,4]. In the sensor cell geometry shown in Figure 2(a), the optical path is definitely open without the (4 mm thick) compensating glass plate while the optical path is well-closed with it on the input/output port of the Faraday cell. For both cases, magnitudes of the cross-talk with an external current were measured as a function of the distance of the current from the center of the Faraday cell, and compared with each other in Figure 2(b), together with the theoretical values of equation (3). Figure 2(b) clearly indicates that, in real OCTs, the cross-talk with close currents becomes a significant error if the optical path is not properly closed.

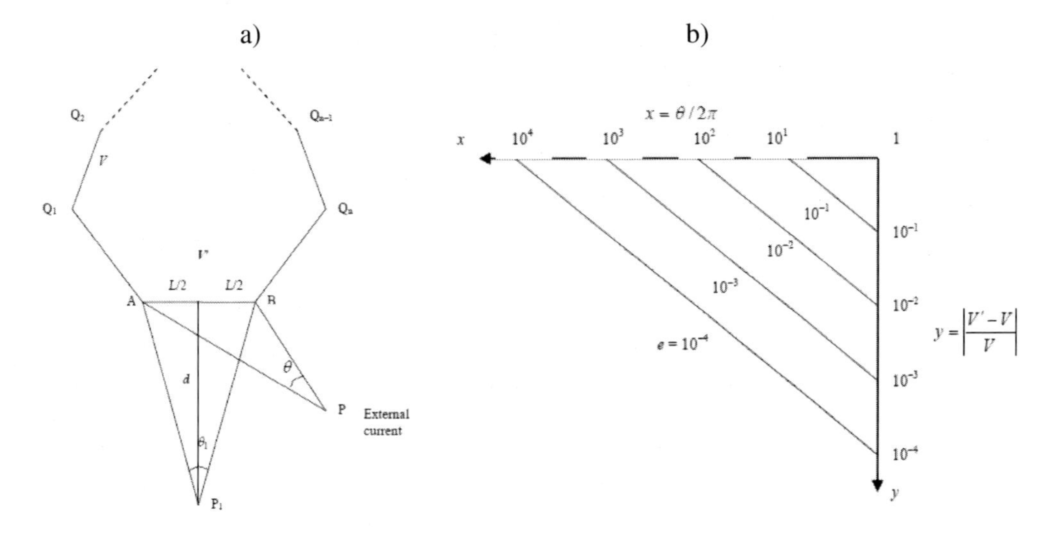

Figure 1. Theoretical value of cross-talk with surrounding current in OCT [3,4]; (a) geometry; (b) dependence of cross-talk e on subtended angle θ and Verdet constant inhomogeneity.

2.3. Signal Interrogating Schemes

Faraday signal can be detected by some different interrogating schemes. In general, a polarimetric interrogation, as follows, is employed. A linearly polarized light is incident on the Faraday material, the output light is received by two polarizers which are oriented at ±45° to the incident polarization direction, and the degree of polarization of the output light is detected as the OCT signal.

As to the light source, a low coherent light is preferable to laser light to avoid its interference noises, and SLD (superluminescent laser diode) is especially used for the coupling to single mode fiber (SMF).

Because the Faraday rotation originates from the magnetically-induced circular birefringence of a substance, interferometric detection of the relevant optical phase shift yields the OCT signal. Though various types of interferometers (e.g., Fabry-Perot, Mach-Zender types) are, in principle, applicable to OCT, a ring interferometer (i.e., Sagnac interferometer) is the most suited one for the Faraday signal interrogation from a closed optical path in particular. The phase difference between the clockwise and the counter-clockwise traveling light beams in the interferometer yields twice the Faraday rotation [5], and then the undesirable reciprocal phase shifts due to temperature changes or vibration are cancelled. Similarly, a ring laser, or an active Sagnac interferometer, can also be used for the Faraday signal interrogation. Interference of the clockwise and the counter-clockwise traveling light beams going out of the ring laser generates an optical beat whose frequency is proportional to the Faraday rotation angle. The Faraday rotation angle is then interrogated in the frequency domain with a linear scale within the free spectral range of the laser [6,7].

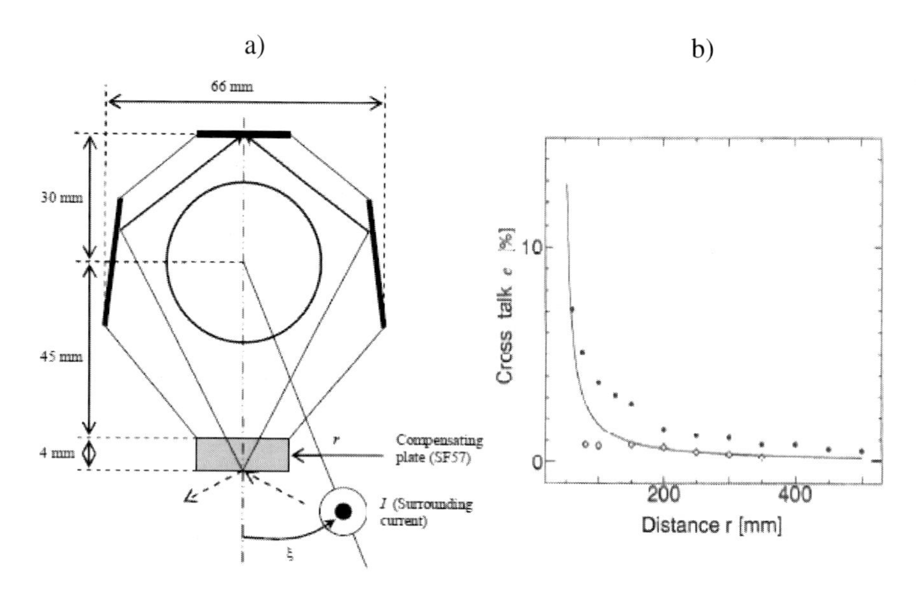

Figure 2. Demonstration of cross-talk with surrounding current in OCT [3,4]; (a) experimental arrangement, (b) dependence of cross-talk e on external current distance r at $\xi = 0°$ with (◦) and without (•) compensator. The solid line represents the calculation by equation (3).

In general, the Faraday signal detection is accompanied by optical noises. For dc current sensing, in particular, the drift of light power becomes a serious noise. To reduce it, the heterodyne detection using two frequencies of laser light [5,8] or a phase-modulated (or ac) Sagnac interferometer is very much suited.

3. All-Fiber Type OCT

An SMF made of silica is an ultra-low loss, thin, long, flexible, cheap and intrinsically isotropic optical waveguide, and therefore, its application to OCT started immediately after the invention of the fiber [9,10]. In reality, however, it took a long time to arrive at practical OCTs mainly because their development was much hindered by the involvement of linear birefringence (LB) in real fibers.

3.1. Magneto-Optical Properties of SMFs

The general theory for Faraday effect in isotropic SMFs is presented in reference [11]. In the simultaneous presence of LB and Faraday effect in a medium, when a linear polarization is input into to medium along the principal axes of LB, the amplitude of the electric field of the output light is given by [12]

$$F = E_0 HVL\mathrm{sinC}\{[(\delta/2)^2 + VH]^{1/2} L\}, \tag{4}$$

where δ is the linear retardation per meter, L is the optical path length, H is the applied magnetic field, E_0 is the amplitude of the electric field of the input light, and $\mathrm{sinC}x = \sin x/x$. Obviously, for $\delta = 0$, equation (4) becomes $F = E_0\sin(VHL)$, which is the known relationship for Faraday effect in isotropic mediums. In the case that $|\delta| >> |VH|$, equation (4) becomes $F = E_0VHL\mathrm{sinC}(\delta L/2)$, so that $|F| \leq 2E_0|VH/\delta|$. For silica fibers subjected to a bending of e.g. 50 mm radius of curvature, $|\delta|$ amounts to the order of 1 radian/m while $|VH|$ is typically 10^{-5} I rad.A^{-1}m^{-1}; it then follows that $|HV/\delta| \leq 10^{-5}$ I A^{-1}, and hence, the field conversion efficiency $|F|/E_0$ is smaller than the order of 10^{-5} I A^{-1}, which is usually very small except for especially heavy currents.

Therefore, some approaches to eliminate LB in silica fibers were conducted. One of them is to induce circular birefringence (CB) by providing a post-drawing fiber with twist [13,14] or an in-drawing fiber with spin (i.e., spun fiber) [15]. For such a fiber, equation (4) becomes

$$F = E_0 H(VH + \delta_T/2)L\mathrm{sinC}\{[(\delta/2)^2 + (VH + \delta_T/2)^2]^{1/2} L\}, \tag{5}$$

where δ_T is the provided CB/m. In the case when CB is much larger than LB (i.e., $|\delta_T| >> |\delta|$), equation (5) becomes

$$F = E_0\sin[(VH + \delta_T/2)L], \tag{6}$$

which indicates that the effect of LB is neglected against CB, and F becomes the same as that for the fiber without LB case except involving a constant bias polarization rotation. When a twist rate of 1 turn/m is applied to a post-drawing silica SMF, $|\delta_T|$ amounts to 1 rad/m.

The study of the thermal relaxation of fiber stress has been attempted, and it has been shown that the bend-induced LB in silica fiber could be reduced by annealing [16]. The annealed fiber, however, is still sensitive to newly applied stress so that, in its practical use, a careful remedy is needed to avoid the application of new stress.

Naturally, it is most desirable to invent an SMF which has inherently no photo-elastic effect. Such fiber was developed by using flint glass as the fiber material [17]. Compared with silica, flint glass has about 100-times smaller photo-elastic constant (Brewster coefficient) and, of additional advantage, its Verdet constant is 6-times larger ($V = 1.5\times10^{-5}$ rad/A at a wavelength of $\lambda = 850$ nm) and little temperature dependent. The transmission loss is, however, pretty high. The flint glass SMF is now commercially available (HOYA Co., Tokyo, Japan).

One more point to be noted in the application of SMF for OCT sensors is the geometrical rotation of the polarization plane of the transmitted light from the fiber. As a general feature of light transmission in optical fibers, if the fiber is not in-plane, the light is then subjected to a purely geometrical rotation of the polarization plane, depending on the geometry of the fiber path [18]. In all-fiber OCTs, this geometrical polarization rotation can cause polarization instability.

Fortunately, this geometrical polarization rotation is a reciprocal effect so that it can be cancelled by making the light beam to go forth and back in the same fiber (i.e., by the use of the reflection scheme) [19].

In the transmission type OCTs, however, this geometrical rotation can introduce uncertainty in measurement due to polarization instability. As such, the sensing fiber coil needs to be fixed on a rigid plane in a stable manner.

3.2. Practical All-Fiber OCTs

Various types of all-fiber OCTs have been developed so far and several of them have been put into either test or practice for the electric power systems including power generators in power plants, railway systems, nuclear devices etc. Regarding flint-glass fiber OCTs, both the transmission and reflection types with polarimetric interrogation have been developed [20]. The transmission-type, in which a flint glass fiber coil is fixed on a durable metallic film, could measure currents up to 300 kA (which corresponds to a Faraday rotation of 45°) within a measurement error of 2 kA to meet the Japanese industrial standard [20]. Figure 3(a) shows the optical diagram of the reflection-type OCT. A low coherent light from SLD (λ = 1550 nm) is remotely sent (via a polarization maintaining fiber) to the sensing part, and coupled into the flint glass fiber after being passed through an optical unit that serves as both a polarizer and a polarization bias element.

The reflected light from the sensing fiber is again passed through the same optical unit and ±45° polarization components of the emerging light are received (via an SMF) by two photo detectors 1 and 2. Figure 3(b) shows the application of the reflection-type OCT for fault current detection. The sensing fiber could, on account of the reflection-type, freely encircle the three phases of electric cables, and accurate phase current detection (summation of the three phase currents) is carried out [21]. The annealed-silica-fiber OCTs with polarmetric interrogation have been put into for testing. In order to reduce the effects of thermal expansion and vibration, the annealed fiber coil is installed in the glass epoxy compound package [22].

A Sagnac interferometer OCT using annealed silica SMF has also been tested for applications, and such OCTs are found to have a measurement accuracy of better than 0.5% for the environmental temperature range from −35 °C to 85 °C [23]. A Sagnac interferometer OCT using spun silica SMF has also been developed for the application in D.C. railway transportation system [24].

All of the all-fiber OCTs reported so far, unfortunately, did not describe about the closing feature of the sensing fiber loop so that there are ambiguities about the possible cross-talk errors with very close surrounding currents.

a)

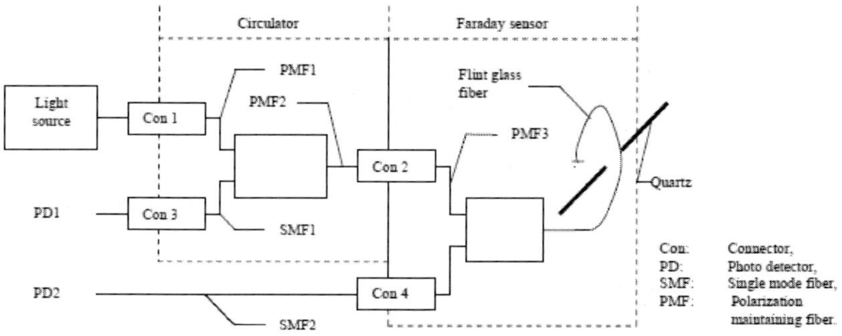

b)

Figure 3. Practical flint glass OCT of the reflection type [21] ; (a) optical diagram, (b) fault section locating system for 22 kV power cable lines.

4. Bulk-Optic OCT

In order to get rid of the problem of fiber LB or to make it possible to use a variety of Faraday materials, it is important to study the use of non-fiber bulk-optic materials as the Faraday element. In this type of OCT (or the bulk-optic type of OCT), the light beam travels in a zigzag path within a Faraday cell. The key requirement for the bulk-optic OCT is to achieve the polarization-maintaining 100% reflectance of the traveling light at each reflecting surface of the Faraday cell. To this end, the oldest scheme applied the crossed right-angle total-reflection glass prisms for the corner reflection [25,26] but it, in principle, could not well fulfill the HICOC condition. In order to fulfill the HICOC condition precisely, new schemes have been developed, which applied the thin-film coating technology. Two different approaches have been conducted. The first approach extended the well-known quarter-wavelength multilayer coating technology to perform the polarization-maintaining 100% reflectance. The theory required the specific conditions of "above Brewster's angle and below critical-angle" of light on every boundary of the multilayer thin films. The theoretical requirement was experimentally demonstrated by the use of conventional coating materials.

The HICOC condition was thus for the first time very exactly fulfilled [4]. The resulting form of the Faraday cell was a polygon of specific form.

The second approach used the coating of the bi-layer dielectric thin films on the total reflection surface to achieve the polarization-maintaining 100% reflectance for light beam at oblique angle of incidence [27]. The optical design was conducted by the trial-and-error numerical calculation. The sensor system developed is shown in Figure 4(a) [27]. Each of the three reflecting surfaces of a square block Faraday cell made of flint glass was coated by the suitably thick SiO_2 and Ta_2O_5 films to produce no polarization-retardation and 100% reflectance on the internal reflection for 45° oblique incidence of light. The input/output port of the Faraday cell was provided by attaching a low-Verdet-constant (SF7) tiny prism on one uncoated surface of the cell. This Faraday cell had good tolerances with respect to light beam angle, wavelength and film thickness. The current sensing characteristics of this OCT are shown in Figure 4(b). The polarization extinction ratio was better than 2×10^{-4} and high isolation from the surrounding current was demonstrated [Figure 4(c)]. Figure 4(d) shows the ac Faraday signal and the dc light power as a function of applied weight to the cell; it indicates that the OCT is very stable against the mechanical disturbances applied to the Faraday cell.

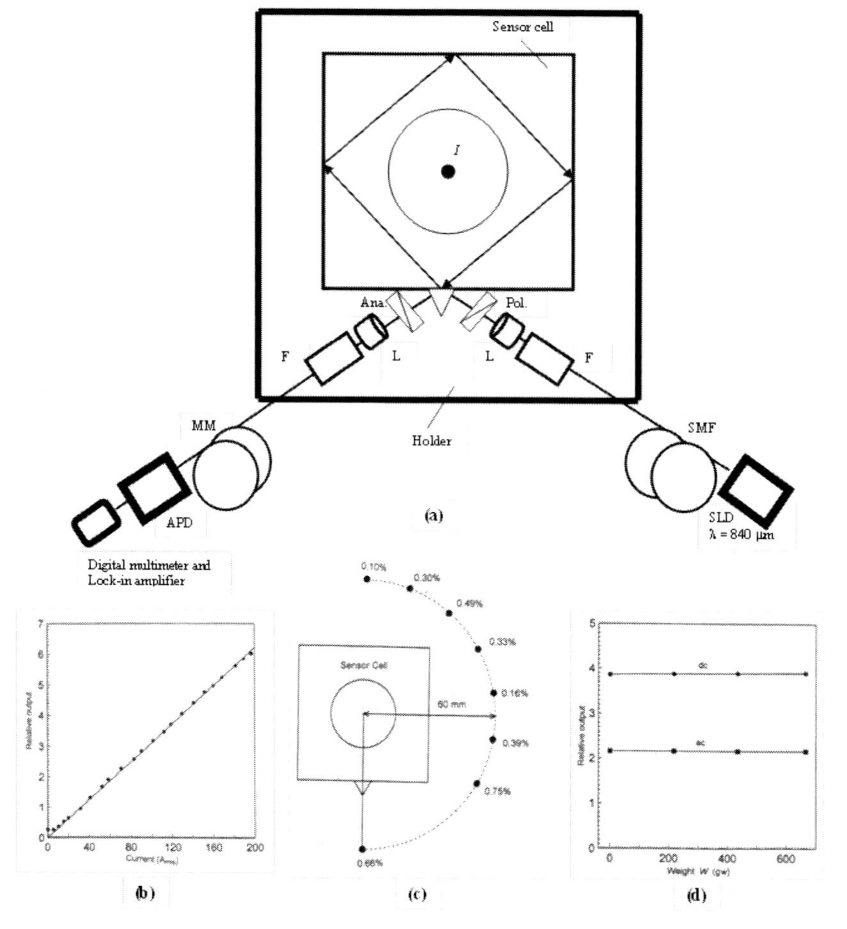

Figure 4. Precise and stable bulk-optic OCT [27]; (a) current sensor system, (b) measured current characteristics, (c) measured cross-talk at different positions of the external current, (d) measured weight characteristics of dc and ac Faraday signals, indicating high mechanical stability of the OCT.

5. HYBRID TYPE OCT

The third type of OCT is fabricated by a simple modification of the traditional electric CT. This type of OCT, or the hybrid OCT, is composed of a high-permeability (metallic) magnetic core with a small gap, and a Faraday element inserted in the core gap. In principle, the hybrid type OCT cannot rigorously fulfill Ampere's circuital law, and therefore, cannot be a much exact OCT.

Moreover, the use of metallic magnetic cores as the field concentrator reduces the insulating feature of OCT, and saturates the Faraday signal against large currents.

Despite these drawbacks, the hybrid type OCT is an attractive one because, compared with the bulk-optic type OCT, the hybrid one allows the use of various Faraday materials, which include crystals and thin films, and does not need specific internal refection schemes of the Faraday cell (unlike the bulk-optic OCT).

Also, the hybrid type OCT can be used as a clamp-type OCT, like the electric CT. In order to achieve high isolation from surrounding currents and to improve the accuracy of OCT, it is required that the gap length of the core is possibly small. This is the key requirement for developing high performance hybrid type OCT. Earlier studies developed the multiple-reflection Faraday cell made of a thin FR5 glass plate, as shown in Figure 5, which is suitable for the insertion into a small gap of a ring core [28–30].

An especially attractive application of the hybrid type OCT is the small current measurement, which is very difficult with the all-fiber or bulk-optic type OCT. The achievable sensitivity is governed by the performances of Faraday materials and magnetic core systems.

As for high Verdet constant materials, transparent magnetic materials such as rare earth iron garnet (RIG), e.g., yttrium iron garnet (YIG), are attractive. Improvements on their magneto-optical properties such as temperature characteristics [31], linearity [32] and response speed [33] were conducted.

Especially, as for the temperature characteristics, Bi-doped YIG with high temperature stability has been developed, and is now commercially available.

A magnetic material (MM) has the following inherent features as the Faraday material:

I Verdet constant is typically 100 times higher than non-magnetic materials,

II when MM is applied in the hybrid OCT, because of its high permeability, the total magnetic resistance of the core system becomes small, thereby enhancing the current-sensitivity,

III it is possible to magnetize MM in the other directions than that of the applied magnetic field by virtue of demagnetization and/or magnetic-anisotropy (such as uniaxial anisotropy) effects peculiar to MM,

IV the magnetic response of MM is subjected to nonlinearity, hysteresis and saturation against the applied magnetic field, unlike non-magnetic materials,

V the produced magnetization is not uniform in MM because of the domain structure, which scatters or diffracts the light beam, and hence, causes the measurement uncertainty. This domain effect, however, is averaged and can be reduced by the illumination of a broad-width light beam as the probe light.

a)

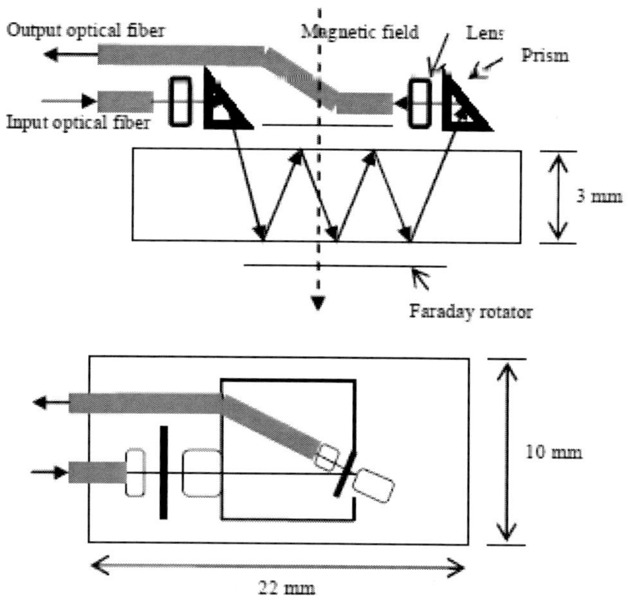

b)

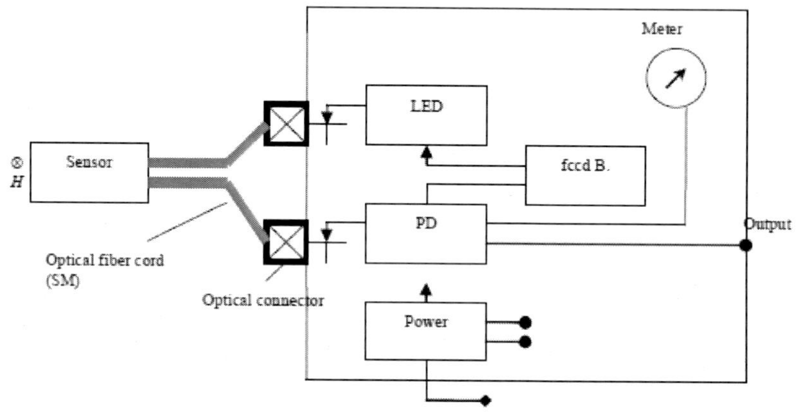

Figure 5. Optical magnetic-field sensor of the multiple-reflection type [30]; (a) sensor head, (b) instrument system.

It results from the above features of MM that a possibly smaller core gap and a broad-beam illumination are desirable for the hybrid OCT using MM. Usually, in the core gap, the probe light is passed through the Faraday element in the direction parallel to the core gap (longitudinal configuration). Then, because the gap length is very limited, special optical designs such as a core system with holes [34] or a micro-optic system [35] have been introduced.

It has also been demonstrated that the hybrid OCT using MM functions well even in the transverse configuration where the probe light is passed through the Faraday element in the direction perpendicular to the core gap [36,37]. The operating principle of this unusual hybrid OCTs is particularly based on the feature no. (v) of MM as mentioned above.

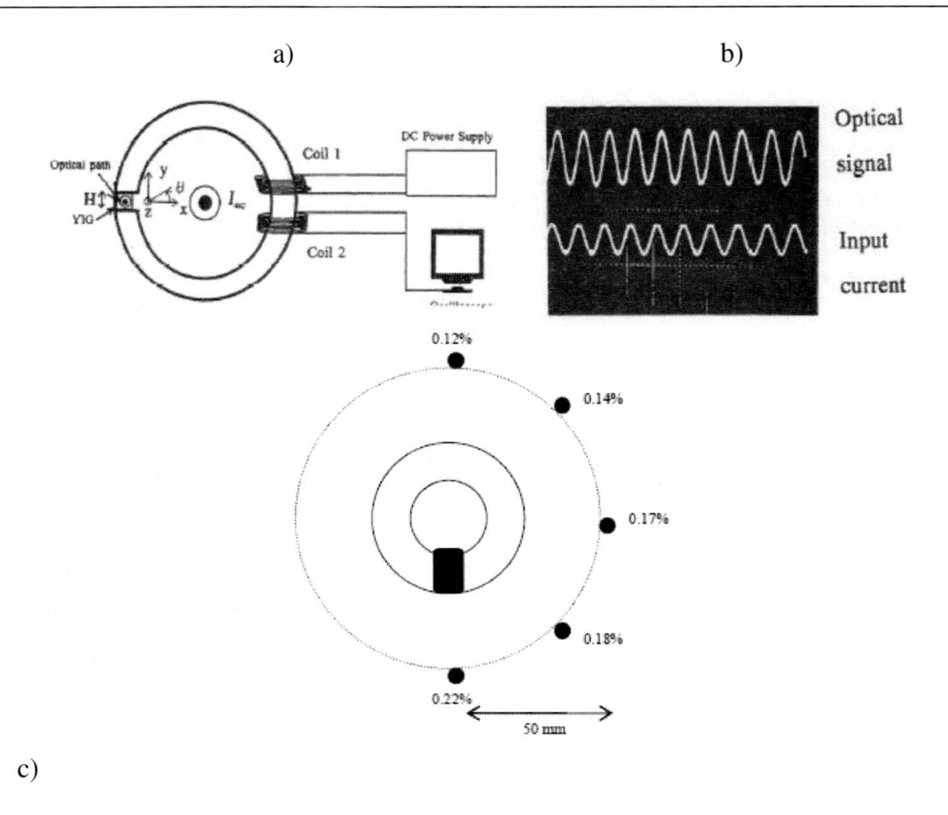

Figure 6. Fiber-linked YIG/ring-core Faraday effect optical current sensor in the transverse configuration [36]; (a) entire system, (b) oscilloscope traces of optical signal for 50Hz-100mA$_{rms}$ current, (c) measured cross-talk from the surrounding current at various angular positions.

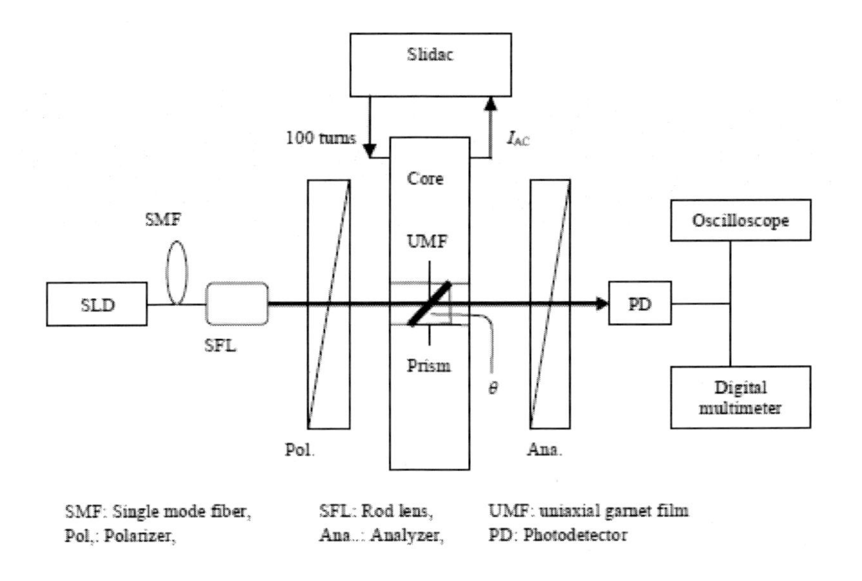

Figure 7. Schematic diagram for magnetic garnet film/ring core Faraday effect optical current sensor in the transverse configuration [37].

Figure 6(a) shows the entire system of the hybrid OCT using a YIG bulk crystal Faraday cell in the transverse configuration [36]. The YIG crystal (3×10×10 mm^3) was tightly inserted in a small gap (of 3 mm) of a permalloy core (of the size 40 mm outer diameter × 10 mm thickness) to overcome the problem of air gap.

Under the illumination of a light beam (from 1.3 μm SLD) on the crystal in the transverse direction to the core, a very high current sensitivity of a modulation depth of 20% per Ampere was obtained, which has been the highest sensitivity of OCT ever reported and about 10^4 times higher than that of the flint-glass fiber OCT.

Figure 6(b) shows the typical optical response signal. Good linearity for currents from 0 to 1 Ampere was obtained and the isolation from surrounding currents was very high, as shown in Figure 6(c).

Figure 7 shows the entire system of the hybrid type OCT using a uniaxial Bi-substituted garnet thin film (650 μm thick) in the transverse configuration [37]. The garnet film was obliquely (typically at 45° inclination angle) inserted into the 3.5 mm air gap of a permalloy ring core, and was illuminated by a collimated optical beam (1.3 μm SLD) from the transverse side of the core. The high sensitivity of about 0.08% per Ampere modulation-depth, good linear response from 0 to 1 Ampere and good reproducibility were demonstrated.

6. Other OCTs and Remarks

The OCT technology is closely related to the optical magnetic field sensing technology. Our descriptions are limited to the OCT which basically relies on the Ampere's circuital law for the sensing principle. However, if there are no other currents than a measuring current, and moreover, the magnetic field distribution due to the measuring current is a priori known, then other techniques than the ones based on the optical Ampere's law can also be applied for the current sensing. One of the useful techniques for this purpose is the introduction of a solenoid coil to transform a current into a uniform magnetic field within the solenoid coil, as long as the resistance and reactance of the solenoid do not matter. Various types of solenoid-combined OCTs were studied in connection with the optical sensing of electric power, and their actual performances are described in refs. [38–41].

Conclusion

The OCT based on Faraday effect has the invaluable features of non-contact, highly insulating and high-speed measurements. Studies on OCT have been made to fulfill the HICOC condition satisfactorily. The currently-known most effective HICOC technologies use ultra-low photo-elastic coefficient fibers, dielectric-thin film coatings and magnetic Faraday materials in the transverse configuration, for the all-fiber, bulk-optic and hybrid type of OCTs, respectively. Such OCTs can provide very simple and flexible sensor head, accomplish the fulfillment of the HICOC condition very precisely and achieve very high current sensitivity. OCTs are now mostly applied for the electric power industries, but can find other wide applications too in both engineering and science. It is very much expected that the OCT technology will improve more with the progress of optoelectronic materials and devices.

About the Authors

Toshihiko Yoshino was born in Tokyo, Japan, on April 27, 1939. He received the B.S., M.S., and Ph.D. degrees in Applied Physics from Tokyo University (Tokyo) in 1963, 1965 and 1968, respectively. From 1968 to 1988, he was with the Institute of Industrial Science, Tokyo University as a Research Associate, Lecturer and successively an Associate Professor. From 1988 to 2002, he was a Professor in the Department of Electrical Engineering, and also, from 2000 to 2002, the Director of the Venture Business Laboratory, Gunma University (Kiryu, Japan). Since 2002, he has been the Principal of the Kaisei Academy (Arakawa-ku, Tokyo). His research area of interest has been Applied Optics. From 1971 to 1973, he was a Guest Researcher with the Hanover University (Germany), where he worked in the area of Laser Spectroscopy. In 1985, he was a Visiting Professor with the IBM Research Laboratory (San Jose, CA), where he worked on optical sensing. He is a Consulting Professor of the North China Electric Power University (Beijing, China).

Dr. Yoshino is a member of the Japan Society of Applied Physics, the Optical Society of America, and a Fellow of Institute of Physics, UK.

Masayuki Yokota was born in Gunma, Japan, on January 31, 1967. He received the B.S., M.S., and Ph.D. degrees in Electronic Engineering from Gunma University in 1989, 1991, and 1998, respectively. Currently, he is a research associate at the Department of Electronic Engineering, Gunma University, where he has been working on optical engineering.

Dr. Yokota is a Member of the Japan Society of Applied Physics (JSAP) and the Optical Society of America (OSA).

REFERENCES

[1] Yoshino, T. *Proc. SPIE* 1987, 798, 258–266.

[2] Day, G. W., Rose, A. H. *Proc. SPIE* 1988, 985, 138–150.

[3] Yoshino, T.; Takahashi, Y.; Gojyuki, M. *Opt. Rev.* 1997, 4, 108–110.

[4] Yoshino, T.; Gojyuki, M.; Takahashi, Y.; Shimoyama, T. *Appl. Opt.* 1997, 36, 5566–5573.

[5] Arditry, H. J.; Bourbin, Y.; Mapuchon, M.; Puech, C. *Tech. Dig. of the 3rd Int. Conf. on Integrated Optics and Optical Fiber Communication (IOOC'3, San Francisco)* 1981, 128–129.

[6] Kueng, A.; Nicati, P. A.; Robert, P. A. *Opt. Rev.* 1997, 4, 56–57.

[7] Takahashi, Y.; Yoshino, T. *J. Light. Tech.* 1999, LT-17, 591–597.

[8] Yoshino, T.; Hashimoto, T.; Nara, M.; Kurosawa, K. *J. Light. Tech.* 1992, LT-10, 503–513.

[9] Smith, A. M. *Appl. Opt.* 1978, 17, 52–56.

[10] Rapp, A.; Harms, H. *Appl. Opt.*1980, 19, 3729–3747.

[11] Yoshino, T. *J. Opt. Soc. Am. B* 2005, 22, 1856–1860.

[12] Yoshino, T. (ed.) Optical fiber sensor technologies (2nd ed.), Dai-ichi International Co.: Tokyo, 1986, 15–160.

[13] Ulrich, R.; Simon, A. *Appl. Opt.* 1979, 18, 2241–2251.

[14] Rashleigh, S. C.; Ulrich, R. *Appl. Phys. Lett.* 1979, 34, 769–770.

[15] Laming, R. I.; Payne, D. N.; *J. Light. Tech.* 1989, 7, 2084–2094.

[16] Tang, D.; Rose, A. H.; Day, G. W., Etzel, S. M. *J. Light. Tech.* 1991, 9, 1031–1037.

[17] Kurosawa, K.; Yoshida, S.; Sakamoto, K. *J. Light. Tech.* 1995, 13, 1378–1384.

[18] Ross, J. N. *Opt. Qunatum Electron.* 1984, 16, 455–461.

[19] Yoshino, T.; Iwama, M. *Opt. Lett.* 1999, 24, 1626–628.

[20] Kurosawa, K. *Opt. Rev.* 1997, 4, 38–44.

[21] Kurosawa, K.; Shirakawa, K.; Saito, H.; Itakura, E.; Sowa, T.; Hiroki, Y. Kojima, T. *Proc. of the 16th Int. Conf. on Optical Fiber Sensors* (OFS'16, Nara) 2003, 316–319.

[22] Willsch, M.; Bosselmann, T. *Proc. of the 15th Int. Conf. on Optical Fiber Sensors* (OFS'15, Boulder) 2002, 407–410.

[23] Bohnert, K., Gabus, P.; Brandle, H., *Proc. of the 16th Int. Conf. on Optical Fiber Sensors* (OFS'16, Nara) 2003, 752–755.

[24] Hayashiya, H.; Kumagai, T.; Hino, M.; Endo, T.; Ando, M.; Negishi, H. *Proc. of the 32nd Meeting on Lightwave Sensing Technology (Jpn. Soc. Appl. Phys., LST'32, Tokyo)* 2003, 141–146.

[25] Saito, S.; Fujii,Y.; Yokoyama, K.; Hamasaki, J.; Ohno, Y. *IEEE J. Quantum Electron.* 1968, QE-2, 255–259.

[26] Kanai, N.; Takahashi, G.; Sato, T.; Higashi, M.; Okamura, K. *IEEE Trans. Power Del.*, 1986, PWRD-1, 91–97.

[27] Yoshino, T.; Yokota, M.; Aoki, K.; Yamamoto, K.; Itoi, S.; Ohtaka, M. *Appl. Opt.* 2002, 41, 5963–5968.

[28] Yoshino, T. *Jpn. J. Appl. Phys.* 1978, 19, 745–749.

[29] Yoshino, T.; Ohno, Y. *Fiber and Integrated Opt.* 1981, 3, 391–399.

[30] Yoshino, T.; Ohno, Y.; Kurosawa, K. *Tech. Dig. of the 2nd Int. Conf. on Optical Fiber Sensors (OFS'2, Stuttgart)* 1984, 55–58.

[31] Inoue, N.; Yamasawa, K. T. *IEE Japan* 1995, 115-A, 1114–1120.

[32] Numata, T.; Tanakaike, H.; Inokuchi, S.; Sakurai, S. *IEEE J. Mag.* 1990, 26, 1358–1360.

[33] Rochford, K. B.; Rose, A. H.; Deeter, M. N.; Day, G. W. *Opt. Lett.* 1994, 19, 1903–1905.

[34] Yoshino, T.; Hara, H.; Sakamoto, N. *Extended Abstracts of the Spring Meeting of Jpn. Soc. Appl. Phys.*, 1988, 860.

[35] Itoh, N.; Minemoto, H.; Ishiko, D.; Ishizuka, S. *Tech. Dig. of the 11th Int. Conf. on Optical Fiber Sensors (OFS'11, Sapporo)* 1996, 638–641.

[36] Yoshino, T.; Minegishi, K.; Nitta, M. *Meas. Sci. Tech.* 2001, 12, 850–853.

[37] Yoshino, T.; Torihata, S.; Yokota, M.; Tsukada, N. *Appl. Opt.* 2003, 42, 1769–1772.

[38] Li, Y.; Li, C.; Yoshino, T. *Appl. Opt.* 2001, 40, 5738–5741.

[39] Li, C.; Cui, X.; Yoshino, T. *IEEE Tans. Instrum. Meas.* 2001, 50, 1375–1380.

[40] Li, C.; Yoshino, T. *Apl. Opt.* 2002, 41, 5391–5397.

[41] Li, C.; Cui, X.; Yoshino, T. *J. Light. Tech.* 2003, 20, 843–849.

In: Advances in Laser and Optics Research. Volume 10 ISBN: 978-1-62257-795-8
Editor: William T. Arkin

Chapter 11

TWO-DIMENSIONAL PLASMON POLARITON NANOOPTICS BY IMAGING

***Andrey L. Stepanov*[1,*]*, Aurélien Drezet*[2] *and Joachim R. Krenn*[3,†]**
[1]Laser Zentrum Hannover, Hannover, Germany
Kazan Physical Technical Institute, Russian Academy of Sciences, Kazan, Russian Federation,
Kazan Federal University, Kazan, Russian Federation
[2] Instritute Néel CNRS and University Joseph Fourier, Grenoble, France
[3]Institute of Physics, Karl-Franzens University, Graz, Austria

ABSTRACT

A review of the experimental realization of key high efficiency two-dimensional optical elements, built up from metal nanostructures, such as nanoparticles and nanowires to manipulate plasmon polaritons propagating on metal surfaces is reported. Beamsplitters, Bragg mirrors and interferometers designed and produced by elelectron-beam lithography are investigated. The plasmon field profiles are imaged in the optical far-field by leakage radiation microscopy or by detecting the fluorescence of an organic film deposited on the metal structures. It is demonstrated that these optical far-field methods are effectively suited for direct observation and quantitative analysis of plasmon polariton wave propagation and interaction with nanostructures on thin metal films. Several examples of two-dimensional nanooptical devices fabricated and studied in recent years are presented.

INTRODUCTION

The ongoing miniaturization in today's innovative technology has triggered the emergence of *nanotechnology*. Within this field intense research effort in, e.g., material

[*] Corresponding author: E-mail: aanstep@gmail.com.
[†] Joachim R. Krenn: Institute of Physics, Karl-Franzens University, 8010 Graz, Austria.

science, chemistry, electronics or microscopy is conducted, often combined to interdisciplinary research. When it comes to optics, however, we find the nanoscale to be out of reach for conventional optics: diffraction limits the spatial resolution to a value given by about half the light wavelength. To overcome this limitation is of great interest for both basic research and technological applications. The investigation of the wealth of optical phenomena by the broad range of readily available radiation sources and detection devices would clearly profit from an extension of the spatial resolution beyond the diffraction limit. Furthermore the continuous miniaturization of existing electronic devices is expected to meet fundamental limitations in the near future. In particular, the metallic interconnects used to link individual electronic elements within data-processing units as well as interfacing these units with their surrounding electronic infrastructure will not be able to deal with much higher computation frequencies than are used in existing devices. One potential solution to this problem is to substitute conventional electronic devices with *integrated optical elements* working in nanoscale dimensions. Nowadays, the drive towards highly integrated optical devices and circuits for use in high-speed communication technologies and in future all-optical photonic chips has generated considerable interest in the field of *optical nanotechnology* [1,2].

Since the beginning of the 1990's new routes towards the control of light wave propagation at the micro- and nano-scale are explored. Relying on *dielectric* structures, photonic band gap materials [3, 4] and high dielectric contrast materials have been explored. For the latter a strong light field confinement permits to reduce the cross section of waveguides and to decrease radiative losses in bends. Recent results have demonstrated that visible light from an evanescent local source can be efficiently propagated through a TiO_2 waveguide only 200 nm wide [5 - 7].

Besides dielectric materials, *metals* have been investigated for their potential in downsizing optics beyond the diffraction limit. Recently *surface plasmon polaritons* (SPPs) excited in metal nanostructures were identified as promising candidates to serve that need. SPPs are resonant electromagnetic surface modes constituted by a light field coupled to a collective oscillation of conduction electrons at the interface of a metal and a dielectric. While SPPs propagate along the metal-dielectric interface, their perpendicular field components decay exponentially into both neighboring media [8]. The corresponding electromagnetic fields are thus strongly localized at the interface. The value of the according wave vectors of optical SPPs is larger than the light wave vector at a given frequency, however, SPP waves can propagate on a two-dimensional surface with almost the speed of light.

Increased interest in optical SPPs comes from recent advances that allow metals to be structured and characterized on the nanometer scale. If metal-dielectric interfaces or surface elements are formed on the nanoscale, it is the spatial dimension of the nanostructures rather than the light wavelength that determines the spatial extension of the SPP field, thus rendering feasible optics beyond the diffraction limit. Therefore, SPPs could allow the realization of novel photonic devices that meet the need of manipulating or guiding electromagnetic fields in nanoscale dimensions. The principal feasibility of *nano-optics* based on metal nanostructures is well documented by recent experiments [9-11]. The existing nano-optical technology and know-how seem complete enough to allow the realization of functional SPP based nanoscale devices.

As a propagating wave, a SPP in an extented metal structure can be used for the direction of light fields, corresponding to *signal transfer* or *optical addressing*. Indeed, recent results on SPP propagation in μm-wide metal stripes and nanowires [12, 13] demonstrate the feasibility of SPP based subwavelength light field transport. Moreover, when a thin metal film acts as a waveguide for SPPs the same structure can be used to carry electrical signals. Propagating SPPs are characterized by a dispersion relation defining the plasmon wavelength to be smaller than the vacuum light wavelength for any given light frequency. Consequently prism, grating or edge coupling [8] have to be applied for optical SPP excitation. The metals of choice are usually gold or silver, as these metals show SPP modes in the visible or near-infrared spectral range, besides rather low ohmic damping [1]. The SPP propagation distance, while depending on material and wavelength is in the order of 10 μm in visible spectral range.

On the other hand, noble metal nanoparticles are known for a long time for spectrally selective absorption and scattering due to particle plasmon excitation [14-16]. The plasmon resonance frequency is a function of particle geometry, the polarization direction of the exciting light wave with respect to the particle axes and the dielectric functions of the particle metal and the surrounding medium. The excitation of particle plasmons gives rise to a spectrally narrow extinction band and to an electromagnetic field strongly enhanced around the nanoparticle with respect to the exciting light field.

As in many applications ensembles rather than individual particles are applied, the mutual interaction of particle plasmons is of high interest. This interaction was recently identified to be particularly strong for regularly arranged nanostructures due to grating effects and can be used for SPP excitation on a surface. Further recent results revealing the properties of particle plasmons indicate that this phenomenon can be efficiently exploited for the local *manipulation* of light fields on the nanoscale. The findings include light field squeezing [11], local plasmon excitation [10], the controlled design of near-field enhancement [17] and spectral selectivity [18], all of which can be applied to manipulate with SPPs on a surface.

In summary, propagating SPPs and particle plasmons can be used for light signal propagation and nanoscale light field manipulation, respectively. A combination of both could obviously allow the realization of nanoscale plasmon-based (*plasmonic*) optical devices. Besides connecting nanoscale metal structures, propagating SPPs can also provide the interface between classical optics (freely propagating or dielectric waveguide bound light) and nanoscale optical and the plasmonic devices.

Lithographic Nanoscale Sample Preparation

Metal nanostructures are available for a long time in the form of island films and colloidal nanoparticles [14-16]. Nevertheless, the detailed investigation of particle plasmon effects proved to be rather difficult as tailoring the geometry and spatial position of the particles within such samples is virtually impossible. These drawbacks can be eliminated by lithographic techniques such as electron beam lithography (EBL), which was recently adapted for the production of metal nanoparticle ensembles well controlled in geometry and position [19].

EBL relies on the local exposure of an electron sensitive resist by a focused electron beam [20]. Usually a standard scanning electron microscope is equipped with commercial hard- and software to extend it to a lithography device. A resist layer (thickness typically 100 nm) can be formed by spin coating substrates such as glass plates with, e.g., polymethylmethacrylate (PMMA). The glass plates are often covered with a few nm of indium tin oxide, which combines the weak electric conductivity needed for EBL with high optical transparency adequate for optical experiments. Electron beam exposure (electron energy ≈ 25 keV) breaks the PMMA molecules into fragments that can be preferentially chemically dissolved. Therewith a PMMA mask for a subsequent deposition process of the material constituting the nanostructures is produced. Deposition is carried out by thermal or electron beam evaporation of materials such as silica, silver or gold in high vacuum. The height of the deposited material, typically ~60 nm is monitored with a quartz microbalance. In a final step, the PMMA mask is removed chemically, leaving the dielectric or metal nanostructures on the substrate. Residual PMMA is ashed by exposing the sample to an argon plasma. When aiming at propagating SPPs, the fabricated discrete nanostructures can be covered by a thin continuous metal film with a typical thickness of 60 nm. A typical structure fabricated by this approach is illustrated in Figure 1 [21]. EBL provides the means to produce metal nanostructures with smallest feature sizes of about 20 nm over areas of typically 100×100 μm.

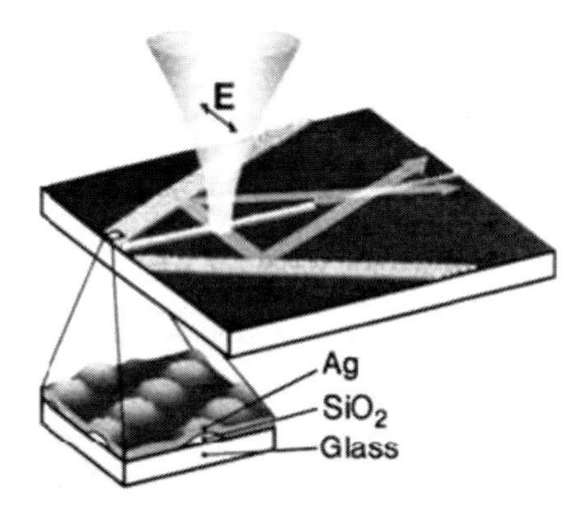

Figure 1. Exemplary plasmonic sample structure: Sketch of sample geometry combined with atomic force microscopy image of the surface [21]. The area of the upper image is 25×25 μm.

LITHOGRAPHIC FLUORESCENCE IMAGING

To couple light to SPPs the light wave vector has to be increased as for a given frequency it is always smaller than the SPP wave vector due to the SPP dispersion relation [8]. By EBL it is possible to create suitable nanoscale surface protrusions on the metal film with geometries giving rise to scattering that provides the Fourier components necessary to match the light and SPP wave vectors. As an example, we launch SPPs by focusing a laser beam onto nanostructure on a gold thin film through a optical microscope objective. A direct way to the SPP field profile is to monitor the fluorescence of a thin molecule layer close to the metal surface by a charge-coupled device (CCD) camera (Figure 2) coupled to an optical microscope. This technique relies on the fluorescence of organic molecules placed in the vicinity of the SPP-carrying metal surface. A 10^{-4} molar solution of DiR or Rhodamine 6G was prepared by dispersing the molecules in a solution of 1 % PMMA in chlorobenzene. The resulting solution was spin cast on the samples yielding a film 30 nm thick when dried. Based on conventional optical microscopy, fluorescence imaging is a quite quick and reliable technique for probing SPP fields.

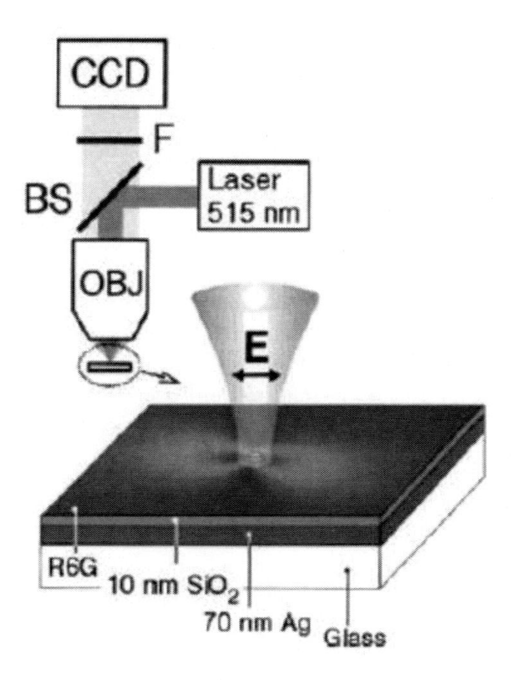

Figure 2. Sketch of the experimental setup for fluorescence imaging of SPPs launched from a nanoscale surface protrusion, combined with an actual SPP fluorescence image. OBJ optical microscope objective (100×, numerical aperture 0.75); BS beam splitter; CCD charge-coupled device camera [22].

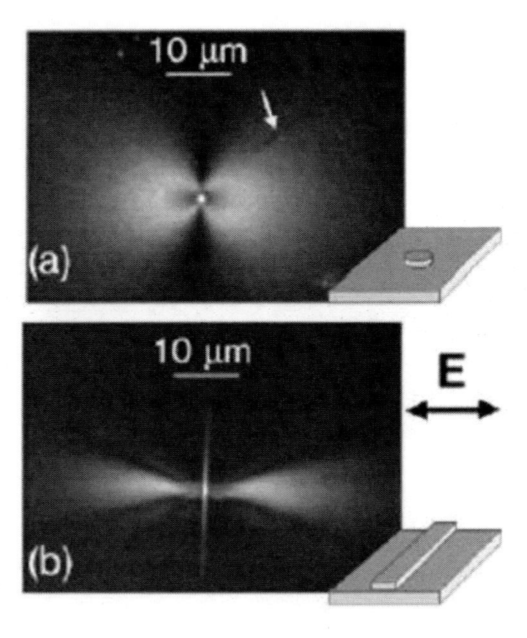

Figure 3. Fluorescence images of SPPs launched from lithographically designed surface features, as sketched in the insets, light wavelength 515nm [21].

As a further example, Figure 3 depicts the fluorescence images of SPPs launched from differently shaped nanostructures, leading to distinctly different plasmon profiles [21].

For comparison, the SPP propagating along the nanowire in Figure 4 has been imaged with a near-field optical microcsope [11].

The present know-how and technology with regard to plasmon effects on both the micro- and nanoscale seem sufficiently advanced to allow the development of functional SPP based optical devices. Therefore basic elements in analogy to conventional optics, as mirrors, beamsplitters, filters, polarizers or resonators have to be realized. Precisely tailored metal nanostructures can serve as the elementary building blocks for these elements. The practical aim of this line of research is the experimental realization and optimization of plasmonic components.

For example, together with the local launch of directed SPPs via a metal nanostructure, SPP mirrors and beamsplitters allow the realization of a SPP interferometer. Additionally, a question of fundamental interest concerns the detailed understanding of how the nanoparticles and -structures that make these SPP optical elements interact with propagating SPPs, i.e., what is the role of their geometries, resonance frequencies, etc. Directed SPPs can be launched as outlined above (Figure 3, [21]).

These SPPs propagating in the flat metal/air interface can be reflected and/or transmitted in-plane by suitable mirrors and beamsplitters.

These elements can be built up and optimized from metal nanostructures. The principle feasibility of this approach to form functional optical elements for a 2D-dimensional SPP optics was recently demonstrated. Figures 5 and 6 display examples of SPPs locally launched on a silver surface and manipulated by silver nanoparticles forming Bragg mirrors and, finally, an interferometer [23].

Based on earlier work [21] we identified in detail the interference conditions for different optical pathlengths in the interferometer arms, based on both lateral position changes of the SPP mirrors and changes in their angular orientation (Figure 7). SPP interferometers constitute thus the first two-dimensional SPP device that is thoroughly characterized.

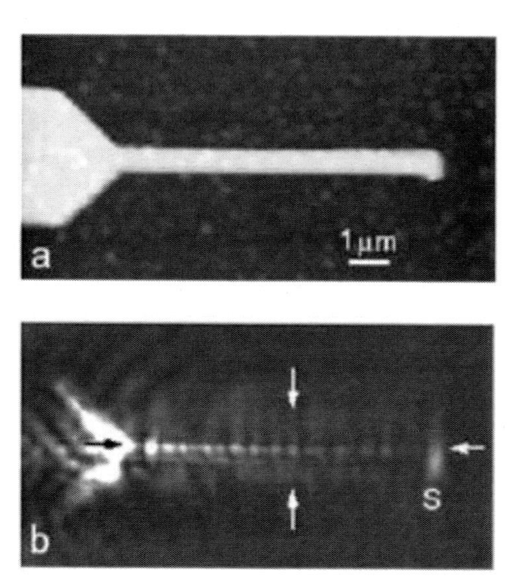

Figure 4. SPP propagation along a gold nanowire (a) topography, (b) near-field optical image acquired with a near field scanning microscopy, light wavelength 800nm [13].

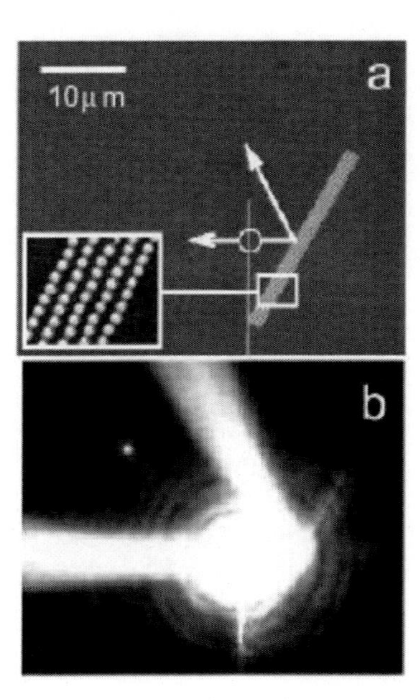

Figure 5. SPP Bragg mirror: (a) Scanning electron microscope image, the inset shows a 6x-magnified view of the area marked by the white box. The white circle and the arrows indicate the laser focus position and the SPP propagation directions, respectively. (b) Corresponding fluorescence image [23].

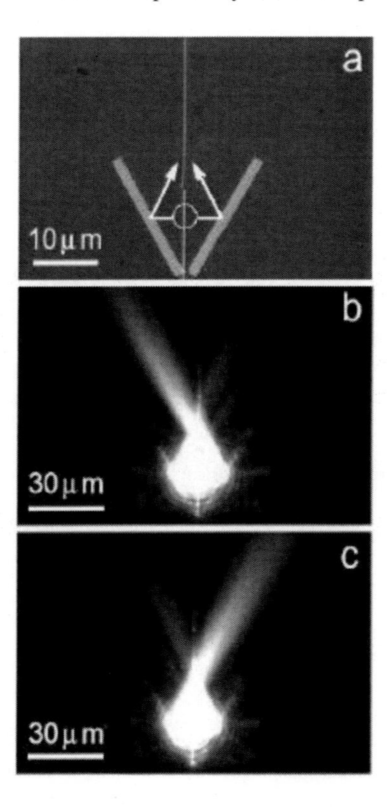

Figure 6. SPP interferometer: (a) Scanning electron microscope image, the white circle and the arrows indicate the laser focus position and the SPP propagation directions, respectively. (b) and (c) show two corresponding fluorescence images with different lateral positions of the right Bragg mirror [23].

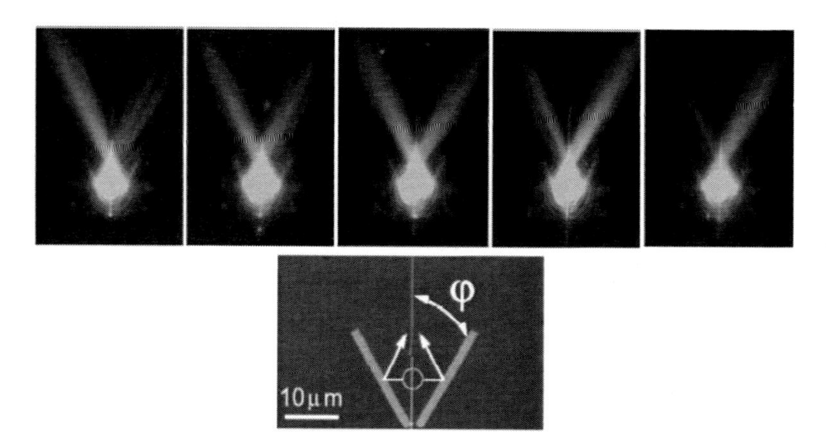

Figure 7. SPP interferometer: (bottom) Scanning electron microscopy image. The white circle and the arrows mark the laser focus position and the SPP propagation directions, respectively, the angle φ defines the SPP mirror orientation. (top) Fluorescence images for different mirror positions.

Besides the realization of a SPP interferometer consisting of mirrors and beamsplitters, the feasibility of a SPP based spectral filter was investigated. Recent results have demonstrated that light extinction from a periodic array of gold nanoparticles deposited onto a planar dielectric waveguide can be controlled by adjusting the array periodicity [24, 25]. Selective suppression of light extinction has been attributed to the coupling of light scattered by the particles with a surface mode of the underneath surface. Despite of the advantages of fluorescence imaging for the visualization of SPP propagation on metal surfaces and their interaction with surface nanostructures we identified restrictions for the practical application of this approach:

- First, fluorescence images have to be recorded within a very limited time (typically a few seconds) after the beginning of laser irradiation because of molecule bleaching. This often restricts the possibilities for a precise optical adjustment and the acquisition of images.
- Second, the fluorescence intensity is in general not proportional to the local SPP field intensity, making quantitative measurements impossible. Furthermore, the efficiency of light/SPP coupling is difficult to assess in *quantitative* terms, as usually no control over the relative contributions of absorbed, transmitted and scattered light field intensities is achieved.

In the following, we discuss and demonstrate that the imaging of SPP fields by means of *leakage radiation* (LR) allows to overcome the drawbacks of fluorescence imaging. We will especially focus on the fact that this approach allows for *quantitative* measurements of the spatial SPP field profile.

Leakage Radiation Imaging

The intensity decay of a plane SPP wave in a perfectly planar metal film between two dielectric media (Figure 8, [26]) defines its intrinsic decay length $L_{int} = 1/2k''$, which is a measure of the "ideality" of the electron gas. k'' is defined as the imaginary part of the complex surface plasmon wave vector $k_{SPP} = k'' + ik''$. *Intrinsic* losses are caused by inelastic scattering of conduction electrons, scattering of electrons at interfaces and LR [8, 27, 28].

LR is emitted from the interface between the metal thin film and a higher-refractive-index dielectric medium (substrate), for example, glass (Figure 8). When the electromagnetic plasmon field crosses the metal film and reaches the substrate, LR appears at a characteristic angle of inclination θ_{LR} [8] with respect to the interface normal.

At this angle the LR wave satisfies $k_{SPP} = nk_0 \sin\theta_{LR}$ where k_{SPP} and nk_0 are the wave vectors of the SPP (real part) and the LR, respectively, with n being the refractive index of glass. For glass with n = 1.5 it follows that $\theta_{LR} \cong 44$ fulfills the phase matching condition, which is larger than the critical angle of total internal reflection, θ_{CRITIC} = 41.8 [8, 27]. It should be mentioned that LR can as well be observed experimentally when SPPs are excited at the metal/air interface by electrons.

Although LR contributes to SPP damping, it permits the direct linearly proportional detection in the far-field of the SPP spatial intensity distribution at the metal/air interface. Indeed, the intensity distribution at the metal/glass interface is proportional to that of the SPPs at the same lateral position and the azimuthal intensity profile equals that of the SPP profile at each point of the interface [8, 27, 28].

A typical experimental setup used for LR imaging is shown in Figure 9 [29]. As mentioned above, LR arises at the interface between the thin metal film and the glass substrate (Figure 8). To avoid total reflection inside the glass, an immersion objective in contact with the bottom part of the sample is required to collect LR images [30]. Note that LR is generated as well in the so-called Kretschmann configuration with a prism [31], where it interferes with the incoming light.

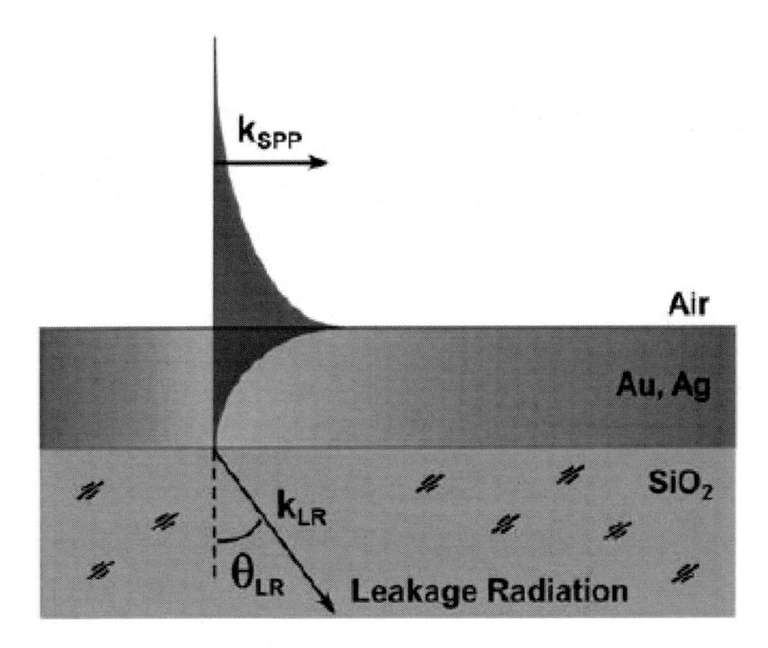

Figure 8. Leakage radiation from SPPs. The SPP field is maximum in the interface and decays exponentially in the perpendicular directions SPPs propagating along at the interface metal/air will couple and transform their energy to electromagnetic fields in the glass substrate [26].

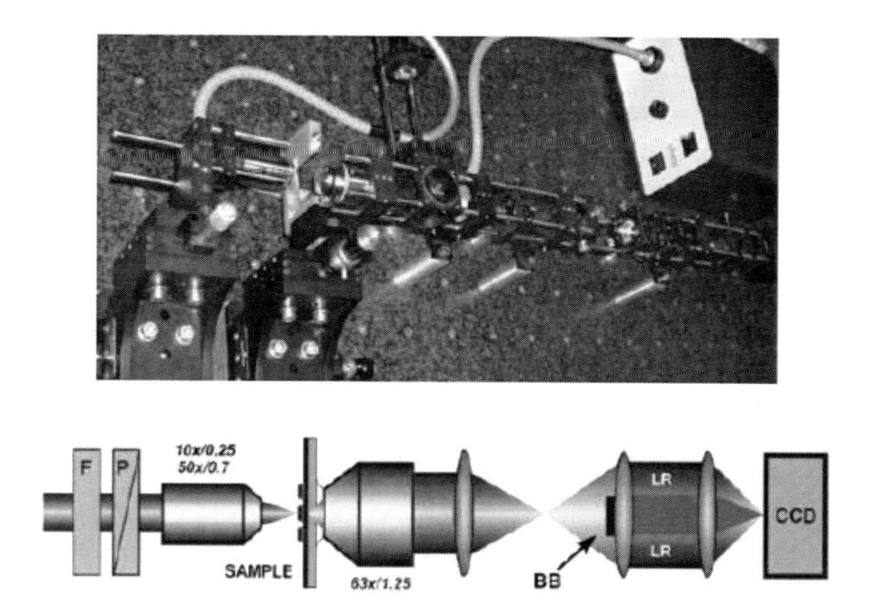

Figure 9. Experimental scheme for LR imaging (bottom). SPPs are exited by laser light focused onto a structured gold film on a glass substrate. LR is emitted into the glass substrate at an angle θ_{LR} (compare Figure 8). F (gray) filter, P polarizer, BB beam block. A picture of the setup is shown on top [29].

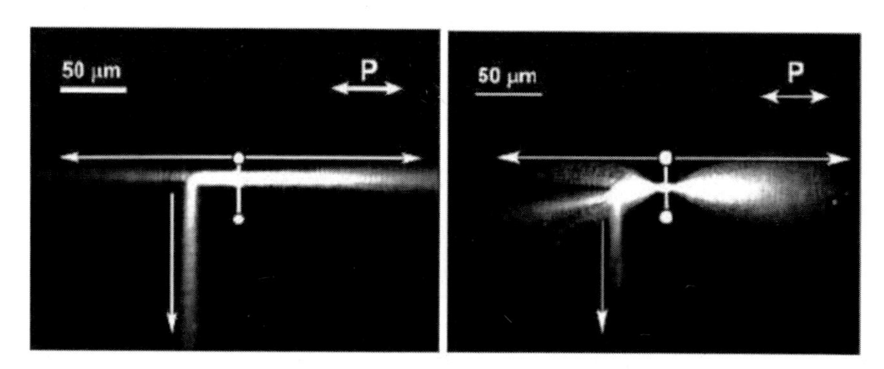

Figure 10. LR images of propagating (and reflected) SPPs acquired with different objectives for focusing the laser beam: Zeiss Achroplan 10x, numerical aperture 0.25 (left) and Zeiss 50×, numerical aperture 0.7 (right). P indicates the laser polarization direction.

The conditions of local excitation have an important impact on the SPP properties, as we now show by applying linearly polarized light from a Titan:Sapphire laser at a wavelength of between 780 and 900 nm focused onto a gold ridge (200 nm wide, 60 nm high) on a 60 nm thick gold film through two different microscope objectives (Zeiss 50×, numerical aperture 0.7 and Zeiss Achroplan 10×, numerical aperture 0.25). With different objectives and thus focus diameters, SPP waves of different divergence angles can be launched as illustrated in Figure 10.

Thus, conventional microscopy based LR imaging proves to be a quick and reliable technique for probing the spatial profile of SPP fields [29]. In the meantime, LR imaging has become a standard technique used by many scientific groups [32 -36].

The quality of LR images is significantly enhanced when the directly transmitted part of exiting laser light is blocked from contributing to the LR image [37].

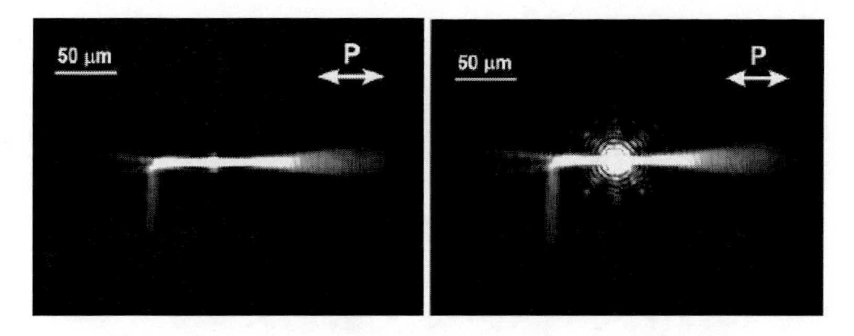

Figure 11. LR images acquired in the configuration with a beam block for the directly transmitted laser beam (left) and without blocking (right). P indicates the laser polarization direction.

Therefore, a reverse diaphragm can be introduced as a central beam block in the first image plane of the sample with a diameter chosen such that the LR appearing under the specific angle θ_{LR} remains unaffected while the directly transmitted laser light is blocked (see Figure 9, „Beam Block"). For illustration, in Figure 11, LR images with and without beam blocked configuration are compared. Large improvements in signal constant and reduced background in the LR image acquired with a beam block are clearly seen.

INTERACTION OF SPPS WITH BEAMSPLITTERS

By applying LR imaging we performed the quantitative analysis of the efficiencies of nanoparticle chains on metal thin films that can be used as SPP beamsplitter in terms of SPP transmittance and reflectance as a function of laser wavelength applied for SPP excitation [26, 29]. Figure 12 depicts the LR image of a structured silver film shown in the scanning electron microscope (SEM) micrograph in the inset. The SPP propagation length is found to be of similar value as that reported from fluorescence imaging [22]. The center of the excitation at the position of the focused laser spot is overexposed due to the limited dynamic range of the CCD camera. Actually the SPP excitation intensity was adjusted to a level beyond the maximum CCD response near the source point in order to image the decay of the SPP over a decently wide area. As shown in Figure 12, two-counter propagating directed SPPs are locally exited by focusing the laser beam onto a nanowire (200 nm wide, 60 nm high), similar to the previously discussed fluorescence imaging studies [21-23]. The LR image reveals two highly directed SPP beams propagating to the right and left with a quite small divergence angle. The right-bound SPP beam interacts with a beamsplitter consisting of a chain of individual nanoparticles (60 nm high, diameter of 260 nm, and separated by a distance of 400 nm), see Figure 12. The incidence angle of the SPP is 45 with regard to the particle chain direction. The SPP is both partly reflected and transmitted by the the beamsplitter, as depicted by the arrows in the figure. In this particular case the intensity of the transmitted beam is comparably high, with only small contributions from reflectance and scattering. An intensity cross-cut through the intensity distribution along the arrows in Figure 12 is shown in Figure 13. The lateral variations of the SPPs intensity are fit by the expression given in same figure. This behavior is characteristic for the damped radiation from a dipole in two dimensions [27, 28]. To obtain a correct fit, the center of the fit functionas plotted in Figure 13 together with the experimental data has to be accordingly corrected for background ligtht.

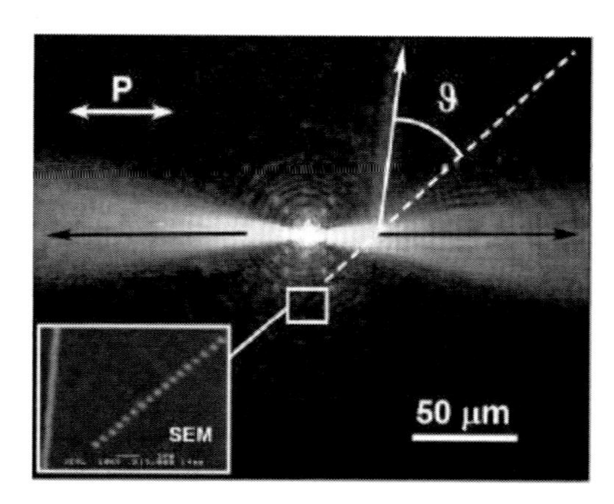

Figure 12. LR image of SPPs interacting with a nanoparticle beamsplitter. The arrows define the directions of SPP propagation. P indicates the polarization direction of the exciting laser beam. The inset shows a magnified SEM image [29].

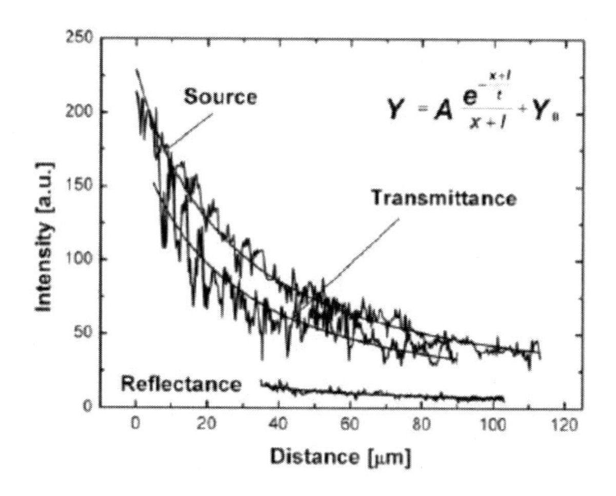

Figure 13. Cross-section intensity profiles along the arrows in Figure 12 together with fitting curves calculated by the inset equation. Y is the intensity; Y_0 is the value of the background intensity in the image; t is the decay constant of SPPs; x is the distance from the focused laser spot and l is the selected distance between laser spot and starting point for profile measurements in the images.

Therefore, a reverse diaphragm can be introduced as a central beam block in the first image plane of the sample with a diameter chosen such that the LR appearing under the specific angle θ_{LR} remains unaffected while the directly transmitted laser light is blocked (see Figure 9, „Beam Block"). For illustration, in Figure 11, LR images with and without beam blocked configuration are compared. Large improvements in signal constant and reduced background in the LR image acquired with a beam block are clearly seen.

The quantitative analysis of the fitting curves yields then all information on transmittance, reflectance and scattering of SPPs at the beamsplitter, which is summarized in Figure 14 [29]. In the near-infrared spectral region for wavelengths longer than ≈ 875 nm the beam splitter is almost transparent for SPPs (T ≈ 85 %), while the according reflectance is very weak and cannot be measured with the present technique. Therefore ≈ 15 % of the incident SPP intensity has to be attributed to SPP scattering to light and absorption by the metal. When moving towards shorter laser wavelengths, the transmittance monotonically decreases and has a minimum at ≈ 800 nm, whereas the reflectance simultaneously increases to a maximum value at the same wavelength. Proceeding further towards lower wavelengths restores the high transmittance and almost negligible reflectance values (Figure 14). SPP transmittance and reflectance thus shows a resonance-like behavior with a resonance wavelength of approximately 800 nm. Although such resonances are well known from metal nanoparticles in dielectric matrices [8], our results show that nanoparticle-on-film systems maintain a resonant behavior in qualitative accordance with theoretical work [38].

In Figure 15 we plot the dependence on the direction of the SPP beam reflected from the nanoparticle beamsplitter (Figure 12) in dependence on the laser wavelength. As parameter for the measurements the angle ϑ between the nanoparticle chain and the direction of the reflectance was selected, as depicted in Figure 12.

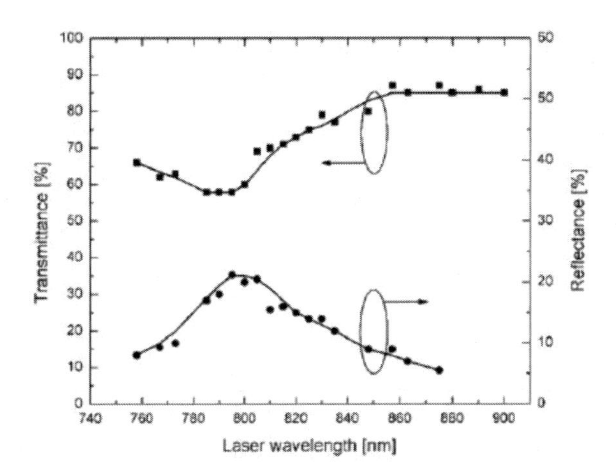

Figure 14. Transmission (T) and reflection (R) efficiencies of the nanoparticle beamsplitter versus laser wavelength. The solid curves are guides to the eye [29].

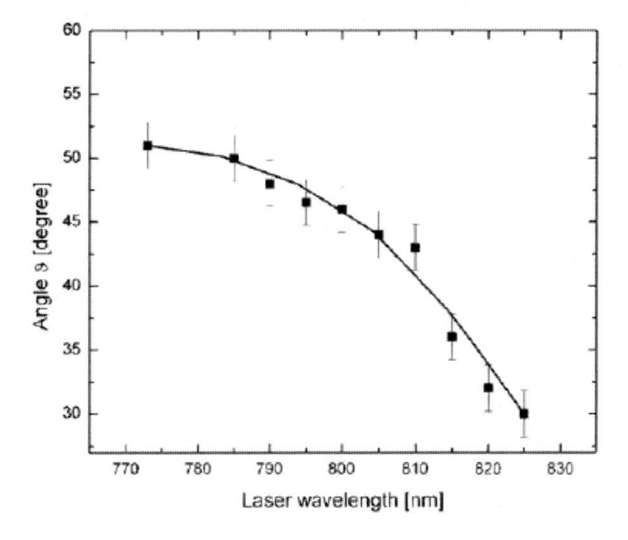

Figure 15. Angle (ϑ) reflectance dependence on the laser wavelength as indicated in Figure 12. The solid curve is a guide to the eye.

Table 1. Beamsplitter efficiency for different incidence angles of SPP incidence

Angle	Transmittance	Reflectance	Scattering
30°	92 %	2 %	6 %
45°	86 %	9 %	5 %
60°	65 %	24 %	11 %
75°	14 %	80 %	6 %

We find that an increase in wavelength leads a decrease in angle ϑ, illustrating the interplay of beamsplitter parameters and the SPP wavelength.

For further analysis we now turn to a nanoparticle beamsplitter with a similar silver nanostructure geometry as the one presented in Figure 12 but with a modified parameters of the nanoparticle chain (height of 70 nm, diameter of 220 nm and separated by a distance of 300 nm), see Figure 16. Again, we analyze SPP transmission and reflection as a function of the incidence angle of the SPP beam. Using the analysis method described above the corresponding values for transmittance, reflectance and scattering were derived and summarized in the Table 1.

We find that increasing the incidence angle of the SPP beam with respect to the particle chain direction increases the SPP reflectance and decreases the according transmittance values. All these results form a solid base for the application-oriented design of components as mirrors, beamsplitters and interferometers [39, 40].

ELLIPTICAL BRAGG REFLECTORS AND INTERFEROMETER

With a specific pattering of, e.g., gold nanoparticles on a thin gold film an elliptical Bragg reflector for SPPs can be realized (Figure 17). In such an elliptical structure, SPPs are generated by laser excitation on a single particle in one focal point F_1. Following reflection from the elliptical Bragg mirror, SPP focusing takes place in the second ellipse focus F_2 [41, 42].

In order to achieve an effective SPP reflectance from the elliptical mirror we used a system of five confocal ellipses arranged by nanoparticles (Figure 17). The distances between the elliptically shaped nanoparticle chains were selected to enable Bragg reflection at a wavelength of 750 nm, just as it was demonstrated before for Bragg mirror constructed by nanoparticles arranged in parallel lines [22].

LR images of the SPP field distributions corresponding to two polarization orientations of the SPP-exciting laser beam are shown in Figures 18a and b. We find that SPP focusing in F_2 is indeed working while the SPP intensity profile within the ellipse depends on the polarization conditions of the exciting laser beam. In Figures 18c and d depict numerical simulations for comparison with the experimental data. Therefore, we characterized the SPP waves by a $\cos^2(\Theta)$ angular intensity distribution Θ being the azimuthal angle between the polarization axis of the laser beam and the direction to the observation point. For modeling we considered the SPP waves from F_1 and the contribution due to SPP reflection from the elliptical mirror. The excellent agreement of measured and calculated images confirms that 2D-elliptical mirrors for SPPs can indeed be used for SPP focussing. The successful experimental realization of the elliptical Bragg reflectors and the effective focalization of SPPs as dicussed in the previous paragraph can be used as the starting point to construct a SPP interferometer based on elliptical Bragg mirrors. The idea of this interferometer is the dependence of the SPP intensity in the focal point F_2 on the polarization angle of the exciting laser beam which launches SPPs in F_1.

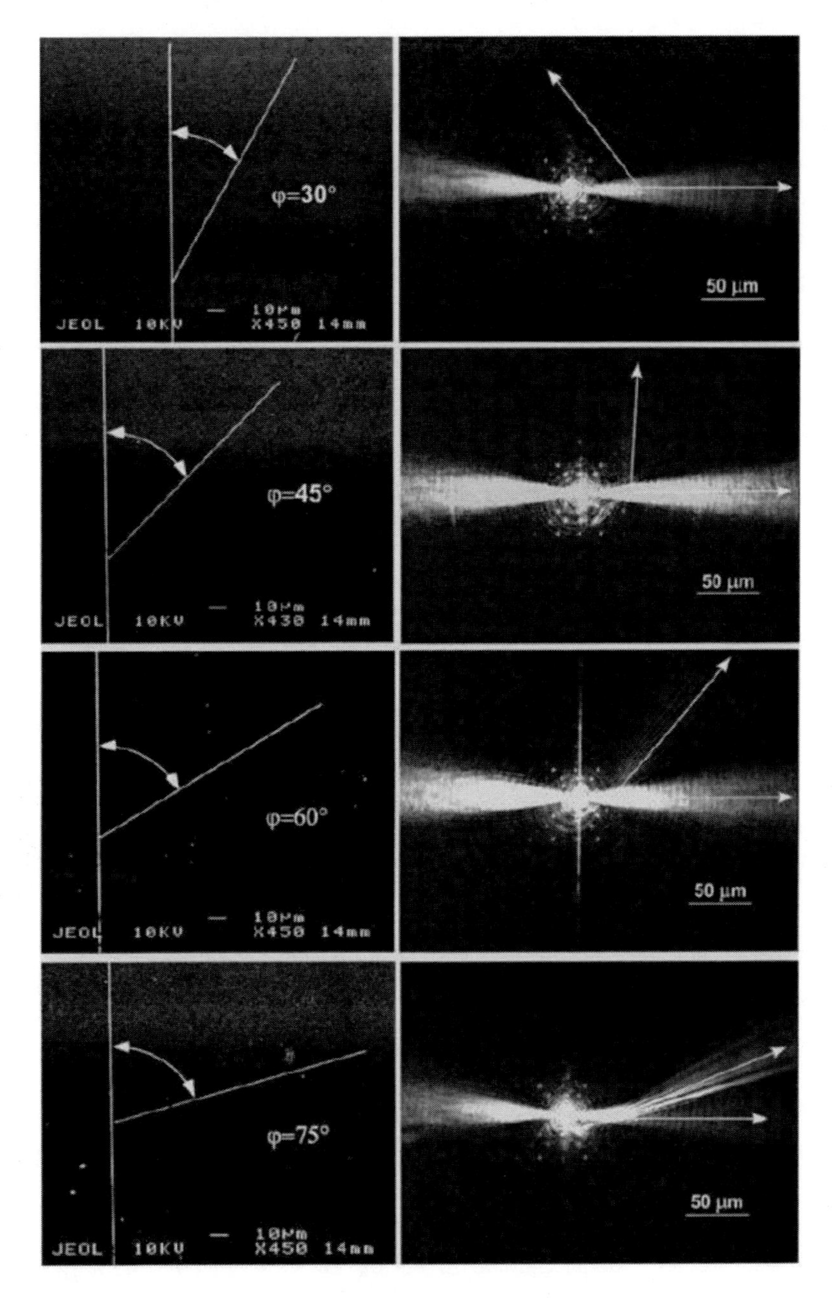

Figure 16. LR images of SPP nanoparticle beamsplitters for different SPP incidence angles. A nanowire as the SPP launching structure and SPP beamsplitters built up from chains of individual nanoparticles are imaged by SEM to the panels on the left hand side. The corresponding LR images are shown on the rigth hand side.

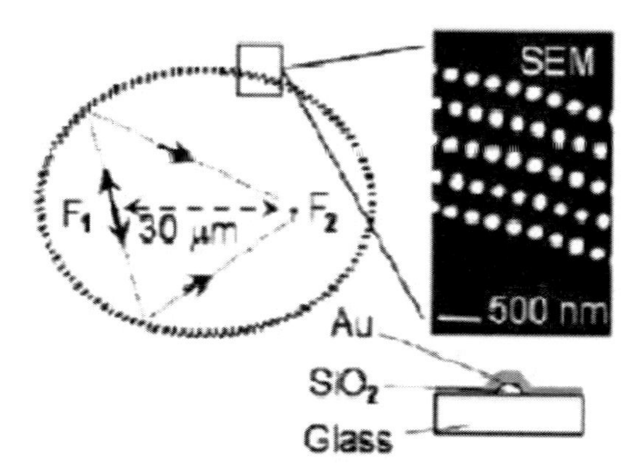

Figure 17. Elliptical SPP nanoparticle reflector. Sketches of the reflector geometry and the cross-section of an individual nanoparticle are combined with a SEM image of a part of the reflector [41].

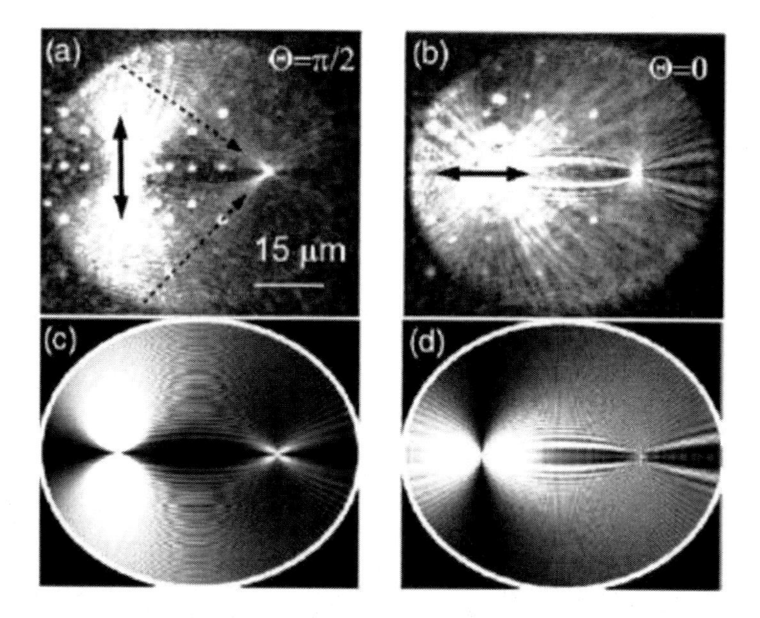

Figure 18. Experimental LR images of locally exited SPP beams upon reflection from elliptical Bragg nanoparticle reflectors. SPP are exited locally in the left focus for (a) vertical and (b) horizontal polarization of the exciting laser beam. Images (c) and (d) show the corresponding theoretical modeling [41]. The black arrows depict the laser polarization direction, the dashed arrows sketch the SPP propagation direction towards the right hand focus after reflection.

The according experimental images are shown in Figure 19 and are found again in excellent agreement with the theoretical modeling for different polarization angles Θ (here defined as the angle between the exciting laser beam polarization and the horizontal), as shown in the same figure [43]. The values of the SPP intensity in F_2 extracted from the experimental measurements are presented in Figure 20 as a function of the angle Θ. This figure quantitatively illustrates the interference condition around F_2 due to SPP reflected from the elliptical Bragg mirror and how this intensity is changed in dependence on laser polarization. Figure 20 is a clear illustration that an efficient angular polarization SPP interferometer can be realized.

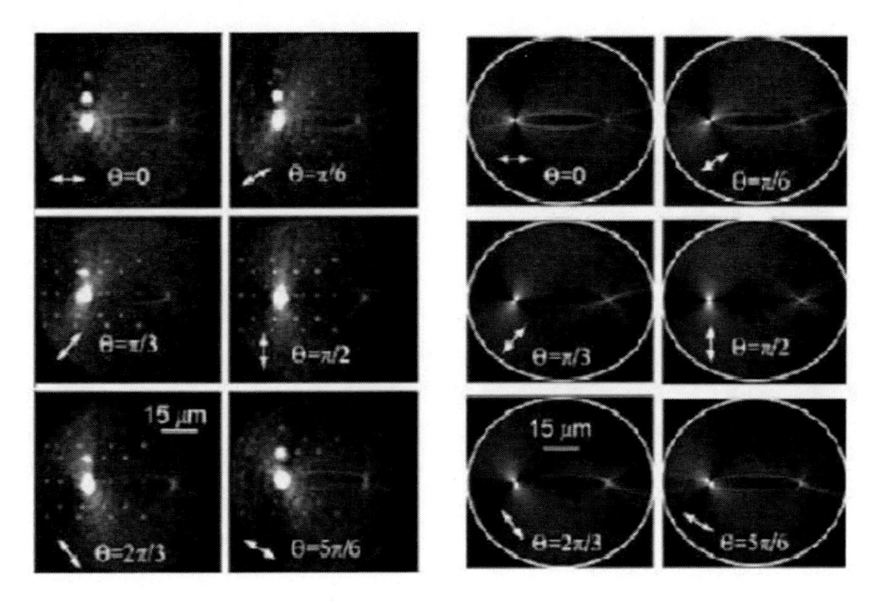

Figure 19. Experimental LR images of SPPs propagating from F_1 to F_2 for different polarization angles Θ, defined as the angle between the exciting laser beam polarization and the horizontal (left set of images). Corresponding theoretical modeling of the LR images (right set of images). The polarization directions are indicated by the white arrows [43].

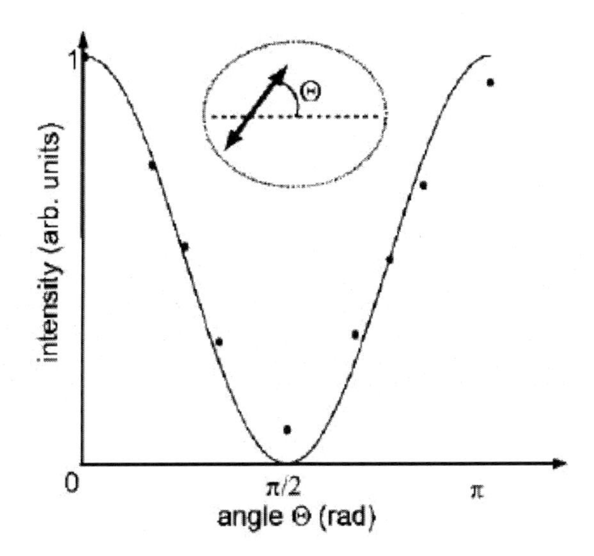

Figure 20. Experimental data (dots) and theoretical model (solid line) for the dependence of the SPP intensity at F_2 on the polarization angle Θ [41].

LINEAR BRAGG REFLECTOR

Besides being built from nanoparticles SPP Bragg mirrors can be constituted as well by arrays of gold nanowires on a gold thin film, which we now apply to demonstrate that LR imaging provides *quantitative* data on the reflection efficiencies of SPPs interacting with surface nanostructures. In Figure 21 the SEM and LR images for different structural parameters of the Bragg mirrors (interline distance) and incidence angles are presented. The images were acquired with different divergence angles of the SPP beam by laser excitation through different microscopy objectives. The quantitative analysis of SPP reflectance was done by the methods described in [29].

Summarizing the main aspects revealed by Figure 21, we conclude:

- The reflectance efficiency is higher for the SPP beam with smaller divergence angle. This finding is readily explained by the fact that the SPP intensity is distributed over a larger angular sector for the case wider divergence angles. Using a higher divergence angle allows us however to directly assess the acceptance angle of a Bragg mirror.
- The increase of incidence angle of the SPP beam leads to a decrease of the acceptance angle. This is due a feature of the interference between the incident SPP beam with the SPP beam reflected from individual lines in the Bragg mirror, corrsponding to a respectively wider bandgap for the Bragg mirrors optimized for smaller incidence angles.

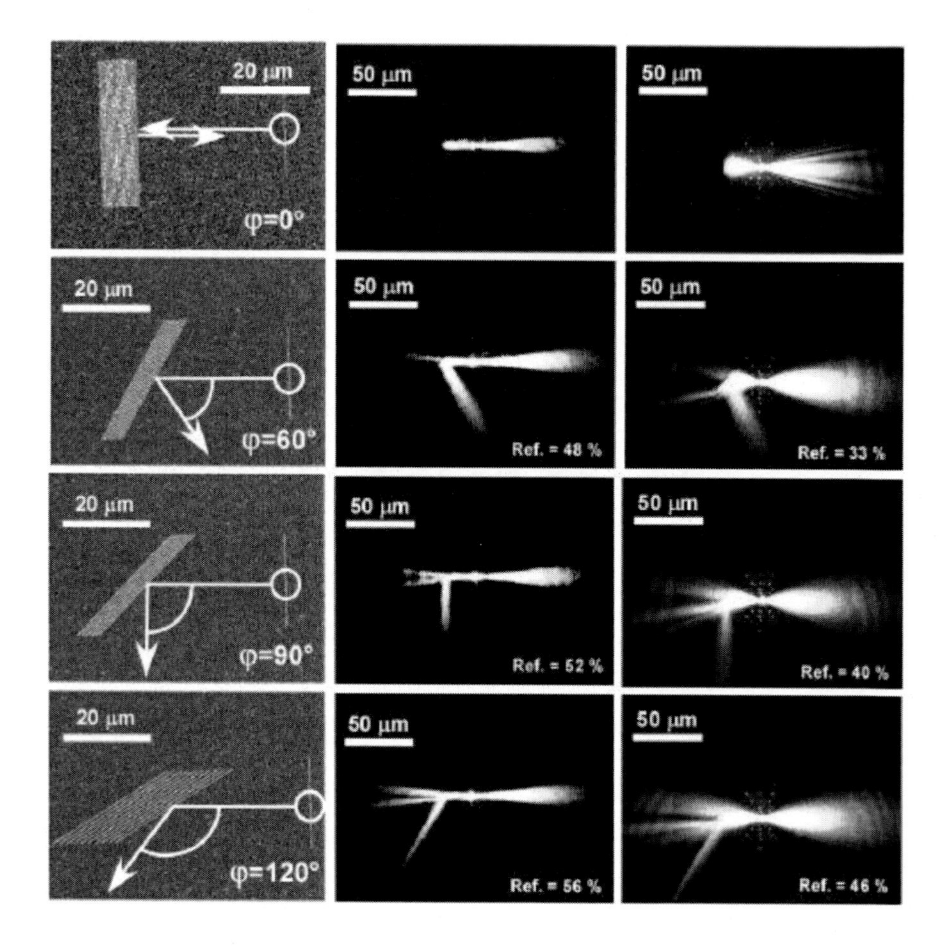

Figure 21. SPP excitation and reflection from Bragg mirrors constituted by gold nanowires. The SEM images (left) show the Bragg mirrors consisting of 10 lines, fabricated for various SPP beam incidence angles. The LR images (right) were acquired with different divergence angles of the SPP beam.

CONCLUSION

Allowing for quantitative analysis leakage radiation imaging has significantly expanded the experimental analysis toolbox with regard to SPP effects on the micro- and nanoscale.

It is an important tool for investigating and further optimizing functional SPP based optical components and devices, aiming at basic elements in analogy to conventional optics, as mirrors, beamsplitters, frequency selective filters etc.

On the fabrication side these developments can rely on advanced lithographic techniques for the precise fabrication of metal nanostructures as elementary building blocks for SPP photonics. For large scale fabrication, nanoimprining lithography migth be a technique enabling both high-resolution and high-throughput production of plasmonic devices.

ACKNOWLEDGMENTS

We wish to thank our partners and co-authors F.R. Aussenegg, H. Ditlbacher, A. Hohenau, A. Leitner, B. Steinberger, N. Galler from the Karl-Franzens University in Graz (Austria) and A. Dereux, J.-C. Weeber, M.U. Gonzalez, A.-L. Baudrion from University of Dijon in France. Also, A.L.S. We wish to thank our partners and co-authors F.R. Aussenegg, H. Ditlbacher, A. Hohenau, A. Leitner, B. Steinberger, N. Galler from the Karl-Franzens University in Graz (Austria) and A. Dereux, J.-C. Weeber, M.U. Gonzalez, A.-L. Baudrion from University of Dijon in France. Also, A.L.S. grateful to the Alexander von Humboldt Foundation and the DAAD in Germany, Austrian Scientific Foundation in the frame of Lisa Meitner Fellowship and the Royal Society in UK for financial support. This work was partly supported by the Ministry of Education and Science of the Russian Federation (FTP "Scientific and scientific-pedagogical personnel of the innovative Russia" No. 02.740.11.0779) and RFBR № 11-02-90420, 11-02-91341 and 12-02-00528.

REFERENCES

[1] Barnes, W. L. Dereux, A. Ebbesen, T. W. *Nature*, 2003, 424, 824-830.

[2] Krenn, J. R., Weeber, J. R., Phil. J. C. *Trans. Roy. Soc.* A 2004, 362, 739-756

[3] Yablonovitch, E. *Phys. Rev. Lett.* 1987, 58, 2059-2062.

[4] Joannopoulos, J. D., Villeneuve, P. R., Fan, S. *Nature*, 1997, 386, 143-149.

[5] Weeber, J. C., Dereux, A., Girard, Ch., Colas des Francs, G., Krenn, J. R. *Phys. Rev. E,* 2000, 62, 7381-7388.

[6] Quidant, R., Weeber, J. C., Dereux, A., Peyrade, D., Colas des Francs, G., Girard, Ch. Chen, Y. *Phys. Rev. E*, 2001, 64, 066607-1-066607-6.

[7] Quidant, R., Weeber, J. C., Dereux, A., Peyrade, D., Chen, Y., Girard, Ch. *Europhys. Lett.,* 2002, 57, 191-197.

[8] Raether, H. In: *Surface Plasmons*; Höhler, G. Ed., Springer: Berlin, 1988.

[9] Smolyaninov, I. I., Mazzoni, D. L., Mait, J., Davi, C. C. *Phys. Rev. B,* 1997, 56, 1601-1611.

[10] Krenn, J. R., Weeber, J. C., Dereux, A., Bourillot, E., Goudonnet, J. P., Schider, B., Leitner, A., Aussenegg, F. R., Girard, C. *Phys. Rev.*, B 1999, 60, 5029-5033.

[11] Krenn, J. R., Dereux, A., Weeber, J. C., Bourillot, E., Lacroute, Y., Goudonnet, J. P., Schider, B., Gotschy, W., Leitner, A., Aussenegg, F. R., Girard, C. *Phys. Rev. Lett.*, 1999, 82, 2590-2593.

[12] Lamprecht, B., Krenn, J. R., Schider, G., Ditlbacher, H., Salerno, M., Felidj, N., Leitner, A., Aussenegg, F. R. *Appl. Phys. Lett*. 2001, 79, 51-53.

[13] Krenn, J. R., Lamprecht, B., Ditlbacher, H., Schider, G., Salerno, M., Leitner, A., Aussenegg, F. R. *Eutophys. Lett.*, 2002, 60, 663-669.

[14] Kreibig, U., Vollmer, M. Optical properties of Metal Clusters, *Springer: Berlin*, 1995.

[15] Stepanov, A. L. In: *Metal-Polymer Nanocomposites*; Nicolais, L. Carotenuto, G. Eds., John Wiley and Sons Publ: London, 2004, pp. 241-263.

[16] Stepanov, A. L. In: *Silver nanoparticles*; Perez, D. P., Ed., In-Tech: Vukovar, 2010.

[17] Lamprecht, B., Schider, G., Lechner, R. T., Ditlbacher, H., Krenn, J. R., Leitner, A., Aussenegg, F. R. *Phys. Rev. Lett.*, 2001, 84, 4721-4724.

[18] Salerno, M., Felidj, N., Krenn, J. R., Leitner, A., Aussenegg, F. R. *Phys. Rev.*, B 2001, 63, 165422-1 - 165422-6.

[19] Gotschy, W., Vonmetz, K., Leitner, A., Aussenegg, F. R. *Appl. Phys.*, B 1996, 63, 381-384.

[20] Handbook of microlithography, micromachining and microfabrication, *Vol. 12 of IEE Materials and Devices Series*, Rai-Choudhury, P. Ed. SPIE: Washington, 1997.

[21] Ditlbacher, H., Krenn, J. R., Schider, G., Leitner, A., Aussenegg, F. R. *Appl. Phys. Lett.*, 2002, 81, 1762-1764.

[22] Ditlbacher, H., Krenn, J. R., Felidj, N., Lamprecht, B., Schider, G., Salerno, M., Leitner, A., Aussenegg, F. R. *Appl. Phys. Lett.*, 2002, 80, 404-406.

[23] Krenn, J. R., Ditlbacher, H., Schider, G., Hohenau, A., Leitner, A., Aussenegg, F. R. *J. Microsc.*, 2003, 209, 167-172.

[24] Linden, S., Kuhl, J., Giessen, H. *Phys. Rev. Lett.*, 2001, 86, 4688-4691.

[25] Linden, S., Christ, A., Kuhl, J., Giessen, H. *Appl. Phys.* B 2001, 73, 311-317.

[26] Stepanov, A. L., Kiyan, R., Reinhardt, C., Chichkov, B. N. In: *Laser Beams: Theory, Properties and Applications;* Thys, M. Desmet, E. Eds., NOVA Sci. Publ.: New York, 2009.

[27] Hecht, B., Bielefeldt, H., Novotny, L., Inouye, Y., Pohl, D. W. *Phys. Rev. Lett.*, 1996, 77, 1889-1892.

[28] Bouhelier, A., Husere, Th. Tamaru, H., Güntherodt, H. J., Pohl, D. W., Baida, F. I., van Labeke, D. *Phys. Rev.*, B 2001, 63, 155404-1 -155404-9.

[29] Stepanov, A. L., Krenn, J. R., Ditlbacher, H., Hohenau, A., Drezet, A., Steinberger, B., Leitner, A., Aussenegg, F. R. *Opt. Lett.*, 2005, 30, 1524-1526.

[30] Hohenau, A., Krenn, J. R., Stepanov, A. L., Drezet, A., Ditlbacher, H., Steinberger, B., Leitner, A., Aussenegg, F. R. *Opt. Lett.*, 2005, 30, 893-895.

[31] Kretschmann, E., Raether, H. Z. *Naturforsch.,* A 1968, 23, 2135-2136.

[32] Kiyan, R., Reinhardt, C., Passinger, S., Stepanov, A. L., Hohenau, A., Krenn, J. R., Chichkov, B. N. *Opt. Express* 2007, 15, 4205-4215.

[33] Radko, I. P., Bozhevolhyi, S. I., Brucoli, G., Martin-Moreno, L., Garcia-Vidal, F. J., Boltasseva, A. *Phys. Rev. B* 2008, 78, 115115-1-7.

[34] Zhang, D., Yuan, X., Bouhelier, A. *Appl. Opt.* 2010, 49, 875-879.

[35] Grandidier, J., Colas, G., Francs, D., Massenot, S., Bouheier, A., Markey, L., Weeber, J.-C., Dereux, A. *J. Microsc.* 2010, 239,167-172.

[36] Wang, J., Zhao, C., Zhang, J. *Opt. Lett.* 2010, 35, 1944-1946.

[37] Drezet, A., Hohenau, A., Stepanov, A. L., Ditlbacher, H., Steinberger, B., Galler, N., Aussenegg, F. R., Leitner, A., Krenn, J. R. *Appl. Phys. Lett.*, 2006, 89, 91117-1 - 91117-3.

[38] Ruppin, R. *Solid State Commun.*, 1981, 39, 903-906.

[39] Gonzalez, M. U., Weber, J.-C., Baudrion, A.-L., Dereux, A., Stepanov, A. L., Krenn, J. R., Devaux, E., Ebbesen, T. W. *Phys. Rev. B* 2006, 73, 155416-1-13.

[40] Gonzalez, M. U., Stepanov, A. L., Weber, J.-C., Hohenau, A., Dereux, A., Quidant, R., Krenn, J. R. *Opt. Lett.* 2007, 32, 2704-2706.

[41] Drezet, A., Stepanov, A. L., Ditlbacher, H., Hohenau, A., Steinberger, B., Aussenegg, F. R., Leitner, A., Krenn, J. R. *Appl. Phys. Lett.*, 2005, 86, 74104-1-74104-3.

[42] Drezet, A., Hohenau, A., Stepanov, A. L., Ditlbacher, H., Steinberger, B., Aussenegg, F. R., Leitner, A., Krenn, J. R. *Plasmonics*, 2006, 1, 141-145.

[43] Drezet, A., Hohenau, A., Koller, D., Stepanov, A. L., Ditlbacher, H., Steinberger, B., Aussenegg, F. R., Leitner, A., Krenn, J. R. *Mater. Sci. Eng.*, B 2008, 149, 220-229.

In: Advances in Laser and Optics Research. Volume 10
Editor: William T. Arkin
ISBN: 978-1-62257-795-8

Chapter 12

NONLINEAR OPTICAL PROPERTIES OF METAL NANOPARTICLES SYNTHESIZED BY ION IMPLANTATION

Andrey L. Stepanov*
Laser Zentrum Hannover, Hannover, Germany
Kazan Federal University, Kazan, Russian Federation
Kazan Physical-Technical Institute, Russian Academy of Sciences,
Kazan, Russian Federation

ABSTRACT

Composite materials containing metal nanoparticles (MNPs) are now considered as a basis for designing new photonic media for optoelectronics and nonlinear optics. Simultaneously with the search for and development of modern technologies intended for nanoparticle synthesis, substantial practical attention has been devoted to designing techniques for controlling the MNP size. One of the promising methods for fabrication of MNPs is ion implantation. Review of recent results on ion-synthesis and nonlinear optical properties of cupper, silver and gold nanoparticles in surface area of various dielectrics as glasses and crystals are presented. Composites prepared by the low energy ion implantation are characterized with the growth of MNPs in thin layer of irradiated substrate surface. Fabricated structures lead to specific optical nonlinear properties for picosecond laser pulses in wide spectral area from UV to IR such as nonlinear refraction, saturable and two-photon absorption, optical limiting. The practical recommendations for fabrication of composites with implanted MNPs for optical components are presented.

1. INTRODUCTION

The search for new nanostructured materials is one of the defining characteristics of modern science and technology [1-6]. Novel mechanical, electrical, magnetic, chemical,

* Corresponding author: E-mail: aanstep@gmail.com.

biological, and optical devices are often the result of the fabrication of new nanostructured materials. The specific interest of this review is recent advantages in optical science and technology, such as development of nonlinear optical random metal-dielectric and metal semiconductor composites based on metal nanoparticles (MNP) synthesized by ion implantation. Simultaneously with the search for and development of novel technologies intended for nanoparticle synthesis, substantial practical attention has been devoted to designing techniques for controlling the MNP size. This is caused by the fact that the optical properties of MNPs, which are required for various applications, take place up to a certain MNP dimension. In this content, ion implantation nanotechnology allows one to fabricate materials with almost any MNP structures, types of metals and their alloys [7-9]; this opens new avenues in engineering nanomaterials with desired properties. Such composites possess fascinating electromagnetic properties, which differ greatly from those of ordinary bulk materials, and they are likely to become ever more important with a miniaturization of electronic and optoelectronic components.

Nonlinear optics plays a key role in the implementation and development of many photonics techniques for the optical signal processing of information at enhanced speed. The fabrication of novel useful nonlinear optical materials with ultrafast time response, high resistance to bulk and surface laser damage, low two-photon absorption and, of course, large optical nonlinearities is a critical for implementation of those applications. In additional, nonlinear materials for optical switching should be manufactured by processes compatible with microelectronics technology. Nonlinear materials with such characteristics are interesting for waveguide applications. The earliest studies of optical analogs to electronic integrated circuits – or integrated optics – were based on the recognition that waveguide geometries allowed the most efficient interaction of light with materials. Optoelectronic devices could be converted to all-optical configurations, with a number of technological advantages, by developing waveguide media with intensity-dependent refractive indices. Nonlinear optical switches must provide conversion of laser signal for pulse duration as short as from nano- to femtoseconds. The nonlinear properties of MNP-containing materials stem from the dependence of their refractive index and nonlinear absorption on incident light intensity. Giant enhancement of nonlinear optical response in a random media with MNPs is often associated with optical excitation of surface plasmon resonances (SPR) that are collective electromagnetic modes and they are strongly dependent on the geometry structure of the composite medium [4]. Therefore, MNP-containing transparent dielectric and semiconductor materials can be effectively applied in novel integrated optoelectronic devices.

Although both classic and quantum-mechanical effects in the linear optical response of MNP composites have been studied for decades [4], the first experimental results on the nonlinear optical effects of MNPs in ruby-glass was obtained quite recently in 1985 by Ricard *et al.* [10]. Driven by the interest in creating nonlinear optical elements with MNPs for applications in all-optical switching and computing devices, variety of experimental and theoretical efforts have been directed at the preparation of composite materials. In practice, to reach the strong linear absorption of a composite in the SPR spectra region, attempts are made to increase the concentration (filling factor) of MNPs. Systems with a higher filling factor offer a higher nonlinear susceptibility, when all other parameters of composites being the same.

MNPs hold great technological promise because of the possibility of engineering their electronic and optical properties through material design. The transition metals of choice are usually gold, silver, or copper, as these metals show SPR modes in the visible or near-infrared spectral range [4]. The advantages of devices based on MNP materials can be understood from the spectacular successes of quantum well materials [11, 12]. The capability of band gap engineering in these structures permits wavelength tuning, while their small size alters the electronic structure of these particles. This provides greater pumping efficiency for applications in optical limiting and switching. The potential advantages of MNP composites as photonic materials are substantial improvement in the signal switching speed. Up to 100 GHz repetition frequencies are expected in communication and computing systems of the 21th century [11]. Figure 1 compares in graphical form the switching speed and switching energies of various electronic, optical materials and devices (adapted from [11, 13]).

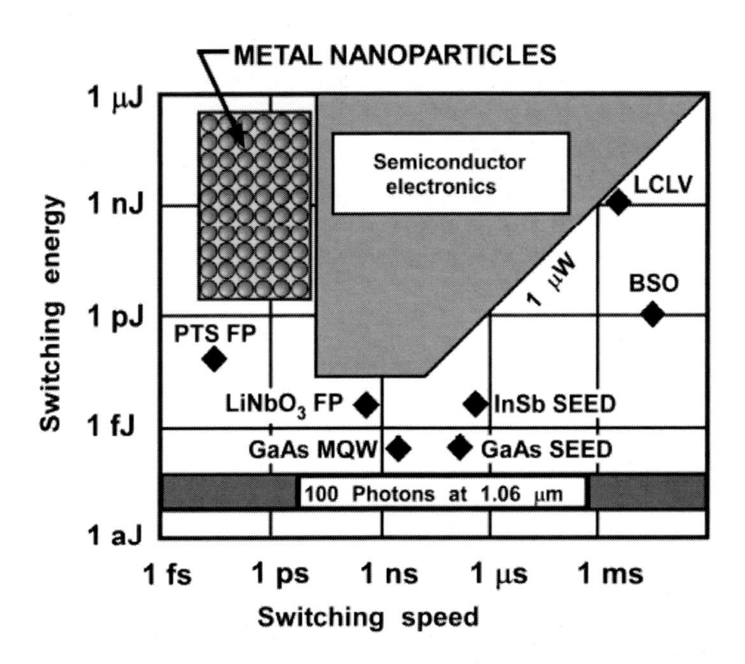

Figure 1. Plot of various photonic materials showing their switching energies and switching speed. Adapted from [11, 13].

Within the broad range of parameters covered by "conventional semiconductor microelectronics", current metal-oxide-semiconductor field-effect transistor devices made in silicon have low switching energies, but switching time in the nanosecond range. Photonic devices based on multiple quantum well (MQW) structures – SEED and GaAs MQW devices and Fabry-Perot (FP) cavities based on ferroelectric such as lithium niobate – have extremely low switching speed in comparison to MNPs [11, 12].

Davenas *et al.* pioneered synthesis of MNPs in dielectrics by ion implantation in 1973 [14, 15], when nanoparticles of various metals (sodium, calcium, etc.) in ionic crystals of LiF and MgO were created. Late in 1975, noble metal nanoparticles such as Au and Ag were fabricated in silicate glasses by Arnold and Borders [16, 17].

As shown in reviews [7-9, 18-23], now developments expanded from the metal implants to the use of compounds, including metal alloys and totally different composition precipitate inclusions. Implanted MNP were fabricated in various materials, as polymers, glass, artificial crystals, and minerals.

Number of publications on nonlinear optical properties of MNPs fabricated in transparent dielectric and semiconductor matrix is increasing every year. There are review articles observed partly this progress [11, 18, 19, 24, 25]. Unfortunately, some of this reviews are already quite old and do not reflect a modern knowledge of the field or restricted to numbers of selected publications only.

However, as followed from a comprehensive list of publications presented in Table 1 by 2011 [26-118], the geography of interest to nonlinear properties of ion-synthesized MNPs covers all world continents. The data in Table 1 includes information on all known types of metal ions and transparent matrices, ion implantation conditions for fabrication of MNPs and for measurements of their optical properties. Nonlinear optical characteristics of composites such as nonlinear refraction (n_2) and absorption (β) coefficients, real ($Re[\chi^{(3)}]$) and imaging ($Im[\chi^{(3)}]$) parts of third order nonlinear susceptibilities ($\chi^{(3)}$) and saturation intensities (I_{sat}) are presented as well. As shown in Table 1, near one hundred articles were already published. It should be mention, that ion implantation technique was first used for ion-synthesis of MNPs in dielectrics to create nonlinear optical materials in 1991 to form copper and gold nanoparticles in silica glass [42, 100]. The present review focuses on advantages in nonlinear optical properties of MNPs fabricated generally by low-energy ion implantation and measured in wide spectral area from ultraviolet to infrared.

2. Optics of Metal Nanoparticle Composites

The nonlinear optical response of medium with MNPs can be described by expanding the i-th component of the polarization P induced by an applied optical field to third order in a power series in this electric field $E=E_0e^{lex}$ [11, 12]

$$P_i = \sum_j \chi_{y}^{(1)} E_j + \sum_{jk} \chi_{yk}^{(2)} E_j E_k + \sum_{jki} \chi_{yki}^{(3)} E_j E_k E_i + \rightleftharpoons , \quad (1)$$

where the summation indices refer to Cartesian conditions in the material-dependent $\chi^{(q)}$ and to the polarization of the applied optical field.

The first-order susceptibility $\chi^{(1)}$ is related to the linear refractive index n_0 and the linear (Lambert-Beer's law) absorption coefficient α_0 through the following equation [13]

$$n_0 = Re\left[1+\chi^{(1)}\right] \text{ and } \alpha_0 = \frac{\omega}{n_0 c} Im\left[\chi^{(1)}\right] , \quad (2)$$

where c – speed of the light, ω - the optical frequency. The value $\chi^{(3)}$ of a centrosymmetric composite has an analogous relationship to the nonlinear coefficients n_2 and β [45]

$$n_2 = \frac{12\pi}{n_0} Re\left[1+\chi^{(1)}\right] \text{ and } \beta = \frac{96\pi^2\omega}{n_0^2 c^2} Im\left[\chi^{(1)}\right] . \quad (3)$$

The susceptibility $\chi^{(3)}$ is a fourth-rank tensor with eighty-one components; however, material symmetries often reduce the number of non-vanishing components substantially. For a MNP with a dielectric constant $\varepsilon(\omega)=\varepsilon_1(\omega)+i\varepsilon_2(\omega)$ occupying a relative volume fraction (filling factor) $p<<1$ in a host of dielectric constant ε_h, the absorption can be presented as [45, 119]

$$\alpha_0 = \frac{\omega}{n_0 c} Im\left[\chi^{(1)}(\omega)\right] = 9p\frac{\omega\varepsilon_h^{3/2}}{c}\frac{\varepsilon_2}{\left(\varepsilon_1+2\varepsilon_h\right)^2+\varepsilon_2^2} = p\frac{\omega}{n_0 c}\left|f_1(\omega)\right|^2\varepsilon_2 , \tag{4}$$

where the $f_1(\omega)$ is the local-field enhancement factor. The absorption coefficient α_0 has a maximum at the SPR frequency [4] where $\varepsilon_1(\omega)+2\varepsilon_h(\omega) = 0$. The effective third-order nonlinear optical susceptibility of the dielectric medium with MNPs $\chi_{eff}^{(3)}$ can be derived by applying Maxwell's equations to the first order in the electromagnetic field to yield [3, 120]

$$\chi_{eff}^{(3)} = p\left|\frac{3\varepsilon_h}{\varepsilon_1+2\varepsilon_h}\right|^2\left(\frac{3\varepsilon_h}{\varepsilon_1+2\varepsilon_h}\right)^2\chi_{met}^{(3)} = p\left|f_1\right|^2 f_1^2\chi_{met}^{(3)} , \tag{5}$$

where f_1 is the same local-field enhancement factor of the polarization describing the $\chi^{(1)}$. This equation shows that the nonlinearity of composites with MNPs comprises two factors: the nonlinearity due to the MNPs itself, and the enhancement contributed by the host matrix. Note that whereas the α_0 varies as $|f_1|^2$, the $\chi_{eff}^{(3)}$ varies as $|f_1|^2 f_1^2$. Hence, it is expected a significantly greater enhancement due dielectric nonlinearity in the $\chi_{eff}^{(3)}$.

3. Nonlinear Optical Properties of Ion-Synthesized Nanoparticles Near IR-Area (1064 Nm)

3.1. Nonlinear Absorption of MNPs Studied by Z-Scan

In numerous studies, the nonlinear optical characteristics of the composite materials with MNPs fabricated by various methods were generally studied using lasers operating at frequencies that correspond to the spectral range of the SPR in particles [120] and Table 1. One other hand, one should take into account that, when used in practice as optical switches, optical limiters, and so on, these nonlinear materials have to operate at the wavelengths of the most frequently used industry available lasers, such as Nd:YAG (λ = 1064 nm), Ti:Al_2O_3 (λ = 800 nm), and so on.

Table 1. Types optically transparent dielectric and semiconductor matrixes with metal nanoparticles synthesized by ion implantation

Metal (Ion)	Matrix	Synthesis conditions: Energy (E), keV Dose (D), ion/cm^2, Current density (J), μA/cm^2 Annealing temper. (T), °C and time	Study nonlinear optical method	Laser parameters: Wavelength (λ), nm Pulse duration (τ), ps Repetition rate (ν), Hz Intensity (I_0), W/cm^2 Pulse energy (P), mJ	Nonlinear parameters: Refract. coef. (n_2), cm^2/W Absorption coef. (β), cm/W Satur. intensity (I_{sat}), W/cm^2 $Re[\chi^{(3)}]$, $Im[\chi^{(3)}]$, $\lvert\chi^{(3)}\rvert$, esu	Authors
Co	SiO_2	$E = 50$ $D = 4 \cdot 10^{16}$ $J = 2$	Z-scan	$\lambda = 770$ $\tau = 0.13$ $\nu = 76 \cdot 10^9$ $I_0 = 11.4 \cdot 10^9$	$n_2 = 1.8 \cdot 10^{-9}$	Cattaruzza *et al.* 1998 [26]
Ni	SiO_2	$E = 100$ $D = 6 \cdot 10^{16}$	Z-scan	$\lambda = 770$ $\tau = 0.13$ $\nu = 76 \cdot 10^9$ $I_0 = 9.8 \cdot 10^9$	$n_2 = 1.7 \cdot 10^{-10}$	Falconieri *et al.* 1998 [27] Cattaruzza *et al.* 2002 [28]
Cu	Al_2O_3	$E = 40$ $D = (0.5 - 1.0) \cdot 10^{17}$ $J = 2.5 - 12.5$	RZ-scan	$\lambda = 1064$ $\tau = 55$ $\nu = 2$ $I_0 = 7.7 \cdot 10^9$	$n_2 = -(1.3 - 1.7) \cdot 10^{-11}$ $Re[\chi^{(3)}] = -(1.0 - 1.4) \cdot 10^{-9}$	Ganeev *et al.* 2005 [29] 2006 [30] Ryasnyanskiy *et al.* 2005 [31]
Cu	Al_2O_3	$E = 60$	Z-scan	$\lambda = 500 - 700$ $\tau = 0.2$ $\nu = 10^3$ $I_0 = 1.1 \cdot 10^7$	$\beta_{500-580} < 0$ $\beta_{580-700} > 0$	Plaksin *et al.* 2008 [32]
Cu	ITO	$E = 40$ $D = (0.5 - 7.5) \cdot 10^{16}$ $J = 4$	Z-scan	$\lambda = 532$ $\tau = 7.0 \cdot 10^3$ $\nu = 10$ $I_0 = (6.2 - 15.5) \cdot 10^9$	$n_2 = (5.2 - 8.3) \cdot 10^{-8}$ $Re[\chi^{(3)}] = (5.4 - 7.4) \cdot 10^{-6}$ $\beta = -(3.5 - 3.6) \cdot 10^{-3}$ $Im[\chi^{(3)}] = -(1.2 - 1.3) \cdot 10^{-6}$ $\lvert\chi^{(3)}\rvert = (5.5 - 7.5) \cdot 10^{-6}$	Ryasnyanskiy *et al.* 2006 [33]
Cu	LiNbO3	*$E = 60$* *$D = (0.3 - 2.0) \cdot 1017$* *$J = 10$*	PPTNS	*$\lambda = 574$* *$\tau = 0.2$* *$\nu = 103$* *$P = 16$*	*Bleaching absorption in 590 - 620 nm*	Takeda et al. 2002 [34-36] Kishimoto et al. 2003 [37] Plaksin et al. 2005 [38] 2006 [39]

Metal (Ion)	**Matrix**	***Synthesis conditions:*** ***Energy (E), keV*** ***Dose (D), ion/cm2,*** ***Current density (J), μA/cm2*** ***Annealing temper. (T), °C and time***	**Study nonlinear optical method**	***Laser parameters:*** ***Wavelength (λ), nm*** ***Pulse duration (τ), ps*** ***Repetition rate (ν), Hz*** ***Intensity (I0), W/cm2*** ***Pulse energy (P), mJ***	**Nonlinear parameters:** **Refract. coef. (n2), cm2/W** **Absorption coef. (β), cm/W** **Satur. intensity (Isat), W/cm2** **Re[χ(3)], Im[χ(3)], |χ(3)|, esu**	**Authors**
Cu	$MgAl_2O_4$	$E = 60$ $D = 3.0 \cdot 10^{16}$ $J = 10$	DFWM	$\lambda = 532$	$\lvert\chi^{(3)}\rvert = (1.0 - 3.0) \cdot 10^{-8}$	Kishimoto *et al.* 2000 [40]
Cu	$MgO_2 \cdot Al_2O_4$	$E = 60$ $D = 3.0 \cdot 10^{16}$ $J = 1\text{-}100$	PPTNS	$\lambda = 574$ $\tau = 0.2$ $\nu = 10^3$ $P = 16$	Bleaching absorption in 590 - 620 nm	Takeda *et al.* 2002 [34-36] 2001 [41]
Cu	SiO_2	$E = 160$ $D = 6.0 \cdot 10^{16}$ $J = 2 - 7.5$	Z-scan	$\lambda = 532$ $\tau = 100$ $\nu = 76 \cdot 10^9$ $I_0 = 5.0 \cdot 10^6$	$n_2 = 2.0 \cdot 10^{-15}$	Becker *et al.* 1991 [42] Haglund *et al.* 1992 [43]
Cu	SiO_2	$E = 160$ $D = 1.2 \cdot 10^{17}$ $J = 2.5$	Z-scan	$\lambda = 570 - 600$ $\tau = 6$ $I_0 = 5.0 \cdot 10^8$	$n_2 = (2.0 - 4.2) \cdot 10^{-10}$ $Re[\chi^{(3)}] = 2.4 \cdot 10^{-8}$ $\beta = -(0.1 - 1.0) \cdot 10^{-6}$	Haglund *et al.* 1993 [44] 1994 [45] 1998 [46] Magruder *et al.* 1994 [47] Yang *et al.* 1994 [48]
Cu	SiO_2	$E = 160$ $D = 1.2 \cdot 10^{17}$ $J = 0.7 - 7.5$	DFWM	$\lambda = 532$ $\tau = 10$ and 35	$n_2 = 2.0 \cdot 10^{-7}$ $\lvert\chi^{(3)}\rvert = (2.4 - 7.3) \cdot 10^{-8}$	Haglund *et al.* 1994 [45] 1998 [46] 1995 [42] Yang *et al.* 1994 [48] 1996 [49]
Cu	SiO_2	$E = 160$ $D = 1.2 \cdot 10^{17}$ $J = 2.5$	Z-scan	$\lambda = 532$ $\tau = 100$ $\nu = 76 \cdot 10^9$ $I_0 = 1.0 \cdot 10^7$	$n_2 = (2.0 - 4.2) \cdot 10^{-14}$ $Re[\chi^{(3)}] = 2.4 \cdot 10^{-8}$ $\beta = -(3.0 - - 8.0) \cdot 10^{-3}$	Magruder *et al.* 1994 [47]

Table 1. (Continued)

Metal (Ion)	Matrix	*Synthesis conditions: Energy (E), keV Dose (D), ion/cm2, Current density (J), μA/cm2 Annealing temper. (T), °C and time*	Study nonlinear optical method	*Laser parameters: Wavelength (λ), nm Pulse duration (τ), ps Repetition rate (ν), Hz Intensity (I0), W/cm2 Pulse energy (P), mJ*	**Nonlinear parameters: Refract. coef. (n2), cm2/W Absorption coef. (β), cm/W Satur. intensity (Isat), W/cm2 Re[χ(3)], Im[χ(3)], |χ(3)|, esu**	**Authors**
Cu	SiO_2	$E = 90$ $D = 6 \cdot 10^{16}$	Z-scan	$\lambda = 770$ $\tau = 0.13$ $\nu = 76 \cdot 10^9$ $I_0 = 9.8 \cdot 10^9$	$n_2 = 5.0 \cdot 10^{-11}$	Falconieri *et al.* 1998 [27] Cattaruzza *et al.* 2002 [28]
Cu	SiO_2	$E = 2.0 \cdot 10^3$ $D = (1.0 - 4.0) \cdot 10^{17}$ $J = 2.0$ $T = 1000$, 1 *h*	Z-scan	$\lambda = 532$ and $555 - 600$ $\tau = 4.5$ $\nu = 76 \cdot 10^9$ $I_0 = 8.8 \cdot 10^9$	$n_2 = (4.0 - 6.8) \cdot 10^{-19}$ $\|\chi^{(3)}\| = (0.3 - 4.7) \cdot 10^{-7}$	Ila *et al.* 1998 [50] Sarkisov *et al.* 1998 [51]
Cu	SiO_2	$E = 60$ $D = 3.0 \cdot 10^{16}$ $J = 1\text{-}100$	DFWM	$\lambda = 532$ and 561 $\tau = 7.0 \cdot 10^3$ $I_0 = (0.1 - 1.0) \cdot 10^6$	$\|\chi^{(3)}\| = (0.2 - 2.2) \cdot 10^{-8}$	Takeda *et al.* 1999 [52] 2000 [53]
Cu	SiO_2	$E = 50$ $D = 8.0 \cdot 10^{16}$ $J = 10$	DFWM	$\lambda = 585$ $\tau = 13$ $\nu = 400$ $I_0 = 1.0 \cdot 10^8$	$\|\chi^{(3)}\| = 1.0 \cdot 10^{-7}$	Olivares *et al.* 2001 [54]
Cu	SiO_2	$E = 60$ $D = 3.0 \cdot 10^{16}$ $J = 1\text{-}30$ $T = 800$, 1 *h*	PPTNS	$\lambda = 574$ $\tau = 0.2$ $\nu = 10^3$ $I_0 = 8.0 \cdot 10^{11}$	Bleaching absorption in 590 - 620 nm	Takeda *et al.* 2002 [34-36, 55] 2001 [41] 2004 [56, 57]
Cu	SiO_2	$E = 50$ $D = 8.0 \cdot 10^{16}$ $J = 10$	Z-scan	$\lambda = 354.7$ $\tau = 55$ $\nu = 2$ $I_0 = 4.1 \cdot 10^9$	$n_2 = -0.6 \cdot 10^{-7}$ $Re[\chi^{(3)}] = -1.3 \cdot 10^{-8}$ $\beta = -6.7 \cdot 10^{-6}$ $Im[\chi^{(3)}] = -2.9 \cdot 10^{-9}$ $\|\chi^{(3)}\| = 1.4 \cdot 10^{-8}$	Ganeev *et al.* 2003 [58] 2004 [59]

Metal (Ion)	Matrix	*Synthesis conditions:* *Energy (E), keV* *Dose (D), ion/cm2,* *Current density (J), μA/cm2* *Annealing temper. (T), °C and time*	Study nonlinear optical method	*Laser parameters:* *Wavelength (λ), nm* *Pulse duration (τ), ps* *Repetition rate (ν), Hz* *Intensity (I0), W/cm2* *Pulse energy (P), mJ*	**Nonlinear parameters:** **Refract. coef. (n2), cm2/W** **Absorption coef. (β), cm/W** **Satur. intensity (Isat), W/cm2** **Re[χ(3)], Im[χ(3)], |χ(3)|, esu**	**Authors**
Cu	SiO_2	$E = 50$ $D = 8.0 \cdot 10^{16}$ $J = 10$	Z-scan	$\lambda = 532$ $\tau = 55$ $\nu = 2$ $I_0 = 5.4 \cdot 10^{9}$	$\beta = -6.0 \cdot 10^{-6}$ $I_{sat} = 4.3 \cdot 10^{8}$	Ganeev *et al.* 2003 [60] 2004 [61]
Cu	SiO_2	$E = 50$ $D = 8.0 \cdot 10^{16}$ $J = 10$	Z-scan	$\lambda = 1064$ $\tau = 35$ $\nu = 2$ $I_0 = 1.0 \cdot 10^{10}$	$n_2 = -1.4 \cdot 10^{-7}$ $Re[\chi^{(3)}] = -3.2 \cdot 10^{-8}$ $\beta = -9.0 \cdot 10^{-6}$ $Im[\chi^{(3)}] = 6.5 \cdot 10^{-9}$ $\|\chi^{(3)}\| = 3.3 \cdot 10^{-8}$	Ganeev *et al.* 2003 [62, 63] 2004 [64] Stepanov *et al.* 2003 [65]
Cu	SiO_2	$E = 60$ $D = 1.0 \cdot 10^{17}$ $J = 10$ $T = 800$, 1 *h*	Z-scan	$\lambda = 540 - 610$ $\tau = 0.2$ $\nu = 1$ $I_0 = 8.0 \cdot 10^{11}$	$Re[\chi^{(3)}] = -3.1 \cdot 10^{-9}$ $Im[\chi^{(3)}] = 1.7 \cdot 10^{-9}$ $\|\chi^{(3)}\| = -(-1.6 - 3.1) \cdot 10^{-8}$	Takeda *et al.* 2005 [66] Plaksin *et al.* 2008 [32]
Cu	SiO_2	$E = 180$ $D = (0.5 - 2.0) \cdot 10^{17}$ $J = 1.5$ $T = 500 - 900$, 1 *h*	Z-scan	$\lambda = 790 - 800$ $\tau = 0.15$ $\nu = 76 \cdot 10^{9}$ $I_0 = (8.0 - 14.5) \cdot 10^{9}$	$n_2 = -1.6 \cdot 10^{-10}$ $Re[\chi^{(3)}] = (0.9 - 1.4) \cdot 10^{-7}$ $\beta = -(1.6 - 9.0) \cdot 10^{-6}$ $Im[\chi^{(3)}] = (0.8 - 1.7) \cdot 10^{-7}$ $\|\chi^{(3)}\| = (1.2 - 2.3) \cdot 10^{-7}$	Ren *et al.* 2006 [67] Wang *et al.* 2006 [68, 69]
Cu	SiO_2	$E = 100 - 200$ $D = 3.0 \cdot 10^{16}$ $T = 300 - 400$, 1 *h*	Z-scan	$\lambda = 533$ $\tau = 7 \cdot 10^{3}$ $\nu = 0.1$ $I_0 = 0.9 \cdot 10^{9}$	$n_2 = -3.7 \cdot 10^{-15}$ $Re[\chi^{(3)}] = 3.7 \cdot 10^{-12}$ $\beta = (2.8 - 6.4) \cdot 10^{-9}$ $Im[\chi^{(3)}] = 3.7 \cdot 10^{-14}$ $\|\chi^{(3)}\| = 3.7 \cdot 10^{-12}$ $I_{sat} = (2.9 - 5.0) \cdot 10^{7}$	Ghosh *et al.* 2007 [70] 2009 [71]

Table 1. (Continued)

Metal (Ion)	Matrix	*Synthesis conditions:* *Energy (E), keV* *Dose (D), ion/cm2,* *Current density (J), μA/cm2* *Annealing temper. (T), °C and time*	Study nonlinear optical method	*Laser parameters:* *Wavelength (λ), nm* *Pulse duration (τ), ps* *Repetition rate (ν), Hz* *Intensity (I0), W/cm2* *Pulse energy (P), mJ*	**Nonlinear parameters:** **Refract. coef. (n2), cm2/W** **Absorption coef. (β), cm/W** **Satur. intensity (Isat), W/cm2** **Re[χ(3)], Im[χ(3)], |χ(3)|, esu**	**Authors**
Cu	SiO_2	$E = 2.0 \cdot 10^3$ $D = 4.0 \cdot 10^{16}$ $T = 900$, 1 h	VSD	$\lambda = 533$ $\tau = 26$ $P = 16$	$n_2 = -1.1 \cdot 10^{-12}$ $\beta = -2.0 \cdot 10^{-11}$ $\|\chi^{(3)}\| = 8.4 \cdot 10^{-11}$	Torres-Torres *et al.* 2008 [72]
Cu	SiO_2	$E = 2.0 \cdot 10^3$ $D = 4.0 \cdot 10^{16}$ $T = 900$, 1 h	VSD	$\lambda = 533$ $\tau = 7 \cdot 10^3$ $P = 16$	$n_2 = -1.2 \cdot 10^{-11}$ $\beta = -2.0 \cdot 10^{-12}$ $\|\chi^{(3)}\| = 8.8 \cdot 10^{-10}$	Torres-Torres *et al.* 2008 [72]
Cu	SiO_2	$E = 180$ $D = (0.5 - 1.0) \cdot 10^{17}$ $J = 1.5$	Z-scan	$\lambda = 532$ $\tau = 38$ $\nu = 10$ $I_0 = 0.9 \cdot 10^9$	$n_2 = -(1.3 - 0.6) \cdot 10^{-10}$ $\beta = -(458- 151) \cdot 10^{-9}$ $\|\chi^{(3)}\| = (2.1 - 0.8) \cdot 10^{-7}$	Wang *et al.* 2009 [73] Wang *et al.* 2010 [74]
Cu	SiO_2	$E = 180$ $D = (0.5 - 1.0) \cdot 10^{17}$ $J = 1.5$	Z-scan	$\lambda = 1064$ $\tau = 38$ $\nu = 10$ $I_0 = 0.38 \cdot 10^9$	$n_2 = -(1.1 - 0.6) \cdot 10^{-10}$ $\|\chi^{(3)}\| = (1.2 - 0.8) \cdot 10^{-7}$	Wang *et al.* 2009 [73] Wang *et al.* 2010 [74, 75]
Cu	SLSG	$E = 50$ $D = 8.0 \cdot 10^{16}$ $J = 10$	Z-scan	$\lambda = 1064$ $\tau = 35$ $\nu = 2$ $I_0 = 3.0 \cdot 10^{10}$	$n_2 = 3.6 \cdot 10^{-8}$ $Re[\chi^{(3)}] = 0.8 \cdot 10^{-8}$ $\beta = -3.4 \cdot 10^{-6}$ $Im[\chi^{(3)}] = 2.5 \cdot 10^{-9}$ $\|\chi^{(3)}\| = 0.9 \cdot 10^{-8}$	Ganeev *et al.* 2003 [62, 63] 2004 [64] Stepanov *et al.* 2003 [65]
Cu	$SrTiO_3$	$E = 60$ $D = 3.0 \cdot 10^{16}$ $J = 10$	PPTNS	$\lambda = 574$ $\tau = 0.2$ $\nu = 10^3$	Bleaching absorption in 605 - 690 nm and positive absorption in 516 - 506 nm	Takeda *et al.* 2002 [35, 55] 2004 [76]

Metal (Ion)	Matrix	*Synthesis conditions:* *Energy (E), keV* *Dose (D), ion/cm2,* *Current density (J), μA/cm2* *Annealing temper. (T), °C and time*	**Study nonlinear optical method**	*Laser parameters:* *Wavelength (λ), nm* *Pulse duration (τ), ps* *Repetition rate (ν), Hz* *Intensity (I0), W/cm2* *Pulse energy (P), mJ*	**Nonlinear parameters:** **Refract. coef. (n2), cm2/W** **Absorption coef. (β), cm/W** **Satur. intensity (Isat), W/cm2** **Re[χ(3)], Im[χ(3)], |χ(3)|, esu**	**Authors**
Cu	$SrTiO_3$	$E = 60$ $D = 3.0\cdot10^{16}$ $J = 10$ $T = 300$, 1 *h*	Z-scan	$\lambda = 540 - 610$ $\tau = 0.2$ $\nu = 1$ $I_0 = 6.0\cdot10^{11}$	$Re[\chi^{(3)}] = -(0.1 - 2.0)\cdot10^{-9}$ $Im[\chi^{(3)}] = -(0.2 - 1.2)\cdot10^{-9}$ $\|\chi^{(3)}\| = (1.0 - 2.0)\cdot10^{-9}$	Takeda *et al.* 2006 [77]
Cu	$SrTiO_3$	$D = (0.1 - 1.0)\cdot10^{17}$	DFWM Z-scan	$\lambda = 775$ $\tau = 0.25$ $\nu = 1$	$n_2 = (1.8 - 6.2)\cdot10^{-12}$ $\|\chi^{(3)}\| = (1.6 - 5.33)\cdot10^{-10}$	Cetin *et al.* 2010 [77]
Cu	TiO_2	$E = 60$ $D = 3.0\cdot10^{16}$ $J = 10$	PPTNS	$\lambda = 574$ $\tau = 0.2$ $\nu = 10^3$	Bleaching absorption in 585 - 760 nm	Takeda *et al.* 2002 [35]
Cu	ZnO	$E = 160$ $D = (0.1 - 1.0)\cdot10^{17}$ $J = 20$	Z-scan	$\lambda = 532$ $\tau = 55$ $\nu = 2$ $I_0 = 5.0\cdot10^{8}$	$\beta = -(0.4 - 2.1)\cdot10^{-3}$	Stepanov *et al.* 2004 [79] Ryasnyansky *et al.* 2005 [80]
Cu	ZnO	$E = 160$ $D = (0.1 - 1.0)\cdot10^{17}$ $J = 20$	Z-scan	$\lambda = 532$ $\tau = 7.5\cdot10^{3}$ $\nu = 10$ $I_0 = 3.2\cdot10^{7}$	$\beta = -(0.7 - 5.5)\cdot10^{-3}$	Ryasnyansky *et al.* 2005 [80]
Ag	Al_2O_3	$E = 30$ $D = 3.8\cdot10^{17}$ $J = 3 - 10$	RZ-scan	$\lambda = 1064$ $\tau = 55$ $\nu = 2$ $I_0 = 4.3\cdot10^{9}$	$n_2 = (1.1 - 1.8)\cdot10^{-11}$ $Re[\chi^{(3)}] = (0.9 - 1.5)\cdot10^{-9}$	Ganeev *et al.* 2005 [29] 2006 [30] Ryasnyanskiy *et al.* 2005 [81]
Ag	$LiNbO_3$	$E = 1.5\cdot10^{3}$ $D = 2.0\cdot10^{16}$ $T = 500$, 1 *h*	Z-scan	$\lambda = 555 - 600$ $\tau = 4.5$ $\nu = 76\cdot10^{6}$ $I_0 = 8.8\cdot10^{7}$	$n_2 = (0.8 - 1.3)\cdot10^{-8}$	Sarkisov *et al.* 1998 [51]
Ag	$LiNbO_3$	$E = 1.5\cdot10^{3}$ $D = 2.0\cdot10^{16}$ $T = 500$, 1 *h*	Z-scan	$\lambda = 532$ $\tau = 40 - 70$ $\nu = 10$ $I_0 = 1.0\cdot10^{10}$	$n_2 = 5.0\cdot10^{-10}$	Williams *et al.* 1999 [82] Sarkisov *et al.* 2000 [83]

Table 1. (Continued)

Metal (Ion)	Matrix	*Synthesis conditions:* *Energy (E), keV* *Dose (D), ion/cm2,* *Current density (J), μA/cm2* *Annealing temper. (T), °C and time*	Study nonlinear optical method	*Laser parameters:* *Wavelength (λ), nm* *Pulse duration (τ), ps* *Repetition rate (ν), Hz* *Intensity (I0), W/cm2* *Pulse energy (P), mJ*	Nonlinear parameters: Refract. coef. (n2), cm2/W Absorption coef. (β), cm/W Satur. intensity (Isat), W/cm2 Re[χ(3)], Im[χ(3)], |χ(3)|, esu	Authors
Ag	SiO_2	$E = 1.5 \cdot 10^3$ $D = 4.0 \cdot 10^{16}$ $J = 2$ $T = 500$, 1 *h*	Z-scan	$\lambda = 532$ $\tau = 4.5$ $\nu = 76 \cdot 10^6$	$\|\chi^{(3)}\| = 5.0 \cdot 10^{-7}$	Ila *et al.* 1998 [50]
Ag	SiO_2	$E = 60$ $D = 4.0 \cdot 10^{16}$ $J = 10$	Z-scan	$\lambda = 354.7$ $\tau = 55$ $\nu = 2$ $I_0 = 1.3 \cdot 10^9$	$n_2 = -2.7 \cdot 10^{-7}$ $Re[\chi^{(3)}] = -6.0 \cdot 10^{-8}$ $\beta = -1.4 \cdot 10^{-5}$ $Im[\chi^{(3)}] = -6.1 \cdot 10^{-9}$ $\|\chi^{(3)}\| = 6.1 \cdot 10^{-8}$	Ganeev *et al.* 2003 [58] 2004 [63]
Ag	SiO_2	$E = 60$ $D = 4.0 \cdot 10^{16}$ $J = 10$	Z-scan	$\lambda = 532$ $\tau = 55$ $\nu = 2$ $I_0 = (2.5 - 14) \cdot 10^9$	$n_2 = -(6.2 - -0.7) \cdot 10^{-10}$ $Re[\chi^{(3)}] = -(3.5 - -0.4) \cdot 10^{-8}$ $\beta = -(3.6 - -0.5) \cdot 10^{-5}$ $Im[\chi^{(3)}] = -(1.3 - -0.2) \cdot 10^{-8}$	Ganeev *et al.* 2004 [84] Stepanov *et al.* 2010 [85]
Ag	SiO_2	$E = 60$ $D = 4.0 \cdot 10^{16}$ $J = 10$	Z-scan	$\lambda = 1064$ $\tau = 35$ $\nu = 2$ $I_0 = 1.0 \cdot 10^{10}$	$n_2 = 1.5 \cdot 10^{-8}$ $Re[\chi^{(3)}] = 2.5 \cdot 10^{-9}$	Ganeev *et al.* 2003 [62] 2004 [63]
Ag	SiO_2	$E = (1.7 - 2.4) \cdot 10^3$ $D = (4.0 - 7.0) \cdot 10^{16}$ $J = 0.3$ $T = 500$, 1 *h*	Z-scan	$\lambda = 532$ $\tau = 7.0 \cdot 10^3$ $P = 0.14$	Three-photon absorption	Joseph *et al.* 2007 [86]
Ag	SiO_2	$E = 2.0 \cdot 10^3$ $D = 7.0 \cdot 10^{16}$ $T = 600$, 1 *h*	Z-scan	$\lambda = 527$ $\tau = 0.233$ $\nu = \cdot 10^3$	$\beta = -(2.6 - 1.8) \cdot 10^{-6}$ $Im[\chi^{(3)}] = 4.7 \cdot 10^{-10}$ $I_{sat} = (1.2 - 3.5) \cdot 10^7$	Rangel-Rojo *et al.* 2009 [87] 2010 [88]

Metal (Ion)	Matrix	*Synthesis conditions:* *Energy (E), keV* *Dose (D), ion/cm2,* *Current density (J), μA/cm2* *Annealing temper. (T), °C and time*	Study nonlinear optical method	*Laser parameters:* *Wavelength (λ), nm* *Pulse duration (τ), ps* *Repetition rate (ν), Hz* *Intensity (I0), W/cm2* *Pulse energy (P), mJ*	**Nonlinear parameters:** **Refract. coef. (n2), cm2/W** **Absorption coef. (β), cm/W** **Satur. intensity (Isat), W/cm2** **Re[χ(3)], Im[χ(3)], |χ(3)|, esu**	**Authors**
Ag	SiO_2	$E = 200$ $D = 2.0 \cdot 10^{17}$ $J = 2.5$	Z-scan	$\lambda = 532$ $\tau = 38$ $\nu = 10$	$n_2 = -3.0 \cdot 10^{-11}$ $Re[\chi^{(3)}] = 3.0 \cdot 10^{-8}$ $\beta = -7.0 \cdot 10^{-8}$ $Im[\chi^{(3)}] = 2.6 \cdot 10^{-8}$ $\lvert\chi^{(3)}\rvert = 4.0 \cdot 10^{-8}$	Wang *et al.* 2009 [89]
Ag	SiO_2	$E = 200$ $D = 2.0 \cdot 10^{17}$ $J = 2.5$	Z-scan	$\lambda = 1064$ $\tau = 38$ $\nu = 10$	$n_2 = -1.7 \cdot 10^{-10}$ $Re[\chi^{(3)}] = 1.8 \cdot 10^{-7}$ $\lvert\chi^{(3)}\rvert = 1.8 \cdot 10^{-7}$	Wang *et al.* 2009 [89]
Ag	SLSG	$E = 60$ $D = 4.0 \cdot 10^{16}$ $J = 10$	Z-scan	$\lambda = 532$ $\tau = 55$ $\nu = 2$ $I_0 = (2.5 - 14) \cdot 10^9$	$n_2 = -(4.1 - -1.7) \cdot 10^{-10}$ $Re[\chi^{(3)}] = -(2.4 - -1.4) \cdot 10^{-8}$ $\beta = -(6.7 - -1.7) \cdot 10^{-5}$ $Im[\chi^{(3)}] = -(0.6 - 2.4) \cdot 10^{-8}$	Ganeev *et al.* 2004 [84]
Ag	SiO_2	$E = 60$ $D = 4.0 \cdot 10^{16}$ $J = 10$	Z-scan	$\lambda = 1064$ $\tau = 35$ $\nu = 2$ $I_0 = 3.0 \cdot 10^{10}$	$n_2 = 3.5 \cdot 10^{-8}$ $Re[\chi^{(3)}] = 5.7 \cdot 10^{-9}$	Ganeev *et al.* 2003 [63] 2004 [64]
Sn	SiO_2	$E = 400$ $D = 2.0 \cdot 10^{17}$ $J = 1$	DFWM	$\lambda = 460\text{-}540$ $\tau = 5 \cdot 10^3$ $I_0 = 3.0 \cdot 10^{10}$	$\lvert\chi^{(3)}\rvert = 3.0 \cdot 10^{-6}$	Takeda *et al.* 1993 [90] 1994 [91]
Sn	SiO_2	$E = 350$ $D = 8.0 \cdot 10^{17}$ $J = 2$ $T = 200$, 0.5 - 1 *h*	Z-scan	$\lambda = 460\text{-}540$ $\tau = 4.5$ $\nu = 76 \cdot 10^6$	$\lvert\chi^{(3)}\rvert = 1.5 \cdot 10^{-6}$	Ila *et al.* 1998 [50]
Ta	SiO_2	$E = 60$ $D = 3.0 \cdot 10^{16}$ $J = 0.3$ $T = 900$, 1 *h*	PPTNS	$\lambda = 574$ $\tau = 0.2$ $\nu = 10^3$	Bleaching absorption in 502 - 605 nm	Takeda *et al.* 2003 [92]

Table 1. (Continued)

Metal (Ion)	Matrix	*Synthesis conditions:* *Energy (E), keV* *Dose (D), ion/cm2,* *Current density (J), μA/cm2* *Annealing temper. (T), °C and time*	Study nonlinear optical method	*Laser parameters:* *Wavelength (λ), nm* *Pulse duration (τ), ps* *Repetition rate (ν), Hz* *Intensity (I0), W/cm2* *Pulse energy (P), mJ*	**Nonlinear parameters:** **Refract. coef. (n2), cm2/W** **Absorption coef. (β), cm/W** **Satur. intensity (Isat), W/cm2** **Re[χ(3)], Im[χ(3)], |χ(3)|, esu**	Authors
Au	Al_2O_3	$E = (2.75 - 3.0)\cdot 10^3$ $D = 2.2\cdot 10^{16}$ $T = 1100$, 1 *h*	DFWM	$\lambda = 532$ $\tau = 35 - 40$ $\nu = 10$ $I_0 = 1.0\cdot 10^9$	$\|\chi^{(3)}\| = 7.0\cdot 10^{-9}$	White *et al.* 1993 [93]
Au	Al_2O_3	$E = 160$ $D = (0.6 - 1.0)\cdot 10^{17}$ $J = 10$ $T = 800 - 1100$, 1 *h*	RZ-scan	$\lambda = 1064$ $\tau = 55$ $\nu = 2$ $I_0 = (1.8 - 2.3)\cdot 10^9$	$n_2 = -(0.1 - 1.5)\cdot 10^{-10}$ $Re[\chi^{(3)}] = -(0.8 - 1.2)\cdot 10^{-8}$	Ganeev *et al.* 2005 [29] 2006 [30] Stepanov *et al.* 2005 [94] 2006 [95]
Au	Al_2O_3	$E = 60$ $D = 2.0\cdot 10^{17}$ $J = 10$	PPTNS	$\lambda = 568$ $\tau = 0.2$ $\nu = 10^3$	Nonlinear dielectric functions	Takeda *et al.* 2006 [96] 2007 [97]
Au	SiO_2	$E = 2.75\cdot 10^3$ $D = (0.3 - 1.5)\cdot 10^{17}$ $T = 600 - 1100$, 2.2 *h*	DFWM Z-scan	$\lambda = 532$ $\tau = 6$ and 35 $\nu = 3.8$ and 10 $I_0 = 4.5\cdot 10^8$	$n_2 = (1.0 - 8.9)\cdot 10^{-10}$ $\beta = (3.7 - 4.8)\cdot 10^{-5}$ $\|\chi^{(3)}\| = (1.0 - 1.7)\cdot 10^{-10}$	Haglund *et al.* 1994 [45] Yang *et al.* 1996 [49] Magruder *et al.* 1993 [98] White *et al.* 1994 [99]
Au	SiO_2	$E = 1.5\cdot 10^3$ $D = 5.6\cdot 10^{16}$ $J = 0.7$ $T = 700 - 1200$	DFWM	$\lambda = 532$ $\tau = 5\cdot 10^3$ $I_0 = 1.0\cdot 10^9$	$\|\chi^{(3)}\| = (0.12 - 5.0)\cdot 10^{-8}$	Fukumi *et al.* 1991 [100] 1994 [101]
Au	SiO_2	$E = 3.0\cdot 10^3$ $D = 1.2\cdot 10^{17}$ $J = 2.0$ $T = 1200$, 0.5-1 *h*	Z-scan	$\lambda = 532$ $\tau = 4.5$ $\nu = 76\cdot 10^6$	$\|\chi^{(3)}\| = 6.5\cdot 10^{-7}$	Ila *et al.* 1998 [50]
Au	SiO_2	$E = 2.75\cdot 10^3$ $D = 1.5\cdot 10^{17}$ $T = 400$	DFWM	$\lambda = 532$ $\tau = 4\cdot 10^3$ $I_0 = 1.9\cdot 10^6$	$\|\chi^{(3)}\| = (0.3 - 1.3)\cdot 10^{-7}$	Lepeshkin *et al.* 1999 [102] Safonov *et al.* 1999 [103]

Metal (Ion)	Matrix	*Synthesis conditions:* *Energy (E), keV* *Dose (D), ion/cm2,* *Current density (J), μA/cm2* *Annealing temper. (T), °C and time*	Study nonlinear optical method	*Laser parameters:* *Wavelength (λ), nm* *Pulse duration (τ), ps* *Repetition rate (ν), Hz* *Intensity (I0), W/cm2* *Pulse energy (P), mJ*	**Nonlinear parameters:** **Refract. coef. (n2), cm2/W** **Absorption coef. (β), cm/W** **Satur. intensity (Isat), W/cm2** **Re[χ(3)], Im[χ(3)], |χ(3)|, esu**	**Authors**
Au	SiO_2	$E = 60$ $D = (1.0 - 2.0)\cdot 10^{17}$ $J = 10$ -17	PPTNS	$\lambda = 554$ and 568 $\tau = 0.2$ $\nu = 10^3$ $I_0 = (0.7 - 0.9)\cdot 10^9$	Nonlinear dielectric functions, positive absorption in 500 – 650 nm	Takeda *et al.* 2006 [96] 2007 [104]
Au	SiO_2	$E = 2.0\cdot 10^3$ $D = 2.8\cdot 10^{16}$ $T = 1100$, 1 *h*	VSD	$\lambda = 532$ $\tau = 7\cdot 10^3$ $P = 0.14$	$n_2 = -2.0\cdot 10^{-8}$ $Re[\chi^{(3)}] = 1.9\cdot 10^{-9}$ $\beta = -5.0\cdot 10^{-6}$ $Im[\chi^{(3)}] = -1.6\cdot 10^{-9}$ $\lvert\chi^{(3)}\rvert = 2.2\cdot 10^{-9}$	Torres-Torres *et al.* 2007 [105]
Au	SiO_2	$E = 1.5\cdot 10^3$ $D = (0.3 - 1.0)\cdot 10^{17}$ $T = 400$, 1 *h*	Z-scan	$\lambda = 532$ $\tau = 7\cdot 10^3$ $\nu = 0.1$ $I_0 = 1.2\cdot 10^{10}$	$n_2 = -1.5\cdot 10^{-11}$ $Re[\chi^{(3)}] = 1.5\cdot 10^{-8}$ $\beta = -(2.6 - 8.0)\cdot 10^{-8}$ $Im[\chi^{(3)}] = -1.5\cdot 10^{-10}$ $\lvert\chi^{(3)}\rvert = 1.5\cdot 10^{-8}$ $I_{sat} = (1.9 - 2.6)\cdot 10^7$	Ghosh *et al.* 2008 [71] 2009 [106]
Au	SiO_2	$E = 250 - 300$ $D = 1.0\cdot 10^{17}$ $J = 2.5$	Z-scan	$\lambda = 532$ $\tau = 38$ $\nu = 10$ $I_0 = 0.9\cdot 10^9$	$n_2 = -(1.2 - 1.4)\cdot 10^{-10}$ $Re[\chi^{(3)}] = -1.2\cdot 10^{-7}$ $\beta = -(9.7 - 21.0)\cdot 10^{-10}$ $Im[\chi^{(3)}] = -3.6\cdot 10^{-8}$ $\lvert\chi^{(3)}\rvert = (1.3 - 1.6)\cdot 10^{-7}$	Wang *et al.* 2008 [107, 108]
Au	SiO_2	$E = 250 - 300$ $D = 1.0\cdot 10^{17}$ $J = 2.5$	Z-scan	$\lambda = 1064$ $\tau = 38$ $\nu = 10$ $I_0 = 3.8\cdot 10^8$	$n_2 = -0.4\cdot 10^{-10}$ $Re[\chi^{(3)}] = -(4.3 - -1.2)\cdot 10^{-8}$ $\lvert\chi^{(3)}\rvert = (0.1 - 4.3)\cdot 10^{-8}$	Wang *et al.* 2008 [107, 108]

Table 1. (Continued)

Metal (Ion)	Matrix	*Synthesis conditions: Energy (E), keV Dose (D), ion/cm2, Current density (J), μA/cm2 Annealing temper. (T), °C and time*	Study nonlinear optical method	*Laser parameters: Wavelength (λ), nm Pulse duration (τ), ps Repetition rate (ν), Hz Intensity (I0), W/cm2 Pulse energy (P), mJ*	**Nonlinear parameters: Refract. coef. (n2), cm2/W Absorption coef. (β), cm/W Satur. intensity (Isat), W/cm2 Re[χ(3)], Im[χ(3)], |χ(3)|, esu**	Authors
Cu-Ni	SiO_2	$E = 90$ and 100 $D = 6\cdot10^{16}$ and $6\cdot10^{16}$	Z-scan	$\lambda = 770$ $\tau = 0.13$ $\nu = 76\cdot10^9$ $I_0 = 9.8\cdot10^9$	$\|\chi^{(3)}\| = 6.8\cdot10^{-10}$	Falconieri *et al.* 1998 [27] Cattaruzza *et al.* 2002 [28]
Cu-Ni	SiO_2	$E = 90$ and 100 $D = 6\cdot10^{16}$ and $6\cdot10^{16}$	Z-scan	$\lambda = 532$ $\tau = 6$ $\nu = 0.5 - 1$ $I_0 = 2.0\cdot10^9$	$n_2 = 1.5\cdot10^{-10}$ $\|\chi^{(3)}\| = 5.0\cdot10^{-12}$	Cattaruzza *et al.* 2002 [28] Battaglin *et al.* 2000 [109]
Cu-Ag	SiO_2	$E = 160$ and 305 $D = 1.2\cdot10^{16}$ $J = 1.3 - 3$	Z-scan	$\lambda = 570$ $\tau = 6$ $\nu = 3.8\cdot10^6$ $I_0 = 4.0\cdot10^8$	$n_2 = (0.1 - 1.6)\cdot10^{-9}$ $\beta = -(1.4 - -3.8)\cdot10^{-5}$	Magruder *et al.* 1994 [110]
Cu-Ag	SiO_2	$E = 30$ and 43 $D = (1.0 - 2.0)\cdot10^{17}$	Z-scan	$\lambda = 790$ $\tau = 0.15$ $\nu = 76\cdot10^9$ $I_0 = 8.8\cdot10^9$	$n_2 = -(3.8 - 4.3)\cdot10^{-12}$ $\beta = (1.0 - 2.2)\cdot10^{-6}$ $\|\chi^{(3)}\| = (0.8 - 1.5)\cdot10^{-8}$	Wang *et al.* 2007 [111]
Cu-Ag	SiO_2	$E = 180$ and 200 $D = (1.0 - 2.0)\cdot10^{17}$ $J = 1.5 - 2.5$	Z-scan	$\lambda = 1064$ $\tau = 38$ $\nu = 10$	$n_2 = (0.6 - 3.0)\cdot10^{-10}$ $\|\chi^{(3)}\| = (0.6 - 2.1)\cdot10^{-7}$	Wang *et al.* 2008 [112] 2010 [113, 114
Ag-Au	SiO_2	$E = 130$ and 190 $D = 9.0\cdot10^{16}$ $J = 2$ $T = 800$, 1 *h*	Z-scan	$\lambda = 572$ $\tau = 5$ $\nu = 1$ $I_0 = (1.6 - 5.0)\cdot10^9$	$n_2 = -1.6\cdot10^{-10}$ $\beta = 1.3\cdot10^{-5}$ $\|\chi^{(3)}\| = (0.9 - 1.7)\cdot10^{-8}$	Cattaruzza *et al.* 2003 [115] 2005 [116]

Metal (Ion)	Matrix	*Synthesis conditions:* *Energy (E), keV* *Dose (D), ion/cm2,* *Current density (J), μA/cm2* *Annealing temper. (T), °C and time*	**Study nonlinear optical method**	*Laser parameters:* *Wavelength (λ), nm* *Pulse duration (τ), ps* *Repetition rate (ν), Hz* *Intensity (I0), W/cm2* *Pulse energy (P), mJ*	**Nonlinear parameters:** **Refract. coef. (n2), cm2/W** **Absorption coef. (β), cm/W** **Satur. intensity (Isat), W/cm2** **Re[χ(3)], Im[χ(3)], |χ(3)|, esu**	**Authors**
Ag-Au	SiO_2	$E = 130$ and 190 $D = 3.0 \cdot 10^{16}$ $T = 800$, 1 *h*	Z-scan	$\lambda = 525$ $\tau = 6$ $\nu = 1$ $I_0 = 0.8 \cdot 10^9$	$\beta = -(3.42 - -1.7) \cdot 10^{-4}$ $I_{sat} = (0.1 - 3.2) \cdot 10^8$	Cesca *et al.* 2010 [117]
Ti-Au	SiO_2	$E = 320$ and $1.1 \cdot 10^3$ $D = (0.6 - 2.0) \cdot 10^{16}$ $T = 900$, 2 *h*	Z-scan	$\lambda = 532$ $\tau = 6$ $\nu = 3.8 \cdot 10^6$ $I_0 = 4.0 \cdot 10^8$	$n_2 = (0.6 - 1.2) \cdot 10^{-9}$ $\beta = 5.3 \cdot 10^{-6}$	Magruder *et al.* 1995 [118]

Abbreviations: soda-lime silicate glass (SLSG), indium-tin oxide (ITO), degenerate four wave mixing (DFWM), pump-probe transient nonlinear spectroscopy (PPTNS), Z-scan and RZ-scan by reflection and vectorial self-diffraction (VSD)

For comparison some data for ion synthesized nontraditional MNPs are also presented.

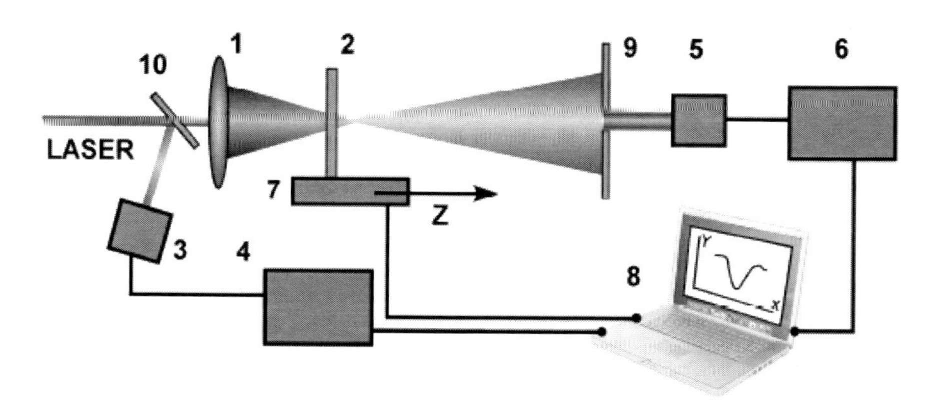

Figure 2. Z-scan setup. (1) focal length lens; (2) sample; (3) and (5) photodiodes; (4) and (6) digital voltmeters; (7) translation stage; (8) computer; (9) aperture.

Hence, in order to create new materials promising for practical use in laser systems and integral optics and to optimize their characteristics, one should study the nonlinear optical properties of these materials not only in the SPR spectral region, but also at the frequencies specific for industrial lasers. Materials characterized with nonlinear properties in near IR are now searching for applications in the field of telecommunication.

Here, recent results on nonlinear optical properties of copper and silver nanoparticles [62-65], synthesized by ion implantation in glass host matrices studied by the classical Z-method [121, 122] at the wavelength of a picosecond mode-locked Nd:YAG laser (λ = 1064 nm) are presented.

The used Z-scan setup is shown in Figure 2. Laser pulse duration was 35 ps, pulse energy 1 mJ and 2-Hz pulse repetition rate. The radiation had a spatial distribution close to Gaussian and was focused by a 25-cm focal length lens (1) onto the samples (2). The beam-waist diameter of the focused radiation was measured to be 90 μm using a CCD camera. The samples were transferred in steps of 2 mm along Z-axis when scanning focal region. The maximum laser intensity at the focal point was $3 \cdot 10^{10}$ W/cm^2, whereas the intensities of optical breakdown were $6 \cdot 10^{10}$ W/cm^2 and $8 \cdot 10^{10}$ W/cm^2 for the glasses with copper and silver nanoparticles, respectively. The fluctuations of the laser energy from pulse to pulse did not exceed 10 %. The energy of single laser pulses was measured by a calibrated photodiode (3). The samples were moved by a translation stage (7) along the Z-axis. A 1-mm aperture (9) with 1 % transmittance was fixed at the distance of 100 cm from the focusing plane (closed-aperture scheme). A photodiode (5) was kept behind the aperture. The radiation energy registered by photodiode (5) was normalized relative to the radiation energy registered by photodiode (3) in order to avoid the influence of non-stability of laser parameters. The experimental data accepted as normalized transmittance $T(z)$. The closed-aperture scheme allowed the determination a value of n_2 and the open-aperture scheme was used for the measurements a value of β.

The samples with copper were prepared by Cu-ion implantation into amorphous SiO_2 and soda lime silicate glasses (SLSG) as described in details [65]. The energies of 50 keV and dose $8 \cdot 10^{16}$ ion/cm^2 and at a beam current density of 10 μA/cm^2 were used. The penetration depth of the MNPs in the glasses for given energy of implantation was not exceed 80 nm.

Optical transmittance spectra of implanted samples Cu:SiO_2 and Cu:SLSG are presented in Figure 3. MNPs such as Cu in dielectric medium show optical absorption determined by SPR [4]. The spectra are maximized near 565 nm Cu:SiO_2 and 580 nm for Cu:SLSG that gives evidence for formation of the Cu nanoparticles in the glasses. Depending on the ion implantation conditions, the incorporation of accelerated ions into silicate glasses leads to the generation of radiation-induced defects, which can initiate reversible and irreversible transformations in the glass structure [6].

This can result in structural imperfections of different types, such as the generation of extended and point defects, local crystallization and amorphization, the formation of a new phase either from atoms involved the glass structure or from implanted ions, etc. In particular, the formation of MNPs in the glass brings about an increase in its volume and the generation of internal stresses within an implanted layer. The radiation-induced defects are responsible for an increase in the absorption in the range of the UV fundamental absorption edge in the spectrum of the glass. In our case, this effect can be observed in the short-wavelength range of the optical transmittance spectra displayed in Figure 3. It should be noted that, presented here nonlinear optical study were performed upon exposure of the samples to laser radiation at a wavelength of 1064 nm, which lies far from the UV spectral range of the linear absorption attributed to the SPR and interband transitions in MNPs and glasses. For this reason, in what follows, the contributions associated with interband transitions and radiation-induced defects will be eliminated from the analysis of the experimental results.

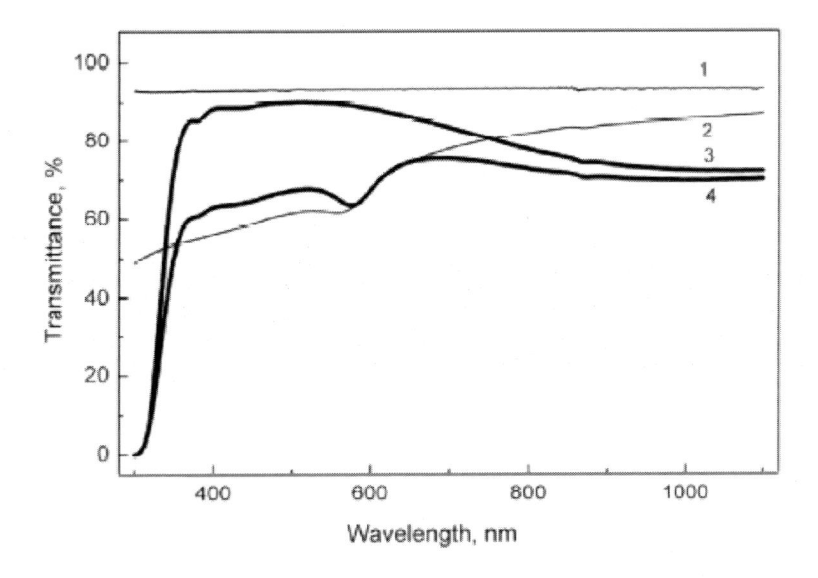

Figure 3. Transmittance spectra of (1) SiO_2 and (3) SLSG before and after Cu-ion implantation with energy of 50 keV and dose of $8{\cdot}10^{16}$ ion/cm^2 (2) Cu:SiO_2 and (4) Cu:SLSG [65].

Figure 4 shows the experimental dependences of the $T(z)$ measured for both glasses containing copper nanoparticles in the Z-scan scheme with an open aperture [65]. Recall that the measurements carried out in this scheme make it possible to determine the nonlinear coefficient β. Silicate glasses did not demonstrate the nonlinear absorption at applied laser intensities. It can be seen from Figure 4 that the experimental dependences $T(z)$ exhibit specific features inherent in nonlinear absorption: the $T(z)$ decreases as the focal point is approached and reaches a minimum at $Z = 0$. Each point in the graphs was obtained by averaging over the values measured for 40pulses. Some distribution of the experimental points in the graphs is caused, to some extent, by the energy instabilities and, for the most part, by the time instabilities of laser radiation. The nonlinear coefficient β of composite materials can be determined from the relationship for the $T(z)$, which, in the case of the scheme with an open aperture, is written as [122, 123]

$$T(z) = q(z)^{-1}\ln(1+q(z)). \quad (6)$$

here, $q(z) = \beta I(z)L_{\mathrm{eff}}$ is the laser beam parameter, $L_{eff} = (1-e^{\alpha_0 L})/\alpha_0$ is the effective optical path with MNPs in the sample, L is the sample thickness, and $I(z)$ is the intensity of the light passed through the sample as a function of its position along the Z-axis. The parameter $q(z)$ describes the propagation of the laser beam in the material, because the following relationship holds

$$1/q(z)=1/G(z)-2\lambda\Delta\phi/\pi w^2-i\lambda/\pi w^2, \quad (7)$$

where $G(z) = z[1+z_0^2/z^2]$ is the radius of the wave front curvature in the Z-direction, $z_0 = kw^2/2$ is the diffraction length of the beam, $k = 2\pi/\lambda$ is the wave vector, $\Delta\phi = \Delta\Phi_0/(1+ z_0^2/z^2)$, $\Delta\Phi_0 = (2\pi/\lambda)n_2 I_0 L_{\mathrm{eff}}$ - phase shift of frequency gained by the radiation passed through the sample, $w(z) = w_0(1 + z_0^2/z^2)^2$ is the beam radius at the point Z, and w_0 is the beam radius at the focal point (at a level of $1/e^2$).

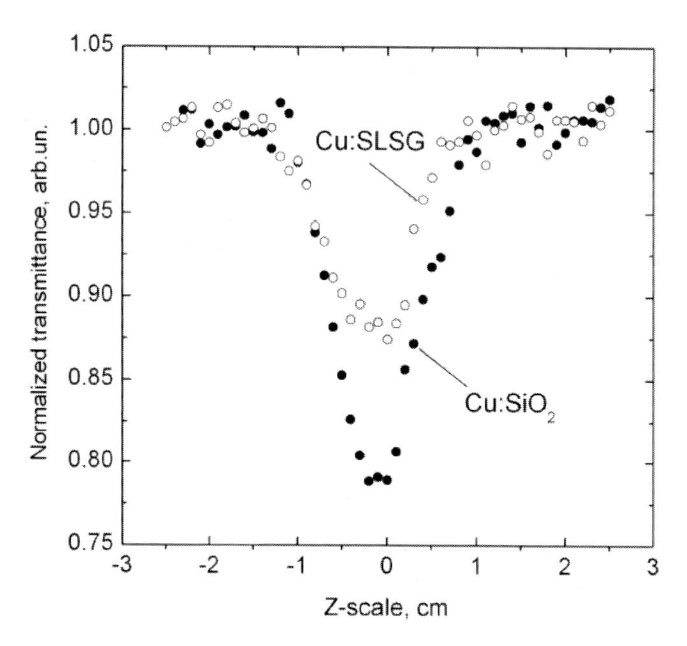

Figure 4. Normalized transmittance as a function of the Z-position of $Cu:SiO_2$ and Cu:SLSG composites in the z-scan scheme with an open aperture. Samples fabricated by Cu-ion implantation with energy of 40 keV and dose of $8\cdot10^{16}$ ion/cm^2. Laser intensity is $8\cdot10^9$ W/cm^2 and pulse duration is 35 ps [65].

Table 2. Nonlinear optical parameters of silicate glasses with ion-synthesized copper and silver nanoparticles measured at the wavelength of 1064 nm

Sample	n_2, 10^{-8} esu	β, 10^{-6} cm/W	$Re\chi^{(3)}$, 10^{-9} esu	$Im\chi^{(3)}$, 10^{-9} esu	$\lvert\chi^{(3)}\rvert$, 10^{-8} esu
Cu:SiO_2	-13.7	9	-32.0	6.5	3.28
Cu:SLSG	3.6	3.42	8.3	2.5	0.87
Ag:SiO_2	1.5		2.5		
Ag:SLSG	3.5		5.7		
SLSG	$8.1\cdot10^{-6}$		$1.4\cdot10^{-6}$		$1.4\cdot10^{-6}$

At $Z = 0$ (focal plane), the parameter $q(0) = q_0$ is defined by the expression

$$q_0 = \beta I_0 L_{\mathrm{eff}}, \tag{8}$$

where $I_0 = I(0)$ – intensity at the focal point.

Using formulas (6) and (8), it is possible to write

$$T_0 = q_0^{-1}\ln(1+q_0), \tag{9}$$

where T_0 is the minimum of $T(z)$ in the focal plane in the scheme with an open aperture. Expression (9) permits to determine the nonlinear absorption coefficient β. The values of β calculated in this way from the experimental data for the Cu:SiO_2 and Cu:SLSG composites are equal to $9.0\cdot10^{-6}$ and $3.42\cdot10^{-6}$ cm/W, respectively (Table 2).

As can be seen, the nonlinear coefficients β for these composites differ by a factor of 2.63. However, to compare correctly the coefficients β for the Cu:SiO_2 and Cu:SLSG, it is necessary to take into account the individual linear coefficients α_0 for layers with copper nanoparticles in different glasses ($\alpha_0^{SiO_2} = 9340$ and $\alpha_0^{SLSG} = 5800$ cm^{-1}). By assuming that the thicknesses of the layers with MNPs in the implanted glasses are virtually identical (~80 nm) [124], the nonlinear coefficient β to the linear coefficient α_0 for the relevant composite ($U = \beta/\alpha_0$) can be normalized. As a result, the normalized values of $U^{SiO_2} = 9.64\cdot10^{-10}$ and $U^{SLSG} == 6.73\cdot10^{-10}$ cm^2/W, which differ by a factor of 1.432, were obtained.

To account for this discrepancy between the parameters U for different samples, proper allowance must be made not only for the difference in the linear absorption coefficients but also for the specific features in the location of the SPR peaks attributed to copper nanoparticles. As can be seen from Figure 3, the SPR peaks assigned to MNPs is observed at 565 nm (ω_p = 17699.1 cm^{-1}) for the Cu:SiO_2 composite and at 580 nm (ω_p = 17241.4 cm^{-1}) for the Cu:SLSG composite. In such MNP systems a two-photon resonance related to the SPR can be assumed [3, 11]. On the other hand, it is known that, in the range of excitations and their associated transitions in nonlinear medium, the optical nonlinearities become more pronounced with a decrease in the detuning of the frequency from the resonance (in a case, two-photon) excitation [125]. In present experiment, the frequency detuning should be treated as the difference between the SPR frequency and the frequency of two photons of the laser radiation used ω_{20} = 18797 cm^{-1} (532 nm). The difference in the location of the SPR peaks for copper nanoparticles in the SiO_2 and SLSG composites can be estimated from the following ratio:

$$M = \left(\omega_{20} - \omega_{\mathrm{p}}^{\mathrm{Cu:SiO_2}}\right)^{-1} / \left(\omega_{20} - \omega_{\mathrm{p}}^{\mathrm{Cu:SLSG}}\right)^{-1} = 1.42 \quad (10)$$

This value is in qualitative agreement with a ratio of 1.432 between the nonlinear coefficients β normalized to the linear coefficients α_0 for different glasses.

The most interesting feature in the nonlinear optical properties of glasses with copper nanoparticles irradiated at a wavelength of 1064 nm is the fact that the doubled frequency of the laser radiation is close to the SPR peak frequency of copper particles. This is illustrated by the diagram in Figure 5, which shows the spectral positions of the SPR peaks of the MNPs in the samples studied and the spectral positions of the fundamental and doubled frequencies of the laser radiation [63, 64]. Thus, from the above date, it can draw the following conclusions: (1) In the near-IR range, the large nonlinear absorption coefficients determined experimentally for glasses containing copper nanoparticles are explained by the SPR in MNPs; (2) The efficient nonlinear absorption in the composites with copper nanoparticles considered is associated with both the linear absorption of the material and the effect of two-photon resonance at the SPR frequency for copper nanoparticles, which leads to the two-photon transition at the wavelength 1064 nm.

The theoretical realization of two-photon absorption associated with the SPR of colloidal metal (silver) particles in a solution was previously discussed in [126, 127], but in these experiments the authors were difficult to analyze this possibility due to the experimental problems related to the efficient aggregation of colloidal silver under laser irradiation, which changed the nonlinear optical properties of the samples in time.

When analyzing the results presented here, it is expedient to dwell on the possible fields of practical application of the studied composites. As is known, media with nonlinear (in particular, two-photon) absorption are very promising as materials for optical limiters, which can serve, for example, for the protection of eyes and highly sensitive detectors against intense optical radiation [128].

Early, the majority of studies in this field have been performed using nanosecond laser pulses. In this case, the main mechanisms responsible for nonlinear effects are associated with the reverse saturable nonlinear absorption (fullerenes and organic and metalloorganic compounds) and nonlinear scattering (solutions of colloidal metal aggregates). Picosecond and subpicosecond laser pulses have been used only to examine the optical limiting in media belonging primarily to semiconductor materials (two-photon absorption and strong nonlinear refraction).

Since two-photon absorption at a wavelength of 1064 nm is observed in the $Cu:SiO_2$ and Cu:SLSG samples, it is of interest to investigate the optical limiting effect in these composites in the scheme with an open aperture. Here, it was assumed that the sample is located in the region corresponding to a minimum transmittance, i.e., in the focal plane of a beam (Z = 0).

The nonlinear absorption was studied experimentally at an operating intensity from 10^7 to 10^{11} W/cm^2 [62, 63].

With the use of the linear and nonlinear (two-photon) absorption coefficients, the dependences of the normalized *T*(*z*) on the laser radiation intensity presented in Figure 6. It can be seen from Figure 6 that, at the maximum intensity, the Cu:SiO_2 composite is characterized by an approximately fifteen-fold limiting, whereas the Cu:SLSG composite exhibits an approximately three-fold limiting. Consequently, these composites can serve as nonlinear materials for optical limiting. It is clear that the Cu:SiO_2 composite is more preferable from the practical standpoint.

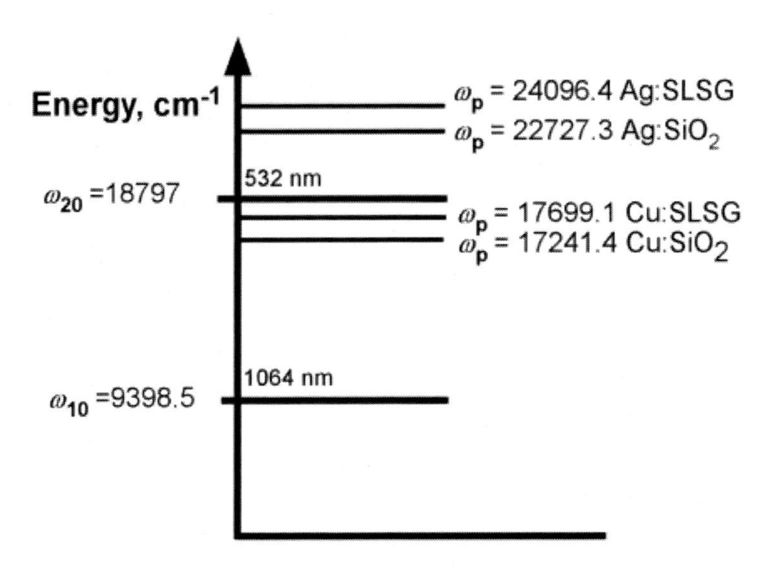

Figure 5. Diagram of the SPR frequencies for the glasses with implanted copper and silver nanoparticles [63, 64].

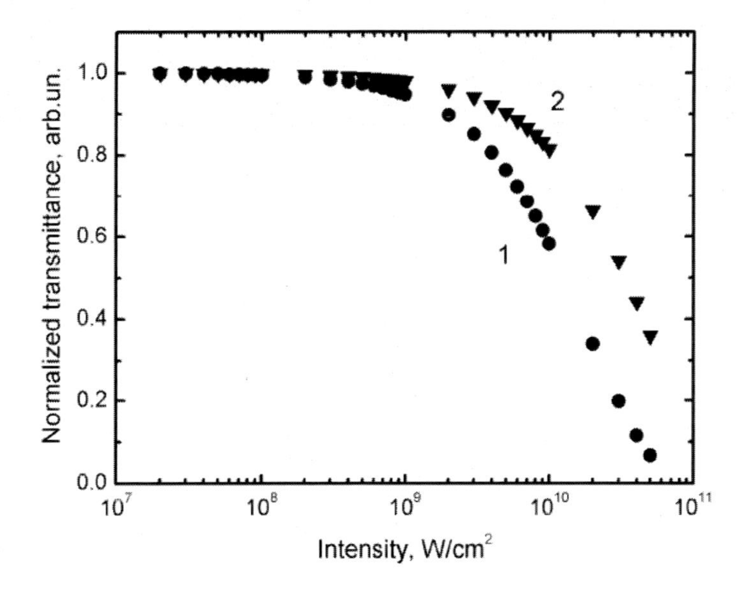

Figure 6. Calculated curves *T*(*z*) as a function of the incident radiation intensity for (1) Cu:SiO_2 and (2) Cu:SLSG composites [62].

3.2. Nonlinear Refraction of MNPs Studied by Z-Scan

Consider the nonlinear optical refraction of Cu:SiO_2 and Cu:SLSG composites [63, 65]. Figure 7 shows the $T(z)$ dependences of the samples in the Z-scan scheme with a closed aperture. Each point on the plot corresponds to a value averaged over 40 pulses. A specific feature of the samples with copper nanoparticles is that the plots for different types of substrate glass demonstrate opposite signs of nonlinear refraction under the same implantation conditions.

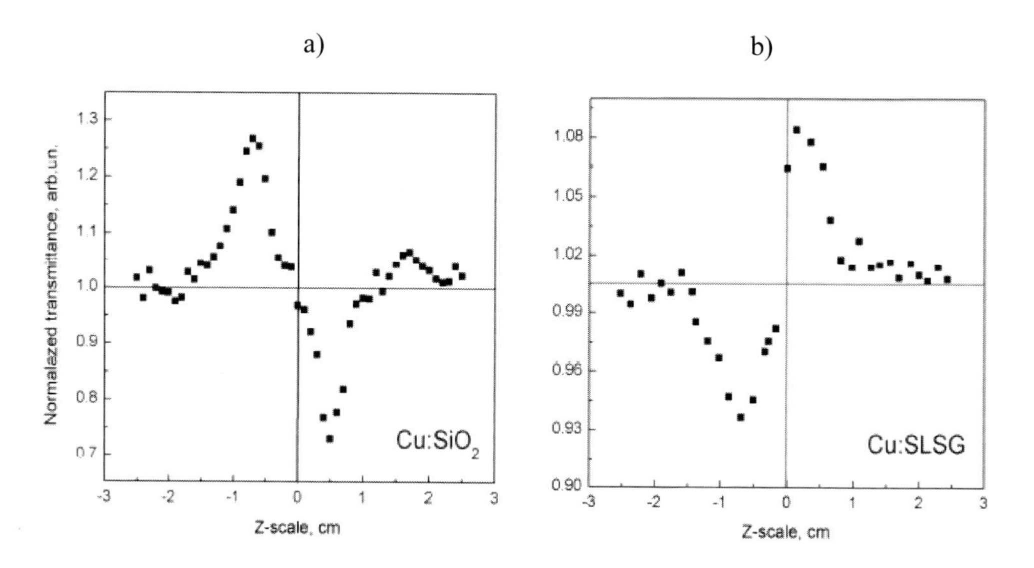

Figure 7. Dependence of the curves $T(z)$ for (a) Cu:SiO_2 and (b) Cu:SLSG composites in the Z-scan scheme with a close aperture. Samples fabricated by Cu-ion implantation with energy of 40 keV and dose of $8{\cdot}10^{16}$ ion/cm^2. Laser intensity is $8{\cdot}10^{9}$ W/cm^2 and pulse duration is 35 ps [65].

From the position of the $T(z)$ peak in the positive or negative Z-scan region (the nonlinearity sign), it can be can conclude that the Cu:SiO_2 sample are characterized by self-defocusing of the laser beam ($n_2 < 0$), while the Cu:SLSG sample demonstrate a self-focusing effect ($n_2 > 0$).

To recognize the reason for the different signs of nonlinear refraction in the glasses with copper nanoparticles, the observed self-action effects has to by analyzed.

Since the wavelength of the laser radiation used in this study considerably exceeds the size of the MNPs synthesized by the implantation [6, 9], the optical properties of the nanoparticles can be considered within the framework of the effective medium theory [4]. Such an approach allows to consider the composites as optically homogeneous materials, disregarding the presence of MNPs in them.

In general, among nonlinear optical processes contributing to the nonlinear part of the refractive index, it should be taken into account the optical Kerr effect, caused by the electronic response of atoms and molecules [125] and associated with the presence of resonance transitions in the medium [129]. The nonlinear index n_2 may vary considerably depending on the type of interaction (resonance or nonresonance). Nonresonance contributions to the n_2 of such medium as glasses are usually positive [130]. In the case of resonance interactions involving one-photon or two-photon processes, the sign of the nonlinear index n_2 is determined from the difference between the frequency of an incident electromagnetic laser wave ω_{10} (or a multiple frequency ω_{i0}) and the intrinsic resonance frequency of the material (ω_p in the case of MNPs).

In particular, the nonlinear index n_2 is negative only for frequencies that are slightly below the one-photon resonances or slightly above the two-photon resonances [130].

For a homogeneous condensed medium characterized by the occurrence of resonance transitions, one can consider the standard two-level energy model [125]. Then, the corresponding equation for the nonlinear index n_2 will have the form

$$n_2 = -2\pi N \frac{|\mu_{i0}|^4}{n_0 \eta (\omega_{io} - \omega_p)^3}, \tag{11}$$

where ω_p and ω_{i0} correspond to the frequencies of the SPR of MNP and the laser radiation, respectively; the subscript i denotes one- and two-photon processes; N is the concentration of active excitation centers considered to be dipoles (virtually equal to the number of nanoparticles in the sample); and μ_{i0} is the transition dipole moment at the frequencies ω_{i0}.

As follows from equation (11), N and μ_{i0} have no effect on the sign of the nonlinear index n_2 (the sign of the nonlinearity), which is determined only by the detuning from the resonance Δ_{i0}. Keeping in mind that $\chi^{(3)}$ depends linearly on n_2 [125], it can be written the relation determining the sign of the nonlinearity as

$$\operatorname{sgn} Re\left[\chi^{(3)}\right] \propto -(\omega_{20} - \omega_p)^{-3} = -\operatorname{sgn} \Delta_{i0}, \tag{12}$$

and analyze it only for the frequency ω_{20}, i.e., for the doubled frequency of the laser radiation, which lies in the vicinity of the SPR of the samples with copper particles. In other words, it is considered the effect that the frequency detuning between the sum frequencies of two laser photons and the SPR frequency exerts on nonlinear optical processes.

As was mentioned (Figure 7) the samples show different spectral positions of the SPR maxima for MNPs in glass host matrices of different types. For example, the SPR maximum for the Cu:SLSG samples is in the vicinity of 580 nm, while the SPR peak of the Cu:SiO_2 samples lies near 565 nm.

Substituting the frequencies of the SPR of copper nanoparticles in SiO_2 and SLSG (Figure 5) and the frequency $\omega_{20} \sim 18797$ cm^{-1} (the frequency of the two-photon excitation of the laser radiation used) into equation (12), a negative sign of the detuning is obtained, which points to a negative contribution to the nonlinear susceptibility for the matrices of both types. These conditions correspond to the self-defocusing of laser radiation, which was experimentally observed for the Cu:SiO_2 sample (Figure 7a). Thus, the two-level model used in this paper gives the proper sign of nonlinearity in the Cu:SiO_2 system excited by laser radiation at a frequency lying outside the SPR region of its particles, namely, at a frequency about two times lower than the SPR frequency.

A noticeable contribution to the n_2 can also be made by the thermal effect, i.e., by the heat transfer from MNPs and defects of a dielectric host matrix heated by laser radiation [131]. The rise time τ_{rise} of n_2 variations is determined by $\tau_{rise} = R_{beam}/V_s$, where R_{beam} is the beam-waist radius and V_s is the sound velocity in the lattice. In present case (R_{beam} = 75 μm, $V_s \approx$ 5500 m/s) the time necessary for both the distribution of the density of a material and its n_2 to reach their stationary values — t_{relax} ~ 13–15 ns — is three orders of magnitude longer than the pulse duration (35 ps). This allows one to exclude from consideration the influence of the thermal effect on the nonlinear optical properties of the composites at present experimental conditions and regard the electronic optical Kerr effect in MNPs as the main factor.

At On the other hand, for the Cu:SLSG sample, the self-focusing of the laser radiation (Figure 7b) was observed [63], which contradicts the conclusions derived from equation (12). To reveal the reasons for the different self-action effects in the glasses, it is necessary to consider the influence of the substrate on the nonlinear optical properties of the composites. For this purpose, the dependences of the $T(z)$ for both types of glasses without MNPs was measured.

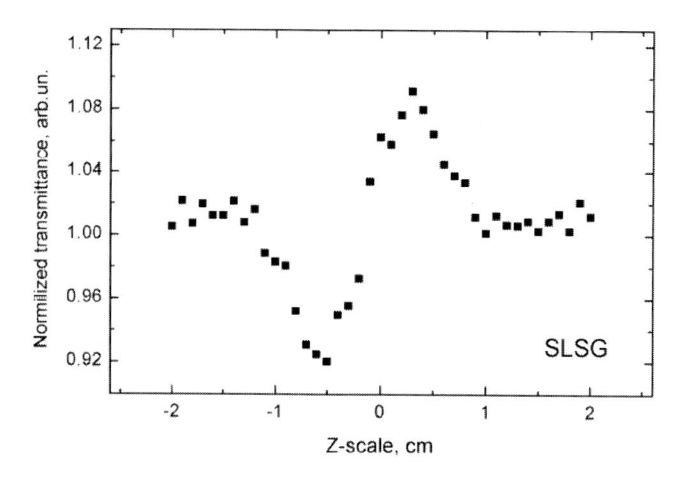

Figure 8. Dependence of the curves $T(z)$ for SLSG in the z-scan scheme with a close aperture [63].

The SiO_2 substrate shows no noticeable changes in the character of the $T(z)$ under irradiation with the intensities used in this study; i.e., this glass does not demonstrate nonlinear refraction and the nonlinearities observed in the Cu:SiO_2 samples are evidently caused by the copper nanoparticles. At the same time, the SLSG matrix exhibits a self-focusing effect (Figure 8). To estimate the contribution of the glass substrate to the optical refraction of the Cu:SLSG sample, it was determined and compared the values of $\chi^{(3)}$ for the SLSG and Cu:SLSG samples. In the general case, when a material simultaneously exhibits both nonlinear refraction and nonlinear absorption, the nonlinear susceptibility is a complex quantity

$$\chi^{(3)} = Re\left[\chi^{(3)}\right] + iIm\left[\chi^{(3)}\right] \quad (13)$$

where the real part is related to the nonlinear index n_2 and the imaginary part is related to the nonlinear coefficient β. As was mentioned the used glasses, contrarily to the samples with nanoparticles, have no nonlinear absorption and, hence, $\chi^{(3)}$ for the SLSG substrates is directly real-valued and can be expressed in terms of n_2 as

$$Re\left[\chi^{(3)}\right]=\frac{n_0}{3\pi}n_2 \quad (14)$$

This parameter can be experimentally estimated using the known Z-scan relations [122]

$$\Delta T_{m-v}=0.404(1-S)^{0.25}\left|\Delta\Phi_0\right| \quad (15)$$

where $\Delta T_{\text{m-v}}$ is the difference between the maximum and the minimum (valley) of the measured $T(z)$, S is the percent of radiation passing through the aperture and reaching the photodiode.

Applying relations (15) to the experimental data (Figure 8) obtained for the SLSG substrate, a value of $n_2 = 8.1\cdot10^{-14}$ esu was estimated, while for the Cu:SLSG sample this parameter was detected to be of $n_2 = 3.6\cdot10^{-8}$ esu. Using relation (14) for the $Re[\chi^{(3)}]$, it was obtained that, for SLSG, which shows no nonlinear absorption, $|\chi^{(3)}|$ is equal to $1.4\cdot10^{-14}$ esu, while $Re[\chi^{(3)}]$ for Cu:SLSG is equal to $8.3\cdot10^{-9}$ esu (Table 2).

Taking into account that the nonlinear susceptibility in Cu:SLSG is a complex-valued parameter, whose imaginary part is expressed via the nonlinear coefficient β as

$$Im\left[\chi^{(1)}\right]=\frac{n_h\varepsilon_h c^2}{\omega}\beta \quad (16)$$

Using equation (13) and the value of $Im[\chi^{(3)}]$ for the given medium, then the nonlinear susceptibility is $|\chi^{(3)}| = 8.7\cdot10^{-9}$ esu (Table 2) [63].

Since the $Re[\chi^{(3)}]$ is responsible for the nonlinear refraction in a material, it was compared to such nonlinear parameters (Table 2) for the SLSG and Cu:SLSG samples. In order to eliminate the influence of both the linear absorption and the difference in the thicknesses of the samples, in practice, one does not compare directly the values of $Re[\chi^{(3)}]$, but rather their normalized values $Re[\chi^{(3)}]\ L_{\text{eff}}$ (in this case, L_{eff} includes a correction for the linear absorption). Using L_{eff} of the wavelength 1064 nm for the samples of both types, it is possible to get $Re[\chi^{(3)}]\ L_{\text{eff}} = 4.92\cdot10^{-14}$ esu·cm for Cu:SLSG, $Re[\chi^{(3)}]\ L_{\text{eff}} = 3.45\cdot10^{-15}$ esu·cm for SLSG. Therefore, the nonlinear parameters of pure SLSG are lower by an order of magnitude than the same parameters for the glasses containing nanoparticles.

Discussing the reasons for the self-focusing observed in the experiment (Figure 7), i.e., the reasons for the positive contribution to the nonlinear susceptibility of Cu:SLSG, it should be take into account the considerable (~30%) linear absorption of SLSG in the spectral region of the laser radiation (Figure 3).

If a material exhibits the effect of saturation at the laser wavelength, the total absorption will decrease upon the laser irradiation. However, if a material is characterized by nonlinear absorption, the total absorption will increase, as is observed for the glasses with nanoparticles. The nonlinear absorption of Cu:SLSG causes an additional decrease in the intensity of the transmitted light in the focal plane by approximately 10–12%, while for Cu:SiO_2, whose linear absorption is ~15%, this value is equal to 18%. Irrespective of the conditions of the laser radiation which was choose here, the increase in the total absorption of Cu:SLSG can be caused by the nonlinear thermal effect predicted early [132], and observed in [133] for silicate glasses containing radio-frequency sputtered copper nanoparticles with size of 2.2±0.6 nm irradiation at the wavelength 1064 nm, when positive nonlinear susceptibility was recorded using trains of picosecond pulses (100 pulses in a train).

Consider nonlinear refraction of glass containing silver nanoparticles. The composites with silver were prepared by Ag-ion implantation into amorphous SiO_2 and soda lime silicate glasses (SLSG) as described [23]. The energies of 60 keV and dose $4{\cdot}10^{16}$ ion/cm^2 and at a beam current density of 10 μA/cm^2 were used. The penetration depth of the MNPs in the glasses for given energy of implantation did not exceed 80 nm [124]. Optical transmittance spectra of implanted samples Ag:SiO_2 and Ag:SLSG are presented in Figure 9.'

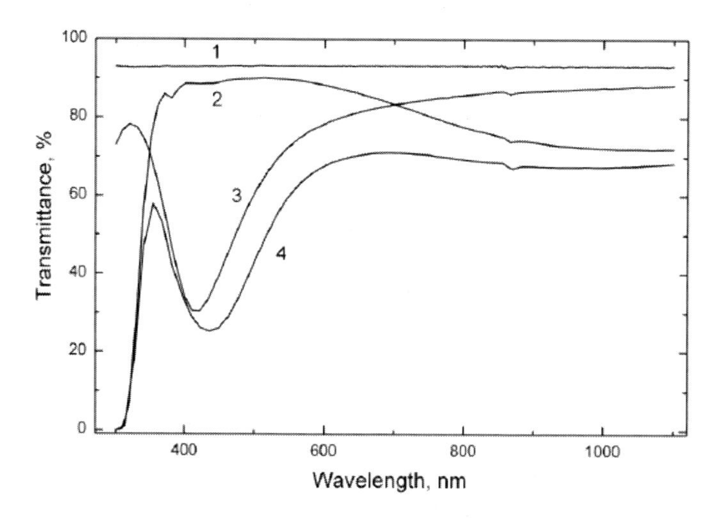

Figure 9. Transmittance spectra of (1) SiO_2 and (2) SLSG before and after Ag-ion implantation with energy of 60 keV and dose of $4{\cdot}10^{16}$ ion/cm^2 (3) Ag:SiO_2 and (4) Ag:SLSG [23].

MNPs such as Ag in dielectric medium show optical absorption determined by SPR with maximum near 415 – 440 nm [4]. Figure 10 shows the experimental dependences of the $T(z)$ of the samples Ag:SiO_2 and Ag:SLSG during Z-scanning in the scheme with a closed aperture. As follows from Figure 10, both types of glasses with silver nanoparticles demonstrate self-focusing of laser radiation. The values n_2 and $Re[\chi^{(3)}]$ are presented in Table 2. The nonlinear absorption was not detected for these samples.

Estimate the nonlinear optical contributing to both the magnitude and the sign of the nonlinear susceptibility $\chi^{(3)}$. Similarly to the case of sample with copper nanoparticles, the spectral positions of the SPR bands in the glasses with silver nanoparticles depend on the type of substrate (Figure 9).

The SPR maximum lies at about 415 nm (ω_p = 24096.4 cm^{-1}) for the Ag:SiO_2 and at 440 nm (ω_p = 22727.3 cm^{-1}) for the Ag:SLSG samples. As can be seen from the diagram in Figure 5, the frequency of the sum of two photons of the laser radiation is lower than the SPR frequency for nanoparticles in either matrix, which corresponds to a positive sign of the detuning and, as a consequence, leads to a positive contribution to the nonlinear susceptibility. Hence, no two-photon absorption occurs in the samples with silver nanoparticles.

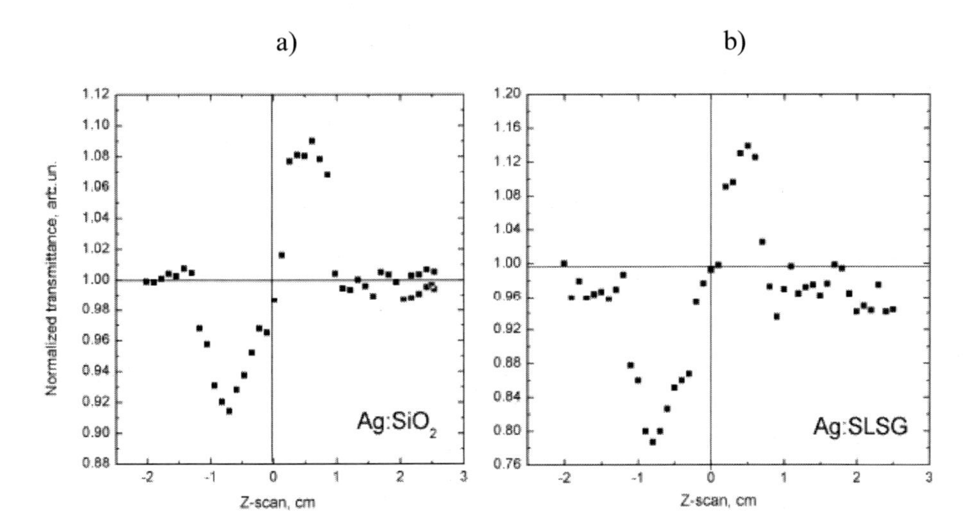

Figure 10. Dependence of the curves $T(z)$ for (a) $Ag:SiO_2$ and (b) Ag:SLSG composites in the Z-scan scheme with a close aperture. Samples fabricated by Ag-ion implantation with energy of 60 keV and dose of $4{\cdot}10^{16}$ ion/cm^2. Laser intensity is $8{\cdot}10^9$ W/cm^2 and pulse duration is 35 ps [63].

3.3. RZ-Scan Technique

There are different approaches for the study of nonlinear optical properties of various materials, for example, degenerate four-wave mixing [136], nonlinear optical interferometry [137], Z-scan [121, 122]). As was mentioned the latter technique allows determining both the value and the sign of nonlinear optical indexes n_2 and β. There are several modifications of the Z-scan technique, such as transmission Z-scan (TZ-scan) [121, 122], eclipsing Z-scan [138], two-beams [139], reflection Z-scan (RZ-scan) [140-142], time-resolved Z-scan [143], etc. The RZ-scan has an advantage with comparing to the others that allows studying the optical nonlinearities of materials with a limited optical transparency. This technique is based on the analyze of the surface properties of materials, whereas the others are used for the investigation of bulk characteristics of media. The application of RZ-scan was firstly presented in [14], where the nonlinear refraction of gallium arsenide was studied in at a wavelength of 532 nm at which this semiconductor is fully opaque. On the other hand, this technique can also be applied for transparent materials and can be used for the comparison with conventional TZ-scan. Ricently, the RZ-scan technique was applied for measurement of nonlinear characteristics of low-transparency dielectric layers with MNPs beyond the region of the SPR absorption of particles [29, 94]. Consider some examples with composites based on dielectric with copper, silver and gold nanoparticles synthesized by ion implantation.

Table 3. Ion implantation conditions and nonlinear optical parameters of Al_2O_3 with ion-synthesized silver, copper and gold nanoparticles measured at the wavelength of 1064 nm

Sample	No.	Energy, keV	Current density, $\mu A/cm^2$	Ion dose, 10^{17} ion/cm^2	I_0, 10^9 W/cm^2	n_2, 10^{-11} cm^2/W	$Re[\chi^{(3)}]$, 10^{-9} esu
$Ag:Al_2O_3$	1	30	3	3.75	4.3	3.40	0.94
$Ag:Al_2O_3$	2	30	6	3.75	4.3	3.89	1.07
$Ag:Al_2O_3$	3	30	10	3.75	4.3	5.36	1.48
$Cu:Al_2O_3$	4	40	2.5	0.54	7.7	-3.75	-1.04
$Cu:Al_2O_3$	5	40	12.5	1.0	7.7	-4.96	-1.38
$Au:Al_2O_3$	6	160	10	0.6	2.3	-28.15	-7.77
$Au:Al_2O_3$	7*	160	10	0.6	2.8	-32.68	-10.0
$Au:Al_2O_3$	8	160	10	1.0	2.3	-38.76	-10.7
$Au:Al_2O_3$	9*	160	10	1.0	2.8	-44.30	-12.2

*thermal annealing [30].

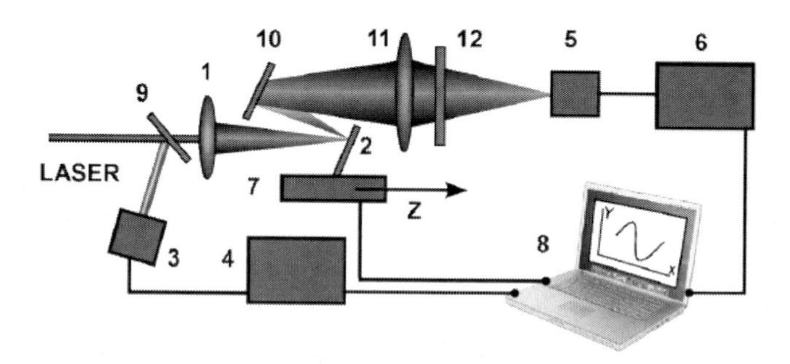

Figure 11. RZ-scan setup. (1) focal lens; (2) sample; (3) and (5) photodiodes; (4) and (6) digital voltmeters; (7) translation stage; (8) computer; (9) beam splitter; (10) mirror; (12) lens.

As a substrate for such model composites an artificial sapphire (Al_2O_3) was used. Whose surface of the sapphire opposite to the implanted surface was frosted, because of which the sample was almost nontransparent in visible and IR-spectral area. Ion implantation was performed with Ag^+, Cu^+ and Au^+ [95, 144]. Experimental conditions of ion implantation used for fabrication of MNPs in Al_2O_3 are shown in Table 3 [30].

The RZ-scan setup for measurement of nonlinear refraction is presented in Figure 11. The Nd:YAG laser (λ = 1064 nm, τ = 55 ps) operated at a 2 Hz pulse repetition rate was applied. Laser radiation was focused by a 25-cm focal length lens (1). The maximum intensity and the beam waist radius in the focal plane were measured to be $I_0 = 7 \cdot 10^9$ W/cm^2 and 72 μm, respectively. The sample (2) was fixed on the translation table (7) and moved along the Z-axis. The angle of incidence of laser radiation on the surface of sample was 30. A part of radiation was reflected from the beam splitter (9) and measured by photo-diode (3) to control the energy of laser pulses.

The radiation reflected from the surface of sample was directed to the mirror (10) and than collected by the lens (11) that allowed registering all the reflected radiation by photo-diode (5). To decrease the influence of the instability of laser radiation, the ratio $R(z)$ between the reflected signal and the incident one was accepted.

In the case of RZ-scan the refractive nonlinearities are measured without aperture (for example, see [145]). In TZ-scan scheme, the phase changes are produced by absorptive nonlinearities and the aperture is needed in this case. The refractive nonlinearities are responsible for the amplitude changes of reflected radiation so there is no need to use an aperture before the detector. The measurements of the refractive nonlinearities of samples (that are the subject of present studies) were carried out without an aperture, thus neglecting the influence of phase changes caused by nonlinear absorption.

The principles of RZ-scan can be described as follows. The sample moves during the experiment through the focal plane of focusing lens. The amplitude and phase of reflected beam change due to the influence of nonlinear refraction and nonlinear absorption. No nonlinear effects appear when the sample is positioned far from the focal plane, so the ratio $R(z)$ of the reflected and incident laser radiation is constant.

When the sample approaches focal plane, the laser intensity becomes higher and the nonlinear effects occur. In the case of positive nonlinear refraction ($n_2 > 0$), the movement of sample close to the focus leads to the growth of $R(z)$. After crossing the focal plane the nonlinear refraction diminishes that leads to a decrease of $R(z)$ down to previous value. In the case of self-defocusing ($n_2 < 0$) the opposite feature will be observed with the valley appearing in the $R(z)$ dependence. One can conclude about the sign of n_2 from the $R(z)$ dependence.

The expression for the intensity of radiation reflected from the surface of a sample can be written in the form [140-142]

$$I_R(z) = I_0 \left| \left(R_0 V_0^{-1}(z) + R_1(\theta)(n_2 - ik_2) I(z) V_1^{-1}(z)(1 - ix') \right) \right|^2 . \tag{17}$$

here, k_2 is the coefficients of nonlinear extinction; R_0 is the linear reflection coefficient, $V_m(z) = g(z) - id/d_m$, $g(z) = d/d_0x$, d is the distance from the sample to the far-field aperture; $d_m = k\omega^2_{m0}/2$, $\omega^2_{m0} = \omega^2(z)/(2m+1)$, $\omega^2(z) = \omega^2(1+x^2)$, $x = z/z_0$, $z_0 = k\omega^2_0/2$ is the diffraction length of the beam; ω_0 is the beam waist radius; z characterizes the sample position:

$$R_1(\theta) = \frac{2n_0^3 \cos(\theta) - 4n_0 \cos(\theta) Sin^2(\theta)}{n_0^4 \cos^2(\theta) - n_0^2 + \sin^2(\theta)} \left[n_0^2 - \sin^2(\theta) \right]^{-1/2} , \tag{18}$$

and θ is the angle of incidence of the beam [145]. Substituting into equation (17) the parameters given above, it will be obtained the following expression for the normalized reflection:

$$R(z,\theta) = 1 - \frac{(4R_1(\theta)/R_0) I_0 k_2 x'}{(x'^2+9)(x'^2+1)} + \frac{(2R_1(\theta)/R_0) I_0 n_2 (x'^2+3)}{(x'^2+9)(x'^2+1)} + \frac{(R_1(\theta)/R_0)^2 I_0^2 (n_2^2 + k_2^2)}{(x'^2+9)(x'^2+1)} . \tag{19}$$

here, the first expression on the right-hand side is responsible for the nonlinear absorption, the second expression describes the nonlinear refraction, and the third expression characterizes their joint effect. It should be noted that equation (18) was derived without taking into account the effect of thermal processes, which are characteristic of nanosecond pulses [131] or of radiation with a high pulse repetition rate [131]. To determine the real part of the third-order nonlinear susceptibility, the expression (14) was used.

For practical purpose the $R(z)$ power could be presented as follows [144]:

$$P(z)=1+2Re\left[R(n_2+ik_2)\right]\frac{\int_0^{\infty}|E(\rho,z)|^4\,\rho d\rho}{\int_0^{\infty}|E(\rho,z)|^2\,\rho d\rho}, \quad (20)$$

where ρ the radial coordinates, and $E(\rho,z)$ is the incident beam amplitude.

This equation describes the general case, when both nonlinear refraction and nonlinear absorption appear simultaneously during the reflection from the sample. However, the application of open-aperture RZ-scan allowed neglecting the influence of nonlinear absorption for the measurements of nonlinear refraction [146].

3.3.1. RZ-Scan Study of Ag:Al_2O_3

The spectra of linear optical reflection for both virgon Al_2O_3 and Ag:Al_2O_3 composites obtained by ion implantation under different conditions presented in Figure 12. Samples 1 - 3 Ag:Al_2O_3 were implanted at fixed doses and energies but at different ion current densities, which increases with the sample number (Table 3). The thickness of a substrate layer containing silver nanoparticles was about 50 nm [124].

As is seen from Figure 12, in contrast to nonimplanted Al_2O_3, all the implanted samples are characterized by the presence in the visible spectral region of a broad selective reflection band with a maximum near 460 nm, whose intensity is slightly higher for the samples obtained at higher ion currents. This reflection band appears due to the formation of silver nanoparticles in the implanted Al_2O_3 and corresponds to the SPR absorption in MNPs [4]. A sharp increase in the reflection intensity in the shorter wavelength region beginning from approximately 380 nm (beyond the SPR band) is caused by the absorption of light by the Al_2O_3 matrix and by interband transitions in metal NPs.

As was shown recently [147], an increase in the ion current density during the implantation of silver ions into SiO_2 leads to an increase in the portion of the metal phase (MNPs) in the sample. This is explained by an increase in the temperature of the dielectric irradiated by high ion currents and, hence, by a higher diffusion mobility of implanted silver ions and thus their more efficient incorporation into MNPs. In the general case, this may result in an increase both in the number of MNPs and in their sizes, which leads to a higher SPR absorption of MNPs. Therefore, the rising in the SPR reflection intensity in Figure 12 should be related to a larger portion of metallic silver in Al_2O_3 implanted at higher currents.

Since the spectral positions of the SPR band maxima almost do not change, hence it is possible to conclude that an increase in the ion current density under these conditions of silver implantation into Al_2O_3 results in a higher concentration of MNPs rather than in their dimensions, which would immediately cause a spectral shift of the SPR reflection maximum.

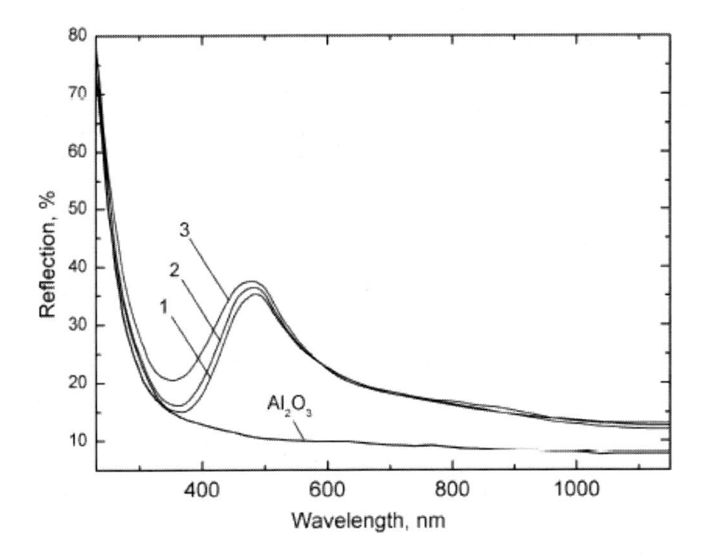

Figure 12. Reflectance spectra of Al_2O_3 before and after Ag-ion implantation with energy of 30 keV, dose of $3.75 \cdot 10^{17}$ ion/cm^2 and different current densities (1) 3; (2) 6 and (3) 10 μA/cm^2 [30].

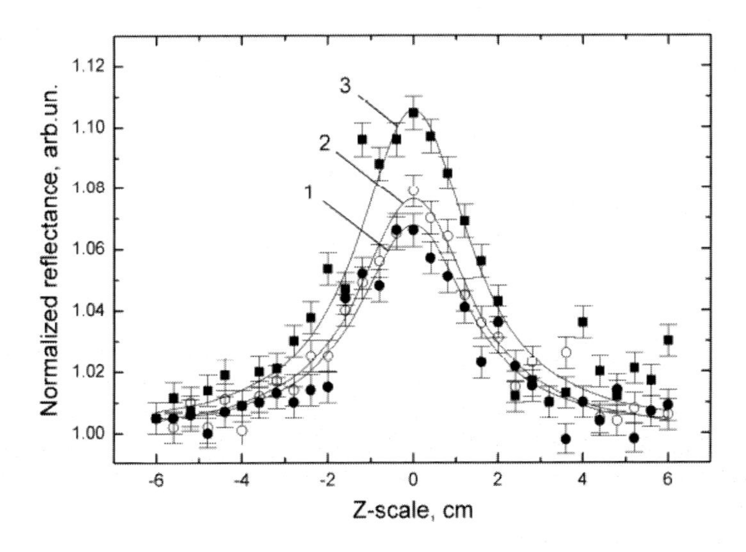

Figure 13. Dependence of the curves $R(z)$ for Ag:Al_2O_3 composites implanted with energy of 30 keV, dose of $3.75 \cdot 10^{17}$ ion/cm^2 and different current densities (1) 3; (2) 6 and (3) 10 μA/cm^2. Laser intensity is $4.3 \cdot 10^9$ W/cm^2 and pulse duration is 55 ps. Solid line is a fitting [30].

The experimental $R(z)$ dependences for the Ag:Al_2O_3 samples measured by RZ-scanning at the wavelength 1064 nm at the laser radiation intensity $I_0 = 4.3 \cdot 10^9$ W/cm^2 are presented in Figure 13. Dependences $R(z)$ for all the samples have the shape of a bell with the top directed upward, symmetrical with respect to $Z = 0$, and by the positive n_2.

It should be noted that the tops of the bell-shaped dependences are higher for the samples implanted at higher ion currents, i.e., for the samples with a higher content of metallic silver.

Virgin Al_2O_3 shows no such optical nonlinearity in experiments with laser radiation intensity up to the optical breakdown.

Thus, the nonlinear optical effects shown in Figure 13 are caused by the presence of silver nanoparticles in Al_2O_3. It is also interesting that the optical nonlinearities of silver particles are observed at laser irradiation at a wavelength outside the SPR absorption of the MNPs.

Using modeled $R(z)$ dependences (Figure 13) and fitting by them the experimental data, values of n_2 and $Re[\chi^{(3)}]$ in each sample were estimated and presented in Table 3. The analysis of the results shows that the samples with a higher concentration of silver nanoparticles have higher values of n_2 and $Re[\chi^{(3)}]$.

3.3.2. RZ-Scan Study of Cu:Al_2O_3

The second type of samples is the Al_2O_3 with copper nanoparticles. In contrast to the previous series of Ag:Al_2O_3 samples (1–3) presented in Table 3, in which the content of MNPs was varied by using different ion current densities, one of the samples with copper nanoparticles (sample 4) was obtained by implantation with a small dose ($0.54 \cdot 10^{17}$ ions/cm^2) and a low ion current (2.5 μA/cm^2), while the other sample (sample 5) was implanted at a larger dose (10^{17} ions/cm^2) and a higher current (12.5 μA/cm^2). The energy of ion implantation was equal to 40 keV for both samples (Table 3).

The choice of ion implantation regimes (Table 3) in the case of copper nanoparticles allowed to obtained samples with a noticeably different filling factor of the metal phase and, in particular, with different sizes of MNPs. This is illustrated in Figure 14, which shows the linear reflection spectra of Cu:Al_2O_3 with the SPR absorption bands of copper NPs [148], whose maxima take clearly different positions. Sample 5, which has a higher concentration of copper nanoparticles, exhibits the band at a longer wavelength (with the maximum at ~ 650 nm) than sample 4 (~ 610 nm), which points to the presence of larger MNPs in sample 5 [21, 148]. A pronounced difference in the portion of the metal phase and in the size of MNPs also manifests itself in the intensity of the SPR bands. The reflection intensity for sample 5 is noticeably higher than that for sample 4.

The experimental and calculated dependences $R(z)$ for the Cu:Al_2O_3 samples are shown in Figure 15. Since the efficiency of the electronic SPR excitation in copper nanostructures (in particular, in films, wires, etc.) is known to be noticeably lower than in silver particles [4], it was chosen a somewhat higher intensity of laser radiation for measuring their nonlinear optical properties $I_0 = 7.7 \cdot 10^9$ W/cm^2 (Table 3) than for Ag:Al_2O_3.

The $R(z)$ dependences obtained are also bell-shaped and symmetrical with respect to the point Z=0, which points to self-focusing in the Cu:Al_2O_3 samples due to the presence of copper nanoparticles. However, in contrast to the case with Ag:Al_2O_3, when the tops of the bells are directed downward, which clearly testifies to the self-defocusing of the laser beam in the samples with copper nanoparticles, i.e., to a negative n_2 (Table 3).

Such a different behavior of the nonlinear optical properties of Ag:Al_2O_3 and Cu:Al_2O_3 samples (a difference in the signs of n_2) was also observed in transmission Z-scan measurements at the same wavelength of 1064 nm in SiO_2 with silver and copper nanoparticles (Table 2).

Estimated values of n_2 and $Re[\chi^{(3)}]$ in Cu:Al_2O_3 by simulating the $R(z)$ dependences and comparing them with experimental data are presented in Table 3. It is found that the sample with a higher content of the metal phase (sample 5) has higher $|n_2|$ and $|Re[\chi^{(3)}]|$.

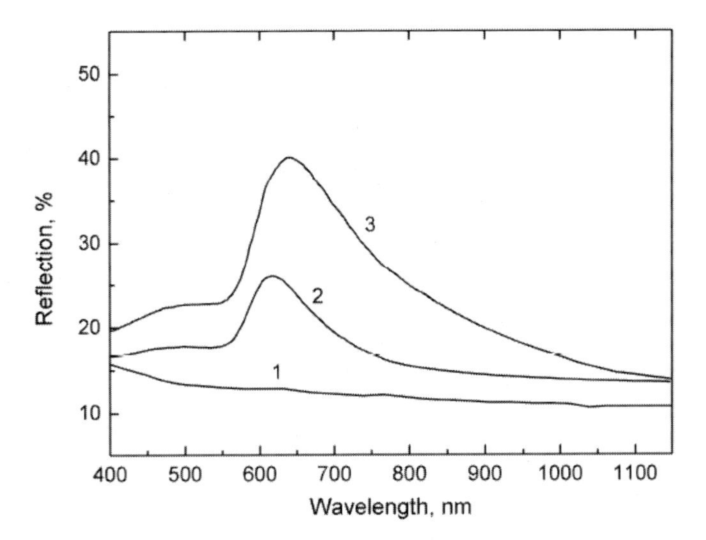

Figure 14. Reflectance spectra of Al_2O_3 before (1) and after Cu-ion implantation with energy of 40 keV and different parameters: dose of $0.54 \cdot 10^{17}$ ions/cm^2 and 2.5 μA/cm^2 (2) and dose of $1.0 \cdot 10^{17}$ ion/cm^2 and 12.5 μA/cm^2 (3) [147].

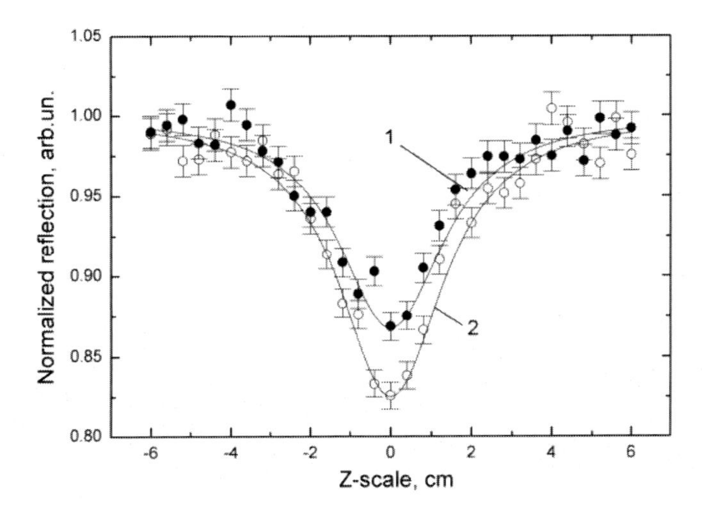

Figure 15. Dependence of the curves $R(z)$ for Cu:Al_2O_3 composites implanted with energy of 40 keV (1) dose of $0.54 \cdot 10^{17}$ ions/cm^2 and current density 2.5 μA/cm^2 and (2) dose of $1.0 \cdot 10^{17}$ ion/cm^2 and current density of 12.5 μA/cm^2. Laser intensity is $7.7 \cdot 10^9$ W/cm^2 and pulse duration is 55 ps. Solid line is a fitting [30].

3.3.3. RZ-Scan Study of Au:Al_2O_3

Samples (6–9) described in Table 3 are the Al_2O_3 containing ion-synthesized gold nanoparticles, which are also characterized by efficient SPR absorption in the visible spectral region. As in the case with copper ions, two different implantation doses, $0.6 \cdot 10^{17}$ and $0.1 \cdot 10^{17}$ ions/cm^2, but higher irradiation energies, 160 keV (samples 6 and 8, Table 3) [94, 95]. Since, at such high energies, the implanted impurity accumulates in a thicker subsurface layer of the irradiated dielectric [124], the impurity concentration necessary for the nucleation of MNPs accumulates over a longer time. In order to increase the size of MNPs, some of samples 7 and 9 were annealed in a furnace for 1 *h* at a temperature of 800°C (Table 3) [94,

95]. The spectra of linear optical reflection from the Au:Al_2O_3 samples 6–9 are presented in Figure 16. The formation of gold nanoparticles by ion implantation is proved by the presence of SPR reflection bands peaked at about 610 nm. Comparing samples 6 and 8 (curves 1 and 3, Figure 16), obtained directly by ion implantation, it is possible to see that an increase in the implantation dose (sample 8) results in a slight shift of the maximum of the SPR reflection (to ~ 620 nm), which is accompanied by a noticeable increase in the intensity. This fact, as in the previous case with the implantation with copper ions (samples 4 and 5), points to a higher filling factor of the metal phase in sample 8 (Au:Al_2O_3). Subsequent thermal treatment of these samples almost does not change the positions of the maxima of the SPR bands but leads to a sharp increase in the reflection in the long-wavelength spectral range. A broad reflection shoulder (sample 9, curve 4 in Figure 16) or even an additional maximum (sample 7, curve 2 in Figure 16) appearing near the SPR reflection bands can be associated with redistribution of the metal phase in the dielectric volume due to the high temperature of the material and, hence, with the formation of aggregates of MNPs. Similar spectral behavior was observed in experiments with fractal aggregates of silver particles but in solutions [149].

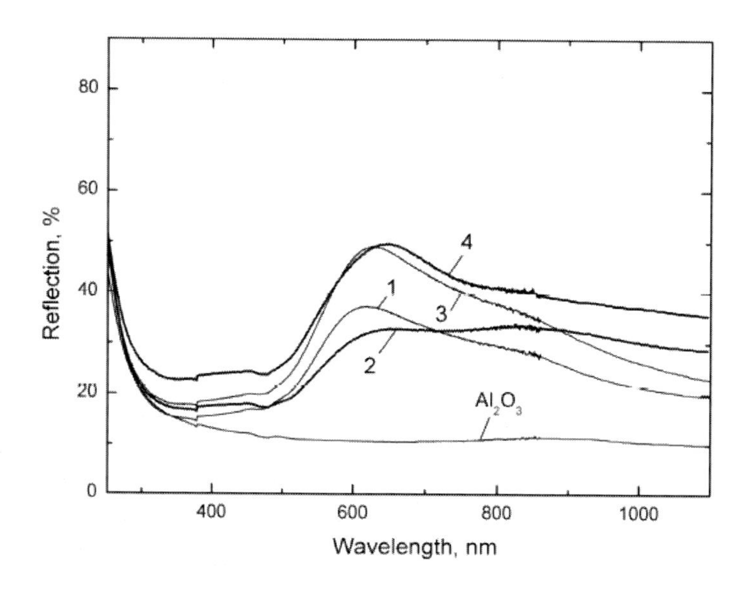

Figure 16. Reflectance spectra of Al_2O_3 before and after Au-ion implantation with energy of 160 keV, current density of 10 μA/cm^2 with different doses of $0.6{\cdot}10^{17}$ ions/cm^2 (3, 4) and $1.0{\cdot}10^{17}$ ion/cm^2 (1, 2). Samples annealed after ion implantation during 1 *h* at temperature of 800°C (2, 4) [94, 95].

At present, there are no data in the literature on the nonlinear optical properties of gold nanoparticles in a solid matrix in the near-IR region (1064 nm). The first experimental results for Au:Al_2O3 were given in [94, 95] and presented in Figure 17.

Figure 17a shows the curves for samples 6 and 7, which were obtained directly by ion implantation, while Figure 17b presents the dependences for samples 8 and 9, obtained by ion implantation and subsequent thermal treatment. As is seen, the $R(z)$ dependences have a bell-like shape symmetrical with respect to the point $Z = 0$, with the tops directed downward. Thus, samples (6-9) are characterized by self-defocusing, which corresponds to a negative n_2 and $Re[\chi^{(3)}]$ which values presented in Table 3.

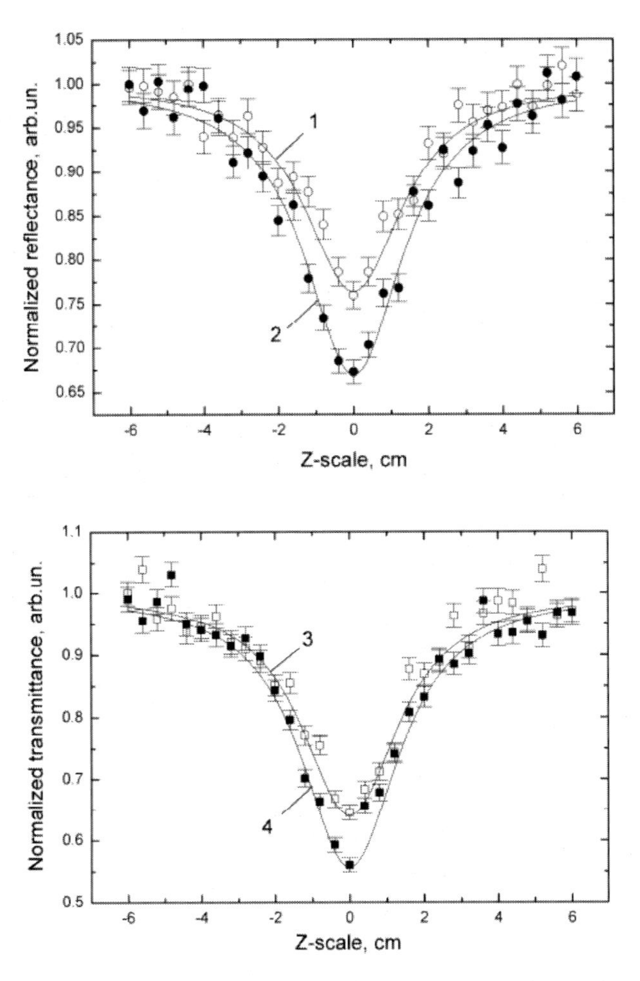

Figure 17. Dependence of the curves $R(z)$ for $Au:Al_2O_3$ composites implanted with energy of 160 keV with doses $0.6\cdot10^{17}$ ions/cm^2 (1, 2) and $1.0\cdot10^{17}$ ion/cm^2 (3,4) at current density of 10 μA/cm^2 with different doses of $0.6\cdot10^{17}$ ions/cm^2 (3,4) and $1.0\cdot10^{17}$ ion/cm^2 (1, 2). Samples created by ion implantation (1, 3) and samples created by ion implantation with subsequent thermal annealing during 1 *h* at temperature of 800°C (2, 4). Laser intensity is $2.3\cdot10^9$ W/cm^2 and pulse duration is 55 ps. Solid line is a fiting [94,95].

3.3.4. Nonlinear Refraction of MNPs Studied by RZ-Scan

For the composites based on Al_2O_3, the frequencies ω_p are equal to 22222 cm^{-1} (~ 450 nm, Figure 12) for silver nanoparticles, 16393 – 15384 cm^{-1} (~ 610 – 650 nm, Figure 14) for cupper nanoparticles, and 16129 cm^{-1} (~ 620 nm, Figure 16) for gold nanoparticles. Consider relation (12) for the case of a one-photon process ($I = 1$), i.e., for the frequency $\omega_{10} = 9398$ cm^{-1} ($\lambda = 1064$ nm), which is the fundamental frequency of the laser radiation used. Substituting the values of ω_{10} and ω_p into relation (12), positive values of $Re[\chi^{(3)}]$ and a negative detuning Δ_{10} for all the samples can be obtained. The positive $Re[\chi^{(3)}]$ and n_2 correspond to the self-focusing of the laser radiation in the sample, which was observed experimentally for the $Ag:Al_2O_3$ samples (samples 1–3, Table 3).

On the other hand, the positive values of the signs of $Re[\chi^{(3)}]$ and n_2 contradict the self-defocusing experimentally detected in the samples with cupper and gold nanoparticles (Figures 15 and 17). Hence, the description of nonlinear processes in the approximation of one-photon excitation for the $Cu:Al_2O_3$ and $Au:Al_2O_3$ systems is incorrect. Therefore, consider the case of two-photon excitation for all the composite systems studied using again expression (12) but with the doubled frequency of the laser radiation, $\omega_{20} = 18797$ cm^{-1} (λ = 532 nm). This frequency lies in the vicinity of the SPR frequencies of MNPs in $Cu:Al_2O_3$ and $Au:Al_2O_3$. Determine the signs of Δ_{20} for composites and see how they correlate with the nonlinear optical processes observed experimentally. In this case, the signs of $Re[\chi^{(3)}]$ for the $Cu:Al_2O_3$ and $Au:Al_2O_3$ systems are negative, which agrees with the self-defocusing detected in experiments and suggests the occurrence of two-photon absorption in these samples. For the $Ag:Al_2O_3$ samples, this sign turns out to be positive again, as in the case of one-photon excitation, and correlates with the self-focusing observed in experiments. However, for silver nanoparticles in Al_2O_3, it is difficult to choose between the one-photon and two-photon excitation mechanisms. Probably, the two mechanisms are simultaneously realized in this type of MNPs and their manifestation depends on the dominant frequency of laser excitation. Thus, the two-level model correctly predicts the sign of the nonlinearity in the $Cu:Al_2O_3$ and $Au:Al_2O_3$ systems in the case of excitation by laser radiation at a frequency divisible by the doubled SPR frequency.

As was mention a change in n_2 of a composite material can be caused by the thermal effect due to heat transfer from MNPs or defects of the dielectric host matrix heated by laser radiation [131]. Despite the duration of laser pulses used was rather short (τ = 55 ns), the influence of the thermal effect on the nonlinear refraction can be analyzed. Estimate how large a change in the refractive index of crystalline sapphire $\Delta n_{Al_2O_3}$ caused by heating can be.

The change in the refractive index due to the thermal effect can be represented in the form [131]

$$\Delta n(r,z,t) = \frac{1}{C_h p_h} \frac{dn}{dT} \Delta E(r,z,t) \quad , \qquad (21)$$

where C_h and ρ_h are, respectively, the heat capacity and the density of the host matrix with MNPs (in the case of sapphire, C_h = 0.419 J/g·K and ρ_h = 3.97 g/cm^3); dn/dT is the thermo-optic coefficient, equal to $13.7 \cdot 10^{-6}$ 1/K; and $\Delta E(r, z, t)$ is the energy of the radiation absorbed in a unit volume of the material over a time t. The thermo-optic coefficient for sapphire is positive, and, hence, the thermal effect should lead to the self-focusing of laser radiation in all the samples.

Since the self-focusing of laser radiation was experimentally observed only for the samples with silver nanoparticles the thermal effect for such samples were analyzed using present experimental conditions of nonlinear measurements [94, 95]. Thus, in the case of $Ag:Al_2O_3$, the energy of the absorbed radiation is $3.87 \cdot 10^{-6}$ J. For the layer with silver nanoparticles of 50 nm thickness and the beam waist radius 72 μm, the analyzing volume is $4.88 \cdot 10^{-10}$ cm^3. In this case, the energy $\Delta E(r, z, t)$ is $7.92 \cdot 10^3$ J/cm^3. Substituting these values into equation (21), the value of $\Delta n(r, z, t) \approx 6.53 \cdot 10^{-2}$ will be obtained. In same time, the experimental values are from $4.33{\cdot}10^{-2}$ to $6.83{\cdot}10^{-2}$. Hence, the thermal effect may manifest itself in the case of samples with silver nanoparticles. However, as it was mentioned the time τ_{rize} necessary for a change in the medium density and a corresponding change in the refractive index is determined by the ratio of the beam waist radius to the speed of sound. Taking into account our experimental conditions ($\omega_0 = 72$ μm at the wavelength 1064 nm and $V_s \sim 5000–5500$ m/s), $\tau_{rize} \approx 13–15$ ns will be again estimated. This time is three orders of magnitude longer than the pulse duration used (55 ps), and, hence, the thermal effect caused by the propagation of an acoustic wave can be excluded from consideration in our case.

In conclusion, RZ-scan method is suited for study of nonlinear refraction of samples based on dielectrics with MNPs. Although the sensitivity of the RZ-scan method is slightly lower than that of the classical transmission Z-scan, the RZ-scan method allows one to extend the spectral range of study to the region of low transparency of composite materials. The sign of the $Re[\chi^{(3)}]$ is analyzed on the basis of the two-level model, and it is shown that $Re[\chi^{(3)}]$ of the samples with copper and gold nanoparticles are determined by the two-photon process. It is difficult to make similar conclusion for samples with silver nanoparticles.

4. Nonlinear Optical Absorption of Ion Synthesized Silver Nanoparticles in Visible Range

The silver nanoparticles doped in different dielectrics demonstrate variable nonlinear optical properties in visible range [120]. The interest on such structures is based on the prospects of the elaboration of optical switchers with ultrafast response, optical limiters, and intracavity elements for mode locking.

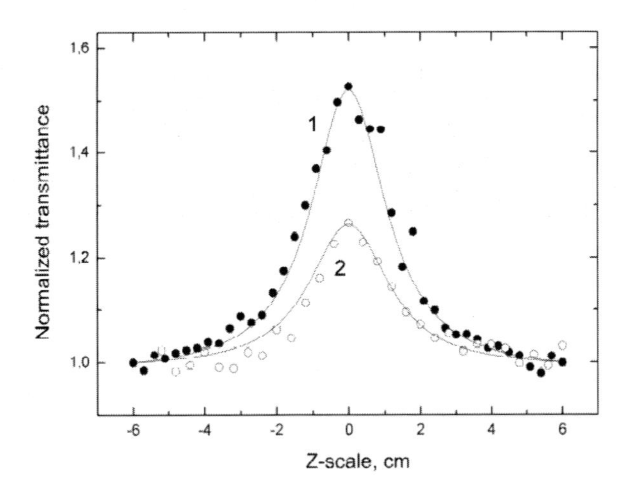

Figure 18. Normalized transmittance Ag:SLSG (1) and $Ag:SiO_2$ (2) samples at laser radiation intensity of $I_0 = 2.5{\cdot}10^9$ W/cm^2. Solis lines show is a fittings [84].

Silver nanoparticles have an advantage over another metal nanoparticles (i.e., gold and copper) from the point of view that the surface plasmon resonance energy of silver is far from the interband transition energy. So, in the silver nanoparticle system it is possible to investigate the nonlinear optical processes caused solely by SPR contribution.

It should be noted that previous studies of nonlinear optical parameters of silver nanoparticles-doped glasses were mostly focused on determination of third-order nonlinear susceptibility $\chi^{(3)}$. The saturated absorption in silicate glasses doped with ion-synthesized Ag nanoparticles at wavelength of 532 nm and their dependence on laser radiation intensity are considered at present review. As was shown the ion-synthesized silver nanoparticles in Ag:SLSG and Ag:SiO_2 demonstrate the SPR band with minimum transmission in the range of 410–440 nm (Figure 9). Early, it was predicted that glasses with silver-doped nanoparticles could possess a nonlinear saturated absorption [150]. The spectral dispersion of the imaginary part of susceptibility $Im[\chi^{(3)}]$ of such glass was detected as negative value in the spectral SPR range of 385 – 436 nm. The nonlinear coefficient β is also negative in the case of saturated absorption. The $T(z)$ dependences of Ag:SLSG and Ag:SiO_2 samples measured using open aperture Z-scan scheme at laser radiation intensity of I_0=2.5·10^{9}W/cm^2 and pulse duration of 55 ps is presented in Figure 18 [84]. The transmission of samples was increased due to nonlinear saturated absorption as they approached close to the focal plane.

Experimentally estimated nonlinear parameters of the β are -6.7·10^{-5} cm/W in Ag:SLSG and -3.6·10^{-5} cm/W in Ag:SiO_2. Note that the coefficient β can be presented as $\beta = \alpha/I_s$ where I_s is saturated intensity. Then the values of I_s are 1.1·10^9 and 1.4·10^9 W/cm^{-2} and the Im$\chi^{(3)}$ are -2.4·10^{-8} and -1.3·10^{-8} esu in Ag:SLSG and Ag:SiO_2, respectively.

In Figures 19 and 20 values of β in dependence of laser intensity varied from 10^9 to 2·10^{10} W/cm^2 are presented. As seen from the figures there are a decrease β of for higher intensities. In particularly, a 21- and 12-fold decrease of β was measured at I_0 = 1.15·10^{10} W/cm^2 for Ag:SLSG and Ag:SiO_2, respectively, compared to β detected at I_0 = 1·10^9 W/cm^2.

The variations of nonlinear transmission in MNP structures in dielectrics were early attributed in some cases to the fragmentation, or fusion of nanoparticles following the their photothermal melting [151, 152]. It was reported about the alteration of the sign of nonlinear refractive index of small Ag clusters embedded in SLSG [153]. They noted that thermal effects could change the properties of nanoclusters. The transparency in these samples was associated with oxidation of Ag nanoparticles. However, no irreversible changes of transmittance were observed in present experiments. The reverse saturated absorption can be responsible for the decrease of negative nonlinear absorption of Ag nanoparticles and it could be assume that in the case of picosecond pulses the reverse saturated absorption starting to play an important role in the overall dynamics of nonlinear optical transmittance of MNPs contained compounds, taking into account the saturation of intermediate transitions responsible for saturated absorption. Thus, saturated absorption in Ag:SLSG and Ag:SiO_2 was dominated at small intensities and decreased with the growth of intensity due to influence of competing effects, whereas the selfdefocusing at low intensities was changed to self-focusing at high intensities. The possible mechanism of the decrease of $Im[\chi^{(3)}]$ is the influence of nonlinear optical processes with opposite dependences on laser intensity, also such as two-photon absorption [122]. The wavelength range corresponded to the interband transitions in Ag is located below 320 nm, so the two-photon absorption connected with interband transitions can be involved in the case of 532 nm radiation.

The possibility of two-photon absorption due to interband transition of photoexcited electrons was previously demonstrated for Ag particles [154]. The three-photon absorption connected with interband transition for Ag nanoparticles was analysed in (Kyoung et al., 1999). Thus, saturated absorption in Ag:SLSG and Ag:SiO_2 was dominated at small intensities and decreased with the growth of intensity due to influence of competing effects, whereas the self-defocusing at low intensities was changed to self-focusing at high intensities.

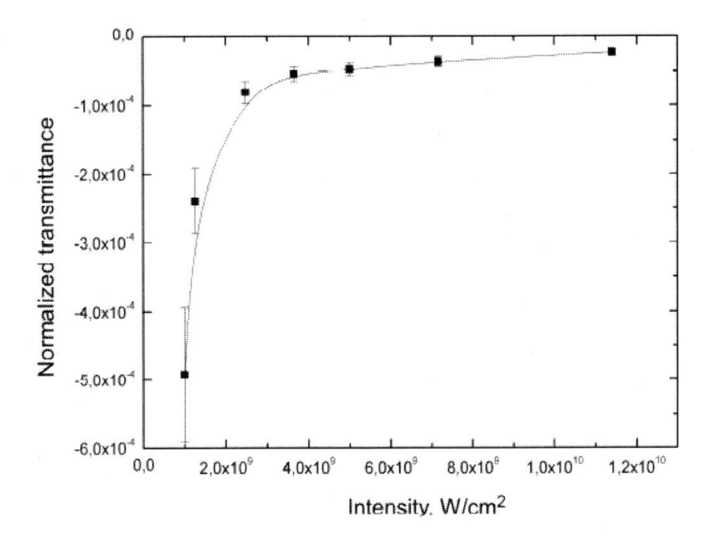

Figure 19. Coefficient β of Ag:SLSG in dependence of laser intensity.

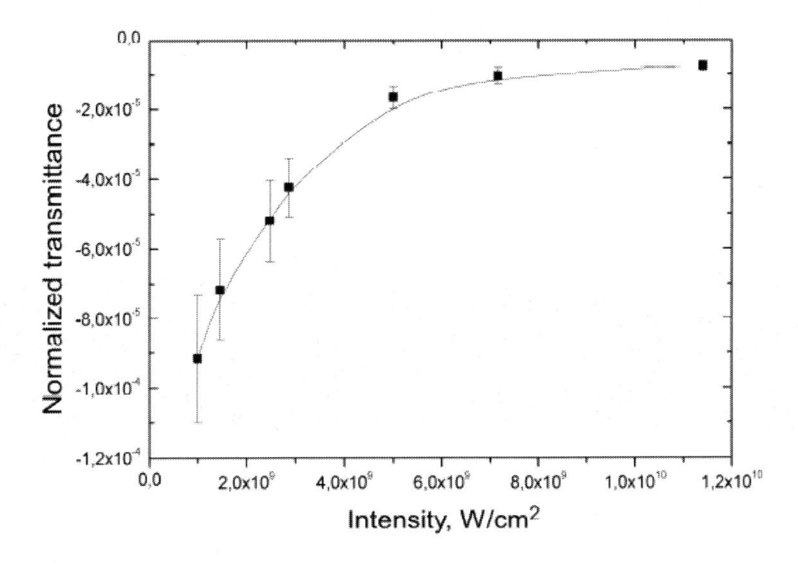

Figure 20. Coefficient β of Ag:SiO_2 in dependence of laser intensity.

ACKNOWLEDGMENTS

I wish to thank my partners and co-authors from different countries D. Hole, P.D. Townsend, I.B. Khaibullin, V.I. Nuzhdin, V.F. Valeev, Yu.N. Osin, R.A. Ganeev, A.I. Ryasnyanskiy, T. Usmanov, M.K. Kodirov, V.N. Popok, and U. Kreibig, Also, I grateful to the Alexander von Humboldt Foundation and the DAAD in Germany, Austrian Scientific Foundation in the frame of Lisa Meitner Fellowship and the Royal Society in UK for financial support. This work was partly supported by the Ministry of Education and Science of the Russian Federation (FTP "Scientific and scientific-pedagogical personnel of the innovative Russia" No. 02.740.11.0779) and RFBR № 11-02-90420, 11-02-91341 and 12-02-00528.

REFERENCES

[1] Zhang, J. Z. Optical properties and spectroscopy of nanomaterials; *Wold Sci. Pub.*: London, 2009.

[2] Optical properties of nanostructured random media; Shalaev, V. M.; Ed,: Springer: Berlin, 2002.

[3] Flytzanis, C.; Hache, F.; Klein, M. C.; Ricard, D.; Rousignol, P. *Nonlinear optics in composite materials*; Elsevier Science: Amsterdam, 1991.

[4] Kreibig, U.; Vollmer, M. *Optical properties of metal clusters*; Springer: Berlin, 1995.

[5] Stepanov, A. L. In: High-power and Femtosecond Lasers: Properties, Materials and Applications; Barret, P.-H.; Palmer M.; Eds., NOVA Sci. Publ.: New York, 2009, pp. 27-70.

[6] Townsend, P. T.; Chandler, P. J.; Zhang, L. *Optical effects of ion implantation*; Cambridge Univ. Press: Cambridge, 1994.

[7] Stepanov, A. L. In: *Metal-Polymer Nanocomposites*; Nicolais, L.; Carotenuto, G.; Eds.; John Wiley and Sons Publ.: London, 2004; pp. 241-263.

[8] Gonella, F.; Mazzoldi, P. In: *Handbook of nanostructured materials and nanotechnology*; Nalwa, H. S.; Ed.; Academic Press: London, 2000.

[9] Stepanov, A. L. Ion-synthesis of metal nanoparticles and their optical properties; NOVA Sci. Publ.: New York, 2011.

[10] Ricard, D.; Roussignol, P.; Flytzanis, C. *Opt. Lett.* 1985, 10, 511-513.

[11] Haglund Jr., R. F. In: *Handbook of optical properties*. Vol. II. Optics of small particles, interfaces, and surfaces; Hummel, R. F.; Wismann, P.; Eds.; CRC Press: London, 1997; pp. 198-231.

[12] Haglund Jr., R. F.; Osbone, D. H.; Magruder III, R. H.; White, C. W.; Zuhr, R. A.; Hole, D. E.; Townsend; P. D. Proc. Conf. Science and technology of atomically engineered materials; Jena, P.; Khanna, S. N.; Rao, B. K.; Eds.; *World Sci.*: Singapure, 1995; pp. 411-418.

[13] Von Neumann, J. In: *Fundamentals of photonics*, Saleh B. E. A.; Teich, M. C.; Eds.; Wiley: New York, 2001; p. 856-903.

[14] Davenas, J.; Perez, A.; Thevenard, P.; Dupuy, C. H. S. *Phys. Stat. Sol.* A. 1973, 19, 679-686.

[15] Treilleux, M.; Thevenard, P.; Ghassagne, G.; Hobbs, L. H. *Phys. Stat. Sol.* A. 1978, 48, 425-430.

[16] Arnold, G. W. *J. Appl. Phys.* 1975, 46, 4466-4473.

[17] Arnold, G. W.; Borders, J. A. *J. Appl. Phys.* 1977, 48, 1488-1496.

[18] Mazzoldi, P.; Arnold, G. W.; Battaglin, G.; Bertoncello, R.; Gonella, F. *Nucl. Instr. Meth. Phys. Res.* B. 1994, 91, 478-492.

[19] Mazzoldi, P.; Arnold, G. W.; Battaglin, G.; Gonella, F.; Haglund Jr., R. F. J. *Nonlinear Opt. Phys. Mater.* 1996, 5, 285-230.

[20] Chakraborty, P. *J. Mater. Sci.* 1998, 33, 2235-2249.
[21] Stepanov, A.; Khaibullin, I. B. *Rev. Adv. Mater. Sci.* 2005, 9, 109-129.
[22] Meldrum, A.; Lopez, R.; Magruder III, R. H.; Boatner, L. A.; White, C. W. In: Material science with ion beam; arnes, H.; Ed.; Springer: Berlin, 2010; pp. 255-285.
[23] Stepanov, A. L. *Rev. Adv. Mater. Sci.* 2010, 26, 1-29.
[24] Ganeev, R. A. *J. Opt. A.: Pure Appl. Opt.* 2005, 7, 717-733.
[25] Ganeev, R. A.; Usmanov, T. *Quant. Electr.* 2007, 37, 605-622.
[26] Cattaruzza, E.; Gonella, F.; Mattei, G.; Mazzoldi, P.; Gatteschi, D.; Sangregorio, S.; Falconieri, M.; Salvetti, G.; Battaglin, G. *Appl. Phys. Lett.* 1998, 73, 1176-1778.
[27] Falconieri, M.; Salvetti, G.; Cattarruzza, E.; Gonella, F.; Mattei, G.; Mazzoldi, P.; Piovesan, M.; Battaglin, G.; Polloni, R. *Appl. Phys. Lett.* 1998, 73, 288-290.
[28] Cattaruzza, E.; Battaglin, G.; Gonella, F.; Polloni, R.; Mattei, G.; Maurizio, C.; Mazzoldi, P.; Sada, C.; Montagna, M.; Tosello, C.; Ferrari, M. *Phil. Mag.* B. 2002, 82, 735-744.
[29] Ganeev, R.; Ryasnyanskiy, A. I.; Stepanov, A. L.; Marques, C.; da Silva, R. C.; Alves, E. *Opt. Comm.* 2005, 253, 205-213.
[30] Ganeev, R.; Ryasnyanskiy, A. I.; Stepanov, A. L.; Usmanov, T.; Marques, C.; da Silva, R. C., Alves, E. *Opt. Spectr.* 2006, 101, 615-622.
[31] Ryasnyanskiy, A. I. *Nonlin. Opt. Quant. Opt.* 2005, 33, 17-28.
[32] Plaksin, O.; Takeda, Y.; Amekura, H.; Kishimoto, N.; Plaksin, S. J. *Appl. Phys.* 2008, 103, 114302-1 - 114302-5.
[33] Ryasnyansky, A. I.; Palpant, B.; Debrus, S.; Khaibullin, R. I.; Stepanov, A. L. *J. Opt. Soc. Am.* B. 2006, 23, 1348-1352.
[34] Takeda, Y.; Lee, C. G.; Kishimoto, N. *Nucl. Instr. Meth. Phys. Res.* B. 2002, 191, 422-427.
[35] Takeda, Y.; Lee, C. G.; Bandourko, V. V.; Kishimoto, N. *Proc. SPIE,* 2002, 4628, 46-54.
[36] Takeda, Y.; Bandourko, V. V.; Lee, C. G.; Kishimoto, N. *Mater. Trans.* 2002, 43, 1057-1061.
[37] Kishimoto, N.; Takeda, Y.; Umeda, N.; Okubo, N.; Faulkner, R.G. *Nucl. Instr. Meth. Phys. Res.* B. 2003, 206, 643-638.
[38] Plaksin, O. A.; Takeda, Y.; Kono, K.; Umeda, N.; Fudamoto, Y.; Kishimoto, N. *Mat. Sci. Eng.* B. 2005, 120, 84-87.
[39] Plaksin, O. A.; Kishimoto, N. *Phys. Solid State*, 2006, 48, 1933-1939.
[40] Kishimoto, N.; Takeda, Y.; Umeda, N.; Gritsyna, T.; Lee, C. G.; Saito, T. *Nucl. Instr. Meth. Phys. Res.* B. 2000, 166-167, 840-844.
[41] Takeda, Y.; Umeda, N.; Gritsyna, V. T.; Kishimoto, N. *Nucl. Instr. Meth. Phys. Res.* B. 2001, 175-177, 463-467.
[42] Becker, K.; Yang, L.; Haglund Jr., R. F.; Magruder III, R. H.; Weeks, R. A.; Zuhr, R. A. *Nucl. Instr. Meth. Phys. Res.* B. 1991, 59-60, 1304-1307.
[43] Haglund Jr., R. F.; Magruder III, R. H.; Morgen, S. H.; Henderson, D. O.; Weller, R. A.; Yang, L.; Zuhr, R. A. *Nucl. Instr. Meth. Phys. Res.* B. 1992, 65, 405-411.
[44] Haglund Jr., R. F.; Yang, L.; Magruder III, R. H.; Wittig, J. E.; Becker, K.; Zuhr, R. A. *Opt. Lett.* 1993, 18, 373-375.
[45] Haglund Jr., R. F.; Yang, L.; Magruder III, R. H.; Wittig, C. W.; Zuhr, R. A.; Yang, L.; Dorsinville, R.; Alfano, R. R. *Nucl. Instr. Meth. Phys. Res.* B. 1994, 91, 493-504.

[46] Haglund Jr., R. F. *Mat. Sci. Eng.* A. 1998, 253, 275-283.
[47] Magruder III, R. H., Haglund Jr., R. F.; Yang, L.; Wittig, J. E.; Zuhr, R. A. *J. Appl. Phys.* 1994, 76, 708-715.
[48] Yang, L.; Becker, K.; Smith, F. M.; Magruder III, R. H.; Haglund Jr., R. F.; Yang, L.; Dorsinville, R.; Alfano, R. R.; Zuhr, R. A. *J. Opt. Soc. Am.* B. 1994, 11, 457-461.
[49] Yang, L.; Osbone, D. H.; Haglund Jr., R. F.; Magruder III, R. H.; Wittig, C. W.; Zuhr, R.A.; Hosono, H. *Appl. Phys. A.* 1996, 62, 403-415.
[50] Ila, D.; Williams, E. K.; Sarkisov, S.; Smith, C. C.; Poker, D. B.; Hensley, D. K. *Nucl. Instr. Meth. Phys. Res. B.* 1998, 141, 289-293.
[51] Sarkisov, S.; Williams, E. K.; Curley, M.; Ila, D.; Venkateswarlu, P.; Poker, D. B.; Hensley, D. K. *Nucl. Instr. Meth. Phys. Res. B.* 1998, 141, 294-298.
[52] Takeda, Y.; Gritsyna, V.T.; Umeda, N.; Lee, C. G.; Kishimoto, N. *Nucl. Instr. Meth. Phys. Res.* B. 1999, 148, 1029-1033.
[53] Takeda, Y.; Zhao, J. P.; Lee, C. G.; Gritsyna, V. T.; Kishimoto, N. *Nucl. Instr. Meth. Phys. Res.* B. 2000, 166-167, 877-881.
[54] Olivares, J.; Requejo-Isidro, J.; del Coso, R.; de Nalda, R.; Solis, J.; Afonso, C. N.; Stepanov, A. L.; Hole, D.; Townsend, P. D.; Naudon, A. *J. Appl. Phys.* 2001, 90, 1064-1066.
[55] Takeda, Y.; Lee, C. G.; Kishimoto, N. *Nucl. Instr. Meth. Phys. Res.* B. 2002, 190, 797-801.
[56] Takeda, Y.; Lu, J.; Plaksin, O. A.; Amekura, H.; Kono, K.; Kishimoto, N. Nucl. *Instr. Meth. Phys. Res.* B. 2004, 219-220, 737-741.
[57] Takeda, Y.; Lu, J.; Okubo, N.; Plaksin, O. A.; Suga, T.; Kishimoto, N. *Vacuum*, 2004, 74, 717-721.
[58] Ganeev, R.; Ryasnyanskiy, A. I.; Stepanov, A. L.; Usmanov, T. *Phys. Stat. Sol.* B. 2003, 238, R5-R7.
[59] Ganeev, R.; Ryasnyanskiy, A. I.; Stepanov, A. L.; Usmanov, T. *Phys. Solid State,* 2004, 46, 351-356.
[60] Ganeev, R.; Ryasnyanskiy, A. I.; Stepanov, A. L.; Usmanov, T. *Quant. Electr.* 2003, 33, 1081-1084.
[61] Ganeev, R.; Ryasnyanskiy, A. I.; Stepanov, A. L.; Usmanov, T. *Phys. Stat. Sol.* B. 2004, 241, R1-R4.
[62] Ganeev, R.; Ryasnyanskiy, A. I.; Stepanov, A. L.; Usmanov, T. *Phys. Solid State*, 2003, 45, 1355-1359.
[63] Ganeev, R.; Ryasnyanskiy, A. I.; Stepanov, A. L.; Kodirov, M. K.; Usmanov, T. *Opt. Spectr.* 2003, 95, 967-975.
[64] Ganeev, R.; Ryasnyanskiy, A. I.; Stepanov, A. L.; Usmanov, T. *Phys. Stat. Sol.* B. 2004, 241, 935-944.
[65] Stepanov, A. L.; Ganeev, R.; Ryasnyanskiy, A. I.; Usmanov, T. *Nucl. Instr. Meth. Phys. Res.* B. 2003, 206, 624-628.
[66] Takeda, Y.; Plaksin, O. A.; Kono, K.; Kishimoto, N. *Surf. Coat.Technol.* 2005, 196, 30-33.
[67] Ren, F.; Jiang, C. Z.; Wang, Y. H.; Wang, Q. Q.; Wang, J. B. *Nucl. Instr. Meth. Phys. Res.* B. 2006, 245, 427-430.
[68] Y. H. Wang, C. Z. Jiang, F. Ren, Q. Q. Wang, D. J. Chen D. J. Fu *Physica* E. 33 (2006) 444-248.

[69] Wang, Y. H.; Ren, F.; Wang, Q. Q.; Chen, D. J.; Fu, D. J.; Jiang, C. Z. *Phys. Lett.* A. 2006, 357, 364-368.

[70] Ghosh, B.; Chakraborty, P.; Mohapartra, S.; Kurian, P. A.; Vijayan, C.; Deshmukh, P. C.; Mazzoldi, P. *Mat. Lett.* 2007, 61, 4512-4515.

[71] Ghosh, B.; Chakraborty, P.; Singh, B. P.; Kundu, T. *Appl. Surf. Sci.* 2009, 256, 389-394.

[72] Torres-Torres, C.; Reyes-Esqueda, J. R.; Cheng-Wong, J. C.; Crespo-Sosa, A.; Rodriguez-Fernandez, L.; Oliver, A. *J. Appl. Phys.* 2008, 104, 14306-1 -14306-5.

[73] Wang, Y. H.; Wang, Y. M.; Lu, J. D.; Ji, L. L.; Zang, R. G.; Wang, R. W. *Physica* B. 2009, 404, 4295-4298.

[74] Wang, Y. H.; Wang, Y. M.; Lu, J. D.; Ji, L. L.; Zang, R. G.; Wang, R. W. *Opt. Comm.* 2010, 283, 486-489.

[75] Wang, Y. H.; Wang, Y. M.; Han, C. J.; Lu, J. D.; Ji, L. L. *Physica* B 2010, 405, 2664-2667.

[76] Takeda, Y.; Lu, J.; Plaksin, O. A.; Kono, K.; Amekura, H.; Kishimoto, N. *Thin Solid Films*, 2004, 464-456, 483-486.

[77] Takeda, Y.; Plaksin, O. A.; Lu, J.; Kono, K.; Amekura, H.; Kishimoto, N. *Nucl. Instr. Meth. Phys. Res.* B. 2006, 250, 372-376.

[78] Cetin, A.; Kibar, R.; Hatipoglu, M.; Karabulut, Y.; Can, N. *Physica* B. 405 (2010) 2323-2325.

[79] Stepanov, A. L.; Khaibullin, R. I.; Can, N.; Ganeev, R. A.; Ryasnyanski, A. I.; Buchal, C.; Uysal, S. *Tech. Phys. Lett.* 2004, 30, 846-849.

[80] Ryasnyanskiy, A. I.; Palpant, B.; Debrus, S.; Ganeev, R.; Stepanov, A. L.; Can, N.; Buchal, C.; Uysal, S. *Appl. Opt.* 2005, 44, 2839-2845.

[81] Ryasnyanskiy, A. I. *J. Appl. Spect.* 2005, 72, 712-715.

[82] Williams, E. K.; Ila, D.; Darwish, A.; Poker, D. B.; Sarkisov, S. S.; Curley, M. J.; Wang, J.-C.; Svetchinkov, V. L.; Zandbergen, H. W. *Nucl. Instr. Meth. Phys. Res.* B. 1999, 148, 1074-1078.

[83] Sarkisov, S. S.; Curley, M. J.; Williams, E. K.; Ila, D.; Svetchinkov, V. L.; Zandbergen, H. W.; Zykov, G. A.; Banks, C.; Wang, J.-C.; Poker, D. B.; Hensley, D. K. *Nucl. Instr. Meth. Phys. Res.* B. 2000, 166-167, 750-757.

[84] Ganeev, R.; Ryasnyanskiy, A. I.; Stepanov, A. L.; Usmanov, T. *Opt. Quant. Electron.* 2004, 36, 949-960.

[85] Stepanov, A. L. In: *Silver nanoparticles*; Perez, D. P.; Ed.; In-tech: Vukovar, 2010; p. 93-120.

[86] Joseph, B.; Suchand Sandeep, C. S.; Sekhar, B. R.; Mahapatra, D. P.; Philip, R. *Nucl. Instr. Meth. Phys. Res.* B. 2007, 265, 631-636.

[87] Rangel-Roja, R.; McCarthy, J.; Bookey, H. T.; Kar, A. K.; Rodriguez-Fernandez, L.; Cheang-Wong, J.-C.; Crespo-Soso, A.; Lopez-Suarez, A.; Oliver, A.; Rodriguez-Iglesias, V.; Silva-Pereryra, H. G. *Opt. Comm.* 2009, 282, 1909-1912.

[88] Rangel-Roja, R.; Reyes-Esqueda, J. A.; Torres-Torres, C.; Oliver, A.; Rodriguez-Fernandez, L.; Crespo-Soso, A.; Cheang-Wong, J. C.; McCarthy, J.; Bookey, H. T.; Kar, A. K. In: *Silver nanoparticles*, Perez, D. P.; Ed.; In-tech: Vukovar, 2010; p. 35-62.

[89] Wang, Y. H.; Peng, S. J.; Lu, J. D.; Wang, R. W.; Cheng Y. G.; Mao, Y. I. *Vacuum*, 2009, 83, 412-415.

[90] Akeda, Y.; Hioki, T.; Motohiro, T.; Noda, S. *Appl. Phys. Lett.* 1993, 63, 3420-3422.

[91] Takeda, Y.; Hioki, T.; Motohiro, T.; Noda, S.; Kurauchi, T. *Nucl. Instr. Meth. Phys. Res.* B. 1994, 91, 515-519.
[92] Takeda, Y.; Kishimoto, N. *Nucl. Instr. Meth. Phys. Res.* B. 2003, 206, 620-623.
[93] White, C. W.; Thomas, D. K.; Hensley, D. K.; Zuhr, R. A.; McCallum, J. C.; Pogany, A.; Haglund, Jr., R. F.; Magruder, III, R. H.; Yang, L. *Nanostruct. Mat.* 1993, 3, 447-457.
[94] Stepanov, A. L.; Marques, C.; Alves, E.; da Silva, R. C.; Silva, M. R.; Ganeev, R.; Ryasnyanskiy, A. I.; Usmanov, T. *Tech. Phys. Lett.* 2005, 31, 702-705.
[95] Stepanov, A. L.; Marques, C.; Alves, E.; da Silva, R. C.; Silva, M. R.; Ganeev, R.; Ryasnyanskiy, A. I. *Tech. Phys.* 2006, 51, 1474-1481.
[96] Takeda, Y.; Plaksin, O. A.; Wang, H.; Kono, K.; Umeda, N.; Kishimoto, N. *Opt. Rev.* 2006, 13, 231-234.
[97] Takeda, Y.; Plaksin, O. A.; Wang, H.; Kishimoto, N. *Nucl. Instr. Meth. Phys. Res.* B. 2007, 257, 47-50.
[98] Magruder, III, R. H.; Yang, L.; Haglund, Jr., R. F.; White, C. W.; Yang, L.; Dorsinville, R.; Alfano, R. R. *Appl. Phys. Lett.* 1993, 62, 1730-1772.
[99] White, C. W.; Zhou, D. Z.; Budai, J. D.; Zhur, R. A.; Magruder, III, R. H.; Osbone, D. H. *Mat. Res. Soc. Proc.* 1994, 316, 499-507.
[100] Fukumi, K.; Chayahara, A.; Kadono, K.; Sakaguchi, T.; Horino, Y.; Miya, M.; Fujii, K.; Hayakawa, J.; Satou, M. *Jpn. J. Appl. Phys.* 1991, 30, L742-L744.
[101] Fukumi, K.; Chayahara, A.; Kadono, K.; Sakaguchi, T.; Horino, Y.; Miya, M.; Fujii, K.; Hayakawa, J.; Satou, M. *J. Appl. Phys.* 1994, 75, 3075-3080.
[102] Lepeshkin, N. N.; Kim, W.; Safonov, V. P.; Zhu, J. G.; Armstrong, R. L.; White, C. W.; Zuhr, R. A.; Shalaev, V. M. J. Nonlin. *Opt. Phys. Mat.* 1999, 8, 191-210.
[103] Safonov, V. P.; Zhu, J. G.; Lepeshkin, N. N.; Armstrong, R. L.; Shalaev, V. M.; Ying, Z. C.; White, C. W.; Zuhr, R. A. *Proc. SPIE*, 1999, 3788, 34-41.
[104] Takeda, Y.; Plaksin, O. A.; Kishimoto, N. *Opt. Express*, 2007, 10, 6010-6018.
[105] Torres-Torres, C.; Khomenko, A. V.; Cheang-Wong, J. C.; Rodriguez-Fernandez, L.; Crespo-Soso, A.; Oliver, A. *Opt. Express*, 2007, 15, 9248-9253.
[106] Ghosh, B.; Chakraborty, P.; Sundaravel, B.; Vijayan, C. *Nucl. Instr. Meth. Phys. Res.* B. 2008, 266, 1356-1361.
[107] Wang, Y. H.; Lu, J. D.; Wang, R. W.; Peng, S. J.; Mao, Y. I.; Cheng, Y. G. *Physica* B. 2008, 403, 3399-3402.
[108] Wang, Y. H.; Lu, J. D.; Wang, R. W.; Mao, Y. I.; Cheng, Y. G. *Vacuum*, 2008, 82, 1220-1223.
[109] Battaglin, G.; Calvelli, P.; Cattaruzza, E.; Polloni, P.; Borsella, E.; Cesa, T.; Mazzoldi, P. J. *Opt. Soc. Am.* B. 2000, 17, 213-218.
[110] Magruder III, R. H.; Osbone, D. H.; Zuhr, R. A. J. *Non.-Cryst. Solids*, 1994, 176, 299-303.
[111] Wang, Y. H.; Jiang, C. Z.; Ren, F.; Wang, Q. Q.; Chen, D. J.; Fu, D. J. *J. Mat. Sci.* 2007, 42, 7294-7298.
[112] Wang, Y. H.; Jiang, C. Z.; Xiao, X. H.; Cheng, Y. G. *Physica* B 2008, 403, 2143-2147.
[113] Wang, Y. H.; Wang, Y. M.; Han, C. J.; Lu, J. D.; Ji, L. L.; Wang, R. W. *Physica* B. 2010, 405, 2848-2851.
[114] Wang, Y. H.; Wang, Y. M.; Han, C. J.; Lu, J. D.; Ji, L. L.; Wang, R. W. *Vacuum*, 2010, 85, 207-210.

[115] Cattaruzza, E.; Battaglin, G.; Gonella, F.; Calvelli, P.; Mattei, G.; Maurizio, C.; Mazzoldi, P.; Padovani, S.; Polloni, R.; Sada, C.; Scremin, B. F.; D'Acapito, F. *Composites Sci. Technol*. 2003, 68, 1203-1208.

[116] Cattaruzza, E.; Battaglin, G.; Gonella, F.; Calvelli, P.; Mattei, G.; Maurizio, C.; Mazzoldi, P.; Polloni, R.; Scremin, B. F. *Appl. Surf. Sci*. 2005, 247, 390-395.

[117] Cesca, T.; Pellegrini, G.; Bello, V.; Scian, C.; Mazzoldi, P.; Calvelli, P.; Battaglin, G.; Mattei, G. *Nucl. Instr. Meth. Phys. Res*. B. 2010, 268, 3227-3230.

[118] Magruder, III, R. H.; Zuhr, R. A.; Osbone, D. H. *Nucl. Instr. Meth. Phys. Res*. B. 1995, 99, 590-593.

[119] Mie, G. *Ann. Phys*. 1908, 25, 377-420.

[120] Palpant, B. In: *Non-Linear optical properties of matter*; Papadopoulos, M. G.; Ed.; Springer: Amsterdam, 2006; pp. 461-508.

[121] Sheik-Bahae, M.; Said, A. A.; van Stryland, E. W. *Opt. Lett*. 1989, 14, 955-957.

[122] Sheik-Bahae, M.; Said, A. A.; Hagan, D. J.; van Stryland, E. W. *IEEE J. Quan. Elect*. 1990, 26, 760-769.

[123] Kwak, C. H.; Lee, Y. L.; Kim, S. G. J. *Opt. Soc. Am*. 1999, 16, 600-604.

[124] Stepanov, A. L.; Zhikharev, V. A.; Hole, D. E.; Townsend, P. D.; Khaibullin, I. B. *Nucl. Instr. Meth. Phys. Res*. B. 2000, 166-167, 26-30.

[125] Reintjes, J. F. *Nonlinear-optical parametrical processes in liquids and gases*; Academic: Orlando, 1984.

[126] Karpov, S. V.; Popov, A. K.; Slabko, V. V. *JETP Lett*. 1997, 66, 106-111.

[127] Ganeev, R. A.; Ryasnyansky, A. I.; Kamalov, S. R.; Kodirov, M. K.; Usmanov, T. J. *Phys. D: Appl. Phys*. 2001, 34, 56-61.

[128] Tutt, L. W.; Boggess, T. F. *Prog. Quant. Electr*. 1993, 17, 299-338.

[129] Shen, Y. R. *The principles of nonlinear optics*; Wiley: New York, 1989.

[130] Owyoung, A. *IEEE J. Quant. Electr*. 1973, 9, 1064-1069.

[131] Mehendale, S. C.; Mishra, S. R.; Bindra, K. S.; Laghate, M.; Dhami, T. S.; Rustagi, K. C. *Opt. Comm*. 1997, 133, 273-272.

[132] Falconieri, M. J. *Opt. A: Pure Appl. Opt*. 1999, 1, 662-667.

[133] Battaglin, G.; Calvelli, P.; Cattaruzza, E.; Gonella, F.; Polloni, R.; Mattei G.; Mazzoldi, P. *Appl. Phys. Lett*. 2001, 78, 3953-3955.

[134] Mizrahi, V.; DeLong, K. W.; Stegeman, G. I.; Saifi, M. A.; Andejco, M. J. *Opt. Lett*. 1989, 14, 1140-1142.

[135] Rangel-Rojo, R.; Kosa, T.; Hajto, E.; Ewen, P. J. S.; Owen, A. E.; Kar, A. K.; Wherrett, B. S. *Opt. Comm*. 1994, 109, 145-150.

[136] Fribers, S. R.; Smith, P. W. *IEEE J. Quant. Electr*. 1987, 23, 2089-2096.

[137] Moran, M. J.; She, C. Y.; Carman, R. L. *IEEE J. Quant. Electr*. 1975, 11, 259-263.

[138] Xia, T.; Hagan, D. J.; Sheik-Behae, M.; van Stryland, E. W. *Opt. Lett*. 1994, 19, 317-319.

[139] Ma, H.; Gomes, A. S. L.; de Araujo, C. B. *Appl. Phys. Lett*. 1991, 59, 2666-2668.

[140] Petrov, D. V.; Gomes, A. S. L.; de Araujo, C. B. *Appl. Phys. Lett*. 1994, 65, 1067-1069.

[141] Petrov, D. V.; Gomes, A. S. L.; de Araujo, C. B. *Opt. Comm*. 1996, 123, 637-641.

[142] Petrov, D. V. J. *Opt. Soc. Am*. 1996, 13, 1491-1498.

[143] Kawazoe, T.; Kawaguchi, H.; Inoue, J.; Haba, O.; Ueda, M. *Opt. Comm*. 1999, 160, 125-129.

[144] Stepanov, A. L. *Rev. Adv. Mater. Sci*. 2003, 4, 45-60.

[145] Martinelli, M.; Gomes, L.; Horowicz, R. *J. Appl. Opt*. 2000, 39, 6193-6196.

[146] Ganeev, R. A.; Ryasniansy, A. I. *Phys. Stat. Sol*. A. 2005, 202, 120-125.

[147] Stepanov, A. L.; Popok, V. N. *Surf. Sci*. 2004, 566-568, 1250-1254.

[148] Stepanov, A. L.; Kreibig, U.; Hole, D. E.; Khaibullin, R. I.; Khaibullin, I. B.; Popok, V. N. *Nucl. Instr. Meth. Phys. Res*. B. 2001, 178, 120-125.

[149] Karpov, S. V.; Popov, F. K.; Slabko, V. V. *Izv. Akad. Nauk SSSR Ser. Fiz*. 1996, 60, 42-49.

[150] Hamanaka, Y.; Hayashi, N.; Nakamura, A.; Omi, S. J. *Luminesc*. 2000, 87-89, 859-861.

[151] Link, S.; Burda, C.; Mohamed, M. B.; Nikoobakht, B.; El-Sayed, M. A. *J. Phys. Che*. A. 1999, 103, 1165-1170.

[152] Mafune, F.; Kohno, J.-Y.; Takeds, Y.; Kondow, T. *J. Phys. Chem*. B. 2002, 106, 7576-7577.

[153] Osbone Jr., D. H.; Haglund Jr., R. F.; Gonella, F.; Garrido, F. *Appl. Phys. B*. 1998, 66, 517-521.

[154] Kyoung, M.; Lee, M.; *Opt. Comm*. 1999, 171, 145-148.

INDEX

A

B

C

D

E

F

G

H

I

J

K

L

M

N

O

P

Q

R

S

T

U

V

W

Y

Z